MATHEMATICS

Applications and Connections

COURSE 2

GLENCOE

Macmillan/McGraw-Hill

New York, New York   Columbus, Ohio   Mission Hills, California   Peoria, Illinois

Send all inquiries to:
Glencoe Division, Macmillan/McGraw-Hill
936 Eastwind Drive
Westerville, Ohio 43081

ISBN: 0-02-824042-1 (Student Edition)
ISBN: 0-02-824043-X (Teacher's Wraparound Edition)

3 4 5 6 7 8 9 10 RRD-W 01 00 99 98 97 96 95 94 93

Dear Students, Teachers, and Parents,

Middle school students are special! That's why we've written the first and only middle school mathematics program in the United States designed specifically for you. The layout of Mathematics: Applications and Connections will delight your eyes. And the exciting content will hold your interest and show you why you need to study mathematics every day.

Please look carefully as you page through the text. Right away, you'll notice the variety of ways mathematics content is presented to you. You'll see the many connections made among mathematical topics and note how mathematics naturally fits into other subject areas and with technology.

You will note that content for each lesson is clearly labeled up front. And you'll appreciate the easy-to-follow lesson format. It introduces each new concept with an interesting application followed by clear examples.

Each day, as you read the text and complete the activities, you'll see the practical value of mathematics. You'll quickly grow to appreciate how often mathematics is used in real-world situations that relate directly to your life. If you don't already realize the importance of mathematics in your life, you soon will!

Sincerely, The Authors

Kay Balch

Linda Dritsas

David D. Molina

Patricia Frey-Mason

Beatrice Moore-Harris

Jack M. Ott

Ron Pelfrey

Barbara D. Smith

Patricia S. Wilson

Kay Balch teaches mathematics at Mountain Brook Junior High in Birmingham, Alabama. She is also the Mathematics Department Chairperson. Ms. Balch received her B.A. and M.A. from Auburn University in Alabama. She also has an Educational Specialist degree from the University of Montevallo. Ms. Balch is a member of the National Council of Teachers of Mathematics and is active in several other mathematics organizations at the national, state, and local levels.

Linda Dritsas is the Mathematics Coordinator for the Fresno Unified School District in Fresno, California. She also taught at California State University at Fresno for two years. Ms. Dritsas received her B.A. and M.A.(Education) from California State University at Fresno. Ms. Dritsas has published numerous mathematics workbooks and other supplementary materials. She has been the Central Section President of the California Mathematics Council and is a member of the National Council of Teachers of Mathematics and the Association for Supervision and Curriculum Development.

Arthur C. Howard is Consultant for Secondary Mathematics at the Aldine School District in Houston, Texas. He received his B.S. and M.Ed. from the University of Houston. Mr. Howard has taught in grades 7-12 and in college. He is Master Teacher in the Rice University School Mathematics Project in Houston. Mr. Howard is also active in numerous professional organizations at the national and state levels, including the National Council of Teachers of Mathematics. His publications include curriculum materials and articles for newspapers, books, and *The Mathematics Teacher.*

William Collins teaches mathematics at James Lick High School in San Jose, California. He has served as the Mathematics Department Chairperson at James Lick and Andrew Hill High Schools. He received his B.A. from Herbert H. Lehman College and is a Masters candidate at California State University, Hayward. Mr. Collins has been a consultant for the National Assessment Governing Board. He is a member of the National Council of Teachers of Mathematics and is active in several professional mathematics organizations at the state level. Mr. Collins is currently a mentor teacher for the College Board's EQUITY 2000 Consortium in San Jose, California.

Patricia Frey-Mason is the Mathematics Department Chairperson at the Buffalo Academy for Visual and Performing Arts in Buffalo, New York. She received her B.A. from D'Youville College in Buffalo, New York, and her M.Ed. from the State University of New York at Buffalo. Ms. Frey-Mason has published several articles in mathematics journals. She is a member of the National Council of Teachers of Mathematics and is active in other professional mathematics organizations at the state, national, and international levels. Ms. Frey-Mason was named a 1991 Woodrow Wilson Middle School Mathematics Master Teacher.

David D. Molina is a professor at Trinity University in San Antonio, Texas. He received his M.A. and Ph.D. in Mathematics Education from the University of Texas at Austin. Dr. Molina has been a speaker both at national and international mathematics conferences. He has been a presenter for the National Council of Teachers of Mathematics, as well as a conductor of workshops and in services for other professional mathematics organizations and school systems.

Beatrice Moore-Harris is the EQUITY 2000 Project Administrator and former Mathematics Curriculum Specialist for K-8 in the Fort Worth Independent School District in Fort Worth, Texas. She is also a consultant for the National Council of Teachers of Mathematics. Ms. Moore-Harris received her B.A. from Prairie View A & M University in Prairie View, Texas. She has also done graduate work there and at Texas Southern University in Houston, Texas, and Tarleton State University in Stephenville, Texas. Ms. Moore-Harris is active in many state and national mathematics organizations. She also serves on the Editorial Board of NCTM's *Mathematics and the Middle Grades* journal.

Ronald S. Pelfrey is the Mathematics Coordinator for the Fayette County Public Schools in Lexington, Kentucky. He has taught mathematics in Fayette County Public Schools, with the Peace Corps in Ethiopia, and at the University of Kentucky in Lexington, Kentucky. Dr. Pelfrey received his B.S., M.A., and Ed.D. from the University of Kentucky. He is also the author of several publications about mathematics curriculum. He is an active speaker with the National Council of Teachers of Mathematics and is involved with other local, state, and national mathematics organizations.

Barbara Smith is the Mathematics Supervisor for Grades K-12 at the Unionville-Chadds Ford School District in Unionville, Pennsylvania. Prior to being a supervisor, she taught mathematics for thirteen years at the middle school level and three years at the high school level. Ms. Smith received her B.S. from Grove City College in Grove City, Pennsylvania and her M.Ed. from the University of Pittsburgh in Pittsburgh, Pennsylvania. Ms. Smith has held offices in several state and local organizations, has been a speaker at national and state conferences, and is a member of the National Council of Teachers of Mathematics.

Jack Ott is a Professor of Mathematics Education at the University of South Carolina in Columbia, South Carolina. He has also been a consultant for numerous schools in South Carolina as well as the South Carolina State Department of Education and the National Science Foundation. Dr. Ott received his A.B. from Indiana Wesleyan University, his M.A. from Ball State University, and his Ph.D. from The Ohio State University. Dr. Ott has written articles for *The Mathematics Teacher* and *The Arithmetic Teacher* and has been a speaker at national and state mathematics conferences.

Jack Price has been active in mathematics education for over 40 years, 38 of those in grades K-12. He is currently the Co-Director of the Center for Science and Mathematics Education at California State Polytechnic University at Pomona, California, where he teaches mathematics and methods courses for preservice teachers and consults with school districts on curriculum change. Dr. Price received his B.A. from Eastern Michigan University, and has a Doctorate in Mathematics Education from Wayne State University. He is active in state and national mathematics organizations and is a past director of the National Council of Teachers of Mathematics.

Patricia S. Wilson is an Associate Professor of Mathematics Education at the University of Georgia in Athens, Georgia. Dr. Wilson received her B.S. from Ohio University and her M.A. and Ph.D. from The Ohio State University. She has received the Excellence in Teaching Award from the College of Education at the University of Georgia and is a published author in several mathematics education journals. Dr. Wilson has taught middle school mathematics and is currently teaching middle school mathematics methods courses. She is on the Editorial Board of the *Journal for Research in Mathematics Education*, published by the National Council of Teachers of Mathematics.

vi

Elaine Ivey
Mathematics Teacher
Adams Junior High School
Tampa, Florida

Donna Jamell
Mathematics Teacher
Ramsey Junior High School
Fort Smith, Arkansas

Augustus M. Jones
Mathematics Teacher
Tuckahoe Middle School
Richmond, Virginia

Marie Kasperson
Mathematics Teacher
Grafton Middle School
Grafton, Massachusetts

Larry Kennedy
Mathematics Teacher
Kimmons Junior High School
Fort Smith, Arkansas

Patricia Killingsworth
Math Specialist
Carver Math/Science Magnet
  School
Little Rock, Arkansas

Al Lachat
Mathematics Department
  Chairperson
Neshaminy School District
Feasterville, Pennsylvania

Kent Luetke-Stahlman
Resource Scholar Mathematics
J. A. Rogers Academy of Liberal
  Arts & Sciences
Kansas City, Missouri

Dr. Gerald E. Martau
Deputy Superintendent
Lakewood City Schools
Lakewood, Ohio

Nelson J. Maylone
Assistant Principal
Maltby Middle School
Brighton, Michigan

Irma A. Mayo
Mathematics Department
  Chairperson
Mosby Middle School
Richmond, Virginia

Daniel Meadows
Mathematics Consultant
Stark County Local School
  System
Canton, Ohio

Dianne E. Meier
Mathematics Supervisor
Bradford Area School District
Bradford, Pennsylvania

Rosemary Mosier
Mathematics Teacher
Brick Church Middle School
Nashville, Tennessee

Judith Narvesen
Mathematics Resource Teacher
Irving A. Robbins Middle School
Farmington, Connecticut

Raymond A. Nichols
Mathematics Teacher
Ormond Beach Middle School
Ormond Beach, Florida

William J. Padamonsky
Director of Education
Hollidaysburg Area School
  District
Hollidaysburg, Pennsylvania

Delores Pickett
Instructional Supervisor
Vera Kilpatrick Elementary
  School
Texarkana, Arkansas

Thomas W. Ridings
Team Leader
Gilbert Junior High School
Gilbert, Arizona

Sally W. Roth
Mathematics Teacher
Francis Scott Key Intermediate
  School
Springfield, Virginia

Dr. Alice W. Ryan
Assistant Professor of Education
Dowling College
Oakdale, New York

Fred R. Stewart
Supervisor of
  Mathematics/Science
Neshaminy School District
Langhorne, Pennsylvania

Terri J. Stillman
Mathematics Department
  Chairperson
Boca Raton Middle School
Boca Raton, Florida

Marty Terzieff
Secondary Math Curriculum
  Chairperson
Mead Junior High School
Mead, Washington

Tom Vogel
Mathematics Teacher
Capital High School
Charleston, West Virginia

Joanne Wilkie
Mathematics Teacher
Hosford Middle School
Portland, Oregon

Larry Williams
Mathematics Teacher
Eastwood 8th Grade School
Tuscaloosa, Alabama

Deborah Wilson
Mathematics Teacher
Rawlinson Road Middle School
Rock Hill, South Carolina

Francine Yallof
Mathematics Teacher
East Middle School
Brentwood, New York

Table of Contents

High Interest Features

Did You Know?
5, 14, 40, 54, 67, 79

Teen Scene
8, 79

When Am I Ever Going To Use This?
29, 61

Save Planet Earth
31

Cultural Kaleidoscope
74

Journal Entry
10, 13, 35, 53, 66, 77

Chapter

3

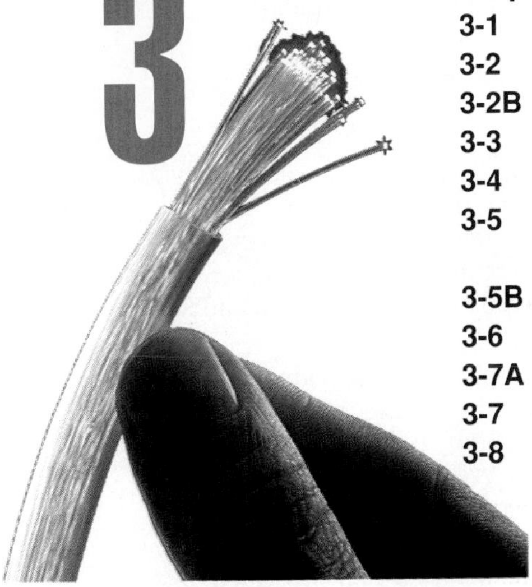

Statistics and Data Analysis

Chapter

4

Patterns and Number Sense

High Interest Features

Teen Scene
90, 158

Did You Know?
93, 102, 109, 113, 136,
150, 164

**When Am I Ever Going
To Use This?**
98, 145

Save Planet Earth
119, 160

Journal Entry
100, 115, 119, 135, 153,
156

Mini-Labs
145, 158, 166

Applications and Connections

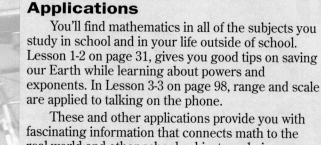

Have you ever asked yourself this question?

"When am I ever going to use this stuff?"

It may be sooner than you think! Here are two of the many ways this textbook will help you answer that question.

Applications

You'll find mathematics in all of the subjects you study in school and in your life outside of school. Lesson 1-2 on page 31, gives you good tips on saving our Earth while learning about powers and exponents. In Lesson 3-3 on page 98, range and scale are applied to talking on the phone.

These and other applications provide you with fascinating information that connects math to the real world and other school subjects and gives you a reason to learn math. Here are some more application topics.

entertainment	science
sports	social studies
smart shopping	music
hobbies	health
ecology	art

Five **DECISION MAKING** features further enable you to connect math to your real-life experiences as a consumer.

Connections

You'll discover that various areas of mathematics are very much interrelated. For example, Lesson 5-8 on page 201 shows one way in which fractions and probability are connected. Example 1 on page 322 connects algebra with making patterns with geometric shapes.

Connections to algebra, geometry, statistics, measurement, probability, and number theory help show the power of mathematics.

The **Mathematics Labs** and **Mini-Labs** also help you connect what you've learned before to new concepts. You'll use counters, measuring tapes, and many other objects to help you discover these concepts.

Chapter 5

Applications with Fractions

Chapter 6

An Introduction to Algebra

High Interest Features

Cultural Kaleidoscope
181

Teen Scene
182, 229

When Am I Ever Going To Use This?
194, 236

Did You Know?
200, 207, 233, 238

Save Planet Earth
222

Journal Entry
177, 185, 196, 227, 235, 245

Mini-Lab
182, 197

Chapter 7

Integers

Chapter 8

Investigations in Geometry

High Interest Features

Did You Know?
259, 263, 297, 303

When Am I Ever Going To Use This?
268, 308

Cultural Kaleidoscope
276

Teen Scene
278, 307

Journal Entry
256, 258, 289, 300, 305, 329

Mini-Labs
298, 314, 321

Chapter 9

Area

Chapter 10

Surface Area and Volume

High Interest Features

Did You Know?
338, 360, 378

Teen Scene
348, 388

When Am I Ever Going To Use This?
366, 383

Cultural Kaleidoscope
386

Journal Entry
340, 353, 358, 380, 400

Mini-Labs
338, 351, 383, 395

Technology

Labs, examples, computer-connection problems, and other features help you become an expert in using computers and calculators as problem-solving tools. You'll also learn how to read data bases, use spreadsheets, and use BASIC and LOGO programs. On many pages, **Calculator Hints** and printed keystrokes illustrate how to use a calculator.

Here are some highlights.

TV Stations
The number of television stations an average U.S. household receives.

27 — 1990
25 — 1988
— 1986
19 —
17 —
1984

$\square$ = 3 stations

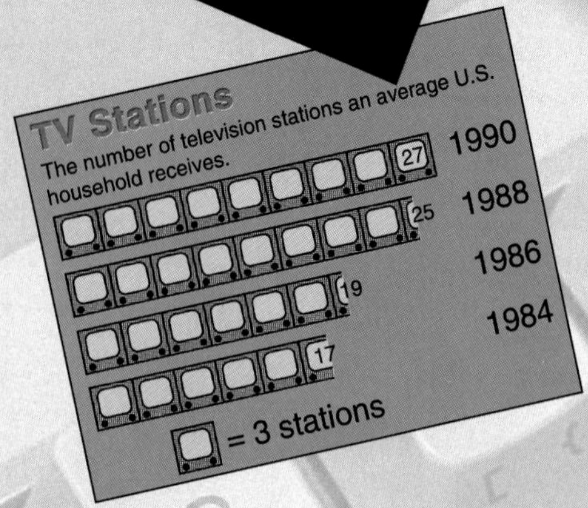

Chapter 11

Ratio, Proportion, and Percent

Chapter 12

Applications with Percent

High Interest Features

Did You Know?
411, 417, 462

When Am I Ever Going To Use This?
426, 465

Save Planet Earth
443

Teen Scene
445, 470

Cultural Kaleidoscope
458

Journal Entry
416, 435, 439, 461, 473, 481

Mini Labs
427, 433, 437, 444, 459, 476

High Interest Features

Teen Scene
497, 536

When Am I Ever Going To Use This?
500, 548

Did You Know?
506, 518, 540, 546

Save Planet Earth
520

Cultural Kaleidoscope
534

Journal Entry
499, 509, 513, 542, 545, 550

Mini-Labs
493, 500

End of Text Materials

Do you learn better when you use counters or tiles to help you work? You'll have the opportunity to work in a cooperative group to find out.

The **Extended Projects Handbook** consists of interesting long-term projects that involve serious issues that impact all citizens in the United States and the world.

If you've ever taken an extended trip, you know how important it is to use a map to guide you safely and surely to your destination. Please look upon the following four pages as a map of what you will learn in mathematics this year. Knowing what's ahead will help you make the most of the text's many features, which are designed to make learning math an interesting and valuable experience.

Objectives tell you exactly what you'll learn in each lesson.

Words to Learn lists the new words you'll encounter.

A variety of features help to guide you through each lesson.

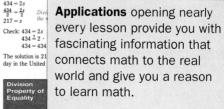

Solving Multiplication and Division Equations

Objective
Solve equations using the multiplication and division properties of equality.

Words to Learn
division
  property of equality
multiplication
  property of equality

How can you tell the difference between fraternal twins and identical twins? Fraternal twins do not necessarily look alike. They are not always the same sex, so you can't always tell they are twins. Identical twins look alike and are of the same sex. In the United States, an average of 434 twin babies are born each day. How many sets of twins are born each day?

You know that twins means two. So two times the number of sets of twins is the number of twin babies. If 434 twin babies are born each day, you can solve the equation $434 = 2s$ to find s, the average number of sets of twins born each day.

Since multiplication and division are inverse operations, equations that involve multiplication can be solved by dividing each side of the equation by the same number. Solve $434 = 2s$ using this method.

$$434 = 2s$$
$$\frac{434}{2} = \frac{2s}{2} \quad \text{Divi...}$$
$$217 = s$$

Check: $434 = 2s$
$434 \stackrel{?}{=} 2 \cdot$
$434 = 434$

The solution is 21... day in the United...

Applications opening nearly every lesson provide you with fascinating information that connects math to the real world and give you a reason to learn math.

Division Property of Equality

$8 = 8$	$a = b$
$\frac{8}{2} = \frac{8}{2}$	$\frac{a}{c} = \frac{b}{c}, c \neq 0$
$4 = 4$	

228   Chapter 6 An Introduction to Algebra

A margin feature called **Teen Scene** shares interesting, math-related tidbits about teens' lifestyles. Other intriguing margin features are **Did You Know?** and **When Am I Ever Going to Use This?**

Equations that involve division can be solved by multiplying each side of the equation by the same number.

Example 1 *Problem Solving*

LOOK BACK
You can review fractions on page 190.

Marketing Did you know that chewing gum loses its flavor after only about 20 minutes ($\frac{1}{3}$ hour)? However, scientists have recently invented chewing gum that will keep its flavor longer, using synthetically derived polymers. If the newly-developed polymer chewing gum keeps its flavor 30 times as long, use the equation $\frac{1}{3} = \frac{h}{30}$ to find h, the number of hours it keeps its flavor.

$$\frac{1}{3} = \frac{h}{30}$$
$$\frac{1}{3} \cdot 30 = \frac{h}{30} \cdot 30 \quad \textit{Multiply each side by 30 to undo the division by 30.}$$
$$10 = h$$

Check: $\frac{1}{3} = \frac{h}{30}$
$\frac{1}{3} \stackrel{?}{=} \frac{10}{30} \quad \textit{Replace h with 10.}$
$\frac{1}{3} = \frac{1}{3}$ ✓

The solution is 10. The newly-developed polymer chewing gum may keep its flavor up to 10 hours.

TEEN SCENE

After World War II, various waxes, plastics, and synthetic rubber virtually replaced chicle in making chewing gum. Artificially sweetened chewing gum found a wide market in the U.S. in the late 20th century, with mint being the favorite flavor.

Multiplication Property of Equality

In words: If each side of an equation is multiplied by the same number, then the two sides remain equal.

Arithmetic	Algebra
$4 = 4$	$a = b$
$4 \cdot 2 = 4 \cdot 2$	$ac = bc$
$8 = 8$	

Example 2

Estimation Hint
••••••••••••
In Example 2, think $360 \div 2 = 180$. The solution is about 180.

Solve $368 = 2.3b$. Check your solution.
$$368 = 2.3b$$
$$\frac{368}{2.3} = \frac{2.3b}{2.3}$$
$$368 \boxed{\div} 2$$
$$160 =$$

Check: 3
3
3

The solut...

Problem Solving gives you the opportunity to use mathematics to find the solution to interesting application problems in marketing and other real-life fields.

Estimation Hints provide clues about when it's best to solve problems using estimation. Some lessons also include **Mental Math Hints, Problem-Solving Hints,** and **Calculator Hints.**

Previewing Your Text

Communicating Mathematics gives you a chance to show what you've learned about a math concept by talking or writing about it, or by drawing a picture or making a model.

Mixed Reviews present problems that help you remember what you've learned. Lesson references tell you exactly where to look in previous lessons to restudy important concepts.

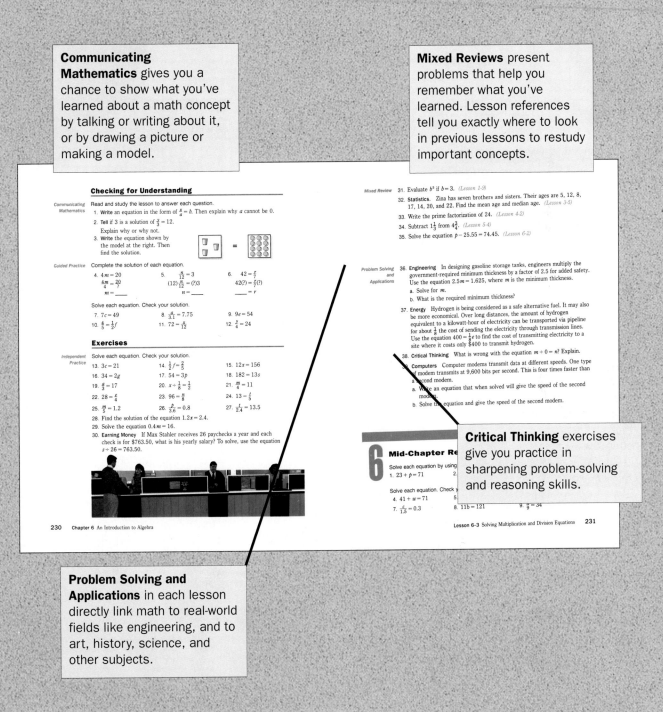

Checking for Understanding

Communicating Mathematics

Read and study the lesson to answer each question.
1. **Write** an equation in the form of $\frac{x}{a} = b$. Then explain why a cannot be 0.
2. **Tell** if 3 is a solution of $\frac{n}{3} = 12$. Explain why or why not.
3. **Write** the equation shown by the model at the right. Then find the solution.

Guided Practice

Complete the solution of each equation.

4. $4m = 20$
$\frac{4m}{4} = \frac{20}{?}$
$m = \underline{\quad}$

5. $\frac{n}{12} = 3$
$(12)\frac{n}{12} = (?)3$
$n = \underline{\quad}$

6. $42 = \frac{r}{7}$
$42(?) = \frac{r}{7}(?)$
$\underline{\quad} = r$

Solve each equation. Check your solution.

7. $7c = 49$
8. $\frac{a}{3.1} = 7.75$
9. $9e = 54$
10. $\frac{4}{5} = \frac{1}{2}f$
11. $72 = \frac{x}{12}$
12. $\frac{y}{4} = 24$

Exercises

Independent Practice

Solve each equation. Check your solution.
13. $3c = 21$
14. $\frac{1}{2}f = \frac{2}{5}$
15. $12x = 156$
16. $34 = 2g$
17. $54 = 3p$
18. $182 = 13s$
19. $\frac{a}{5} = 17$
20. $x \div \frac{1}{8} = \frac{1}{2}$
21. $\frac{m}{4} = 11$
22. $28 = \frac{c}{4}$
23. $96 = \frac{n}{8}$
24. $13 = \frac{t}{5}$
25. $\frac{m}{5} = 1.2$
26. $\frac{p}{3.6} = 0.8$
27. $\frac{y}{2.4} = 13.5$
28. Find the solution of the equation $1.2x = 2.4$.
29. Solve the equation $0.4m = 16$.
30. **Earning Money** If Max Stahler receives 26 paychecks a year and each check is for $763.50, what is his yearly salary? To solve, use the equation $s \div 26 = 763.50$.

230 Chapter 6 An Introduction to Algebra

Mixed Review
31. Evaluate b^5 if $b = 3$. *(Lesson 1-9)*
32. **Statistics.** Zina has seven brothers and sisters. Their ages are 5, 12, 8, 17, 14, 20, and 22. Find the mean age and median age. *(Lesson 3-5)*
33. Write the prime factorization of 24. *(Lesson 4-2)*
34. Subtract $1\frac{1}{3}$ from $4\frac{3}{4}$. *(Lesson 5-4)*
35. Solve the equation $p - 25.55 = 74.45$. *(Lesson 6-2)*

Problem Solving and Applications
36. **Engineering** In designing gasoline storage tanks, engineers multiply the government-required minimum thickness by a factor of 2.5 for added safety. Use the equation $2.5m = 1.625$, where m is the minimum thickness.
a. Solve for m.
b. What is the required minimum thickness?
37. **Energy** Hydrogen is being considered as a safe alternative fuel. It may also be more economical. Over long distances, the amount of hydrogen equivalent to a kilowatt-hour of electricity can be transported via pipeline for about $\frac{1}{8}$ the cost of sending the electricity through transmission lines. Use the equation $400 = \frac{1}{8}e$ to find the cost of transmitting electricity to a site where it costs only $400 to transmit hydrogen.
38. **Critical Thinking** What is wrong with the equation $m \div 0 = n$? Explain.
39. **Computers** Computer modems transmit data at different speeds. One type of modem transmits at 9,600 bits per second. This is four times faster than a second modem.
a. Write an equation that when solved will give the speed of the second modem.
b. Solve the equation and give the speed of the second modem.

6 Mid-Chapter Re

Solve each equation by using
1. $23 + p = 71$
2.

Solve each equation. Check y
4. $41 + w = 71$
5.
7. $\frac{c}{1.5} = 0.3$
8. $11b = 121$
9. $\frac{x}{9} = 34$

Lesson 6-3 Solving Multiplication and Division Equations 231

Critical Thinking exercises give you practice in sharpening problem-solving and reasoning skills.

Problem Solving and Applications in each lesson directly link math to real-world fields like engineering, and to art, history, science, and other subjects.

Getting Into Each Chapter

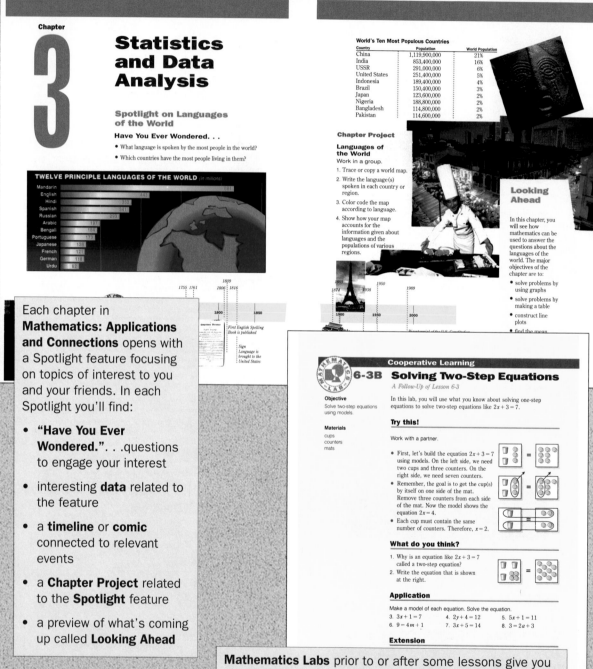

Each chapter in **Mathematics: Applications and Connections** opens with a Spotlight feature focusing on topics of interest to you and your friends. In each Spotlight you'll find:

- **"Have You Ever Wondered."**. . .questions to engage your interest

- interesting **data** related to the feature

- a **timeline** or **comic** connected to relevant events

- a **Chapter Project** related to the **Spotlight** feature

- a preview of what's coming up called **Looking Ahead**

Mathematics Labs prior to or after some lessons give you hands-on experience, with a partner or group, in discovering a math concept on your own. You may also participate in shorter **Mini-Labs** in which you will investigate math concepts within a lesson.

Wrapping Up Each Chapter

A **Study Guide and Review** at the end of each chapter helps you connect the whys and hows of reviewing what you've learned by presenting objectives and examples in the left column and review exercises in the right. This unique format makes it easier for you to retain and apply what you've learned.

Chapter

1 Study Guide and Review

Communicating Mathematics

Choose the correct term to complete the sentence.

1. The first step of the four-step plan is __?__.
2. Front-end estimation is a process that can be used to estimate the __?__ of two values.
3. The expression $3x - 4$ is called an __?__.
4. Order of operations states that __?__ precedes addition and/or subtraction.
5. In the expression 4^3, 4 is called the __?__.
6. In your own words, explain what should be done during the examine step of the four-step plan.

equation
exponent
solution
explore
product
sum
algebraic expression
multiplication
base

Skills and Concepts

Objectives and Examples

Upon completing this chapter, you should be able to:

- solve problems using the four-step plan *(Lesson 1-1)*

  The four steps are *Explore, Plan, Solve* and *Examine.*

- estimate sums and differences using front-end estimation *(Lesson 1-2)*

  Estimate $8,324 + 6,936$.

  $\begin{array}{ccc} 8,324 & & 8,324 \\ +6,936 & \rightarrow & +6,936 \end{array}$

Review Exercises

Use these exercises to review and prepare for the chapter test.

Use the four-step plan to solve.

7. A car traveling 60 mph will travel how far in 7 hours?
8. Ann starts the day with $100. If she spends $35 at the mall, how much is left for groceries?

Estimate.

9. $16,327 + 8,102$
10. $43,212 - 21,605$
11. $864 - 219$
12. $3,986 + 4,627$

Review Exercises

Estimate.

14. $987 \div 11$
15. $12,042 \div 62$
16. $1,100 \div 12$
17. $6,350 \div 92$
18. $254,102 \div 512$

Add or subtract mentally.

19. $37 + 86$
20. $148 - 63$
21. $8,986 + 348$
22. $2,432 - 1,999$
23. $7,345 - 19$
24. $38 + 15,893$

Evaluate each expression.

25. $3 + 7 \cdot 4 - 6$
26. $8(16 - 5) - 6$
27. $12 - 18 \div 9$
28. $83 + 3(4 - 2)$
29. $75 \div 3 + 6(5 - 1)$
30. $10(12 - 2) \div 10$

Evaluate $3(9 + 7) - 4 \div 2 + 3$.

$\begin{aligned} 3(9 + 7) &- 4 \div 2 + 3 \\ &= 3(16) - 4 \div 2 + 3 \\ &= 48 - 2 + 3 \\ &= 49 \end{aligned}$

- evaluate numerical and simple algebraic expressions *(Lesson 1-8)*

  Evaluate $6x - xy + y$ if $x = 10$, $y = 3$.

  $\begin{aligned} 6x - xy + y &= 6(10) - (10)(3) + 3 \\ &= 60 - 30 + 3 \\ &= 33 \end{aligned}$

Evaluate each expression if $p = 12$, $q = 3$, and $r = 5$.

31. $p + q - r$
32. $\frac{p + q}{r}$
33. $3(p + q) - r$
34. $6qr - p$
35. $25 - 2(p - 2q)$

- use powers and exponents in expressions *(Lesson 1-9)*

  $2^5 = 2 \cdot 2 \cdot 2 \cdot 2 \cdot 2 = 32$

Evaluate each expression.

36. 10^3
37. 3^6
38. 15 squared
39. y^4 if $y = 4$

- solve equations using mental math *(Lesson 1-10)*

  Solve $3s = 36$.

  You know that $3 \cdot 12 = 36$. So, the value of s is 12.

Solve each equation using mental math.

40. $t - 12 = 35$
41. $8x = 88$
42. $\frac{m}{4} = 16$
43. $28 + r = 128$

Applications and Problem Solving

Choose the method of computation. Then solve.

44. Use all five even digits to form a two-digit number and a three-digit number which will yield the smallest possible product. *(Lesson 1-5)*

If the problem has enough facts, solve it. If not, write the missing f...

45. **Seating** Anna, Barbara, Conchita, Doris, and Emma sit arou... girls whose names begin with adjacent letters of the alphabe... the table. Emma sits on the right of Conchita. Where do the...

46. **Business** A house painter finds that he needs 30 minutes to... before painting and then 5 minutes for every square foot tha... expression for the total time the painter needs. Let n repres... feet he paints. *(Lesson 1-8)*

Curriculum Connection Projects suggest short research projects and field trips that link math to language arts, health, and other subjects, and to real-world situations.

Curriculum Connection Projects

- **Language Arts** Write ten sentences in which the word *variable* is used as an adjective. Make a crossword puzzle or word search from the ten nouns *variable* describes.
- **Health** Learn an emergency procedure such as CPR or the Heimlich maneuver with a friend. Write steps for demonstrating your procedure to the class.

Read More About It

DeJong, Meindert. *The Wheel on the School.*
Smullyan, Raymond. *Alice in Puzzle-Land.*
Packard, Edward. *The Cave of Time.*

Read More About It lists fiction and nonfiction books that may interest you.

Now that you've looked at your "map" and know what's ahead, you're prepared to begin using the text. You'll soon see that you can do math. Math is for everyone, and you will use it every day. Remember, attitude counts. Textbooks, and your teachers and parents who help you interpret them, are excellent and vital resources. But the most important factor for your success in mathematics is you. Have a wonderful year!

1

Tools for Problem Solving

Spotlight on Skiing

Have you ever wondered . . .

- How much lift tickets would cost for two adults for a three-day weekend if one lift ticket costs $25 per day?

- Why high mountains are covered in snow? Because the temperature falls 41° every time the height increases by 3,280 feet. What is the temperature on top of a mountain that is 10,000 feet high if the temperature at the foot of the mountain is 95°?

SNOW PLACES IN THE U.S.	
1989 Total Snowfall	
Anchorage, Alaska	68.5 inches
Juneau, Alaska	98.2 inches
Flagstaff, Arizona	96.4 inches
Denver, Colorado	59.8 inches
Sault St. Marie, Michigan	116.4 inches
Mt. Washington, New Hampshire	254.8 inches

HI & LOIS

Skiing

Work in a group.

1. Plan a ski weekend with your group. Choose a couple of locations and compare the costs of transportation, lodging, ski-lift tickets, and food at each location. Check with travel agents, airlines, newspapers, ski magazines, and so on.

2. Find out the cost of buying or renting clothes, accessories, and equipment.

3. What factors might affect the cost of your trip?

4. Estimate the total cost of your trip.

5. Choose the most reasonable location.

6. Compare your information with that of other groups. Is there a significant price difference in your ski weekends?

Looking Ahead

In this chapter, you will see how mathematics can be used to answer the questions about skier spending.

The major objectives of this chapter are to:

- solve problems using the four-step plan
- use estimation to solve problems
- use the order of operation
- solve algebraic expressions

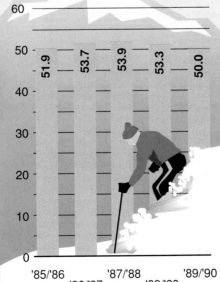

SKIER VISITS

Total number of visits to, or days spent at, U.S. ski areas, in millions.

	51.9	53.7	53.9	53.3	50.0

'85/'86   '87/'88   '89/'90
   '86/'87   '88/'89

3

1-1 A Plan for Problem Solving

Objective

Solve problems using the four-step plan.

In the news, you have often heard of people who are millionaires or billionaires. You know that they are very rich, but just how rich are they?

One way to understand the size of a million is to find how many days it would take you to count to one million. You can use a four-step plan to solve this and other problems. The four steps are described below.

1. **Explore** The plan begins with understanding the problem. You need to know what information you have and need and what is asked.

2. **Plan** After understanding the problem, you should develop a plan to solve it. There are many ideas or strategies you can use. You will learn many of these in this book. It is usually helpful to make an estimate.

3. **Solve** Then you carry out your plan. If the plan does not work, try another—and another.

4. **Examine** Finally, you should look at your answer to see whether it answers the question you were asked. You may also check your answer by solving the problem in another way by using a different strategy. Compare your answer to the estimate. If the answer doesn't make sense, make a new plan and try again.

Example 1

Let's try the four-step plan with the opening problem.

Explore *What is given? What do you know?*
- There are 60 seconds in a minute.
- There are 60 minutes in an hour.
- There are 24 hours in a day.
- Assume that it takes 1 second to say a number.

What is asked?
- How many days will it take you to count to one million?

Plan Since you have assumed it takes 1 second to say a number, it will take 1,000,000 seconds to count to one million. To find the number of days it takes to count to one million, first find out how many seconds there are in a day. To do this, change days to hours, hours to minutes, and minutes to seconds. Then divide to find the number of days it takes to count to one million.

Solve
- Change days to minutes.

$$24 \ \boxed{\times} \ 60 \ \boxed{=} \ \text{1440} \quad \textit{minutes in one day}$$

$$\underset{\textit{hours in a day}}{\uparrow} \qquad \underset{\textit{minutes in an hour}}{}$$

- Change minutes to seconds.

$$1440 \ \boxed{\times} \ 60 \ \boxed{=} \ \text{86400} \quad \textit{seconds in one day}$$

$$\underset{\textit{minutes in a day}}{\uparrow} \qquad \underset{\textit{seconds in a minute}}{}$$

- Divide to find the number of days it takes to count to one million.

$$1000000 \ \boxed{\div} \ 86400 \ \boxed{=} \ \text{11.574074} \quad \textit{days}$$

$$\underset{\textit{count}}{\uparrow} \qquad \underset{\textit{seconds in a day}}{\uparrow}$$

- It would take *about* $11\frac{1}{2}$ days to count to one million. That means not sleeping, not eating, not doing anything except counting.

Examine Is your answer reasonable? You can check division by multiplying.

$$11.574074 \ \boxed{\times} \ 86400 \ \boxed{=} \ \text{999999.99} \quad \checkmark$$

Example 2

The trans-Alaska pipeline can deliver about two million barrels of oil a day. The 810-mile trip from Prudhoe Bay to Valdez in Alaska takes about five days. How many miles a day is this?

Explore You need to find the number of miles traveled in one day. You know the trip is 810 miles long and the trip takes 5 days.

Plan   To find the number of miles the oil travels in one day, divide 810 by 5.

Estimate. $500 \div 5 = 100$
$1,000 \div 5 = 200$

The answer should be between 100 and 200.

Solve   Now solve. $810 \div 5 = 162$

The oil travels 162 miles a day.

Examine   Is your answer reasonable? Remember you can use multiplication to check division: $162 \times 5 = 810$.

You can also examine the problem mentally. If the trip were made in 10 days, the oil would move 81 miles daily. So in five days (half the time), it would move twice as far or 162 miles.

Remember to compare your answer to the estimate to see if your answer is reasonable.

Checking for Understanding

Communicating Mathematics   Read and study the lesson to answer each question.

1. **Tell** why it is important to plan before solving a problem.

2. **Tell** two reasons for including the *Examine* step.

3. **Write** in your own words:

   a. what to do if your first plan does not work.

   b. how you know it did not work.

Guided Practice   4. **Travel**   The Masons traveled 753 miles to the Great Smoky Mountains for their vacation. They took a different route home and traveled 856 miles. How many miles did they travel in all?

5. **Gardening**   Three garden hoses are joined together to reach a flower bed. Two hoses are 25 feet each and one is 50 feet. Will the combined length reach 92 feet?

Exercises

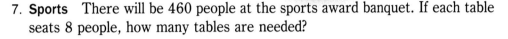

6. **Sports**   Hank Aaron is the all-time home run leader of major league baseball with 755 home runs. In 1966 and 1967 alone, he hit a total of 83 home runs. If he hit 44 home runs in 1966, how many home runs did he hit in 1967?

7. **Sports**   There will be 460 people at the sports award banquet. If each table seats 8 people, how many tables are needed?

8. **Critical Thinking**   Find the least four-digit number that is divisible by 2, 5, and 9.

9. **Smart Shopping**   David saw an advertisement for piano lessons at $15.95 per lesson. How much will 12 lessons cost?

10. **PTA**   The PTA sold 84 tickets to WildWaters Amusement Park in 2 hours. At that rate, how many tickets will they sell in two six-hour days at school?

11. **Transportation**   Nine school buses serve Maplewood Middle School. The buses travel a total of 4,482 miles in one school week. How many miles does each bus travel weekly?

12. **Baseball**   The table at the right gives the highest number of home runs hit in each year during the 1970s in either the National or American Leagues. What is the total number of home runs hit by the leaders during this time?

Year	HR	Year	HR
1970	45	1975	38
1971	48	1976	38
1972	40	1977	52
1973	44	1978	40
1974	36	1979	48

13. **Mathematics and Media**   Read the following paragraph.

> In 1906, R. Fessenden made the first radio transmission of human speech. How is sound transmitted by radio? Sound waves consist of air which is compressed and then expanded. A microphone changes the sound waves into electrical signals, and a transmitter produces radio waves which can carry the sound signals. An aerial radiates the radio waves into the air which can be received by a radio with an antenna. A speaker in the radio converts the radio waves back into sound waves.

In 1989, 343 million radios were in use in the home, 131.4 million in cars, 37.8 million in trucks and vans, and 20.8 million at work. *About* how many radios were in use?

1-2 Front-End Estimation

Objective

Estimate sums and differences using front-end estimation.

Words to Learn

front-end estimation

TEEN SCENE

An average American teenager eats 1,817 pounds of food in one year.

After school, Rob walked to the grocery store for his family. They needed some vegetable oil, chicken, bacon, and a green pepper for the chicken casserole they were making for dinner. The register tape for his purchases is shown below.

```
Lucky Markets
   June 10

oil            2.55
chicken        5.47
bacon          2.15
g. pepper      1.09
             -----
             11.26
```

Rob used **front-end estimation** to determine whether he had been charged the right amount.

He used the "front end" or the left-hand column for a first estimate. He looked at the next column to *adjust* his estimate. Adjusting helps you to get a closer estimate.

$$
\begin{array}{r}
2.55 \\
5.47 \\
2.15 \\
+\ 1.09 \\
\hline
\$10.00
\end{array}
\qquad
\begin{array}{r}
2.55 \\
5.47 \\
2.15 \\
+1.09 \\
\hline
\$1.00
\end{array}
$$

$10.00 + $1.00 = $11.00   Rob's estimate is $11.

Compared to the estimate, the total amount on the register tape is reasonable.

Front-end estimation for addition can be summarized as follows.

Front-End Estimation	Use front-end estimation as follows.
	1. Add the front-end digits.
	2. Adjust by estimating the sum of the remaining digits.
	3. Add the two values.

Front-end estimation can also be used for subtraction.

Example

Estimate the difference of 11,745 and 8,285.

$$
\begin{array}{r}
11{,}745 \\
-8{,}285 \\
\hline
3{,}000
\end{array}
\qquad
\begin{array}{r}
11{,}745 \\
-8{,}285 \\
\hline
500
\end{array}
$$

Subtract the front-end digits. The first estimate is about 3,000. Look at the next column and adjust your estimate. The difference is about 3,500.

$3,000 + 500 = 3,500$

Checking for Understanding

Communicating Mathematics
Read and study the lesson to answer each question.

1. **Write** a subtraction problem and tell what digits are the front-end digits.

2. **Tell** why you should look at the remaining digits after you have used the front-end digit to estimate.

3. **Show** how you would use front-end estimation to estimate the difference of 764 and 149.

Guided Practice   Estimate. Use front-end estimation.

4. $532 + 625 + 419$

5. $927 - 618$

6. $427 + 962 + 520$

7. $8,429 - 6,258$

8. $\$35.80 + \$29.90 + \$41.90$

9. $15,734 - 3,212$

10. $3,624 + 521 + 8,201$

11. $680 - 83$

12. Alma went to the grocery store to buy the ingredients she will need to make Rice Krispie® treats. She was charged $6.04. Use front-end estimation to see if the total amount on the register tape is reasonable.

```
Greene's Grocery
     Aug. 21

Rice Krispies 3.49
Marshmallows   1.10
Margarine      1.45
               -----
Total          6.04
```

Exercises

Independent Practice   Estimate. Use front-end estimation.

13. $4,522 + 6,059$

14. $7,247 - 4,126$

15. $15,923 - 6,782$

16. $8,723 - 5,247$

17. $6,522 + 493 + 8,026$

18. $5,278 + 4,258 + 6,233$

19. $\$6.68 + \$9.00 + \$2.42$

20. $65,378 - 23,753$

21. Estimate the sum of 6,015 and 2,843.

22. Estimate the difference of 4,963 and 519.

23. Estimate the difference of 37,444 and 13,127.

24. Estimate the sum of 3,988 and 12,019.

25. *True* or *False:* In the four-step plan, the *Solve* step comes last. *(Lesson 1-1)*

26. **Travel** A bicyclist planning an 1,800-mile trip decides that he can ride 15 miles per hour for 6 hours each day. How many days will it take for him to complete the trip? *(Lesson 1-1)*

27. **Smart Shopping** At the grocery store, Jeanne is deciding between purchasing a 50-ounce bottle of shampoo priced at 10¢ per ounce and a 30-ounce bottle priced at 20¢ per ounce. Which bottle will cost more? *(Lesson 1-1)*

28. **Music** A popular rock band is putting together a new album. They can use 30 minutes of music for the first side. They have selected 3 songs that are each 4 minutes long and 4 songs that are each 5 minutes long. Will all of their selections fit on the first side? *(Lesson 1-1)*

29. **Smart Shopping** To go back to school, Sam bought 3 pairs of jeans at $31.99 each and 5 shirts at $20.99 each. Estimate the cost of his clothes using front-end estimation.

30. **Clubs** The Band Boosters had 10,420 boxes of cookies to sell in order to buy new band uniforms. After the first week, there were 6,147 boxes left. *About* how many boxes did they sell the first week?

31. **Smart Shopping** Julia bought a pair of shoes for $21.48, including tax. She gave the sales clerk a twenty-dollar bill and a five-dollar bill and received $2.52 in change. Use front-end estimation to see if Julia received the right amount of change.

32. **Critical Thinking** How can you use front-end estimation to estimate the difference of 8,267 and 3,798? Will the exact answer be more or less than the estimate?

33. **Critical Thinking** Given the digits 4, 5, 6, 7, 8, and 9, and using each digit only once, form two three-digit numbers that will give you the greatest sum.

34. **Journal Entry** Write one or more sentences to explain how you know that the sum of 3,659 and 6,510 is about 10,000.

1-3 Compatible Numbers

Objective

Estimate quotients using compatible numbers.

Words to Learn

compatible numbers

The first Mickey Mouse animated film was *Steamboat Willie*. It was released in 1928. A man named Ub Iwerks drew the pictures for it. He drew a total of 14,400 pictures in just 24 days! *About* how many pictures did he draw per day?

You can use **compatible numbers** to estimate the quotient, $14,400 \div 24$. Compatible numbers are two numbers that are easy to divide mentally. They are often members of fact families.

To estimate quotients using compatible numbers, round the divisor to its greatest place-value position. Then replace the dividend with a compatible number.

To estimate $14,400 \div 24$, round 24 to 20. Replace 14,400 with 14,000 since it is compatible with 20.

$$14,000 \div 20 = 700 \quad \leftarrow \quad estimate$$

Ub Iwerks drew *about* 700 pictures a day. That's a lot of pictures!

Mental Math Tip

● ● ● ● ● ● ● ● ● ● ● ●

You can use patterns to find
$14,000 \div 20$.

$14 \div 2 = 7$

$140 \div 20 = 7$

$1,400 \div 20 = 70$

$14,000 \div 20 = 700$

Examples

1 Estimate $5,461 \div 6$.

$5,461 \div 6 \quad \rightarrow \quad 5,400 \div 6 = 900$

5,400 and 6 are compatible numbers.

$5,461 \div 6$ is *about* 900.

2 Estimate $654 \div 72$.

$654 \div 72 \quad \rightarrow \quad 630 \div 70 = 9$

Round 72 to 70. 630 and 70 are compatible numbers.

$654 \div 72$ is *about* 9.

Lesson 1–3 Estimation Strategy: Compatible Numbers **11**

Example 3 *Problem Solving*

Air Travel The first airplane to fly around the world without refueling in midair was the *Voyager*. In 1986, it circled the globe, flying 26,000 miles in 9 days. *About* how many miles did it travel per day?

$$26,000 \div 9 \quad \rightarrow \quad 27,000 \div 9 = 3,000$$

27,000 and 9 are compatible numbers.

The *Voyager* traveled *about* 3,000 miles per day.

Checking for Understanding

Communicating Mathematics

Read and study the lesson to answer each question.

1. **Tell** two numbers that are compatible numbers.

2. **Write** a problem in which estimating with compatible numbers could be used.

3. **Tell** why estimating with compatible numbers is useful.

Guided Practice

Estimate. Use compatible numbers.

4. $731 \div 3$	5. $589 \div 5$	6. $876 \div 9$
7. $498 \div 6$	8. $397 \div 20$	9. $632 \div 30$
10. $221 \div 15$	11. $954 \div 29$	12. $1,418 \div 7$
13. $6,399 \div 80$	14. $3,642 \div 93$	15. $4,328 \div 87$

16. Suppose you drive your moped about 243 miles on 6 gallons of gasoline. *About* how many miles per gallon does your moped get?

Exercises

Independent Practice

Estimate. Use compatible numbers.

17. $619 \div 6$	18. $364 \div 9$	19. $897 \div 3$
20. $744 \div 80$	21. $527 \div 50$	22. $456 \div 80$
23. $432 \div 25$	24. $378 \div 94$	25. $604 \div 78$
26. $2,973 \div 9$	27. $3,740 \div 6$	28. $9,273 \div 3$
29. $3,273 \div 50$	30. $4,927 \div 20$	31. $8,772 \div 30$

32. Estimate the quotient of 6,152 and 58.

33. Estimate the quotient of 8,497 and 45.

34. Estimate the quotient of 7,428 and 27.

35. **Sports** A baseball stadium holds 20,000 people. If 3,650 people can be seated in the bleachers, how many seats are available in the rest of the stadium? *(Lesson 1-1)*

36. **Construction** A new highway under construction requires 4 tons of concrete for every 2 miles of road. How many tons of concrete will be needed to build 100 miles of the highway? *(Lesson 1-1)*

37. Use front-end estimation to estimate the sum of 3,425 and 6,243. *(Lesson 1-2)*

38. **Summer Camp** A total of 486 campers are expected to arrive at Camp Miami. If 269 have already arrived, about how many more campers are still expected? Use front-end estimation. *(Lesson 1-3)*

39. **Banking** Anthony has a savings account that currently has a balance of $2,346.21. He deposits a check in the amount of $521.19. About how much money does he have in his account now? Use front-end estimation. *(Lesson 1-2)*

40. **Art** In a wheat field near Duns, Scotland, a reproduction of Vincent van Gogh's *Sunflowers* was created with 250,000 plants and flowers. The "painting" covered a 46,000-square foot area. *About* how many flowers and plants were planted per square foot?

41. **Travel** Big Ben, located in London's Houses of Parliament, is the name of the clock's largest bell. The bell weighs about 27,000 pounds. This is about 68 times as much as the hammer that strikes the bell weighs. *About* how much does the hammer weigh?

42. **Animals** An elephant in a zoo will eat 3,000 cabbages in a year. *About* how many cabbages does an elephant in a zoo eat in one week?

43. **Critical Thinking** Write a problem in which you could use both front-end estimation and compatible number estimation to solve it.

44. **Journal Entry** Find the definition of *compatible* in a dictionary. Explain how the definition describes the estimation strategy.

1-4 Compensation

Objective

Compute sums and differences using compensation.

Words to Learn

compensation

When you use estimation, you get an approximate answer. When you use compensation, you get an exact answer.

M&M's® were introduced in 1940 as the candy that wouldn't melt in your hand. The name M&M came from the first letters of the last names of the owners of the Mars Company, Forrest Mars and Bruce Murrie. You'll find about 500 candies in 6 colors in a one-pound bag of M&M's®.

Suppose one bag of M&M's® has 153 brown, 98 yellow, 104 red, 56 green, 48 tan, and 52 orange candies. How many yellow and green M&M's® are there?

Before you take out a pencil and paper to solve this problem, let's discuss a strategy for computing exact sums mentally. This strategy is called **compensation.** In compensation, you change a problem to make it easier to solve mentally. Using basic facts and a number line can show you how compensation works.

Addition

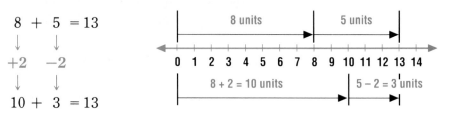

$$8 + 5 = 13$$
$$\downarrow \quad \downarrow$$
$$+2 \quad -2$$
$$\downarrow \quad \downarrow$$
$$10 + 3 = 13$$

The total distance, 13, must stay the same. So if you add units to one section, you must subtract the same number of units from the other section.

Subtraction

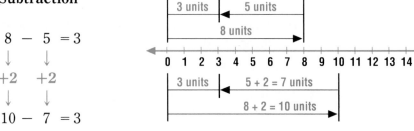

$$8 - 5 = 3$$
$$\downarrow \quad \downarrow$$
$$+2 \quad +2$$
$$\downarrow \quad \downarrow$$
$$10 - 7 = 3$$

The distance, 3, representing the difference of the numbers, must stay the same. So if you add units to one section, you must add the same number of units to the other section also.

Example 1

Find the total number of yellow and green M&M's® using compensation.

$$
\begin{array}{ccccc}
98 & \rightarrow & +2 & \rightarrow & 100 \\
\underline{+56} & \rightarrow & -2 & \rightarrow & \underline{+\ 54} \\
& & & & 154
\end{array}
$$

There are 154 yellow and green M&M's®.

Example 2 *Problem Solving*

Smart Shopping Kim bought a pair of sunglasses for $18.68 and gave the cashier a twenty-dollar bill. How much change should she get back? Solve mentally.

$$
\begin{array}{ccccc}
\$20.00 & \rightarrow & -0.01 & \rightarrow & 19.99 \\
\underline{-18.68} & \rightarrow & -0.01 & \rightarrow & \underline{-18.67} \\
& & & & 1.32
\end{array}
$$

Since you have all 9s from which to subtract, it is easier to subtract mentally.

Kim should get $1.32 in change.

Checking for Understanding

Communicating Mathematics

Read and study the lesson to answer each question.

1. **Tell** how you would use compensation to find $596 + 247$ mentally.

2. **Tell** how you would use compensation to find $303 - 247$ mentally.

3. **Write** the difference between estimating the sum $2,993 + 243$ and using compensation to find the sum.

Guided Practice

Name the numbers you would use to solve each addition and subtraction problem using compensation.

4. $67 + 26$
5. $\$5.97 + \$3.28$
6. $148 + 66$
7. $192 - 58$
8. $152 - 49$
9. $\$7.00 - \$4.28$

Add or subtract mentally.

10. $48 + 43$
11. $157 + 36$
12. $\$4.98 + \$3.27$
13. $74 - 47$
14. $413 - 78$
15. $\$5.01 - \$1.38$

Exercises

Add or subtract mentally.

16. $46 + 24$	17. $78 + 33$	18. $\$2.77 + \$0.43$
19. $45 - 23$	20. $56 - 48$	21. $912 - 89$
22. $108 + 46$	23. $598 + 354$	24. $597 + 398$
25. $500 - 346$	26. $\$6.02 - \$4.29$	27. $366 - 87$
28. $6,992 + 858$	29. $3,987 + 597$	30. $\$26.76 + \$49.25$
31. $\$3.95 - \$2.92$	32. $716 - 486$	33. $4,000 - 2,837$

34. Find the sum of 218 and 45 mentally.

35. Find the difference of 996 and 538 mentally.

36. **Sports**   A triathalon competition consists of running 10 miles, bicycling 35 miles and swimming 3 miles. How many miles will each athlete travel during the competition? *(Lesson 1-1)*

37. **Politics**   A total of 326 members of the U.S. House of Representatives voted on a budget bill. If 212 of them voted against the bill, about how many representatives voted in favor of the bill? Use front-end estimation. *(Lesson 1-2)*

38. **Smart Shopping**   Camille has $75 to spend on a new outfit. She selects a skirt priced at $19.95, a sweater priced at $23.45, and shoes priced at $29.95. Does Camille have enough money to purchase the whole outfit? Use front-end estimation. *(Lesson 1-2)*

39. Estimate $5,426 \div 913$ using compatible numbers. *(Lesson 1-3)*

40. **Employment**   When reviewing his schedule, John finds that he is due to work a total of 213 hours over the next 29 days. *About* how many hours per day will John be working? Use compatible numbers to compute. *(Lesson 1-3)*

41. **Games**   In Scrabble®, there are 98 lettered tiles with assigned point values. On the gameboard, there are 17 pink, 12 dark blue, 24 light blue, 8 red, and 164 gray squares. How many squares are on a Scrabble® gameboard?

42. **Critical Thinking**   If you never backtrack, what is the greatest distance from *A* to *B*? Compute mentally.

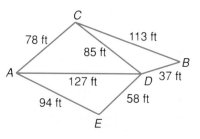

1-5 Choose the Method of Computation

Objective

Solve problems by choosing estimation, mental math, paper and pencil, or calculator.

Ms. Meadows gave her class this problem. Choose five digits. Use all five digits to make a two-digit number and a three-digit number that when multiplied have the greatest product possible.

Which method of computation would you use to solve this problem? You can use estimation, mental math, pencil and paper, or calculator.

Explore   What do you know?

You can use any five digits. You need to make a two-digit and a three-digit number, using all the digits you chose.

What are you trying to find?

Which pair of numbers has the greatest product?

Plan   Use the chart below to help you decide which method of computation to use.

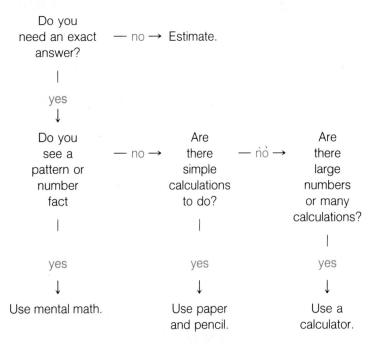

Since you need an exact answer and you need to multiply all the number pairs, use a calculator.

Solve   Let's choose the digits 1, 2, 3, 4, and 5. Place the 5 and 4 in the greatest place-value positions of both numbers. Keep track of the products.

521 ⊠ 43 ⊟ 22403
531 ⊠ 42 ⊟ 22302
532 ⊠ 41 ⊟ 21812
432 ⊠ 51 ⊟ 22032
431 ⊠ 52 ⊟ 22412 ✓
421 ⊠ 53 ⊟ 22313

The two-digit and three-digit numbers that have the greatest product from the numbers 1, 2, 3, 4, and 5 are 431 and 52.

Examine   Products with the 5 and 4 in the same number are less than any of the products in the *Solve* step.

541 ⊠ 32 ⊟ 17312
54 ⊠ 321 ⊟ 17334

Checking for Understanding

Communicating Mathematics

Read and study the lesson to answer each question.

1. **Tell** how to use the chart to help you choose a method of computation.

2. **Write** a sentence explaining how you know when estimation is an acceptable method for solving a problem.

3. **Model** your own chart for choosing a method of computation.

Guided Practice

Read each situation. Write *exact* if the number must be figured exactly and *estimate* if the number can be approximate.

4. amount earned for babysitting

5. attendance at a baseball game

6. grocery bill as you place items in the cart

7. amount of money you give the cashier for groceries

8. amount of change from a purchase

9. average number of cars produced in 5 years

Problem Solving

Practice Choose the method of computation. Then solve.

10. Max needs to buy four markers to make posters for a social studies project. He has $4. Does he have enough money if each marker is 89¢?

11. A trip from Des Moines to Chicago is 360 miles. If 10 gallons of gasoline are used, how many miles per gallon is this?

12. A student council convention had 4,293 students registered. The convention manager had to assign 537 students per hotel. To how many different hotels did the manager have to assign the students?

13. Choose five digits. Use the five digits to form a two-digit and a three-digit number so that their product is the least product possible. Use each digit only once.

14. During a political campaign, each of 256 persons donated $100. How much money was donated?

15. The Mediterranean Sea is an almost completely closed sea of about 900,000 cubic miles of water. Water entering at the Strait of Gibraltar takes about 150 years to flush through the sea. *About* how many cubic miles of water clear the sea in a day?

Mid-Chapter Review

1. **Health** Do you know how much blood is in your body? To find the approximate number of quarts, divide your weight by 30. *(Lesson 1-1)*

Estimate. Use front-end estimation. *(Lesson 1-2)*

2. $4,251 + 822 + 1,089$ 3. $6,743 - 3,221$

Estimate. Use compatible numbers. *(Lesson 1-3)*

4. Estimate the quotient of 4,932 and 79.

5. Estimate the quotient of 5,378 and 61.

Add or subtract mentally. *(Lesson 1-4)*

6. $53 + 27$ 7. $400 - 184$ 8. $\$6.03 + \$0.58$

9. **Sports** The distance between successive bases on a baseball diamond is 27.43 meters. *About* how far does a player run when hitting a home run? *(Lesson 1-5)*

1-6 Classify Information

Objective

Solve problems by classifying information.

The husbands—a doctor, dentist, and artist—and their wives—a physicist, architect, and professor—sit side by side in one row at the theater. No two husbands and no two wives sit together, and no wife sits next to her husband. The names are Jan, Joy, Jill, Bob, Bart, and Brad. (The names and occupations are in no special order.)

The dentist occupies one of the middle seats, and he is sitting next to Joy. Bob sits at one end of the row next to the artist's wife. How are the husbands and wives seated?

Sometimes it helps to classify information when you are deciding how to solve a problem.

Explore What do you know?
- Six adults sit in a row.
- Men and women sit alternately.
- No husband and wife sit together.
- The dentist is in a middle seat next to Joy.
- Bob sits on an end next to the artist's wife.

What are you trying to find?
How are the husbands and wives seated?

Plan Let's start with Bob, the artist's wife, the dentist, and Joy since we know where they are sitting.

Solve Bob sits on one end next to the artist's wife. → Bob, artist's wife

Alternate seating puts the dentist in the middle seat and next to Joy. → Bob, artist's wife, dentist, Joy

Husbands and wives do not sit together. The remaining husband is the artist since his wife is next to Bob. For this same reason, the dentist's wife must be at the end of the row. Joy must be Bob's wife. Therefore, Bob must be the doctor.

The seating arrangement is as follows.

doctor, artist's wife, dentist, doctor's wife, artist, dentist's wife

Examine   Look at the seating arrangement above. No husband or wife sat together and each person is seated according to the clues.

The wives' occupations and the names except for Bob and Joy were extra information.

Checking for Understanding

Communicating Mathematics

Read and study the lesson to answer each question.

1. **Tell** how to identify extra information in a problem.

2. **Write** a sentence explaining what you do if a problem does not have enough information.

Guided Practice

If the problem has enough facts, solve it. If not, write the missing facts.

3. Forty videos rent at the same price. How much will it cost to rent all forty videos?

4. There are 10 CDs in a box. They sell for $13.65 each. How much will 3 CDs cost?

5. Ned says his birthday is on a Tuesday this year. Bill says it's on Monday. Ron says it was on Thursday last year. What day is Ned's birthday?

Problem Solving

Practice

If the problem has enough facts, solve it. If not, write the missing facts.

6. Five girls share a pizza. The large size costs $9.60. The medium size costs $6.25. How much did each girl pay?

7. Dan bought a burger for $2.95 and an apple juice for $0.79. How much change did he get from a $5 bill?

8. First class stamps sell for $0.29. Postcard stamps sell for $0.19. How much do 25 first class stamps cost?

9. Socks cost $4 a pair for striped tops and $3 a pair for solid tops. Paul bought 3 pairs of socks. How much did he spend?

10. Jan has $10.56. Pat has twice as much except for one cent less. How much do they have in all?

Strategies

• • • • • • • • • •

Look for a pattern.

Solve a simpler problem.

Act it out.

Guess and check.

Draw a diagram.

Make a chart.

Work backward.

Planning a Flower Garden

Situation

Suppose the Booster Club at your school is arranging a Spring Surprise.
There are 12 of you in the club and you have $158 in the treasury. Do you
have enough money and people to plant enough bulbs near the school's
front entrance to make a big splash of color in the spring?

Hidden Data

Will you need to add any nutrients to the soil at planting time?
How many bulbs are needed?
What will be the cost of shipping and handling for ordering the bulbs?
How much will the sales tax be, if applicable?

Use this map to determine your zone.

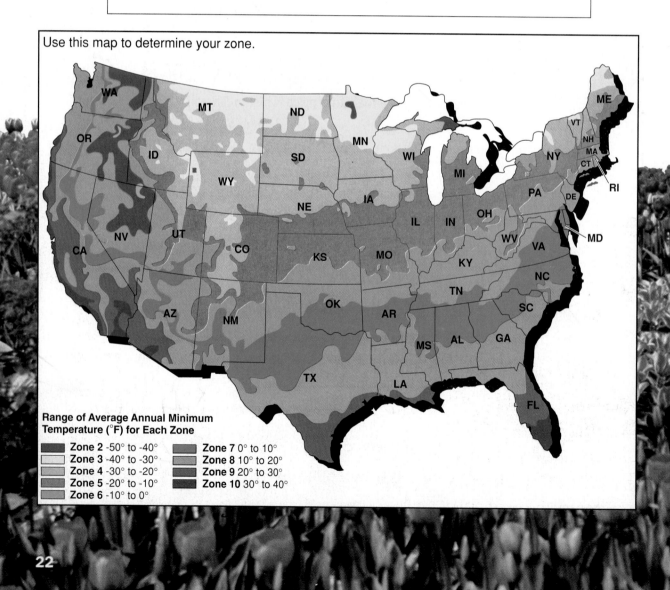

Range of Average Annual Minimum
Temperature (°F) for Each Zone

Zone 2 -50° to -40°   Zone 7 0° to 10°
Zone 3 -40° to -30°   Zone 8 10° to 20°
Zone 4 -30° to -20°   Zone 9 20° to 30°
Zone 5 -20° to -10°   Zone 10 30° to 40°
Zone 6 -10° to 0°

Analyzing the Data

1. What zone is your area in?
2. Which variety of bulbs are compatible with your climate?

Handling & Shipping Charges

Total of Goods		Add:
Up to $20	Calculate	$3.00
$20.01—$40	Charges	$4.00
$40.01—$60	For EACH	$6.00
$60.01—$80	Delivery	$7.00
$80.01—$100	to EACH	$9.00
$100.01—$250	Separate	$13.00
Over $250	Address	$20.00

Narcissus Collection

Bulb size 12 to 15 cm
White, Blush, Yellow; hardy in
zones 4 to 8. Southwinds Mix;
hardy in zones 9 and 10. Shipped
late September through October.
Sets of 50 bulbs

Blush #22...........$30
Yellow #24..........$25
White #28...........$29
Southwinds
    Mix #20..........$28

Crocus for Naturalizing

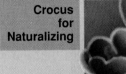

Bulb size 10 cm
Light shade or full sun
Variegated colors; hardy in zones 3 to 8
Shipped in October

50 bulbs #91......$15
100 bulbs #95....$27

Tulip Mixture

SAVE

Bulb size 12 cm
Suitable for cutting
Mix of four colors; hardy in zones 3 to 8
Shipped in October

32 bulbs #71......$18
60 bulbs #75......$32

Lily Bed Low
(plant 20 cm apart)

Bulb size 14 cm
Will not need staking
Multicolored hardy in zones 3 – 10
Shipped in October

40 bulbs #206.....$34
80 bulbs #205.....$60

Making a Decision

3. **Who is** going to weed the garden each year?
4. **How will** the garden be cared for after the bulbs bloom and die each year?
5. For the amount of money the club spends, how many years do you expect the garden to bloom?
6. **Would your club** want to promote this Spring Surprise with a spring bloom for each student? teacher? staff member?

Making Decisions in the Real World

7. **Research** the cost of materials from a nearby nursery for both outside bulbs and inside bulbs.

1-7 Order of Operations

Objective

Evaluate expressions using the order of operations.

Words to Learn

order of operations

Sarah and Erick are photographers on the middle school yearbook. Erick found 2 rolls of film with 36 exposures each and 3 rolls of film with 24 exposures each in the supply cabinet. How many photos can they take before they have to buy more film?

First they make an estimate.

$$2 \times 36 \quad \rightarrow \quad 2 \times 40 \text{ or } 80$$
$$3 \times 24 \quad \rightarrow \quad 3 \times 20 \text{ or } 60$$

They can take about $80 + 60$ or 140 photos before they have to buy more film.

Sarah and Erick then decide to use a calculator to find the exact number of photos. They each use a different calculator. Here are the results.

Sarah: 2 ⊗ 36 ⊞ 3 ⊗ 24 ⊟ **144**

Erick: 2 ⊗ 36 ⊞ 3 ⊗ 24 ⊟ **1800**

Which answer is correct? Sarah reasons that 144 is correct because it is close to the estimate.

To make sure that expressions like $2 \times 36 + 3 \times 24$ have only one value, mathematicians have agreed on the following **order of operations.** Grouping symbols, such as parentheses, are used to change the order of operations.

Order of Operations	1. Do all operations within grouping symbols first.
	2. Do multiplication and division from left to right.
	3. Do addition and subtraction from left to right.

Calculator Hint

• • • • • • • • • • • •

To see whether your calculator follows the order of operations, enter

2 ⊞ 5 ⊗ 3.

If your calculator displays 17, your calculator follows the order of operations.

Some calculators are programmed to follow the order of operations. In the situation described above, Sarah's calculator follows the order of operations. You may want to try the problem shown above with your calculator to determine whether it follows the order of operations.

Example 1

Evaluate $(5 + 4) \div 3$.

$(5 + 4) \div 3 = 9 \div 3$ *Add 5 and 4 since they are in parentheses.*

           $= 3$ *Divide by 3.*

There are other ways to indicate multiplication, besides using the symbol $\times$. One way is to use a raised dot.

$$3 \cdot 5 \quad \boxed{\text{means}} \!\!> \quad 3 \times 5$$

Another way is to use parentheses.

$$2(4 + 5) \quad \boxed{\text{means}} \!\!> \quad 2 \times (4 + 5)$$

Example 2

Evaluate $3(4 + 7) - 5 \cdot 4$.

$$
\begin{aligned}
3(4 + 7) - 5 \cdot 4 &= 3(11) - 5 \cdot 4 && \text{\textit{Add 4 and 7.}} \\
&= 33 - 5 \cdot 4 && \text{\textit{Multiply 3 and 11.}} \\
&= 33 - 20 && \text{\textit{Multiply 5 and 4.}} \\
&= 13 && \text{\textit{Subtract 20 from 33.}}
\end{aligned}
$$

Checking for Understanding

Communicating Mathematics

Read and study the lesson to answer each question.

1. **Show** that $3(4 + 5)$ is equal to $3(4) + 3(5)$.
2. **Write** an expression in which you should add first.
3. **Tell** which operation you should do first in the expression $6(3 + 7)$. Explain.

Guided Practice

Name the operation that should be done first.

4. $3 + 5 \cdot 6$
5. $10 - (3 + 4)$
6. $4 + 2(8 - 6)$
7. $(17 + 3) \div (4 + 1)$
8. $12 - 3(4)$
9. $4(6 + 4) \div 2$

Evaluate each expression.

10. $3 + 5 \cdot 4$
11. $(12 - 4) \div 2$
12. $5 \cdot 8 - 3 \cdot 4$
13. $3 \cdot 4(5 - 3)$
14. $14 - 8 \div 8$
15. $4(10 + 8) \div 6$

Exercises

Independent Practice

Name the operation that should be done first.

16. $12 - 3 \cdot 4$
17. $7 + 3(5 - 2)$
18. $(8 + 3) - 5$
19. $5 + 3(4)$
20. $(8 - 4) \div 2$
21. $7 \times 9 - (4 + 3)$

Evaluate each expression.

22. $7 \cdot 3 + 8 \cdot 2$
23. $(8 - 2) \div 3$
24. $5 - 3 + 1$
25. $16 \div 4 \cdot 2$
26. $12 - 8 \div 4 + 6$
27. $24 \div (7 - 3)$
28. $14 - (19 - 19)$
29. $25 \div (9 - 4)$
30. $84 - 28 \div (4 \cdot 7)$
31. $2(14 - 9) - (17 - 14)$
32. $(26 - 9) - 4 \times 3$
33. $3(24 - 7) - 2 \cdot 13$
34. $16(2 - 1) \div 2 + 2$
35. $(81 + 19) \div 25 + 5$
36. $82 - 43 - 6 \div 6$

Copy each sentence below. Use your calculator to determine where to insert parentheses to make each sentence true. You may use the parentheses keys.

37. $16 + 5 \times 4 \div 2 = 42$

38. $64 \div 8 + 24 - 1 = 1$

39. $36 \div 3 - 9 \div 3 = 1$

40. $18 \div 24 - 18 - 3 = 0$

Mixed Review

41. **Production**   Soft drink cans can be filled by a machine on a production line at a rate of 8 cans per minute. How many cans can be filled during an 8-hour shift? *(Lesson 1-1)*

42. **Consumer Math**   A new car has a sticker price of $16,150. The tax to be paid on this amount is $1,035. About how much is the total cost of the car? Use front-end estimation. *(Lesson 1-2)*

43. Use compatible numbers to estimate $1{,}243 \div 310$. *(Lesson 1-3)*

44. Add $394 + 21$ mentally. *(Lesson 1-4)*

45. **Health**   Stephan weighed 204 pounds before starting a new diet plan. After 4 weeks on the diet, he weighed 186 pounds. Compute mentally to find out how many pounds Stephan lost. *(Lesson 1-4)*

Problem Solving and Applications

46. **Data Search**   Refer to pages 2 and 3. *About* how much more snow fell in Sault St. Marie, Michigan, than in Anchorage, Alaska, in 1989?

47. **Smart Shopping**   At a local coffee house, coffee can be purchased by the pound. Megan buys 3 pounds of Mexican coffee priced at $5 per pound. Nick buys 7 pounds of Spanish coffee priced at $9 per pound. How much more did Nick spend than Megan?

48. **Critical Thinking**   In a collection of nickels and quarters, there are four more nickels than quarters. If the coins are worth $5.60, how many quarters are there?

COMPUTER CONNECTION

49. **Computer Connection**   BASIC is a computer language that can be used to program a computer. The chart at the right shows the BASIC notation for the operations we use in mathematics. The BASIC computer language also follows the order of operations. Evaluate these expressions that are written in BASIC notation.

Operation	Mathematics	BASIC
Addition	$+$	$+$
Subtraction	$-$	$-$
Multiplication	$\times$	$*$
Division	$\div$	$/$

a. $2 * 5 + 9$

b. $25 + 5 - 10$

c. $40/10 * 8$

d. $(19 + 17)/12$

e. $12/(31 - 2 * 15)$

f. $3 * 5 * 15$

1-8A Algebra: Variables and Expressions

A Preview of Lesson 1-8

Objective

Model algebraic expressions.

Materials

counters
cups

The phrase *the sum of four and some number* is an algebraic expression. This phrase contains a *constant* that you know, 4, and an unknown value "some number."

- You can use counters to represent 4 and a cup to represent the unknown value.

- Any number of counters may be in the cup. Suppose you put 3 counters in the cup. Instead of an unknown value, you know the cup has a value of 3. When you empty the cup and count all the counters, the expression has a value of 7.

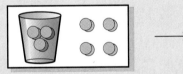

- Consider the phrase *twice some number*. Since you don't know the value of the number, let a cup represent this value. Since it is *twice* some number, you will need to use two cups. The same number of counters should be in each cup.

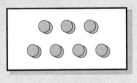

Try this!

Work in groups of three.

Model each phrase with cups and counters. Then put four counters in each cup. How many counters are there in all? Record your answer by drawing pictures of your models.

1. the sum of 3 and a number
2. three times a number
3. 5 more than a number
4. five times a number

What do you think?

5. Write a sentence to describe what the cup represents.
6. Write a sentence that explains why $x + 4$ is called an algebraic expression.

1-8 **Variables and Expressions**

Objective

Evaluate numerical and simple algebraic expressions.

Words to Learn

variable
algebra
algebraic expression
evaluate

Tammy charges $2 per hour for babysitting. If she babysits for one hour, she makes $1 \times \$2$ or $2. If she babysits for two hours, she makes $2 \times \$2$ or $4. The amount she makes increases with the number of hours she babysits.

You can make a table to show the pattern between the number of hours spent babysitting and the amount earned.

Number of Hours	Amount Earned
0	$\$2 \times 0 = \$0$
1	$\$2 \times 1 = \$2$
2	$\$2 \times 2 = \$4$
3	$\$2 \times 3 = \$6$
4	$\$2 \times 4 = \$8$

In the table above, notice that the amount earned per hour is constant, $2, but the number of hours varies. You can use a placeholder, or **variable,** to represent the number of hours spent babysitting. The expression for the amount earned is $\$2 \times$ ▮ or $\$2 \times n$, where n is a variable. This expression can also be written as $2n$, which means 2 times the value of n.

The area of mathematics that involves expressions that have variables is called **algebra.** The expression $2n$ is called an **algebraic expression** because it contains variables, numbers, and at least one operation.

You can **evaluate** an algebraic expression by replacing the variable with a number and then finding the value of the numerical expression. Consider the expression $2n$. How much money will Tammy earn if she babysits for 6 hours?

First use cups and counters to find the solution.

Place six counters in each cup.

The solution is 12.

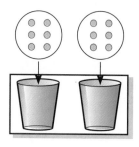

You can also solve this problem algebraically.

$$2n = 2 \times 6 \qquad \textit{Replace n with 6.}$$
$$= 12 \qquad \textit{Tammy will earn \$12.}$$

Example 1

Evaluate $x + y + 4$ if $x = 7$ and $y = 11$.

$$x + y + 4 = 7 + 11 + 4 \qquad \textit{Replace x with 7 and y with 11.}$$
$$= 18 + 4 \qquad \textit{7+ 11+ 18}$$
$$= 22 \qquad \textit{18+ 4= 22}$$

To make sure everyone understands the language of mathematics, mathematicians agree on standard notation for multiplicaton and division with variables.

$3a$ means $\Rightarrow$ $3 \times a$, or $3 \cdot a$, or $(3)(a)$

rs means $\Rightarrow$ $r \times s$

$4cd$ means $\Rightarrow$ $4 \times c \times d$

$\dfrac{m}{2}$ means $\Rightarrow$ $m \div 2$

The expression $110 + \dfrac{A}{2}$ is used to estimate a person's normal blood pressure. A stands for the person's age.

Estimate the normal blood pressure of an 18-year-old.

$$110 + \dfrac{18}{2} = 119$$

The blood pressure of an 18-year-old is *about* 119.

Examples

2 Evaluate $5a + 3b$ if $a = 7$ and $b = 2$.

$$5a + 3b = 5(7) + 3(2) \qquad \textit{Replace a with 7 and b with 2.}$$
$$= 35 + 6 \qquad \textit{Multiply.}$$
$$= 41 \qquad \textit{Add 35 and 6.}$$

3 Evaluate $\dfrac{ab}{3}$ if $a = 7$ and $b = 9$.

$$\dfrac{ab}{3} = \dfrac{(7)(9)}{3} \qquad \textit{Replace a with 7 and b with 9.}$$
$$= \dfrac{63}{3} \qquad \textit{Multiply 7 and 9.}$$
$$= 21 \qquad \textit{Divide 63 by 3.}$$

4 Use a calculator to evaluate $\dfrac{x}{y} + 50$ if $x = 12$ and $y = 2$.

$$\dfrac{x}{y} + 50 = \dfrac{12}{2} + 50 \qquad \textit{Replace x with 12 and y with 2.}$$

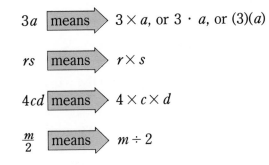

12 ÷ 2 + 50 = 56

Checking for Understanding

Communicating Mathematics

Read and study the lesson to answer each question.

1. **Write** two different expression that mean the same as $4x$.

2. **Write** an expression that means the same as $5 \div a$.

3. **Tell**, in your own words, the difference between numbers and variables.

Guided Practice   Evaluate each expression if $x = 6$, $y = 4$, $a = 3$, $b = 2$, and $c = 7$.

4. $x + 2$ 5. $y - 3$

6. $9 - c$ 7. $14 + y$

8. $x + y$ 9. $a + b$

10. $c + 8 - a$ 11. $y - 1 + x$

12. $x - y + 3$ 13. $a - b + 5$

14. $ab - 1$ 15. $xy - 4$

Exercises

Independent Practice   Evaluate each expression if $a = 6$, $b = 3$, and $c = 2$.

16. $a + b + c$ 17. $3a + b$ 18. $ab - c$

19. $\frac{a}{c} + 4$ 20. $2ab$ 21. $\frac{a}{b} + c$

22. $\frac{2a}{3} - c$ 23. $2(a + b) - c$ 24. $2a - 3b$

25. $5 - \frac{2b}{c}$ 26. $\frac{6(a + c)}{b}$ 27. $c(b + a) - a$

Mixed Review

28. **Consumer Math**   Jackie receives a paycheck stating that her total pay, before taxes, for 52 hours of work amounts to $353. About how much is her hourly wage? Use compatible numbers. *(Lesson 1-3)*

29. **Music**   Tanya owns a vast compact disc collection consisting of 95 CDs. During a sale at a local record store, she purchases 6 more. Compute mentally the number of CDs she now owns. *(Lesson 1-4)*

30. Mr. Jackson plans to walk 14 miles a week. How many more miles does Mr. Jackson have to walk? If the problem has enough facts, solve it. If not, write the missing facts. *(Lesson 1-6)*

31. Evaluate the expression $24 - 12 \div 3 \cdot 5 + 6$. *(Lesson 1-7)*

32. **Nature**   You can estimate the temperature in degrees Fahrenheit by counting the number of times a cricket chirps in one minute, dividing it by 4, and then adding 37.

   a. Write an expression for temperature using this information. Let *c* represent the number of chirps per minute.

   b. Find the temperature if a cricket chirps 104 times in one minute.

33. **Aeronautics**   An aircraft is said to travel at Mach 2 if it travels at twice the speed of sound, Mach 3 if it travels at 3 times the speed of sound, and so on.

   a. If the speed of sound is about 740 mph, write an expression to indicate the speed an aircraft travels in terms of its Mach number. Let *m* represent the Mach number.

   b. Find the speed of an aircraft traveling at 5 times the speed of sound.

34. **Critical Thinking**   Demarcus and Latisha are remodeling the bathroom in their apartment. Latisha is installing 9-inch square ceramic tiles on the wall while Demarcus is installing 12-inch square vinyl tiles on the floor.

   a. How often will the seams line up along an 8-foot wall if they both begin at the left edge of the wall?

   b. How far is each common seam line from the left edge?

Save Planet Earth

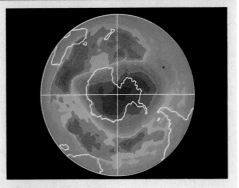

The Ozone Layer
The ozone layer is a thin layer of gas located miles above our heads. It filters out damaging radiation from the sun. Chemicals such as chlorofluorocarbons (CFCs) can destroy the ozone by drifting upward into the atmosphere from our homes, factories, and cities.

Styrofoam, which is often injected with gases used with CFCs, is a threat to the ozone layer, as well as marine life and our already-full landfills.

How You Can Help
Avoid foam packaging in picnic goods, fast food restaurants, and egg cartons.

1-9 Powers and Exponents

Objective

Use powers and exponents in expressions.

Words to Learn

factors
exponent
base
powers
cubed
squared

Do you want to do your part to save Earth? Here is something all Americans can do to help save our planet. We could save 150,000 trees every year, if just 100,000 people would stop delivery of junk mail to their homes.

Numbers such as 100,000 can be written as $10 \cdot 10 \cdot 10 \cdot 10 \cdot 10$. When two or more numbers are multiplied, these numbers are called **factors** of the product. When the same factor is used, you may use an **exponent** to simplify the notation.

$$100,000 = \underbrace{10 \cdot 10 \cdot 10 \cdot 10 \cdot 10}_{5 \ factors} = 10^5 \xleftarrow{} exponent$$

$\underset{base}{\uparrow}$

The common factor is called the **base.** Numbers expressed using exponents are called **powers.**

The powers 2^3, 4^2, and 5^4 are read as follows.

2^3 **two to the third power, or two cubed**
4^2 **four to the second power, or four squared**
5^4 **five to the fourth power**

Examples

---

1   Write 3^4 as a product.

The base is 3. The exponent 4 means 3 is used 4 times.

$3^4 = 3 \cdot 3 \cdot 3 \cdot 3$

2   Write $2 \cdot 2 \cdot 2 \cdot 2 \cdot 2$ using exponents.

2 is a base. Since 2 is a factor 5 times, the exponent is 5.

$2 \cdot 2 \cdot 2 \cdot 2 \cdot 2 = 2^5$

The *order of operations* now must include exponents.

Order of Operations	1. Do all operations within grouping symbols first.
	2. Evaluate all powers before other operations.
	3. Do multiplication and division from left to right.
	4. Do addition and subtraction from left to right.

Evaluate powers as follows.

Example

3 Evaluate 5^3.

$$5^3 = 5 \cdot 5 \cdot 5$$
$$= 125$$

Calculator Hint

Many calculators have a $\boxed{y^x}$ key. This key allows you to compute exponents. Suppose you want to find 12^4. Press

$12 \boxed{y^x} 4 \boxed{=}$.

The answer, 20,736, is displayed immediately.

Examples *Connection*

4 **Algebra** Write x^3 as a product.

The base is x. The exponent 3 means x is used 3 times.

$$x^3 = x \cdot x \cdot x$$

5 **Algebra** Write $a \cdot a \cdot a \cdot a$ using exponents.

a is a base. Since a is a factor 4 times, the exponent is 4.

$$a \cdot a \cdot a \cdot a = a^4$$

6 **Algebra** Evaluate n^4 if $n = 3$.

$$n^4 = 3^4 \qquad \textit{Replace n with 3.}$$

$3 \boxed{y^x} 4 \boxed{=} 81$

Checking for Understanding

Communicating Mathematics

Read and study the lesson to answer each question.

1. **Tell** why exponents are useful.
2. **Tell**, in your own words, what powers are. Use the terms *factor, base,* and *exponent.*
3. **Write** one or two sentences that explain the pattern used when you find the product of 10^5 mentally.

Guided Practice

Write each power as a product of the same factor.

4. 2^4
5. 7^5
6. 12^3
7. 9^7

Write each product using exponents.

8. $6 \cdot 6 \cdot 6$
9. $15 \cdot 15 \cdot 15 \cdot 15$
10. $a \cdot a \cdot a \cdot a \cdot a \cdot a$

Evaluate each expression.

11. 5^4
12. 2^6
13. 4^4
14. 8^3

15. **Fitness** In one year, Americans spent $1,410,000,000 on home exercise equipment. Out of this money, Americans spent about 10^8 dollars on weight sets. About how much money did Americans spend on weight sets?

Exercises

Independent Practice

Write each power as a product.

16. 4^5 17. 9^3 18. 3^4 19. n^9

Write each product using exponents.

20. $7 \cdot 7 \cdot 7 \cdot 7$ 21. $12 \cdot 12$ 22. $x \cdot x \cdot x \cdot x \cdot x$

Evaluate each expression.

23. 7^2 24. 1^{12} 25. 3^5 26. 6^3

27. 3^4 28. 11^2 29. 9^2 30. 13^1

31. 6 squared   32. 8 to the third power   33. 5 cubed

34. Given that $2^5 = 32$, find 2^6 mentally.

Use a calculator to determine whether each sentence is *true* or *false*.

35. $3^7 > 7^3$ 36. $12^4 = 182$ 37. $6^3 < 4^4$

Algebra Evaluate each expression.

38. n^3 if $n = 4$ 39. m^4 if $m = 5$

40. x^6 if $x = 1$ 41. r^2 if $r = 11$

42. s^4 if $s = 3$ 43. t^7 if $t = 2$

Mixed Review

44. Use compatible numbers to estimate how many miles per gallon a car gets if it travels 295 miles and uses 9 gallons of gasoline. *(Lesson 1-3)*

45. Anya was born in 1965. Compute mentally to find her age in 1999. *(Lesson 1-4)*

46. Evaluate the expression $36 + 9 \div 3$. *(Lesson 1-7)*

47. **Smart Shopping** Packages of pencils come with 12 pencils. Pens are packaged 10 to a pack. To start the school year, Susan buys 3 packages of pencils and 2 packages of pens. What is the total number of pens and pencils she buys? *(Lesson 1-7)*

48. **Algebra** Evaluate the expression $2x + 3(x + y) - xy$ if $x = 10$ and $y = 2$. *(Lesson 1-8)*

49. **Animals** It is believed that a dog ages 7 human years for every calendar year that it lives. *(Lesson 1-8)*

   a. Write an expression for determining a dog's age in human years. Let y represent the number of calendar years the dog has lived.

   b. Find the human age of a dog that has lived for 12 calendar years.

Problem Solving and Applications

50. **Geometry** To find the volume of a cube, you find the cube of the length of one edge. Find the volume of a cube with an edge of six inches.

6 in.

51. **Number System** The base-ten number system uses powers of ten to express numbers. For example, 1 quintillion is the number expressed as a 1 followed by 18 zeros. How would this number be expressed as a power with a base of 10?

52. **Critical Thinking** Based on the pattern shown at the right, write a convincing argument that any number, besides 0, raised to the 0 power equals 1. Use the $\boxed{y^x}$ key on a calculator to see what happens.

$$2^4 = 16$$
$$2^3 = 8$$
$$2^2 = 4$$
$$2^1 = 2$$
$$2^0 = ?$$

53. **Critical Thinking** Describe the pattern of the ones digit in successive powers of 3. What is the ones digit in 3^{100}?

Entertainment Industry Use the graph below for Exercises 54–56.

54. a. In which state was the most money spent?

   b. *About* how much money was spent?

55. *About* how much more money was spent filming on location in New York than in Massachusetts?

56. *About* how much money did film and television companies spend shooting on location in these five states?

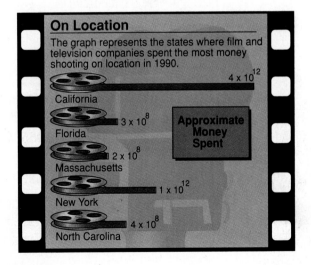

On Location

The graph represents the states where film and television companies spent the most money shooting on location in 1990.

California — 4×10^{12}

Florida — 3×10^8

Massachusetts — 2×10^8

New York — 1×10^{12}

North Carolina — 4×10^8

Approximate Money Spent

57. **Journal Entry** Explain why 2^3 is not equal to 3^2.

1-9B **Spreadsheets**

A Follow-Up of Lesson 1-9

Objective

Describe a spreadsheet.

Words to Learn

spreadsheet
cell

A **spreadsheet** is a computer program that organizes numerical data into rows and columns. It is used for computing calculations and making overall adjustments based on new data.

Most spreadsheets can be as large as 999 rows long and 127 columns wide. The spreadsheet below shows data of test results for four students.

	A STUDENT	B TEST 1	C TEST 2	D TEST 3	E TOTAL	F AVERAGE
1	M. JONES	85	83	90	258	86
2	A. SMITH	91	95	87		
3	C. MAYO	77	89	92		
4	D. WHITE	100	68	93		

Each section of a spreadsheet is called a **cell.** A cell can contain data, labels, or formulas. The first cell in the upper left-hand corner is called cell A1. The cell directly to the right in column B and row 1 is called B1. The cell below that is B2. You can enter information in a spreadsheet one cell at a time.

To perform a calculation, you need to enter a formula in a cell. A formula can contain constant values, operations, and cell locations. Besides + (addition) and − (subtraction), the following operations can also be used: * (multiplication), / (division), and ^ (exponentiation).

What do you think?

Work with a partner.
Use the spreadsheet on
the previous page.

1. What test score is stored in cell C4?
2. How are the totals in column E found?
3. Copy and complete the spreadsheet on page 36.
4. What formula might you enter in cell E2 to find the total test scores for A. Smith?
5. What formulas might you enter in cells F2 to F4 to find the average test scores?
6. What formula might you enter in cell F5 if you wanted to find out the average test score of the four students?

	A	B	C
1	CITY	MILES WIDE	MILES LONG
2	EVERYWHERE		
3	NOWHERE	5	

Use the spreadsheet above to answer each of the following.

7. Cell B2 contains the formula B3 ^ 3. Find the value that would be in cell B2.
8. What value would be in cell C3, if the cell contained the formula (B2 + B3) ^ 2?
9. What value would be in cell C2, if the cell contained the formula (C3/130) ^ 2?

Extension

10. Create a spreadsheet that computes a table of squares and cubes of numbers.

1-10 Solving Equations Mentally

Objective

Solve equations using mental math.

Words to Learn

equation
solve
solution

The __?__ Ocean is on the east coast of the United States. You cannot determine whether this sentence is true or false until you fill in the blank. If you say Atlantic, the sentence is true. If you say Pacific, the sentence is false.

An **equation** is a sentence in mathematics that contains an equal sign.

$45 + 12 = 67$ This sentence is false.

$32 - 10 = 22$ This sentence is true.

The equation $f + 9 = 16$ contains a variable. This equation is neither true nor false until f is replaced with a number. You **solve** the equation when you replace the variable with a number that makes the equation true. Any number that makes the equation true is called a **solution.** The solution to $f + 9 = 16$ is 7 because $7 + 9 = 16$.

Example 1

Which of the numbers 8, 9, or 10 is the solution of $9 + t = 18$?

Replace t with 8.

$9 + t = 18$
$9 + 8 \overset{?}{=} 18$
$17 \neq 18$

This sentence is false.

Replace t with 9.

$9 + t = 18$
$9 + 9 \overset{?}{=} 18$
$18 = 18$

This sentence is true.

Replace t with 10.

$9 + t = 18$
$9 + 10 \overset{?}{=} 18$
$19 \neq 18$

This sentence is false.

The solution is 9.

The equation in the example above was solved by replacing the variable, t, with each number until a true sentence was found.

DID YOU KNOW

The toothbrush is
thought to have been
invented in China in the
late 1400s. It may have
been made of hog's
hair.

Example 2 *Problem Solving*

Health The first toothbrush with nylon bristles was produced in 1938 in the United States. It was called Dr. West's Miracle Tuft Toothbrush. The first electric toothbrush was made by the Squibb Company 23 years later. What year did the first electric toothbrush makes its appearance?

Let y represent the year the electric toothbrush was invented. You need to solve the equation $y - 1938 = 23$ to find y. Start by making a guess. Replace y in the equation with your guess and see if the resulting equation is true. Do this until you find the value of y that makes the equation true.

We guessed 1968 first. Then we guessed 1958 followed by 1961.

$$y - 1938 = 23 \qquad\qquad y - 1938 = 23 \qquad\qquad y - 1938 = 23$$
$$1968 - 1938 \stackrel{?}{=} 23 \qquad 1958 - 1938 \stackrel{?}{=} 23 \qquad 1961 - 1938 \stackrel{?}{=} 23$$
$$30 \neq 23 \qquad\qquad\qquad 20 \neq 23 \qquad\qquad\qquad 23 = 23$$

This sentence   This sentence   This sentence
is false.   is false.   is true.

The solution is 1961. The first electric toothbrush made its appearance in 1961.

Some equations can be solved mentally by using basic facts or arithmetic skills you already know well.

Examples

3 Solve $12m = 120$ mentally.

$$12 \cdot 10 \stackrel{?}{=} 120 \qquad \textit{You know } 12 \cdot 10 = 120.$$
$$120 = 120$$

The solution is 10. The value of m is 10.

4 Solve $b = \frac{56}{7}$ mentally.

$$8 \stackrel{?}{=} \frac{56}{7} \qquad \textit{You know that } \frac{56}{7} = 8.$$
$$8 = 8$$

The solution is 8. The value of b is 8.

Checking for Understanding

Communicating Mathematics

Read and study the lesson to answer each question.

1. **Tell** why it is important to estimate when solving problems mentally.
2. **Write** a definition of *solution*.
3. **Tell** the solution of $4c = 12$.

Guided Practice

Tell whether the equation is *true* or *false* using the given value of the variable.

4. $j + 4 = 14$; $j = 18$
5. $p - 8 = 19$; $p = 27$
6. $10k = 200$; $k = 20$
7. $t \div 7 = 49$; $t = 7$

Name the number that is a solution of the given equation.

8. $q + 8 = 21$; 12, 13, 14
9. $d - 14 = 27$; 39, 40, 41
10. $9 \cdot 9 = r$; 81, 82, 83
11. $x \div 5 = 4$; 10, 20, 30

Solve each equation.

12. $3x = 21$
13. $g + 12 = 30$
14. $13 \cdot 11 = k$
15. $25 + 19 = a$
16. $n \div 4 = 20$
17. $c - 10 = 27$

18. **Travel** If it takes you 5 hours to travel 250 miles in a car, what is the average speed of the car? Use the equation $250 = 5r$ where r is the average speed of the car.

Exercises

Independent Practice

Name the number that is a solution of the given equation.

19. $a + 15 = 19$; 4, 5, 6
20. $b - 13 = 29$; 40, 41, 42
21. $11e = 77$; 6, 7, 8
22. $v \div 10 = 4$; 20, 30, 40
23. $33 + t = 51$; 18, 19, 20
24. $13 \cdot 9 = g$; 107, 117, 127
25. $w \div 12 = 8$; 96, 97, 98
26. $51 - 24 = b$; 17, 27, 37

Solve each equation.

27. $x + 35 = 91$
28. $m + 18 = 24$
29. $c - 15 = 71$
30. $15r = 105$
31. $\frac{n}{8} = 6$
32. $17y = 1,615$
33. $a - 75 = 98$
34. $d + 25 = 80$
35. $z \div 14 = 8$
36. $9g = 108$
37. $r - 29 = 117$
38. $\frac{f}{3} = 61$
39. $\frac{s}{6} = 12$
40. $234 - 89 = h$
41. $43 + z = 65$

42. A number plus four is eight. What is the number? Use the equation $b + 4 = 8$.

43. A number less three is 14. Find the number. Use the equation $p - 3 = 14$.

44. The product of a number and 6 is 84. What is the product? Use the equation $6s = 84$.

45. The quotient of a number and 22 is 7. Find the number. Use the equation $e \div 22 = 7$.

Mixed Review 46. Compute the sum $163 + 75$ mentally. *(Lesson 1-4)*

47. **Decorating** Wallpaper for a bedroom costs $16 per roll for the walls and $9 per roll for the border. If the room requires 12 rolls of paper for the walls and 6 rolls for the border, compute the total cost for the decorating job. *(Lesson 1-7)*

48. *True* or *False:* If $m = 2$ and $n = 3$, then $m + n = mn$. *(Lesson 1-8)*

49. **Food Industry** Pierre runs a pizza parlor. His daily cost of operating the parlor consists of a constant cost of $75 for rent, employee wages, and utilities plus $1 for every pizza he makes.

   a. Write an expression for Pierre's total daily cost. Let n represent the number of pizzas Pierre makes during the day.

   b. Find Pierre's total cost for a day during which he makes 45 pizzas. *(Lesson 1-8)*

50. Write the following product using exponents: $4 \cdot 4 \cdot 4 \cdot 4$. *(Lesson 1-9)*

51. Find the number of seconds in 60 hours. *(Lesson 1-9)*

Problem Solving and Applications 52. **Nutrition** Suppose you eat 2 tablespoons of peanut butter a day. Two tablespoons of peanut butter provides you with 8 grams of protein. If you need 44 grams of protein a day, how many more grams of protein are required?

53. **Geometry** The perimeter of a square is four times the length of one of its sides. What is the perimeter of a square whose side has a length of 21 centimeters?

54. **Critical Thinking** Consider the equation $0 + b = c$. What can you say about the values of b and c?

1 Study Guide and Review

Communicating Mathematics

Choose the correct term to complete the sentence.

1. The first step of the four-step plan is ___?___.

2. Front-end estimation is a process that can be used to estimate the ___?___ of two values.

3. The expression $3x - 4$ is called an ___?___.

4. Order of operations states that ___?___ precedes addition and/or subtraction.

5. In the expression 4^3, 4 is called the ___?___.

6. In your own words, explain what should be done during the examine step of the four-step plan.

equation
exponent
solution
explore
product
sum
algebraic expression
multiplication
base

Skills and Concepts

Objectives and Examples	Review Exercises
Upon completing this chapter, you should be able to:	*Use these exercises to review and prepare for the chapter test.*
• solve problems using the four-step plan *(Lesson 1-1)* The four steps are *Explore, Plan, Solve* and *Examine.*	Use the four-step plan to solve. 7. A car traveling 60 mph will travel how far in 7 hours? 8. Ann starts the day with $100. If she spends $35 at the mall, how much is left for groceries?
• estimate sums and differences using front-end estimation *(Lesson 1-2)* Estimate $8,324 + 6,936$. $\begin{array}{cc} 8,324 & 8,324 \\ +6,936 & +6,936 \\ \hline 14,000 \ + & 1,200 = 15,200 \end{array}$ $ \rightarrow$	Estimate. 9. $16,327 + 8,102$ 10. $43,212 - 21,605$ 11. $864 - 219$ 12. $3,986 + 4,627$ 13. $236,549 + 86,333$

Objectives and Examples

- estimate quotients using compatible numbers *(Lesson 1-3)*

  Estimate $1,457 \div 33$.

  $1,457 \div 33 \rightarrow 1,500 \div 30 = 50$.
  $1,457 \div 33$ is about 50.

- compute sums and differences using compensation *(Lesson 1-4)*

  Add mentally $347 + 28$.

  $$
  \begin{array}{rcccr}
  347 & \rightarrow & +3 & \rightarrow & 350 \\
  28 & \rightarrow & -3 & \rightarrow & +\ 25 \\
  \hline
  & & & & 375
  \end{array}
  $$

- evaluate expressions using the order of operations *(Lesson 1-7)*

  Evaluate $3(9 + 7) - 4 \div 2 + 3$.

  $3(9 + 7) - 4 \div 2 + 3$
  $= 3(16) - 4 \div 2 + 3$
  $= 48 - 2 + 3$
  $= 49$

- evaluate numerical and simple algebraic expressions *(Lesson 1-8)*

  Evaluate $6x - xy + y$ if $x = 10$, $y = 3$.

  $6x - xy + y = 6(10) - (10)(3) + 3$
  $\qquad\qquad = 60 - 30 + 3$
  $\qquad\qquad = 33$

- use powers and exponents in expressions *(Lesson 1-9)*

  $2^5 = 2 \cdot 2 \cdot 2 \cdot 2 \cdot 2 = 32$

- solve equations using mental math *(Lesson 1-10)*

  Solve $3s = 36$.

  You know that $3 \cdot 12 = 36$. So, the value of s is 12.

Review Exercises

Estimate.

14. $987 \div 11$
15. $12,042 \div 62$
16. $1,100 \div 12$
17. $6,350 \div 92$
18. $254,102 \div 512$

Add or subtract mentally.

19. $37 + 86$
20. $148 - 63$
21. $8,986 + 348$
22. $2,432 - 1,999$
23. $7,345 - 19$
24. $38 + 15,893$

Evaluate each expression.

25. $3 + 7 \cdot 4 - 6$
26. $8(16 - 5) - 6$
27. $12 - 18 \div 9$
28. $83 + 3(4 - 2)$
29. $75 \div 3 + 6(5 - 1)$
30. $10(12 - 2) \div 10$

Evaluate each expression if $p = 12$, $q = 3$, and $r = 5$.

31. $p + q - r$
32. $\frac{p + q}{r}$
33. $3(p + q) - r$
34. $6qr - p$
35. $25 - 2(p - 2q)$

Evaluate each expression.

36. 10^3 37. 3^6
38. 15 squared   39. y^4 if $y = 4$

Solve each equation using mental math.

40. $t - 12 = 35$
41. $8x = 88$
42. $\frac{m}{4} = 16$
43. $28 + r = 128$

Applications and Problem Solving

Choose the method of computation. Then solve.

44. Use all five even digits to form a two-digit number and a three-digit number which will yield the smallest possible product. *(Lesson 1-5)*

If the problem has enough facts, solve it. If not, write the missing fact.

45. **Seating** Anna, Barbara, Conchita, Doris, and Emma sit around a circular table. No two girls whose names begin with adjacent letters of the alphabet sit next to each other at the table. Emma sits on the right of Conchita. Where do the others sit? *(Lesson 1-6)*

46. **Business** A house painter finds that he needs 30 minutes to set up his equipment before painting and then 5 minutes for every square foot that he paints. Write an expression for the total time the painter needs. Let *n* represent the number of square feet he paints. *(Lesson 1-8)*

Curriculum Connection Projects

- **Language Arts** Write ten sentences in which the word *variable* is used as an adjective. Make a crossword puzzle or word search from the ten nouns *variable* describes.
- **Health** Learn an emergency procedure such as CPR or the Heimlich maneuver with a friend. Write steps for demonstrating your procedure to the class.

Read More About It

DeJong, Meindert. *The Wheel on the School.*
Smullyan, Raymond. *Alice in Puzzle-Land.*
Packard, Edward. *The Cave of Time.*

1. The Lewis Middle School Band is planning a bus trip to a band competition. There are 138 members in the band and each bus will hold 32 people. How many buses are needed for the trip?

Estimate.

2. $6,394 + 2,412$

3. $13,986 - 2,667$

4. $1,812 \div 94$

5. **Fast Food**   On a given day, a fast-food restaurant sold 1,545 ounces of cola, 1,239 ounces of lemon-lime drink, and 482 ounces of root beer. Estimate the total amount of soft drinks that were sold.

Add or subtract mentally.

6. $85 + 29$

7. $8,799 + 247$

8. $356 - 198$

Choose the method of computation. Then solve.

9. Use all five odd digits to form a two-digit number and a three-digit number which will yield the largest possible product.

If the problem has enough facts, solve it. If not, write the missing fact.

10. **Sales**   Adult tickets for the fair are $5. Bob collected $100 for a total of 30 adult and student tickets. How many of each kind did Bob sell?

11. Name the operation to be done first in the expression $26 + 12 \div 3$.

Evaluate.

12. $13 + 2 \cdot 4 - 6$

13. $5 + \frac{2mn}{4}$ if $m = 6$ and $n = 3$

14. b^3 if $b = 3$

15. **Animals**   Heather left her dog and cat at a kennel for 3 nights each. The kennel charges $8 per night for the dog and $5 per night for the cat. Compute Heather's total bill.

16. **Dessert**   Johnson's Real Ice Cream charges 25¢ for a cone and 75¢ for each scoop of ice cream added to the cone. Write an expression for the cost of an ice cream cone if n represents the number of scoops on it.

Solve.

17. $6a = 72$

18. $\frac{x}{8} = 100$

19. $p + 13 = 25$

20. $35 - m = 19$

Bonus   Find the last digit of 5^{10} without computing.

Chapter Test

Applications with Decimals

Spotlight on Clubs and Recreation

Have You Ever Wondered . . .

- How many people in the United States spend their free time by joining clubs or volunteering?

- How the number of boy scouts compares with the number of girl scouts?

Here are some of the most popular associations that Americans join:	membership in milions
American Association of Retired Persons	28.0
American Automobile Association	29.0
American Bowling Congress	3.3
American Farm Bureau Federation	3.3
Boy Scouts of America	4.8
4-H Program	4.8
Girl Scouts of the U.S.A.	3.1
National Committee to Preserve Social Security and Medicare	5.0
National Geographic Society	10.5

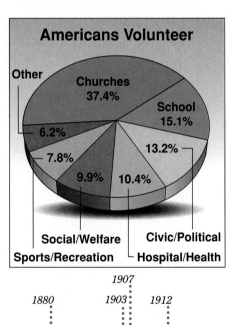

Americans Volunteer

Other

Churches 37.4%

School 15.1%

6.2%

13.2%

7.8%

9.9%   10.4%

Social/Welfare

Civic/Political

Sports/Recreation

Hospital/Health

GIRL SCOUTS

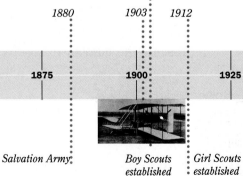

1907

1880   *1903*   *1912*

1850   1875   1900   1925

Salvation Army   *Boy Scouts established*   *Girl Scouts established*

Chapter Project

Clubs and Recreation

Work in a group.

1. Make a list of the organizations or volunteer groups to which you, your friends, and members of your family belong.

2. Prepare a chart showing how many people belong to each organization and how many total hours they spend each week or month with the organization.

3. Figure out how to begin your own club. Make a chart or write a proposal describing the goals of your club, its expenses, and its sources of income.

Looking Ahead

In this chapter, you will see how mathematics can be used to answer the questions about how many people belong to various organizations.

The major objectives of the chapter are to:

- compare and order decimal numbers

- estimate sums, differences, and products of decimals

- multiply and divide decimals

- use decimals to convert units within the metric system

1950s   *1961*   *1991*

1950   **1975**   **2000**

Peace Corps established

Persian Gulf War ends

47

2-1 Comparing and Ordering Decimals

Objective

Compare and order decimals.

Are you hungry for a Quarter Pounder® with Cheese or are you having a Big Mac® attack? A McDonald's® Quarter Pounder® with Cheese contains 76.7 calories per ounce, and a Big Mac® contains 83.3 calories per ounce. The Big Mac® contains more calories per ounce. When you use words like *more, less,* or *equal to,* you are comparing numbers.

You can compare decimals like 76.7 and 83.3 using a number line. To graph these decimals, you need to locate the number on the number line and draw a dot at that point.

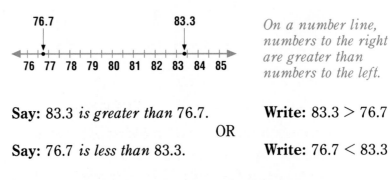

On a number line, numbers to the right are greater than numbers to the left.

Say: 83.3 *is greater than* 76.7.   **Write:** 83.3 > 76.7

OR

Say: 76.7 *is less than* 83.3.   **Write:** 76.7 < 83.3

You can also compare decimals by comparing the digits in each place-value position. The place-value chart below tells the position of each digit in the number 125.0674.

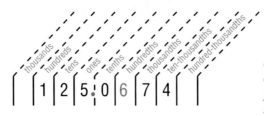

The digit 6 is in the hundredths position. This digit and its place-value position name the number six hundredths, 0.06.

To compare two decimals, align the numbers by their decimal points. Start at the left and compare the digits in each place-value position. Compare as with whole numbers.

1 Compare 24.9 and 25.3.

24.**9** In the tens place, the digits are the same.

2**5**.3 In the ones place, $4 < 5$. So, $24.9 < 25.3$.

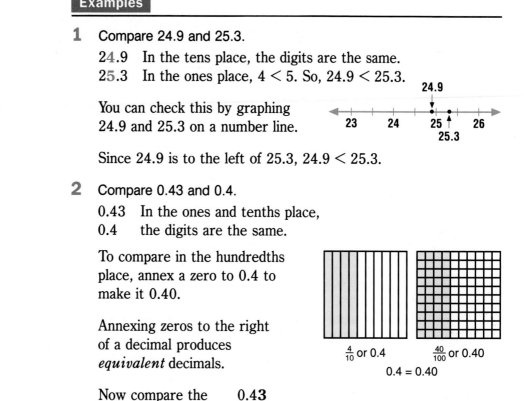

You can check this by graphing
24.9 and 25.3 on a number line.

Since 24.9 is to the left of 25.3, $24.9 < 25.3$.

2 Compare 0.43 and 0.4.

0.43 In the ones and tenths place,

0.4   the digits are the same.

To compare in the hundredths
place, annex a zero to 0.4 to
make it 0.40.

Annexing zeros to the right
of a decimal produces
equivalent decimals.

$\frac{4}{10}$ or 0.4 $\frac{40}{100}$ or 0.40

$0.4 = 0.40$

Now compare the   **0.43**
hundredths place.   **0.40** *In the hundredths place, 3 > 0.*

So, $0.43 > 0.4$.

Checking for Understanding

*Communicating
Mathematics*

Read and study the lesson to answer each question.

1. **Draw** a place-value chart for 840.107.
2. **Tell** a friend how comparing decimals is similar to and different from comparing whole numbers.
3. **Show** how you would compare 0.3 and 0.003.

*Guided
Practice*

Draw a number line to show which decimal is greater.

4. 0.56, 0.51   5. 1.22, 1.02   6. 0.97, 1.06

Replace each ● with $<$, $>$, or $=$.

7. 2.15 ● 2.05   8. 7.9 ● 7.9   9. 1.3 ● 1.31

Exercises

*Independent
Practice*

Draw a number line to show which decimal is greater.

10. 0.31, 0.33   11. 0.23, 0.29   12. 1.4, 1.04

13. 1.037, 1.009   14. 5.23, 5.0066   15. 2.5, 2.49

Replace each ● with <, >, or =.

16. 4.03 ● 4.01   17. 0.77 ● 0.69   18. 8 ● 0.8

19. 0.68 ● 0.680   20. 3.28 ● 3.279   21. 0.23 ● 0.32

22. 0.55 ● 0.65   23. 1.29 ● 1.43   24. 2.6 ● 2.6

25. 0.0034 ● 0.034   26. 1.67 ● 0.48   27. 9.09 ● 9

28. Write a sentence comparing two of the numbers shown on the number line.

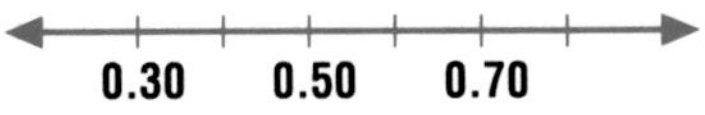

0.30   0.50   0.70

Order each set of numbers from least to greatest.

29. 5.13, 5.07, 5.009   30. 0.9, 0.088, 1.02, 0.98

31. 0.087, 0.901, 2, 1.001   32. 12.3, 12.008, 1.273, 12.54

Mixed Review
33. Use compatible numbers to estimate the quotient of $9,850 and 36. *(Lesson 1-3)*

34. **Algebra**   Evaluate the expression $6mn + n \div m$ if $m = 4$ and $n = 20$. *(Lesson 1-8)*

35. Carla is 6 years old and is 12 years younger than Maria. How old is Maria? *(Lesson 1-10)*

36. Add 82,998 and 3,700 mentally. *(Lesson 1-4)*

Problem Solving and Applications
37. **Critical Thinking**   The numbers 999, 463, 208, and 175 are in order from greatest to least. Place decimal points in each number so that the resulting decimal will be in order from *least* to *greatest*. Do not rearrange the numbers.

38. **Weather**   A barometer is an instrument that measures atmospheric pressure in terms of millimeters of mercury. The higher the mercury rises in the tube, the higher the atmospheric pressure is.
    a. Order the following barometer readings from least to greatest: 29.97 mm, 30.22 mm, 29.13 mm, 30.53 mm, 31.01 mm.
    b. At which barometric reading is the atmospheric pressure the greatest?

39. **Sports**   The chart below shows the winning scores in the 1988 Olympic women's balance beam.
    a. Rank the scores in order from 1 to 4. The highest score gets a rank of 1.
    b. Whose score(s) was higher than 19.875?
    c. Whose score(s) was lower than 19.875?
    d. Whose score was higher, Gabriela's or Phoebe's?

Women's Balance Beam	Points
Phoebe Mills	19.837
Daniela Silivas	19.924
Gabriela Potorac	19.837
Elena Shoushounova	19.875

2-2 **Rounding Decimals**

Objective

Round decimals.

Engineers hope that one day soon they can build a roller coaster that will travel 100 miles per hour. Right now one of the fastest roller coasters in the world is The Beast at King's Island in Ohio. Its top speed is 64.77 miles per hour. What is its top speed rounded to the nearest whole number?

On the number line, the graph of 64.77 is closer to 65 than 64. To the nearest whole number, 64.77 rounds to 65.

64.77

64   65

You can round to any place-value position without using a number line.

Rounding Decimals	Look at the digit to the right of the place being rounded. • The digit remains the same if the digit to the right is 0, 1, 2, 3, or 4. • Round up if the digit to the right is 5, 6, 7, 8, or 9.

Examples

1 Round 1.84 to the nearest tenth.

$$1.84 \;\rightarrow\; \textit{The digit to the right of 8 (tenths place) is 4. So, 8 remains the same.} \;\rightarrow\; 1.8$$

1.84 rounded to the nearest tenth is 1.8.

2 Round 14.295 to the nearest hundredth.

$$14.295 \;\rightarrow\; \textit{The digit to the right of 9 (hundredths place) is 5. So, 9 rounds up.} \;\rightarrow\; 14.30$$

14.295 rounded to the nearest hundredth is 14.30.

Example 3 *Connection*

Geometry There are some numbers that are not exact. Pi (π), which is a number used to find the area of a circle, is such a number. Pi is 3.1415926. . . . Round pi to the nearest hundredth.

First write pi to the thousandths place: 3.141.

$$3.141 \quad \rightarrow \quad \begin{array}{c} \textit{The digit to the right of 4} \\ \textit{(hundredths place) is 1.} \\ \textit{So, 4 remains the same.} \end{array} \quad \rightarrow \quad 3.14$$

3.1415926 . . . rounded to the nearest hundredth is 3.14.

Checking for Understanding

Communicating Mathematics

Read and study the lesson to answer each question.

1. **Tell** why you can ignore all the digits to the right of 7 when you are rounding 6.37215 to the nearest tenth.

2. **Tell,** using the number line below, to what whole number you would round 14.37.

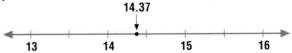

3. **Show** how 9.1651 rounds to 9.2 on a number line.

Guided Practice

Round each number to the place indicated. Draw a number line to support your decision.

4. 0.315 Round to the nearest tenth.
5. 0.2456 Round to the nearest hundredth.
6. 7.0375 Round to the nearest thousandth.
7. 17.499 Round to the nearest tenth.

Round each number to the underlined place-value position.

8. <u>8</u>.2
9. 12.1<u>2</u>56
10. <u>2</u>3.09
11. 0.2<u>3</u>8
12. 16.<u>4</u>875
13. 0.21<u>8</u>
14. 709.0<u>9</u>21
15. 0.08<u>5</u>41

Exercises

Independent Practice

Round each number to the underlined place-value position.

16. <u>2</u>3.48
17. 0.3<u>7</u>
18. 0.7<u>8</u>9
19. 0.9<u>6</u>
20. 1.5<u>7</u>2
21. 0.1<u>6</u>3
22. 0.008<u>4</u>
23. 15.4<u>5</u>1

Round each number to the underlined place-value position.

24. 3.1<u>4</u>53

25. 4.52<u>9</u>88

26. 0.4<u>4</u>5

27. 0.<u>7</u>87

28. <u>3</u>8.56

29. 59.6<u>1</u>

30. 0.<u>5</u>55

31. 1.7<u>0</u>4

32. The height of the world's tallest mountain, Mt. Everest, is 29,067 feet. Round this height to the nearest thousand feet.

33. Draw a number line to show how 3.67 rounds to 4.

Mixed Review

34. Use front-end estimation to estimate the sum of $179 + $213 + $355. *(Lesson 1-2)*

35. Juanita is preparing for her birthday party. She buys 2 boxes of cookies containing 24 cookies each and 3 packages of brownies containing 15 brownies each. Find the total number of dessert items she has bought. *(Lesson 1-7)*

36. Evaluate 4^3. *(Lesson 1-9)*

37. Solve $\frac{x}{6} = 3$ mentally. *(Lesson 1-10)*

38. Order 6.32, 8.75, 9, 10.29, 8.78, and 9.15 from least to greatest. *(Lesson 2-1)*

Problem Solving and Applications

39. **Measurement** Barry is using a ruler marked in centimeters to measure the length of the ribbon at the right.
    a. To the nearest centimeter, what is the length of the ribbon?
    b. To the nearest tenth of a centimeter, what is the length of the ribbon?

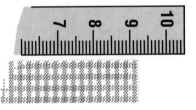

40. **Critical Thinking** In certain everyday situations, it is best to round *all* numbers up. What are some situations in which this might happen?

41. **Measurement** A gill is a unit of capacity that is equal to one-fourth of a pint. It has a volume of 7.2187 cubic inches. To the nearest hundredth of a cubic inch, what is its volume?

42. **Journal Entry** Why do people use rounded numbers, even when they know the exact amount? For example, why would you tell a friend you live about a mile from school, even if you knew that you actually lived 1.1 miles from school?

2-3 Estimating with Decimals

Objective

Estimate with decimals.

Words to Learn

clustering

In a recent year, some of the busiest airports in the United States were Dallas/Fort Worth, San Francisco, Los Angeles, Hartsfield in Atlanta, John F. Kennedy (JFK) in New York, and O'Hare in Chicago. *About* how many millions of passengers use these airports in one year?

Airport	Passengers (in millions)
Dallas/Fort Worth	47.6
Hartsfield, Atlanta	43.3
JFK, New York City	30.3
Los Angeles	45
O'Hare, Chicago	59.1
San Francisco	29.9

To solve this problem, you can use *rounding* to estimate the answer. To estimate by rounding, round each addend to its greatest place-value position. Then complete the operation.

$$
\begin{array}{rcr}
47.6 & \rightarrow & 50 \\
43.3 & \rightarrow & 40 \\
30.3 & \rightarrow & 30 \\
45.0 & \rightarrow & 50 \\
59.1 & \rightarrow & 60 \\
+29.9 & \rightarrow & +\,30 \\
\hline
 & & 260
\end{array}
$$

There are *about* 260 million passengers using these airports in one year.

DID YOU KNOW

260 million passengers ... that's more than the population of the United States. It must be noted, though, that many of these passengers are probably repeat passengers. That is, the same person uses these airports several times a year.

Examples

1 Estimate the difference of 16.295 and 8.762.

$$
\begin{array}{rcr}
16.295 & \rightarrow & 16 \\
-\,8.762 & \rightarrow & -\,9 \\
\hline
 & & 7
\end{array}
$$

The difference is *about* 7.

2 Estimate the product of 34.9 and 48.3.

$$
\begin{array}{rcr}
34.9 & \rightarrow & 30 \\
\times 48.3 & \rightarrow & \times\;50 \\
\hline
 & & 1{,}500
\end{array}
$$

The product is *about* 1,500.

You can also use **clustering** to estimate sums. Clustering is used in addition situations if the numbers seem to be clustered around a common quantity.

Example 3 *Problem Solving*

Homework   Latricia used a calculator to solve a math homework problem. She added 32.8, 29.7, 34.1, 30.9, 27.5, and 33.6 and got 157.9. Check the reasonableness of her answer using clustering.

All the numbers are clustered around 30. There are six numbers. So, the sum is *about* 30 × 6, or 180.

157.9 is not very close to 180. Latricia may have made an error in entering the numbers. She should add the six numbers again.

You can review front-end estimation on page 8.

You can use *front-end* estimation to add or subtract.

Examples

Estimate using front-end estimation.

4   Estimate the sum of 5.82, 2.19, 8.1, and 6.05.

5.82	→	5.82
2.19	→	2.19
8.1	→	8.1
+ 6.05	→	+6.05
21.00		1.00

21 + 1 = 22
The estimate is 22.

5   Estimate 23.9 − 11.4.

23.9	→	23.9
−11.4	→	−11.4
10.0		2.0

10 + 2 = 12
The estimate is 12.

LOOKBACK
You can review compatible numbers on page 11.

You can use *compatible* numbers to estimate quotients.

Example 6

Divide 0.59 by 2.3.

$$2.3\overline{)0.59} \quad \rightarrow \quad 2\overline{)0.6}^{\,0.3}$$ The quotient is *about* 0.3.

Checking for Understanding

Communicating Mathematics

Read and study the lesson to answer each question.

1. **Tell** why estimation is helpful when using a calculator to solve math problems.

2. **Tell** whether you agree or disagree with the following statement and why. *An estimate is a good way to answer the opening question of this lesson.*

3. **Write** a sentence describing when it makes sense to use the clustering method to estimate a sum.

Estimate by rounding.

4. 9.56
    + 5.34

5. 23.84
    + 12.13

6. 6.8
    × 3.7

7. 9.3)‾65.48

Estimate by clustering.

8. 56.9 + 63.2 + 59.3 + 61.1

9. 18.4 + 22.5 + 20.7

Estimate by using front-end estimation.

10. 32.6
     56.2
     + 71.9

11. 13.21
     − 8.23

12. 9.34
     + 3.18

Estimate by using compatible numbers

13. 2.6)‾8.99

14. 38.1)‾984.76

15. 6.8)‾40.79

16. **Games**   You are watching Wheel of Fortune and the letters R, S, T, N, and L are chosen to solve the puzzle. Use the table at the right to estimate what percent of the time at least one of these letters would appear in the puzzle.

Percentage That Each Letter Is Used.					
A	8.2	**J**	0.1	**S**	6.0
B	1.4	**K**	0.4	**T**	10.5
C	2.8	**L**	3.4	**U**	2.5
D	3.8	**M**	2.5	**V**	0.9
E	13.0	**N**	7.0	**W**	1.5
F	3.0	**O**	8.0	**X**	0.2
G	2.0	**P**	2.0	**Y**	2.0
H	5.3	**Q**	0.1	**Z**	0.07
I	6.5	**R**	6.8		

Exercises

Estimate. Use an appropriate strategy.

17. $3.27
     6.75
     + 8.56

18. 19.5
     +56.13

19. 34.3
     −18.9

20. 67.86
     −24.35

21. 7.5
     ×8.4

22. 26.3
     × 9.7

23. 8.1)‾73.8

24. 18.4)‾41.7

25. 121.5
     +487.8

26. $76.22
     − 47.34

27. 32.5
     ×81.4

28. 11.4)‾35.7

29. 50.4 + 51.1 + 48.9 + 49.5

30. 9.9 + 10.0 + 10.3 + 11.1

31. 100.5 + 97.8 + 101.6 + 100.2 + 99.3 + 99.1

32. Estimate the quotient of 119 and 23.

33. Estimate the product of 72 and 99.

34. Estimate the difference of 69.4 and 16.2.

Mixed Review

35. **Sports** Min has a collection of 269 baseball cards. He trades 25 of them to a friend for an autographed baseball. Compute mentally to find the number of cards left in Min's collection. *(Lesson 1-4)*

36. **Algebra** Write an expression that represents a $500 donation plus $5 for every event. Let *n* represent the number of events. *(Lesson 1-8)*

37. Solve $m + 18 = 33$ *(Lesson 1-10)*

38. Replace the ● in the following sentence with $<$, $>$, or $=$. 0.2 ● 0.214 *(Lesson 2-1)*

39. **Outdoor Exercise** A ski resort advertises a new cross country ski trail that is 5.673 miles long. To the nearest 0.1 mile, what is the length of the trail? *(Lesson 2-2)*

Problem Solving and Applications

40. **Critical Thinking** Using rounding, an estimate of $11 + 38$ is 50 and an estimate of $14 + 44$ is 50. Which estimate is closest to the exact sum? Why?

41. **Land Speed** In 1910, the land speed record was 131.72 miles per hour. In 1970, it was 622.29 miles per hour. About how many times faster was the speed in 1970?

42. **Consumer Math** The menu at the right appears at the snack counter of a local movie theater. Use estimation to answer each question.

Popcorn	. . . . sm. $1.69
	. . . . lg. $2.79
Nachos w/cheese	. . . . $2.29
Soft drinks	. . . . sm. $0.95
	. . . . med. $1.35
	. . . . lg. $1.95

    a. *About* how much do a small popcorn and 2 medium soft drinks cost?

    b. Is $4.00 enough to buy nachos with cheese and a small soft drink?

    c. Is $3.21 the correct change from $5.00 for a large popcorn?

43. **Geography** The table at the right shows the length in miles of the three longest rivers in the world. *About* how many miles long are the three rivers?

    Longest Rivers

    This graph shows the longest rivers in the world. Their lengths are shown in thousands of miles.

Nile	4.132
Miss./Mo./Red Rock	3.9
Amazon	3.83

Addition and Subtraction of Decimals

Objective

Add and subtract decimals.

This page and the next provide a review of addition and subtraction of decimals. To add or subtract decimals, align the decimal points. Then start at the right and add or subtract the numbers in each place-value position. Annex zeros when necessary.

Examples

1 Add $0.75 + 0.82$.

Estimate: $0.8 + 0.8 = 1.6$

Align the decimal points.	*Add the hundredths.*	*Add the tenths.*
$\begin{array}{r} 0.75 \\ +0.82 \\ \hline \end{array}$	$\begin{array}{r} 0.75 \\ +0.82 \\ \hline 7 \end{array}$	$\begin{array}{r} {}^{1} \\ 0.75 \\ +0.82 \\ \hline 1.57 \end{array}$ *Compare to the estimate.*

2 Add $1.73 + 24 + 2.236$.

Estimate: $2 + 24 + 2 = 28$

$\begin{array}{r} 1.730 \\ 24.000 \\ +\ 2.236 \\ \hline \end{array}$ *Align the decimal points. Annex zeros so each addend has the same number of decimal places.* $\begin{array}{r} 1.730 \\ 24.000 \\ +\ 2.236 \\ \hline 27.966 \end{array}$

3 Subtract $4.285 - 2.51$.

Estimate: $4 - 3 = 1$

$\begin{array}{r} 4.285 \\ -2.510 \\ \hline \end{array}$ *Align the decimal points. Annex a zero.* $\begin{array}{r} {}^{3\ 12} \\ 4.285 \\ -2.510 \\ \hline 1.775 \end{array}$ *Subtract in each place-value position.*

4 Subtract $2.23 - 0.497$.

Estimate: $2 - 0.5 = 1.5$

$\begin{array}{r} 2.230 \\ -0.497 \\ \hline \end{array}$ *Align the decimal points. Annex a zero. Then subtract.* $\begin{array}{r} {}^{1\ 111210} \\ 2.230 \\ -0.497 \\ \hline 1.733 \end{array}$

Exercises

Independent
Practice

Add or subtract.

1. $\begin{array}{r} 2.3 \\ +4.1 \\ \hline \end{array}$

2. $\begin{array}{r} 0.37 \\ +0.55 \\ \hline \end{array}$

3. $\begin{array}{r} 0.67 \\ -0.43 \\ \hline \end{array}$

4. $\begin{array}{r} 42.76 \\ -31.59 \\ \hline \end{array}$

5. $\begin{array}{r} \$6.78 \\ +\ 4.99 \\ \hline \end{array}$

6. $\begin{array}{r} 8 \\ +6.76 \\ \hline \end{array}$

7. $\begin{array}{r} 8.267 \\ -6.52 \\ \hline \end{array}$

8. $\begin{array}{r} 17.6 \\ -\ 4.739 \\ \hline \end{array}$

9. $\begin{array}{r} 67.4 \\ 8.05 \\ +105.3 \\ \hline \end{array}$

10. $\begin{array}{r} 5.124 \\ 32.45 \\ +\ 8.6 \\ \hline \end{array}$

11. $\begin{array}{r} 18 \\ -\ 9.36 \\ \hline \end{array}$

12. $\begin{array}{r} 7.63 \\ -3.009 \\ \hline \end{array}$

13. $6.6 + 4.58$

14. $5.77 - 2.374$

15. $86.332 - 48.7$

16. $0.563 + 5.8 + 6.89$

17. $23.4 + 9.865 + 18.26$

Solve each equation.

18. $r = 0.32 + 8.99$

19. $32.45 - 18.86 = m$

20. $a = 12 + 7.64$

21. $t = 34.6 - 23.88$

22. $6.2 + 8.57 = y$

23. $c = 26.13 - 13.7$

24. $d = 45.1 + 16 + 8.091$

25. $f = 9.32 + 7.06 + 12.221$

26. Find the difference of 156.003 and 89.42

27. Find the sum of 67.03 and 100.97.

Problem Solving
and
Applications

28. **Critical Thinking**   How is subtraction of decimals similar to subtraction of whole numbers? How is it different?

29. **Algebra**   The equation $5.7 - n = 2.06$ can be solved by solving the related sentence $5.7 - 2.06 = n$. What is the value of n?

30. **Sports**   In the 1984 Olympic games, Edwin Moses of the United States won the 400-meter hurdles with a time of 47.75 seconds. In the 1988 games, another American, Andre Phillips, won the 400-meter hurdles in 47.19 seconds. How much faster was Phillips than Moses?

31. **Statistics**   According to a 1989 projection, there will be 131.2 million males and 137.1 million females in the United States in the year 2000. What will be the total population of the United States in the year 2000?

Review: Addition and Subtraction of Decimals   **59**

2-4A Multiplication with Decimal Models

A Preview of Lesson 2-4

Objective

Multiply decimals using models.

Materials

decimal models
markers

Multiplication of decimals is similar to multiplication of whole numbers. You can multiply decimals using a decimal model.

Try this!

Work in groups of two.

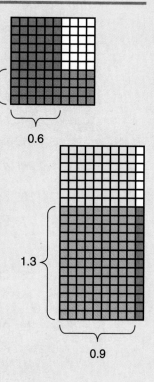

- Model 0.4×0.6 by shading 4 tenths and 6 tenths as shown.

- What is the product? Tell how you figured it out.

- Model 1.3×0.9 by shading 13 tenths and 9 tenths as shown using two decimal models.

- What is the product? Tell how you figured it out.

What do you think?

1. Use decimal models to show each product.
   a. 0.3×3 b. 0.7×0.5 c. 1.4×0.8
   d. 0.2×0.9 e. 0.6×2 f. 1.1×0.1
2. Tell how many decimal places there are in each factor and in each product in Exercise 1.
3. How does the number of decimal places in a product relate to the number of decimal places in the factors?

Extension

4. Use what you have learned in this lab to find the product of 1.3 and 1.8.

2-4 Multiplying Decimals

Objective

Multiply decimals.

Do you know that math and music are related? All sound is caused by vibrations. The number of vibrations per second determines the pitch of the sound. The more vibrations per second the higher the pitch. The number of vibrations per unit of time is called frequency. The frequency of any note multiplied by 1.06 gives the frequency of the note one-half step higher.

These are each a half step.

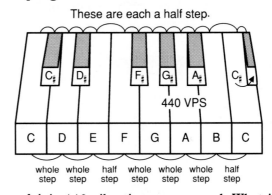

440 VPS

| C | D | E | F | G | A | B | C |

whole step, whole step, half step, whole step, whole step, whole step, half step

Read A♯ as "A sharp."

The frequency of A is 440 vibrations per second. What is the frequency of A♯ (one-half step higher)? To find the frequency of A♯, multiply 440 by 1.06.

Use estimation to help place the decimal point in the product.
Estimate: $1.06 \times 440 \rightarrow 1 \times 440 = 440$

```
     440
  × 1.06
   26 40
  440 0
  466.40
```

Since the estimate is 440, place the decimal point so the answer is in the 400s.
Compared to the estimate of 440, the answer 466.4 is reasonable.

The frequency of A♯ is 466.4 vibrations per second.

There is another way to find the product of two decimals.
Look at the table below.

When am I ever going to use this?

Multiplying decimals is an important concept in the banking industry.

Suppose you deposited $125.50 into an account. After 6 months, you had 1.04 times as much money in the account. How much money is in the account?

$125.50 × 1.04
= $130.52

Factors and Product	Decimal Places in Factors	Decimal Places in Product
5 ⊗ 0.7 ⊜ **3.5**	0, 1	1
0.5 ⊗ 0.7 ⊜ **0.35**	1, 1	2
0.5 ⊗ 0.07 ⊜ **0.035**	1, 2	3

This table gives the number of decimal places in each factor and the number of places in the product. The number of decimal places in the product is the sum of the number of decimal places in the factors.

Example 1 Problem Solving

Animals A snail moves at a speed of about 0.005 kilometers per hour. To find out how far it can travel in a half hour (0.5 hour), multiply 0.005 and 0.5.

$$\begin{array}{r} 0.005 \\ \times\quad 0.5 \\ \hline 0.0025 \end{array}$$ *three decimal places*
one decimal place
four decimal places   *Is the answer reasonable?*

Check with a calculator: 0.005 ⊠ 0.5 ⊟ **0.0025**

A snail can travel 0.0025 kilometers in a half hour.

Example 2 Connection

You can review
evaluating expressions
on page 28.

Algebra Evaluate 2.5 · *a* if *a* = 0.7.

$$2.5 \cdot a = 2.5 \cdot 0.7 \quad \textit{Replace a} \\ = 1.75 \qquad\quad \textit{with 0.7.}$$

$$\begin{array}{r} 2.5 \\ \times\,0.7 \\ \hline 1.75 \end{array}$$ *one decimal place*
one decimal place
Count two decimal places
from the right.

Checking for Understanding

Communicating Mathematics

Read and study the lesson to answer each question.

1. **Tell** in your own words why the number of decimal places in the product of two decimals is equal to the sum of the decimal places in the factors.

2. **Write** the multiplication sentence for the model at the right.

3. **Make a model** to show why $0.1 \times 0.1 = 0.01$.

Guided Practice

Place the decimal point in each product.

4. $1.32 \times 4 = 528$ 5. $0.7 \times 1.1 = 077$ 6. $5.48 \times 3.6 = 19728$

Multiply.

7. $\begin{array}{r} 0.4 \\ \times\,0.7 \\ \hline \end{array}$ 8. $\begin{array}{r} 3.4 \\ \times\,7.8 \\ \hline \end{array}$ 9. $\begin{array}{r} 0.15 \\ \times\,1.23 \\ \hline \end{array}$ 10. $\begin{array}{r} 11.5 \\ \times\,0.47 \\ \hline \end{array}$

11. 0.45×0.02 12. 8.32×0.064 13. 1.9×0.6

14. Evaluate $1.4 \cdot x$ if $x = 0.9$.

15. Evaluate $4.2 \cdot y$ if $y = 3.6$.

Exercises

Independent Practice

Multiply.

16.   0.2
    $\times$ 6

17.   0.3
    $\times$ 0.9

18.   0.45
    $\times$ 0.12

19.   0.0023
    $\times$ 32

20. 10.1×9
21. 4.5×0.34
22. 0.0023×0.35
23. 6.78×1.3
24. 1.5×2.7
25. 0.25×36.3

Solve each equation.

26. $p = 0.45 \times 0.02$
27. $5.1 \times 4.3 = g$
28. $r = 0.08 \times 1.9$
29. $0.25 \times 0.0004 = t$
30. $b = 0.4 \times 3$
31. $m = 1.17 \times 0.09$

Evaluate each expression if $a = 0.6$ and $b = 3.1$.

32. $0.32 \cdot b$
33. $a \cdot 1.2$
34. $12.4 \cdot a$
35. $b \cdot 0.0019$

36. **Animals** A giant tortoise can travel at a speed of about 0.2 kilometers per hour. At this rate, how far can it travel in 1.75 hours?

Mixed Review

37. Use compatible numbers to estimate $\$1.19 \div 16$. *(Lesson 1-3)*
38. Evaluate b^5 if $b = 2$. *(Lesson 1-9)*
39. Draw a number line to show which is greater, 3.77 or 3.7. *(Lesson 2-1)*
40. Round 26.394 miles to the nearest mile. *(Lesson 2-2)*
41. Find the difference of 5.8 and 3.59. *(Lesson 2-3)*

Problem Solving and Applications

42. **Biology** The Marshall Island goby, the world's smallest fish, measures 0.47 inch. The striped bass, an American sport fish, is about 25.5 times longer. How long is the bass?

43. **Currency** On October 2, 1991, the Japanese yen was worth 0.0075 United States dollars. At that time, how much was 450 yen worth in United States dollars?

44. **Critical Thinking** Make up a problem in which the factors each have two decimal places, but the product has only three.

45. **Mathematics and Time** Read the following paragraphs.

> Throughout history, people have measured the passing of time by observing natural events like the succession of days and nights, different positions of the sun in the sky, the phases of the moon, or the tides. In 3500 B.C., the first measuring device appeared in Egypt. It was called a *clepsydrae* ('klep-sə-drə), or water clock.
>
> Today, time is measured with great precision. We use watches with hands, digital watches, and chronometers to regulate time.

Earth revolves around the sun in 365.24 days. How many days does it take Earth to revolve 12 times?

2-5 Powers of Ten

Objective

Multiply decimals mentally by powers of ten.

How far away is the moon? Scientists have calculated it to be 2.39×10^5 miles from Earth. You can multiply 2.39 and 10^5 to find the number of miles.

$$2.39 \times 10^5 = 2.39 \times 100,000$$

2.39 $\boxed{\times}$ 100000 $\boxed{=}$ **239000.**

The moon is 239,000 miles from Earth.

How can you find the product of a power of 10, like 10^5, and another number without using a calculator or paper and pencil?

Consider the following products. Look for a pattern.

Decimal		Power of Ten		Product
2.39	$\times$	10^0 (or 1)	=	2.39
2.39	$\times$	10^1 (or 10)	=	23.9
2.39	$\times$	10^2 (or 100)	=	239
2.39	$\times$	10^3 (or 1,000)	=	2,390
2.39	$\times$	10^4 (or 10,000)	=	23,900

Note that as the decimal is multiplied by greater powers of 10, the decimal point in the product is farther to the right of the original position. Study the last two columns. The exponent in the power of 10 and the number of places the decimal point moved to the right are the same.

You can use this pattern to multiply mentally.

Mental Math Hint

• • • • • • • • • • • •

You can also determine the number of places the decimal point moves to the right by counting the number of zeros in the power of ten. That is, in Example 2, you can see that there are three zeros in 1,000, so you will move the decimal point three places to the right.

Examples

1 Multiply 0.34 and 10^4 mentally.
$0.34 \times 10^4 = 3,400.$

Move the decimal point 4 places to the right.

The solution is 3,400.

2 Solve $c = 13.1 \times 1,000$.
$c = 13.1 \times 1,000$ *Rename*
$= 13.1 \times 10^3$ *1,000 as 10^3.*
$= 13,100.$

Move the decimal point 3 places to the right.

The solution is 13,100.

Checking for Understanding

Communicating Mathematics

Read and study the lesson to answer each question.

1. **Tell** a classmate how you would solve $x = 2.378 \times 100$ mentally.

2. **Show** the steps you would take to multiply 100×0.28 using paper and pencil. Compare it to solving the problem mentally.

Guided Practice

Choose the correct product mentally.

3. 2.34×100; 0.0234 or 234

4. 0.8×10^0; 8 or 0.8

5. 0.028×10^2; 2.8 or 28

6. 1.4×10^5; 140,000 or 14.07

Multiply mentally.

7. 12.53×10

8. $0.605 \times 1,000$

9. 3.159×100

Solve each equation.

10. $n = 2.031 \times 10^4$

11. $a = 0.78 \times 10^2$

12. $1.32 \times 10^3 = x$

Exercises

Independent Practice

Multiply mentally.

13. 0.05×100

14. 4.527×10^0

15. $2.78 \times 1,000$

16. 5.492×10^4

17. 0.925×100

18. 99.44×10^2

Solve each equation.

19. $c = 0.78 \times 1,000$

20. $m = 11.23 \times 10^5$

21. $6.894 \times 1,000 = k$

22. $28.1 \times 100 = b$

23. $t = 9.3 \times 1,000$

24. $3.76 \times 10^6 = y$

25. Suppose a can of soup costs 73¢. If you purchase 100 cans to donate to the Food Bank, what will the cost be?

Mixed Review

26. Estimate the sum of 1,237 and 896. *(Lesson 1-2)*

27. Order 3, 0.3, 3.33, 0.33, 0.03, 3.03 from least to greatest. *(Lesson 2-1)*

28. Round 63.2765 to the nearest whole number. *(Lesson 2-2)*

29. **Travel** At one point during Manuel's summer trip, he drove 356.5 miles in 6.3 hours. Estimate Manuel's speed. *(Lesson 2-3)*

Problem Solving and Applications

30. **Geology** In 1872, Yellowstone National Park became the first national park in the United States. It is noted for its hot springs that erupt through the surface of Earth forming geysers. The park has about 2,000 hot springs. If 0.1 of these are geysers, how many geysers are in the park?

31. **History** During the seventeenth century, the Spanish used a coin called the *real*. A real was equal to $0.125 in today's U.S. currency. If you had 100 reals, how much money did you have in U.S. currency?

32. **Critical Thinking** Explain why the decimal point moves right when multiplying by powers of ten and why the decimal point moves left when dividing by powers of ten.

33. **Computer Connection** The BASIC computer language uses a special notation, E, for powers of 10. For example, the notation 2.4E3 means 2.4×10^3. Write the number represented by each of the following.
    a. 1E6   b. 6.07E2   c. 3.9256E7

34. **Journal Entry** Write a rule that states how to multiply by a decimal number and a power of ten.

Mid-Chapter Review

Replace each ● with $<$, $>$, or $=$. *(Lesson 2-1)*
1. 0.28 ● 0.028   2. 1.32 ● 1.320   3. 3.25 ● 3.2

Round each number to the underlined place-value position. *(Lesson 2-2)*
4. 0.4<u>9</u>   5. 6.<u>5</u>2   6. 0.9<u>9</u>7   7. 3.<u>1</u>24

Estimate. *(Lesson 2-3)*

8. 43.67
   $+\,17.32$

9. $\$6.23$
   $-\ \ 4.09$

10. 2.8
    $\times\,8.2$

11. $5.7\overline{)43.9}$

12. Estimate the sum of 71.28, 68.4, 70.73, 69.45, and 73.21. *(Lesson 2-3)*

Multiply. *(Lesson 2-4)*

13. 0.7
    $\times\,0.9$

14. 6.4
    $\times\,5.8$

15. 5.32
    $\times\,4.1$

16. 0.87
    $\times\ \ 16$

17. **Postage** Mr. Jackson buys 1,000 stamps for his business. If each stamp cost 29¢, how much did he spend altogether? *(Lesson 2-5)*

2-6 Scientific Notation

Objective

Express numbers greater than 100 in scientific notation and vice versa.

Words to Learn

scientific notation

DID YOU KNOW

Mercury takes only 88 Earth days to make one trip around the sun. However, it rotates slowly on its axis. One day on Mercury is about 59 Earth days long.

Did you know that the planet Mercury was named after the Roman wing-footed messenger of the gods? The planet was so named because of the speed it travels. Mercury orbits the sun at 30 miles per second while Earth orbits the sun at only 18.5 miles per second. Mercury is about 36,000,000 miles from the sun and is the closest planet to the sun.

Large numbers like 36,000,000 can be expressed in scientific notation.

A number in **scientific notation** is written as the product of a number greater than or equal to 1 and less than 10 and a power of ten. The power of ten is written with an exponent. To find the exponent, count the number of places the decimal point was moved in the original number.

Write 36,000,000 in scientific notation.

3.6000000   *Move the decimal point to get a number between 1 and 10.*

3.6×10^7 *The decimal point was moved 7 places.*

Mercury is about 3.6×10^7 miles from the sun.

Examples

1 Write 347,000 in scientific notation.

3.47000   *Move the decimal point to get a number between 1 and 10.*

3.47×10^5 *The decimal point was moved 5 places.*

2 The planet Mars is an average distance of 141,710,000 miles from the sun. Express this number in scientific notation.

1.41710000   *Move the decimal point to get a number between 1 and 10.*

1.4171×10^8 *The decimal point was moved 8 places.*

Notice that the number in Example 2 has more digits to the right of the decimal point than the numbers in other examples. Usually, the decimal part of a number written in scientific notation is rounded to the hundredths place.

1.4171×10^8 → The distance from Mars to the Sun is about 1.42×10^8 miles.

A number that is in scientific notation can be written in standard form when necessary.

Example 3 *Problem Solving*

Astronomy   The diameter of Jupiter is 1.43×10^5 kilometers. The diameter of Earth is 1.28×10^4 kilometers. How much greater is Jupiter's diameter?

Write the numbers in standard form.
$1.43 \times 10^5 = 1.43 \times 100,000 = 143,000$
$1.28 \times 10^4 = 1.28 \times 10,000 = 12,800$

The difference is $143,000 - 12,800$ or $130,200$ kilometers. Jupiter's diameter is 130,200 kilometers greater than Earth's.

Checking for Understanding

Communicating Mathematics

Read and study the lesson to answer each question.

1. **Show** how you would determine which is greater: 25,200 or 1.75×10^5.
2. **Show** a classmate how to write 5,280, the number of feet in a mile, in scientific notation.

Guided Practice

Write each number in scientific notation.

3. 890   4. 4,300   5. 6,235
6. 52,000   7. 820,000   8. 126,400,000

Write each number in standard form.

9. 9.87×10^3 10. 6×10^2 11. 1.75×10^4
12. 2.3×10^4 13. 4.95×10^8 14. 0.57×10^6

Exercises

Independent Practice

Write each number in scientific notation.

15. 7,500   16. 8,450   17. 40,700
18. 630,000   19. 400,000   20. 32,000,000
21. 7,900,000   22. 558,000   23. 160,000,000

Write each number in standard form.

24. 5×10^3 25. 1.42×10^4 26. 4.2×10^2
27. 5.47×10^5 28. 9.5×10^6 29. 2.71×10^7
30. 8.08×10^3 31. 6.024×10^8 32. 5.75×10^4

33. The circulation of *'Teen* magazine is about 1,100,000. Write this number in scientific notation.

34. **History**   In 1989, 4.056×10^5 people immigrated to the United States from Mexico. Write this number in standard form.

Mixed Review   35. **Hobby**   Last year Gene had 237 baseball cards. He collected another 78 cards this year. Compute mentally to find the total. *(Lesson 1-4)*

36. Evaluate $100 \div 10 + 2 \cdot 6 \div 4$. *(Lesson 1-7)*

37. The quotient of a number and 8 is 14. Find the number. *(Lesson 1-10)*

38. **Math**   Carissa solved a difficult math problem in 7.9 minutes. Ted took 2.3 times longer to solve the same problem. How long did it take Ted to solve the problem? *(Lesson 2-4)*

39. Multiply 1,000 and 18.7. *(Lesson 2-5)*

Problem Solving and Applications   40. **Critical Thinking**   Order from least to greatest: 5.29×10^4; 8.35×10^2; 9.05×10^3; 5.29×10^3.

41. **Geography**   Mauna Kea, a Hawaiian mountain, would be 3.35×10^4 feet tall if its height was measured from the ocean floor. Mt. Everest, the highest mountain in the world, is about 29,000 feet tall.
   a. If Mt. Everest were next to Mauna Kea on the ocean floor, which would be taller?
   b. How much taller would it be?

42. **Astronomy**   The closest approach of a planet to the sun is called its *perihelion*. In 1991, Pluto was near its perihelion, which is 2,762,000,000 miles. At the same time, Neptune's perihelion was 2.766×10^9 miles. At that time in history, which of these planets was the ninth (and outermost) planet?

DATA SEARCH

43. **Data Search**   Refer to page 46. Order from greatest to least the nine organizations that have the most members.

2-7A Division with Decimal Models

A Preview of Lesson 2-7

Objective

Divide decimals using models.

Materials

decimal models
markers

Division can be thought of as putting a collection of objects into equal groups. You can show division of decimals using a decimal model.

Try this!

Work with a partner.

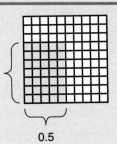

0.5

- Model $0.35 \div 0.5$ by shading 35 squares such that 5 columns of squares are shaded.

- How can you write the 35 squares as a decimal?

- How many squares are in each column?

- Write the squares in each column as a decimal.

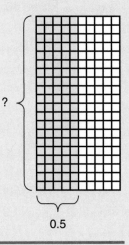

- Model $1 \div 0.5$ by shading 100 squares such that 5 columns of squares are shaded. To do this, you will need two decimal models.

- What is the quotient? Tell how you figured it out.

0.5

What do you think?

1. True or False? The examples show that decimal division is similar to whole-number division.

2. Explain how the decimal models show each quotient.
   a. $0.35 \div 0.05$ b. $1 \div 0.05$

3. Use decimal models to show each quotient.
   a. $0.25 \div 0.5$ b. $2 \div 0.8$ c. $0.6 \div 0.2$

4. When a whole number is divided by a decimal less than 1, as in $2 \div 0.5$, is the quotient greater or less than the divisor? Why?

Extension

5. Predict the quotient of $1 \div 0.25$.

2-7 Dividing Decimals

Objective

Divide decimals.

Leonardo Da Vinci was an artist, scientist, and inventor. But he is best known as an artist. He was born in Italy in 1452. His most famous painting is the Mona Lisa. The painting hangs in the Louvre Museum in Paris, France. It has a width of 0.5 meters and an area of 0.4 square meters. What is the height of the Mona Lisa?

You can use a decimal model to find the height. Shade 40 squares so that 5 columns of squares are shaded. The height is 0.8 meters.

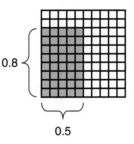

You can also solve this problem by dividing the area by the width: $0.4 \div 0.5$.

Estimate: $\frac{1}{2} \div \frac{1}{2} = 1$

0.4 $\boxed{\div}$ 0.5 $\boxed{=}$ 0.8

The height of the painting is 0.8 meters.

To divide two decimals using paper and pencil, it is best to change the divisor to a whole number by moving the decimal point to the right. You must also move the decimal point in the dividend the same number of places to the right. Study the models on the next page to see how this works.

Change 0.5 to 5 and 0.4 to 4 by multiplying by 10 (moving the decimal point one place to the right.)

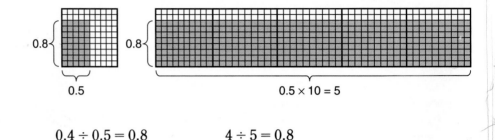

0.8 { 0.5   0.8 { 0.5 × 10 = 5

$$0.4 \div 0.5 = 0.8 \qquad\qquad 4 \div 5 = 0.8$$

Notice that the answer is the same for both models. So dividing 0.4 by 0.5 has the same result as dividing 4 by 5.

When you are using paper and pencil to find a quotient, the divisor is usually changed to a whole number. Of course, the dividend must be changed in the same way.

Examples

1 $9.6\overline{)199.68}$

$$
\begin{array}{r}
20.8 \\
9.6\overline{)199.68} \\
-192 \\
\hline
7\ 68 \\
-7\ 68 \\
\hline
0
\end{array}
$$

Change 9.6 to 96 and 199.68 to 1,996.8 by moving both decimal points one place to the right.

2 Solve $n = 0.7 \div 0.05$.

$$
\begin{array}{r}
14 \\
0.05\overline{)0.70} \\
-\ 5 \\
\hline
20 \\
-20 \\
\hline
0
\end{array}
$$

Annex a zero. Why?

Estimation Hint

• • • • • • • • • • • • •

In Example 1, use compatible numbers to estimate
199.68 ÷ 9.6.
200 ÷ 10 = 20

Example 3 *Problem Solving*

Sports In 1991, American sprinter Carl Lewis set a world record of 9.86 seconds for the 100-meter dash. A honeybee can fly the same distance in 20.706 seconds. How many times faster than a honeybee is Carl Lewis?

We need to divide 20.706 by 9.86.
Estimate: $20 \div 10 = 2$.

$$
\begin{array}{r}
2.1 \\
986.\overline{)2070.6} \\
-1972 \\
\hline
98\ 6 \\
-98\ 6 \\
\hline
0
\end{array}
$$

Change 9.86 to 986 and 20.706 to 2,070.6.

Carl Lewis is 2.1 times faster than a honeybee.

Checking for Understanding

Read and study the lesson to answer each question.

1. **Write** one or two sentences explaining why you should estimate before using a calculator.

2. **Write** the division sentence for the model at the right.

0.9

3. **Make a model** to show $0.56 \div 0.7$.

4. **Tell** a friend why $4.4 \div 0.8$ and $44 \div 8$ have the same quotient.

Guided Practice

Without finding or changing each quotient, change each problem so that the divisor is a whole number.

5. $0.36 \div 0.4$	6. $1.05 \div 0.7$	7. $4.4 \div 1.1$
8. $2.94 \div 0.084$	9. $1.89 \div 0.9$	10. $50.4 \div 0.56$

Divide.

11. $3 \div 0.6$	12. $4.2 \div 1.2$	13. $0.084 \div 0.056$
14. $0.287 \div 0.035$	15. $51 \div 0.8$	16. $0.245 \div 0.7$

Exercises

Independent Practice

Without finding or changing each quotient, change each problem so that the divisor is a whole number.

17. $0.82 \div 0.4$	18. $68.13 \div 0.003$	19. $2.6 \div 1.3$
20. $0.00945 \div 0.021$	21. $1.488 \div 3.1$	22. $14.42 \div 0.206$

Divide.

23. $0.6\overline{)4.8}$	24. $0.7\overline{)0.21}$	25. $0.5\overline{)35}$
26. $1.6\overline{)0.768}$	27. $0.53\overline{)74.2}$	28. $0.075\overline{)0.345}$
29. $9\overline{)8.19}$	30. $1.2\overline{)108}$	31. $7.5\overline{)0.345}$

32. Find the quotient of 6.51 and 0.7.

33. Find the quotient of 0.89 and 1.78.

Solve each equation.

34. $3.68 \div 0.92 = p$	35. $f = 0.4664 \div 5.3$
36. $1.25 \div 5 = s$	37. $a = 7.56 \div 0.63$
38. $0.42 \div 3.5 = w$	39. $n = 2.04 \div 0.6$
40. $17.94 \div 2.3 = m$	41. $c = 2.665 \div 4.1$

42. **Mileage** Susannah has 2 gallons of gasoline left in her car. Her car averages 15 miles per gallon. If Susannah's home is 32 miles away, will she make it home before she runs out of gasoline? *(Lesson 1-1)*

43. **Algebra** Evaluate $3x - y \div 6$ if $x = 4$ and $y = 12$. *(Lesson 1-8)*

44. **Consumer Math** A survey of the weekly average amount spent on groceries for a family of four is $147.2653. Find the weekly average to the nearest cent. *(Lesson 2-2)*

45. Estimate 21.7×6.3. *(Lesson 2-3)*

46. Write 635,000 in scientific notation. *(Lesson 2-6)*

47. **Currency** All currency bills (paper money) in the United States, no matter what their value is, are the same size. A currency bill has an area of about 15.86 square inches and a length of about 6.1 inches. What is the width of a currency bill?

48. **Animals** A blue whale, the largest creature ever to live on Earth, can weigh 153.26 tons.
    a. An eighteen-wheel semitractor trailer fully loaded, can weigh 48.5 tons. How many times heavier is the blue whale?
    b. How many times heavier than *you* is the blue whale?

49. **Science** The bacterium *E. coli* has a diameter of 0.001 millimeter. The head of a pin has a diameter of 1 millimeter. How many bacteria *E. coli* could fit across the head of a pin?

50. **Critical Thinking** Place a decimal point in each of the numbers 11,008 and 256 such that their quotient will be 430.

CULTURAL KALEIDOSCOPE

Benjamin Banneker

Benjamin Banneker (1731–1806) was a mathematician, an astronomer, a compiler of almanacs, an inventor, a writer, and a very important African-American intellectual.

In 1761, he attracted attention by building a wooden clock that kept precise time. In 1773, Banneker began astronomical calculations and accurately predicted a solar eclipse which occurred in 1789.

Appointed to the District of Co-lumbia Commission by President George Washington in 1790, he helped survey Washington D.C. He published almanacs annually from 1791 to 1802 and sent his first one to Thomas Jefferson, then United States Secretary of State. Banneker, who opposed war and slavery, worked very hard to try to bring about better conditions for African-Americans.

2-8 Rounding Quotients

Objective

Round decimal quotients to a specified place.

Marisa and three of her friends ordered a large pizza for $13.89. To find out how much each of them owes, they divide $13.89 by 4.

Estimate: $12 ÷ 4 = $3

13.89 ⌹ 4 ⌹ **3.4725**

Since the smallest unit of money is a penny ($0.01), they round the quotient to two decimal places.

3.4725 → 3.48   *When a quotient involves money, it is usually rounded up.*

Each person should pay $3.48.

There are other situations, besides money, when it is useful to round quotients.

Calculator Hint
●●●●●●●●●●●●●
Some calculators round and some calculators truncate results. *Truncate* means to cut off at a certain place-value position, dropping the digits that follow. Does your calculator round or truncate results?

Example 1   *Problem Solving*

Sports   A table tennis table has an area of 4.165 square meters, while a tennis court has an area of 260.76 square meters. How many times larger is the tennis court than the table to the nearest hundredth?

To solve, divide 260.76 by 4.165.

Estimate: 240 ÷ 4 = 60

260.76 ⌹ 4.165 ⌹ **62.607443**

≈ 62.61   *Round to the nearest hundreth.*

≈ *means "is approximately equal to"*

The tennis court is about 62.61 times larger than the table tennis table.

Sometimes, it is helpful to round a quotient that is very large to the greatest place-value position of the whole number.

Example 2 *Problem Solving*

Astronomy Ceres, one of the largest known asteroids, is 690.4 kilometers in diameter. Jupiter, the first planet beyond the asteroid belt, has a diameter of 142,748.8 kilometers. How many times longer is Jupiter's diameter than Ceres' diameter?

To find out, divide 142,748.8 by 690.4.

Estimate: $140,000 \div 700 = 200$

$142748.8 \boxed{\div} 690.4 \boxed{=} 206.762457$

≈ 200 *Round to the nearest hundred.*

Jupiter's diameter is *about* 200 times longer than Ceres'.

Checking for Understanding

Communicating Mathematics

Read and study the lesson to answer each question.

1. **Write** one or two sentences that explain how to round the quotient of a money problem.

2. **Write** an example of a division problem involving very large quantities that requires rounding to the greatest place-value position.

Guided Practice

Divide. Round to the nearest tenth.

3. $0.7\overline{)29.3}$ 4. $2.6\overline{)56.38}$ 5. $0.04\overline{)15.999}$

Divide. Round to the greatest place-value position of the quotient.

6. $0.14\overline{)56,382.9}$ 7. $1.8\overline{)26,788.13}$ 8. $0.09\overline{)25,788.89}$

Divide. Round up to the next cent.

9. $6\overline{)\$8.25}$ 10. $15\overline{)\$168.57}$ 11. $23\overline{)\$48.98}$

12. Round the quotient of 14.85 and 0.023 to the nearest hundredth.

Exercises

Independent Practice

Divide. Round to the nearest hundredth.

13. $0.03\overline{)0.00775}$ 14. $2.2\overline{)6.329}$ 15. $13\overline{)56.728}$

Divide. Round to the greatest place-value position of the quotient.

16. $0.9\overline{)32{,}567.1}$ 17. $2.31\overline{)123{,}678.55}$ 18. $5.6\overline{)5{,}498.21}$

Divide. Round up to the next cent.

19. $0.05\overline{)\$45.67}$ 20. $6.3\overline{)\$89.73}$ 21. $0.4\overline{)\$74.23}$

22. Round the quotient of $7.69 and 5 to the nearest cent.

23. Round the quotient of 7,180,000 and 2.05 to its greatest place-value position.

Mixed Review

24. Estimate the number of inches in 2,573 feet. *(Lesson 2-3)*

25. **Pets** Steve has two dogs. Rex is 1.7 years old and Lady is 2.5 times older. How old is Lady? *(Lesson 2-4)*

26. Write 3.075×10^4 in standard form. *(Lesson 2-6)*

27. **Smart Shopping** A jar of peanut butter holds 39 ounces and costs $1.95. What is the cost per ounce in cents? *(Lesson 2-7)*

Problem Solving and Applications

28. **Buildings** Each story in an office building is about 3.66 meters tall. Find the height in stories of each of the structures.
    a. Great Pyramid of Cheops
    b. Gateway Arch
    c. Statue of Liberty

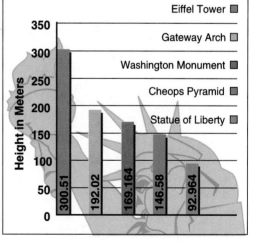

Height of Various Structures

Height in Meters

Eiffel Tower ▪ 300.51
Gateway Arch ▪ 192.02
Washington Monument ▪ 169.164
Cheops Pyramid ▪ 146.58
Statue of Liberty ▪ 92.964

29. **Geography** Alaska, the largest state in the United States, has an area of 1,478,458 square kilometers. Rhode Island, the smallest state in the United States, has an area of 2,732 square kilometers. How many times larger is Alaska?

30. **Critical Thinking** Write a division problem where the quotient rounded to the nearest tenth and the quotient truncated to the tenths place are the same. Write a division problem where they are different.

31. **Journal Entry** Name three situations that require division and state how you would round each quotient. Explain.

2-9 The Metric System

Objective

Change metric units of length, capacity, and mass.

Sound travels at about 343 meters per second in air. In water, sound travels about 1.435 kilometers per second. Does sound travel faster in air or water? One way to find out is to change 1.435 kilometers per second to meters per second.

Words to Learn

meter
metric system
gram
liter

Both the measurements above are based on the **meter** (m), which is the basic unit of length in the **metric system.** A meter is about the distance from the floor to a doorknob.

All units of length in the metric system are defined in terms of the meter. A prefix is added to indicate the decimal place-value position of the measurement. Study the chart below.

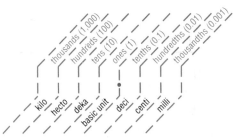

Notice that each place value is 10 times the place value to its right.

Notice that the value of each metric prefix is 10 times the value of the prefix to its right.

Mental Math Hint

•••••••••••••

To multiply or divide by a power of ten, you can move the decimal point.

One way to solve the problem above is to change 1.435 kilometers to meters. Since 1 km = 1,000 m, multiply by 1,000.

$1.435 \times 1,000 = 1,435$

Sound travels 1,435 meters per second in water.

Since $1,435 > 343$, sound travels faster in water than in air.

This diagram can help you change metric units.

MULTIPLY to change from longer units to shorter units.

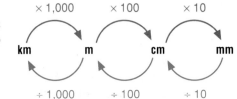

DIVIDE to change from shorter units to longer units.

Examples

1   0.9 cm = ▧ mm

*To change from centimeters
to millimeters, multiply by
10 since 1 cm = 10 mm.*

$0.9 \times 10 = 9$
0.9 cm = 9 mm

2   4,500 m = ▧ km

*To change from meters to
kilometers, divide by 1,000
since 1 km = 1,000 m.*

$4,500 \div 1,000 = 4.5$
4,500 m = 4.5 km

The **gram** (g) is the basic unit of mass in the metric system. *Mass* is
the amount of matter that an object contains. A thumbtack has a
mass of about one gram. You will notice that kilogram, gram, and
milligram are related in a manner similar to kilometer, meter, and
millimeter.

Examples

3   2,647 g = ▧ kg

*To change from grams to
kilograms, divide by 1,000
since 1 kg = 1,000 g.*

$2,647 \div 1,000 = 2.647$
2,647 g = 2.647 kg

4   6 g = ▧ mg

*To change from grams to
milligrams, multiply by
1,000 since 1 g = 1,000 mg.*

$6 \times 1,000 = 6,000$
6 g = 6,000 mg

The **liter** (L) is the basic unit of capacity in the metric system.
Capacity is the amount of dry or liquid material an object can hold.
Soft drinks often come in a 2-liter plastic container. You will notice
that kiloliter, liter, and milliliter are related in a manner similar to
kilometer, meter, and millimeter.

Examples

5   0.8 L = ▧ mL

*Multiply by 1,000 since
1 L = 1,000 mL.*

$0.8 \times 1,000 = 800$
0.8 L = 800 mL

6   862 L = ▧ kL

*Divide by 1,000 since
1 kL = 1,000 L.*

$862 \div 1,000 = 0.862$
862 L = 0.862 kL

Checking for Understanding

Read and study the lesson to answer each question.

1. **Tell** how you know when to multiply or divide when you are converting measures.

2. **Tell** how the metric system and decimals are similar.

Complete.

3. 550 mm = ▓ cm
4. 43.8 km = ▓ m
5. 814 g = ▓ kg

6. 16.5 g = ▓ mg
7. 5 L = ▓ mL
8. 32 L = ▓ mL

9. 89 km = ▓ m
10. 67.1 kg = ▓ g
11. 0.6 L = ▓ mL

12. How many grams are in 1.01 kilograms?

13. How many centimeters are in 0.56 meter?

Exercises

Complete.

14. 234 mm = ▓ cm
15. 5.8 m = ▓ cm
16. 13.2 cm = ▓ m

17. 0.9 cm = ▓ mm
18. 46 km = ▓ m
19. 6,700 m = ▓ km

20. 567 mg = ▓ g
21. 80 g = ▓ kg
22. 0.62 kg = ▓ mg

23. 73.8 kg = ▓ g
24. 24.7 g = ▓ mg
25. 8.1 L = ▓ mL

26. 329 mL = ▓ L
27. 47 L = ▓ kL
28. 0.52 kL = ▓ mL

29. How many milliliters are in 0.07 liters?

30. How many centimeters are in 6.302 kilometers?

31. How many milligrams are in 0.014 kilograms?

32. Evaluate 2^8. *(Lesson 1-9)*

33. Which is greater, 3.19 or 3.1? *(Lesson 2-1)*

34. Solve the equation $16.2 \div 2.5 = n$. *(Lesson 2-7)*

35. **Advertising** A 30-second advertisement on a local television station during prime time costs the advertiser $1,280. To the nearest cent, how much does the ad cost per second? *(Lesson 2-8)*

36. **Research** Look up the *metric system* in your school library. What are the meanings of the prefixes: micro-, tera-, giga-, and nano-?

37. **Critical Thinking** Order from least to greatest.
    0.0031 km   3.4 cm   53.25 mm   0.49 m

38. **Geometry** In order for three numbers to represent the measures of the sides of a triangle, the sum of any two numbers must be greater than the third. Could 1.3 centimeters, 9.5 millimeters, and 0.127 meters be the lengths of the sides of a triangle? Draw a diagram to explain your answer.

2-10 Determine Reasonable Answers

Objective

Determine whether answers are reasonable.

Megan has $80 to spend on clothes for school. After looking at sale ads in the newspaper, she decides that she will buy 2 pairs of jeans for $29.99 each and 2 belts for $8.18 each. She thinks that she would have $10.00 left to buy hair accessories. Does this seem reasonable?

Explore   You know Megan has $80 to spend. You know that she wants to buy 2 pairs of jeans for $29.99 and 2 belts for $8.18 each.

You want to find out whether it is reasonable for Megan to have $10 left to buy hair accessories.

Plan   Estimate to find the total amount Megan will spend. Then determine if she has enough money left for hair accessories.

Solve
$$\begin{array}{r} \$30.00 \\ 30.00 \\ 10.00 \\ + \ 10.00 \\ \hline \$80.00 \end{array}$$ *Round $29.99 to $30.00.*

Round $8.18 to $10.00

$$\begin{array}{r} \$80.00 \\ - \ 80.00 \\ \hline \$ \ \ 0.00 \end{array}$$ It is not reasonable for Megan to have $10 left for hair accessories.

Examine   You can use a calculator to see how much Megan will actually have left.

2 ⨯ 29.99 = STO 2 ⨯ 8.18 + RCL = ⸀76.34

C/CE 80 − 76.34 = ⸀3.66

Checking for Understanding

Communicating
Mathematics

Read and study the lesson to answer each question.

1. **Tell** another strategy you could have used to solve this problem.

2. **Write** a problem with an unreasonable answer and ask a classmate to explain why they think the answer is unreasonable.

Guided Practice

Solve.

3. Mr. Eldridge eats food that has 2,755 calories in an average day. When he multiplied the number of calories he eats per day by the number of days in a week, the calculator showed 192,850. Is this answer reasonable? Explain.

4. At the Book Fair, Brian wants to buy 2 science fiction books for $2.95 each, 3 magazines for $2.95 each, and 1 bookmark for $0.39. Does he need to bring $15 or $20 with him?

5. Vinney's Video Haven is selling 3 blank video tapes for $14.96. Carlene says she can get 9 tapes for under $40. Is her answer reasonable?

Problem Solving

Practice

Solve using any strategy.

Strategies
●●●●●●●●●●
Look for a pattern
Solve a simpler problem.
Act it out.
Guess and check.
Draw a diagram.
Make a chart.
Work backwards.

6. New car carriers deliver new cars from the loading dock at the auto plant to car dealerships. Each truck can carry about 20,000 pounds of weight. If an economy-sized car weighs about 2,330 pounds, what is a reasonable number of cars that could be transported on one truck?

7. The trail up to the top of Pike's Peak is about 22 miles long. The Arnold family drove about one third of the way. Did they drive about 7 miles or 17 miles?

8. Felicia's vacation lasted 8 days and 7 nights. She spent $95 per night for the hotel and $30 per day for food. How much did she spend on food and lodging?

9. In the 1988 Olympics, Steven Lewis ran the 400-meter dash in 43.87 seconds. Round this decimal to the nearest tenth of a second.

10. Suppose a flask contains 750 milliliters of an acid. A chemist pours 0.5 liters of the acid into a solution. How many milliliters are left in the flask?

11. The width of the continental shelf along West Africa varies. It is about 4.5 miles wide in the Gulf of Guinea and only 2.5 miles wide off Angola. What is the average width of those two readings?

12. **Photography**   A paramecium that is 0.25 millimeters long is magnified to 60 millimeters for a science book photograph. How many times larger is the paramecium in the photo than the actual paramecium?

13. During one 20-game span of the Los Angeles Lakers, Magic Johnson scored 498 points and James Worthy scored 425 points. Find the total points scored together by both athletes.

14. Suppose a relative matches your age with dollars on your birthday. You are 13. How much money have you been given over the years by this relative?

15. Al Johnson is measuring an exterior wall of the family room in order to paint it. There is one large window and a door on that wall. Al multiplies the width times the height of the wall and gets 240 square feet. The paint label says a quart covers 100 square feet. How much paint should he buy?

16. Ms. Francis drove her car 427 miles on 15.8 gallons of gasoline.
    a. To the nearest mile, how many miles per gallon does her car get?
    b. What was the cost of gasoline at $1.439 per gallon?

17. Suppose a grocery store advertises 3 Snickers® bars for $1.00. If you buy only 1 Snickers® bar, how much will you pay?

18. Peta places a long distance phone call to her grandparents in California and talks for 45 minutes. The phone company bills the call at a rate of $0.10 per half-minute. How much does the call cost Peta?

19. The average adult male weighs 162 pounds. The average adult cat weighs 10 pounds. *About* how many times heavier is a man than his cat?

2 Study Guide and Review

Communicating Mathematics

Choose the correct term or number to complete the sentence.

1. The number 0.04 is (less, greater) than 0.041.

2. When rounding decimals, the digit in the place being rounded should be rounded up if the digit to the right is a (4, 7).

3. The number of decimal places in the product when multiplying decimals is the (sum, product) of the number of places in the factors.

4. In scientific notation, a number is written as a (sum, product) of a decimal number and a power of ten.

5. The basic unit of mass in the metric system is the (gram, meter).

6. In your own words, explain the relationship between a meter and a centimeter.

Skills and Concepts

Objectives and Examples

Upon completing this chapter, you should be able to:

- compare and order decimals
  (Lesson 2-1)

  Order the following decimals from least to greatest. 3.2, 0.4, 0.43, 3.5, 4

  0.4, 0.43, 3.2, 3.5, 4

- round decimals *(Lesson 2-2)*

  Round 237.359 to the nearest tenth.

  The digit to the right of the 3 in the tenths place is 5, so round up.
  237.359 → 237.4

Review Exercises

Use these exercises to review and prepare for the chapter test.

Order each set of decimals from least to greatest.

7. 4.2, 3.9, 3.15, 3.04, 3.7

8. 15.91, 1.59, 0.159, 0.06, 1.4

9. 0.15, 0.149, 0.105, 0.015, 0.501

10. 16.3, 16.03, 16, 15.99, 15.09

11. 26.04, 25.7, 0.257, 2.046, 2.04

Round each number to the underlined place-value position.

12. 5.75

13. 13.274

14. 129,342

15. 0.076

16. 81.349

17. 257.196

18. 200,543

19. 0.0015

Objectives and Examples

• estimate with decimals *(Lesson 2-3)*

You can use rounding, clustering, front-end estimation, or compatible numbers to estimate sums, differences, products, and quotients of decimals.

• multiply decimals *(Lesson 2-4)*

$$\begin{array}{r} 3.2 \\ \times 0.6 \\ \hline 1.92 \end{array}$$ *1 decimal place*
   1 decimal place
   Count 2 decimal places from the right.

• multiply decimals mentally by powers of ten *(Lesson 2-5)*

$100 \times 2.3 = 230$

• express numbers greater than 100 in scientific notation and vice versa *(Lesson 2-6)*

$256,000 = 2.56 \times 10^5$
$6.79 \times 10^7 = 67,900,000$

• divide decimals *(Lesson 2-7)*

$$\begin{array}{r} 12.6 \\ 3.6\overline{)45.36} \\ -36 \\ \hline 9\,3 \\ -7\,2 \\ \hline 2\,16 \\ -2\,16 \\ \hline 0 \end{array}$$

Review Exercises

Estimate.
20. $13.72 + 12.07$
21. $243.46 + 112.3$
22. $25.73 - 2.19$
23. 11.75×3.13
24. $72.4 \div 9.3$
25. $150.96 \div 4.76$

Multiply.
26. 2.6×3.7
27. 0.13×2
28. 12.5×0.0017
29. 7.5×3.03
30. 1.001×0.4

Multiply.
31. 13.7×10^3
32. $0.0065 \times 10,000$
33. 6.37×10
34. 128.63×10^2

Write each number in scientific notation.
35. $6,000$
36. $459,000,000$

Write each number in standard form.
37. 1.37×10^4
38. 9.99×10^6

Divide.
39. $12 \div 1.2$
40. $8.4 \div 0.2$
41. $0.0036 \div 0.9$
42. $1.47 \div 0.06$
43. $5 \div 0.005$
44. $0.002 \div 0.01$

Objectives and Examples	*Review Exercises*

• round decimal quotients to a specific place *(Lesson 2-8)*

$1.25 \div 4 = 0.3125$ Rounded to the nearest tenth, 0.3125 is 0.3.

Divide. Round to the indicated place-value position.

45. $3.5 \div 1.3$ to the nearest tenth.

46. $14.78 \div 2.6$ to the nearest whole number.

• change metric units of length, capacity, and mass *(Lesson 2-9)*

$1.39 \text{ kg} = $ 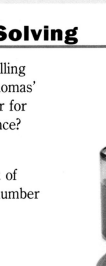 g
Multiply by 1,000 since 1 kg = 1,000 g.
$1.39 \text{ kg} = 1,390 \text{ g}$

Complete.

47. 27 mm = ▮ m   48. 3.9 mg = ▮ g
49. 6.85 km = ▮ m   50. 3.3 mL = ▮ L
51. 16 cm = ▮ mm   52. 0.04 kL = ▮ L
53. 43 g = ▮ kg   54. 3.9 kL = ▮ mL

Applications and Problem Solving

55. **Smart Shopping** Thomas' Apple Orchard is selling apple cider in 34.7-ounce bottles for $2.08. Thomas' competitor is selling 24.6-ounce bottles of cider for $1.99. Who is selling at the lower price per ounce? Round to the nearest cent. *(Lesson 2-8)*

56. **Remodeling** Anna needs 1.2×10^3 square feet of carpeting to carpet her new house. Write this number in standard form. *(Lesson 2-6)*

57. When Kevin divided 78,278.2 by 1,547, the calculator displayed 506. Is this a reasonable answer? *(Lesson 2-10)*

Curriculum Connection Projects

• **Geography**   List the populations of the ten largest and ten smallest countries in the world. Rewrite each population as a power of ten.

• **Physical Education**   Work with two friends to measure, to the nearest tenth of a meter, how far a person travels while performing a cartwheel. Find how many cartwheels must be performed to travel 1.5 kilometers.

Read More About It

McGraw, Eloise. *The Money Room.*
Simon, Seymour. *The Paper Airplane Book.*
Sachar, Louis. *Sideways Arithmetic from Wayside School.*

Chapter

2 Test

Order each set of decimals from least to greatest.

1. 12.6, 4.3, 8.7, 4, 12.06
2. 0.07, 0.7, 0.71, 1.07, 1.71

Round each number to the underlined place-value position.

3. 13.2$\underline{7}$5
4. 0.0$\underline{7}$6
5. 1$\underline{2}$,436
6. $\underline{0}$.995
7. The average attendance at a game during the football season was 2,176.34. Round to the nearest person.

Estimate.

8. $27.34 + 12.95$
9. $236.95 - 107.07$
10. 23.6×2.95
11. $11.1\overline{)142.6}$
12. 91.6×7.999

13. **Earning Money** During a two-week period, Hiroko worked 89.7 hours. Her hourly wage is $6.85. Estimate the amount earned.

Multiply or divide.

14. 0.3×8
15. 5.5×0.004
16. $6.7 \times 1,000$
17. 0.0047×10^5
18. $3,003 \times 100$
19. $0.4\overline{)4.8}$
20. $0.24\overline{)0.0072}$
21. $0.0016\overline{)64}$

22. **Pets** Alicia's cat Felix weighed 0.95 pounds at birth. On his first birthday, Felix weighed in at 9.2 times his birth weight. What was Felix's weight on his first birthday?

23. How many pennies are there in $36.00?

Change from standard form to scientific notation or vice versa.

24. 23,000
25. 632
26. 8.03×10^4
27. 1.6349×10^3

28. **Child Care** The Wests spend $3,575 per year on day care for their daughter Lisa. Lisa actually is in day care 245 days per year. What is the cost per day? Round to the nearest cent.

Complete.

29. $1.62 \text{ L} = \blacksquare \text{ mL}$
30. $243 \text{ g} = \blacksquare \text{ kg}$
31. $0.09 \text{ km} = \blacksquare \text{ mm}$

32. **Party Time** To mix a punch, Yolanda starts with 4 liters of ginger ale and adds 2,650 milliliters of cranberry juice.
    a. How many liters are in the punch bowl when the punch is complete?
    b. How many 200-mL glasses of punch can be served?

33. **Smart Shopping** Teresa bought 3 pounds of apples for $2.89, 2 pounds of carrots for $1.79, and 4 avocados for $0.99 cents each. Should she expect to pay about $5 or $10 at the checkout?.

Bonus How many times greater is 1.73×10^{20} than 1.73×10^6?

Chapter Test

Statistics and Data Analysis

Spotlight on Languages of the World

Have You Ever Wondered. . .

- What language is spoken by the most people in the world?
- Which countries have the most people living in them?

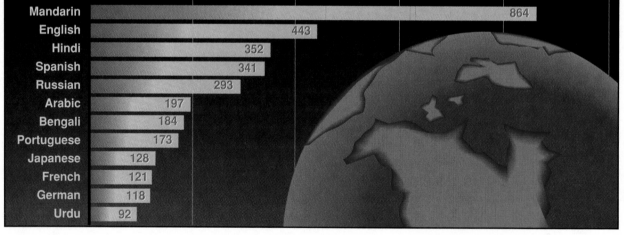

TWELVE PRINCIPAL LANGUAGES OF THE WORLD (in millions)

Language	Millions
Mandarin	864
English	443
Hindi	352
Spanish	341
Russian	293
Arabic	197
Bengali	184
Portuguese	173
Japanese	128
French	121
German	118
Urdu	92

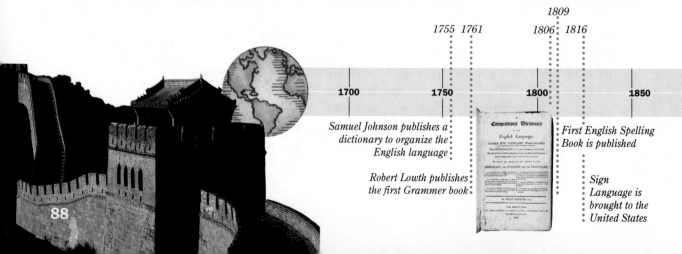

1755 1761

1809

1806 1816

1700   **1750**   **1800**   **1850**

Samuel Johnson publishes a dictionary to organize the English language

Robert Lowth publishes the first Grammer book

First English Spelling Book is published

Sign Language is brought to the United States

World's Ten Most Populous Countries, 1990

Country	Population	Percent of World Population
China	1,119,900,000	21%
India	853,400,000	16%
USSR	291,000,000	6%
United States	251,400,000	5%
Indonesia	189,400,000	4%
Brazil	150,400,000	3%
Japan	123,600,000	2%
Nigeria	118,800,000	2%
Bangladesh	114,800,000	2%
Pakistan	114,600,000	2%

Chapter Project

Languages of the World

Work in a group.

1. Trace or copy a world map.

2. Write the language(s) spoken in each country or region.

3. Color code the map according to language.

4. Show how your map accounts for the information given about languages and the populations of various regions.

Looking Ahead

In this chapter, you will see how mathematics can be used to answer the questions about the languages of the world. The major objectives of the chapter are to:

- solve problems by using graphs

- solve problems by making a table

- construct line plots

- find the mean, median, and mode of a set of data

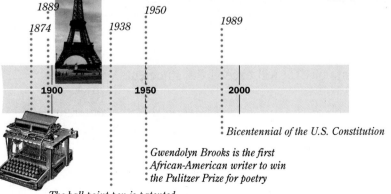

1889
1874
1950
1938
1989

1900
1950
2000

Bicentennial of the U.S. Constitution

Gwendolyn Brooks is the first African-American writer to win the Pulitzer Prize for poetry

The ball-point pen is patented

3-1 Use a Graph

Objective

Solve problems by interpreting bar graphs, line graphs, and circle graphs.

Statistics is a branch of mathematics that consists of collecting, organizing, and summarizing numerical facts. When the data is collected and displayed in a graph, you can look for trends and make predictions based on these facts.

The double-line graph shows the number of United States shipments of compact discs and cassette tapes. By looking at the graph, when do you think United States shipments of CDs will exceed that of cassettes?

In 1975, over 257,000 shipments of record albums were sent to retailers. By 1989, that number hit a 15-year low of 34,600.

Explore What do you know?
You know the number of cassette tapes and CD shipments as shown on the graph.
What are you trying to find?
You are trying to predict when shipments of CDs will exceed that of cassettes.

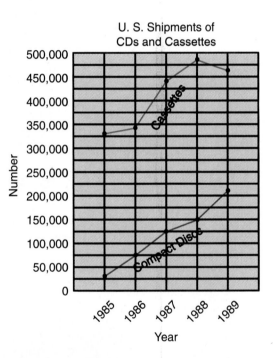

U. S. Shipments of CDs and Cassettes

Number

Year

Plan Study the graph. Determine what each of the lines represents. Look for trends in the number of shipments from 1985 to 1989. When did cassette shipments start to decline?

Next extend the lines mentally to predict when the shipments of CDs will exceed the shipments of cassettes.

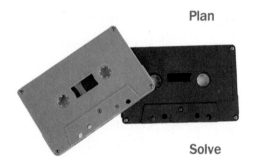

Solve By looking at the graph, you can see that the shipments of cassettes begins to decline in 1988 and the shipments of CDs have steadily risen since 1985. You can predict that CD shipments will exceed the shipment of cassettes in about 3 years or in 1992.

Examine   Copy the graph. Extend the lines and the horizontal and vertical scales to see when the lines cross.

Example

The double-graph shows the number of degrees earned for various years during a 27-year period. Were there more advanced degrees in 1969–70 or 1986–87?

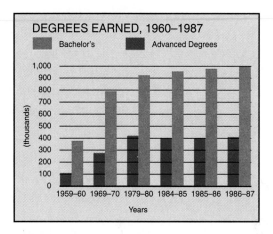

The key, below the title of the graph, shows that the blue bars represent the number of advanced degrees earned. Locate the bars for 1969–70 and 1986–87. In 1969–70, about 275,000 advanced degrees were earned. In 1986–87, about 400,000 advanced degrees were earned. So, about 125,000 more advanced degrees were earned in 1986–87 than in 1969–70.

Checking for Understanding

Communicating Mathematics   Read and study the lesson to answer each question.

1. **Tell** why different colors are used to show data in a double-line graph.
2. **Write** a sentence explaining why you might display data on a graph rather than in a chart.

Guided Practice   Solve. Use the line graph.

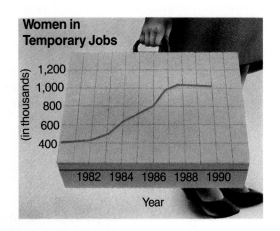

3. About how many more women worked as temporary employees in 1986 than in 1982?
4. In what year were there the greatest number of temporary employees?
5. The economic recession was the cause of the drop in temporary jobs in 1990. Given the steady growth since 1982, do you think the number of temporary jobs will continue to decline? Why?

Problem Solving

Solve. Use the circle graph.

6. What part of the budget is spent on printing?

7. How much more money is spent on photography than design?

8. If the total budget is increased by $200 next year, about how much could be spent on printing?

Publishing Budget for School Newspaper

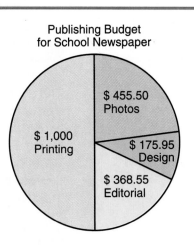

$1,000 Printing

$455.50 Photos

$175.95 Design

$368.55 Editorial

Strategies

• • • • • • • • • •

Look for a pattern.

Solve a simpler problem.

Act it out.

Guess and check.

Draw a diagram.

Make a chart.

Work backward.

Solve. Use any strategy.

9. Melanie made a display with 66 cat food cans. She put 1 less can in each row than was in the row below. How many cans were in the bottom row?

10. Ilsa bought 6 rolls of film on sale for $28.13. She took 216 photos on her vacation. If each roll had the same number of exposures, how many photos did she get from each roll of film?

11. Danny bought a new pair of athletic shoes for $55.59, a new sweatshirt for $24.87, and sweat pants for $34.99. Did he spend more or less than $100?

12. Alaska, the largest U.S. state, has an area of 1,478,458 square kilometers. Rhode Island, the smallest, has an area of 2,732 square kilometers. How many times larger is Alaska?

13. A 64 ounce carton of orange juice costs $0.05 per ounce. What is the cost of the juice?

14. There were 12 more seventh graders at the pep rally than sixth graders. There were 178 sixth and seventh graders at the rally. How many seventh graders went to the pep rally?

3-2 Make a Table

Objective

Solve problems by organizing data in a table.

Words to Learn

frequency table

Team 7-A is studying a unit on communication. They learned that Alexander Graham Bell patented the telephone in 1876. They decided to take a survey to find out how many classmates on Team 7-B knew the year that Bell patented the telephone. The class wanted to organize the data in a way that made it easy to study the results.

Mr. Kim's class organized the data in a **frequency table.**

Year	Tally	Frequency
1825					3
1850	⧸⧸⧸⧸ ||||	9
1854	⧸⧸⧸⧸		6
1862	⧸⧸⧸⧸ |||	8
1876				2
1898	⧸⧸⧸⧸	5

DID YOU KNOW

In 1884, Bell Telephone Company set up the first long distance telephone line between Boston and New York. It used copper wire instead of iron, which allowed the signals to travel farther.

Explore   What do you know?
You know how each person responded.
What are you trying to find?
You are trying to find the number of people that knew the year the telephone was patented.

Plan   Study the frequency table that Mr. Kim's class made. A tally has been marked for each response, and the frequency is the sum of the tally marks.

Solve   The frequency table shows that only 2 people surveyed knew the year the telephone was patented.

Examine   Read down the column marked Year until you find 1876. Read across the row, 1876, until you find the frequency column that shows the number of responses is 2.

You can make a frequency table using data from a survey or questionnaire.

Example

Rosita asked all the students in her seventh-grade math class to vote for their favorite amusement park out of Disneyland (DL), Dollywood (DW), or Six Flags (SF.)

DL	DW	SF	DL	DW
SF	DL	DW	DL	DL
DW	SF	DL	DW	DW
DL	DW	DW	SF	DW

To make a frequency table:
- Draw a table with three columns.
- In the first column, list the items in the set of data.
- In the second column, tally the data.
- In the third column, write the frequency or number of tallies.

Park	Tally	Frequency
Disneyland |卌 ||	7
Dollywood	卌 ||||	9
Six Flags	||||	4

Checking for Understanding

Communicating Mathematics

Read and study the lesson to answer each question.
1. **Tell** why it is important to be able to organize data.
2. **Write** a sentence about what each column represents in the frequency table above.

Guided Practice 3. Copy the table and complete the frequency column.

Heights of Junior High Students

HEIGHT(cm)	TALLY	FREQUENCY
145	|	
150	卌	
155	卌 卌	
160	|	
165	卌 |	
170					

4. The ages of guests at a birthday party are given at the right.
   a. Make a frequency table.
   b. What age was the most common at the party?

25	21	30	24
26	26	21	30
25	25	26	25
30	25	25	26

Problem Solving

Practice Solve using any strategy.

5. Make a frequency table for the data at the right.
   a. Which time appeared the most often?
   b. Which time appeared the least often?

Lengths of TV Commercials (in seconds)						
30	30	10	20	60	10	10
30	60	10	20	20	30	30
20	10	60	20	30	30	30

Strategies
• • • • • • • • • •
Look for a pattern.
Solve a simpler pattern.
Act it out.
Guess and check.
Draw a diagram.
Make a chart.
Work backward.

6. **Collect Data** Take a survey of your classmates' favorite television show.
   a. Organize the data in a frequency table.
   b. Which television show was picked as a favorite the most often?

7. The price of a T-shirt is $15.00 plus $0.62 sales tax. How much would 6 shirts cost?

8. Dina rented 20 movies at $2.95 each. How much did she spend on rentals?

9. An elevator sign reads "DO NOT EXCEED 2,500 POUNDS." How many people each weighing about 150 pounds can be in the elevator at the same time?

10. Marco bought 6 tickets to the circus. He gave the cashier $170 and received $8 in change. How much did one ticket cost?

11. Choose five digits. Use the five digits to form a two-digit and a three-digit number so that their product is the least product possible.

12. **Reading** Chuck is reading a book that has 16 chapters. Each chapter has 28 pages. How many pages does the book have?

13. **Mathematics and Telecommunications** Read the following paragraphs.

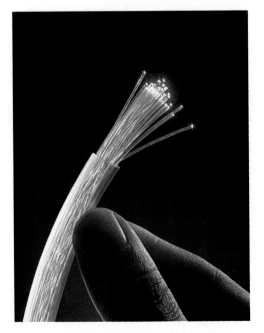

In 1966, it was first suggested to use fiber optic cables instead of copper wires to carry telephone conversations. Pulses of light are used to transmit calls down the fiber optic cables.

Fiber optic cable is made up of strands of glass. Each strand is the width of a human hair. Light is beamed into the inner core, bouncing along it. Thousands of telephone calls can be carried at the same time in each strand of glass.

The fibers in a cable are bundled. If each fiber is 0.0005 inch thick, how thick is a bundle of 3,000 fibers?

3-2B Data Base

A Follow-Up of Lesson 3-2

Objective

Work with a data base.

Words to Learn

data base
file
record
fields

A **data base** is a collection of data organized for rapid search and retrieval, usually by a computer. A company's data base is one of its most valuable resources. A data base computer program can assist a company in analyzing the past and making projections for the future.

A data base is organized into files. A **file** is a collection of data about a particular subject. The subject, or sub unit within the file, is called a **record.** For example, in a sales file that contains data about a company's sales, you may find records about the individual salesperson or company.

The first step in creating a data base file is to specify its structure. Name the file and then describe the elements, or **fields** within each record. A field name can be up to 10 characters. It must start with a letter and include no spaces between the characters. For each field, you must specify its name, the type of data it will contain (character, C, or numeric, N), the size of each field, and, if numeric, how many decimal places the field contains.

Field	Specifications
001	Name,C,20
002	Number,C,5
003	Profit,N,8,2

The file structure specification above means that the first field (001) in each record is the company's name up to 20 characters. The second field (002) is the company's number up to 5 characters. The company's number is coded as character since there are no calculations done with numbers. The profit field size (003) is specified as 8, and number of decimal places as 2. This indicates that there will be positions for a 5-digit dollar amount, a decimal point, and two positions for cents.

Try this!

Work with a partner.

1. What does a file in a data base contain?
2. What information would you expect to find in a sales file?
3. How do you enter a field structure?
4. Explain why you cannot use the format, 001 Customer Name,C,20 to specify a field?

Consider the following data base.

CUSTOMER NAME	CUSTOMER NUMBER	TERRITORY	DATE	ORDER NUMBER	AMOUNT
NEWPORT RENTAL	54213	15	112091	11340	8000.00
DOLLAR FURNITURE	15682	10	111891	11289	792.00
SERVICE STAR	38900	12	112091	11300	3000.00
NEWPORT RENTAL	54213	15	111791	11250	400.00
CITY LIMITS	80087	11	111891	11280	10.00
EVERY STEP	18851	13	112191	11450	350.00
DOLLAR FURNITURE	15682	10	112291	11560	1000.00
DOLLAR FURNITURE	15682	10	112091	11301	850.00
EVERY STEP	18851	13	111991	11299	985.00
DOLLAR FURNITURE	15682	10	112391	11600	50.00

What do you think?

5. How many fields does this data base contain? What are they?
6. How many records does the data base contain?
7. What is the largest number that can be entered in the amount field?
8. How would you use this data base to predict sales for each company over a period of time?
9. How would you use the data base to predict sales for a specific territory?

Extension

10. Create your own data base. Write a short paragraph describing the type of information it contains, how many fields there are, and what the field specifications are.

3-3 Range and Scales

Objective

Choose appropriate scales and intervals for data.

Words to Learn

range
scale
interval

Telecommunications Professional

Telecommunications have brought the world closer together by electronically communicating information by fax, telex, cellular phone, electronic mail, and on-line data bases.

Careers for computer programmers and systems analysts are expected to grow by 50% within the next decade. A technical background in mathematics and computer programming is essential in this field.

For more information, contact the National Telecommunications and Information Administration, Dept of Commerce, 14th St., Washington, DC 20230.

How much time do you spend talking on the phone in one week? Is it hours, minutes, or seconds? The average number of hours a teenager spends talking on the telephone is two hours a day. Girls are likely to talk a half hour more than boys.

Susan Ching asks 11 of her classmates how many hours they spent on the phone last week. She records the data in the table at the right. To analyze this data, she needs to find an appropriate scale and intervals for the data.

Number of Hours on the Telephone			
Sue	15	Lyn	17
Bob	11.5	Kathy	16
Tim	14	John	12
Joe	11.5	Carole	14
Pete	13	Ida	13.5
Jill	16.5		

First, Susan finds the range of the data. The **range** is the difference between the greatest number and the least number in the set of data.

$$\text{The range is } 17 - 11.5 \text{ or } 5.5$$
greatest number ↗   ↖ least number

Next she chooses a scale for the number line. The **scale** must include numbers from 11.5 to 17; that is all data points. So, Susan decides on a scale of 11 to 18. This scale will allow her to plot all the data she has collected.

She then decides on the interval. The **intervals** separate the scale into equal parts. Since the data is grouped closely together, she decides on an interval of 1.

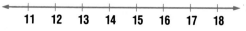

There is more than one correct way you can choose the scale and interval for the same data.

Example 1

Choose a different scale and interval for the data above. Draw a number line to show them.

The least number is 11.5, and the greatest number is 17. Another appropriate scale for this data is 10 to 20.

Since the scale is larger, a good choice for the interval is 2.

Checking for Understanding

Communicating Mathematics Read and study the lesson to answer each question.

1. **Tell**, in your own words, how to find the range, the scale, and the interval for a set of data.

2. **Tell** the advantages and disadvantages of the two different scales and intervals used for the data on p. 98.

3. **Draw** a number line that shows a scale of 0 to 50 and intervals of 5.

Guided Practice Name the scale and intervals of each number line.

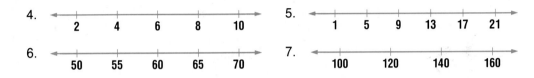

4. (number line: 2, 4, 6, 8, 10)

5. (number line: 1, 5, 9, 13, 17, 21)

6. (number line: 50, 55, 60, 65, 70)

7. (number line: 100, 120, 140, 160)

Find the range for each set of data. Choose two different scales and intervals for each set of data. Draw two number lines to show each.

8. 3, 7, 1, 9, 3, 5

9. 14, 19, 4, 0, 13, 8, 2

10. 25, 75, 50, 34, 56

11. 785, 900, 456, 832, 678

12. 4.5, 2.3, 4.5, 7.8, 5.5, 5.1, 3.9

Exercises

Independent Practice Find the range for each set of data. Choose an appropriate scale and intervals. Draw a number line to show the scale and intervals.

13. 2, 6, 8, 9, 12, 4

14. 9, 0, 18, 19, 2, 9, 8, 13, 4

15. 20, 60, 30, 80, 90, 120, 40

16. 6.4, 4.2, 3.6, 2.4, 5.0

17. 200; 600; 300; 800; 900; 1,200; 400

18. 14.5, 18.2, 21.6, 18.8, 17.3, 14.1

19. Scott has been taking bowling lessons. His scores for the first ten games are 36, 54, 72, 89, 90, 110, 146, 134, 140, and 145.
    a. Find a scale and intervals to graph the scores.
    b. Draw a number line to show them.

Mixed Review 20. **Home Economics** To make a new dress, Melissa buys 4 yards of cotton fabric at $5 a yard and 2 yards of lace at $3 a yard. What is the total cost for the fabric and lace she purchases? *(Lesson 1-7)*

21. **Physical Fitness** Ed completes an obstacle course in 6.9 minutes. It takes Andre 1.4 times longer to complete the course. How long does it take Andre to complete the course? *(Lesson 2-4)*

22. Complete the sentence 2.33 km = ■ m. *(Lesson 2-9)*

Problem Solving and Applications

23. **Critical Thinking** Given a set of data, can there be more than one range? More than one scale? More than one interval? Explain your reasoning for each answer.

24. **History** The table at the right shows the length of reign of the 11 most recent rulers of England and Great Britain.

    a. Which ruler had the shortest reign?

    b. Draw a number line with an appropriate scale and intervals for this data.

Ruler	Reign (years)
George I	13
George II	33
George III	59
George IV	10
William IV	7
Victoria	63
Edward VII	9
George V	25
Edward VIII	1
George VI	15
Elizabeth II	41*

* as of 1993

25. **Television** The table at the right shows the longest running national TV series, as of 1989.

    a. Which show had the longest running time?

    b. Draw a number line with an appropriate scale and intervals for the number of seasons.

Program	Number of Seasons	Years
Walt Disney	33	1954–86
Ed Sullivan	24	1948–71
60 Minutes	21	1968–
Gunsmoke	20	1955–75
Red Skelton	20	1951–71
Meet the Press	18	1947–65
What's My Line?	18	1950–67
Lassie	17	1954–71
Lawrence Welk	17	1955–71

26. **Journal Entry** Take a survey of ten of your classmates to find the number of hours each person spends on the phone in a week.

    a. Find the range.

    b. Choose an appropriate scale and intervals.

    c. Draw a number line.

    d. How does your number line compare with Susan Ching's number line on page 98?

3-4 Line Plots

Objective

Construct line plots.

Words to Learn

line plot
outlier
cluster

Have you ever curled up with a good book on a rainy day? For many people, reading is very enjoyable and relaxing.

Meredith conducted a survey of her classmates to determine how many books each student read last month. The results are shown at the right.

Number of Books	
Emilio—1	Marna—2
Jeremy—4	Bruce—3
Alicia—2	Jana—1
Wai—2	Rick—5
Daniel—0	Scott—2
Charo—3	Ellen—10
Sarah—1	Enrique—2
Bret—3	Mark—0
Mika—1	Kathleen—2
Julie—5	Sei—2

One way to organize this data is to present it on a number line. A **line plot** is a picture of information on a number line.

- First draw a number line. Find the range of the data and determine a scale and intervals.

  The least number of books is 0, and the greatest number is 10. The range is $10 - 0$ or 10. You can use a scale of 0 to 10 and an interval of 2 to draw the number line. Other scales and intervals could also be used.

- Put an "x" above the number that represents the number of books each student read. If the number is odd, place the "x" halfway between the appropriate notches.

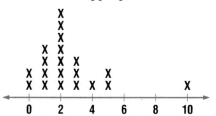

This statistical graph is called a line plot.

You can make some observations about the data from the line plot above.

- The number that occurs most frequently is 2.
- The number 10 is far apart from the rest of the data. It is called an **outlier.**
- There seems to be a **cluster** of data between 1 and 3. Data that are grouped closely together are called a cluster.

Example 1

The table at the right shows the voter turnout for presidential elections from 1928 to 1988. Draw a line plot of this data.

The least percent is 50.2, and the greatest is 62.8. You can round to the nearest whole percent to graph them more easily. An appropriate scale for this graph is 50 to 63 with an interval of 1.

Unlike a line graph, a line plot does not need to start at zero.

Year	Voter Turnout
1928	51.8%
1932	52.6
1936	56.8
1940	58.8
1944	56.1
1948	51.1
1952	61.6
1956	59.4
1960	62.8
1964	61.9
1968	60.9
1972	55.2
1976	53.5
1980	52.6
1984	53.1
1988	50.2

```
                    X
                    X
   X   X   X   X   X   X   X   X        X           X
                                        X           X   X
  ──┼───┼───┼───┼───┼───┼───┼───┼───┼───┼───┼───┼───┼──
   50  51  52  53  54  55  56  57  58  59  60  61  62  63
```

Checking for Understanding

Communicating Mathematics

Read and study the lesson to answer each question.

1. **Tell** the advantage of using a line plot instead of a table to display data.

2. **Write** the definition of an outlier.

3. **Write** the outliers and clusters, if any, for the line plot below.

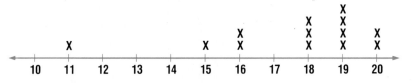

```
                                                    X
                                           X        X
                                   X       X        X    X
              X            X       X       X        X    X
  ──┼────┼────┼────┼────┼────┼────┼────┼────┼────┼────┼──
   10   11   12   13   14   15   16   17   18   19   20
```

4. **Draw** a line plot that shows ten items of data. The scale should be 10 to 50 with an interval of 5. Eight of the ten items should be from 25 through 35.

Guided Practice

Make a line plot for each set of data. Circle any outliers on the line plot.

5. 3, 5, 4, 5, 9, 10, 2, 4, 3, 12, 6, 4

6. 50, 45, 35, 40, 40, 30, 55, 35, 45, 35

7. 110, 115, 114, 106, 101, 119, 108, 102, 111, 114

8. Make a line plot of the test scores: 100, 89, 88, 84, 90, 97, 100, 89, 90, 90, 73, 91, 83, 95. Name any outliers and clusters.

Exercises

Independent Practice

Make a line plot for each set of data. Circle any outliers on the line plot.

9. 36, 45, 42, 16, 41, 30, 38, 52, 32, 33
10. 340, 500, 600, 640, 730, 520, 600, 560, 490, 670
11. 1983, 1980, 1976, 1985, 1984, 1989, 1990, 1985, 1987, 1976, 1988, 1986
12. 3.2, 3.6, 3.7, 4.0, 3.8, 3.3, 3.2, 3.0, 4.0, 3.6, 3.2

13. Ruth wanted the best price on her favorite shampoo. She found the following prices at different stores: $2.40, $2.35, $2.50, $2.25, $3.00, $2.75, $2.40, $2.10, $2.50.

    a. Make a line plot of the data.
    b. What did Ruth find out about the prices?

Mixed Review

14. **Travel** The Jones' drove 274 miles one day and 304 miles the next day. *About* how far did they travel in two days? Use front-end estimation. *(Lesson 1-2)*

15. Find the range and appropriate scale and interval for the following data:
    3, 17, 21, 19, 36, 15, 12, 9.
    Draw a number line to show the scale and interval.
    (Lesson 3-3)

Problem Solving and Applications

16. **Collect Data** Conduct a survey of your classmates to determine how many books each student read last month.
    a. Make a line plot of the data.
    b. Write a sentence that compares this line plot to the one on page 101.

17. **Critical Thinking** Compare a line plot to a bar graph.
    a. How is a line plot like a bar graph?
    b. How is it different?
    c. Which is easier to construct?

18. **Computer Connection** Marcos asked 15 of his classmates how many times they bought food from a fast-food restaurant last month. He entered the data in a computer and made a line plot of the data. What conclusions can you make from this line plot? *Hint: Look for outliers and clusters.*

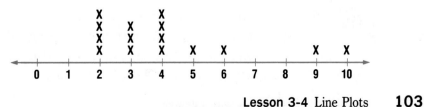

Mean, Median, and Mode

Objective

Find the mean, median, and mode of a set of data.

Words to Learn

mean
median
mode
average

Suppose you are watching the movie *Home Alone* on TV and it is interrupted by commercials that seem to go on forever. You decide to time the length of the next 15 commercials to find out the average length of a commercial. The times, written to the nearest 10 seconds, are listed below.

10 60 20 20 30 20 40 20
40 50 30 10 60 20 50

In mathematics, there are three common ways to describe the data: the mode, the median, and the mean.

Mode	The mode of a set of data is the number or item that appears most often.

A line plot of the commercial lengths can quickly give you the mode.

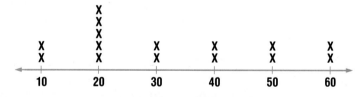

The mode is 20 because 20 occurs most often.

Median	The median is the middle number in a set of data when the data are arranged in numerical order.

The median can also be found by looking at the line plot. Since there are 15 numbers in the data set, the eighth number is the median. The median is 30.

Mean	The mean of a set of data is the arithmetic average.

The mean, or arithmetic **average,** is what people usually are talking about when they say "average." It is found by adding the numbers in the data set and then dividing by the number of items in the set.

$$\text{mean} = \frac{(10+10+20+20+20+20+20+30+30+40+40+50+50+60+60)}{15}$$
$$= \frac{480}{15} \text{ or } 32$$

The mean is 32. Note that the mode, the median, and the mean are *not* the same for this set of data.

Example 1

Joe's mathematics test scores for the first half of the year are 60, 93, 99, 72, 80, 96, 95, 91. Find the mode, median, and mean.

There is no mode because each score occurs only once. To find the median, arrange the numbers in numerical order: 60, 72, 80, 91, 93, 95, 96, 99. Since there are 8 numbers, the median is the mean of the fourth and fifth numbers, 91 and 93.

$$\text{median} = \frac{91+93}{2} \qquad \text{The median is 92.}$$

To find the mean, calculate the arithmetic average.

$(\; 60 \; + \; 72 \; + \; 80 \; + \; 91 \; + \; 93 \; + \; 95 \; + \; 96 \; + \; 99 \;) \; \div \; 8$

$= 85.75 \qquad$ The mean is about 86.

In Example 1, if Joe wants to brag about his scores, he would probably use the median, 92. When his mathematics teacher computes his grade, she will probably use the mean, 86.

Example 2 *Problem Solving*

Business A survey was conducted by a bedding firm to determine the average number of hours people sleep each night. The bedding firm wants to use this information to convince people that they should buy a good mattress since they spend a lot of time sleeping.

Number of Hours	Number of People
5	4
6	10
7	35
8	35
9	16

a. Determine the mode, median, and mean of the data.
b. Which number would the bedding firm probably use? Why?

Since the 35 people sleep 7 hours and 35 people sleep 8 hours, there are two modes, 7 and 8.

One hundred people were surveyed, so the median is the mean of the two middle numbers, 8 and 8. The median is 8 hours.

$$\text{mean} = \frac{4(5)+10(6)+35(7)+35(8)+16(9)}{100}$$
$$= \frac{749}{100} \text{ or } 7.49 \qquad \text{The mean is 7.49 hours.}$$

The bedding firm would use the median because it shows the greatest number of hours spent sleeping.

Checking for Understanding

Read and study the lesson to answer each question.

1. **Tell**, in your own words, how you would find each of the following from a set of data.
   a. the median, if there is an odd number of items.
   b. the median, if there is an even number of items
2. **Show** a set of data that meets each condition.

   a. 1 mode   b. 2 modes   c. no mode

Guided Practice

Order the data from least to greatest. Then find the mode(s), median, and mean.

3. 13, 17, 20, 18, 17, 15, 12
4. 3, 4, 7, 6, 5, 7, 8, 2
5. 90, 92, 94, 91, 90, 94, 95, 98

Exercises

*Independent
Practice*

Find the mode(s), median, and mean for each set of data.

6. 1, 5, 8, 3, 5, 4, 6, 2, 3, 7
7. 56, 75, 65, 57, 76, 66, 65, 64
8. 8.9, 8.0, 9.0, 9.1, 9.3, 9.4
9. 1,755; 1,780; 1,755; 1,805; 1,805

10. Your math teacher makes a frequency table of students' test scores on the last test. Find the mode, median, and mean of these scores.

Test Score	Tally	Number
100	卌	5
95					3
90				2
85	卌	5

Mixed Review

11. Express 32,500 in scientific notation. *(Lesson 2–6)*
12. Make a line plot for the following data: 21, 19, 16, 21, 20, 19, 18, 21, 19. Circle any outliers. *(Lesson 3–4)*

*Problem Solving
and Applications*

13. **Sports**   The scores of the winning teams of the Super Bowl from 1981 to 1990 are 27, 26, 27, 38, 28, 46, 39, 42, 20, and 55.
    a. Make a line plot.
    b. Find the mode, median, and mean.

14. **Critical Thinking**   Insert another number in the set 9, 12, 17, 15, 13 so that the mean of the resulting set is 14.

15. **Weather** The high temperatures for the first week of school were 84, 87, 89, 81, 83, 80, and 89.
    a. Find the mode, median, and mean.
    b. Which number would you use to emphasize how hot it was?

16. **Business** The manager of The Foot Locker keeps a record of the sizes of each athletic shoe sold. Which number is probably most useful: the mean, median, or mode? Explain.

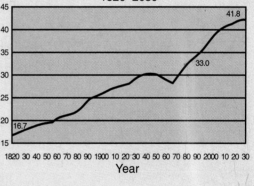

17. **Critical Thinking** Create a set of data with five numbers for each of the following situations.
    a. The mode, median, and the mean are the same.
    b. The mean is greater than the median.
    c. The mean is not equal to one of the numbers in the set.

DATA SEARCH

18. **Data Search** Refer to pages 88 and 89. What is the total number of Arabic-speaking people and Portuguese-speaking people?

Mid-Chapter Review

3

Use the line graph to answer the question.

1. Was the median age of the U.S. population greater or less than 25 in 1970? *(Lesson 3-1)*

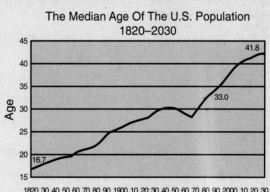

The Median Age Of The U.S. Population 1820–2030

The scores on a 20-point quiz are given at the right.

2. Make a frequency table for this set of data. *(Lesson 3-2)*

3. What is the range of the data? *(Lesson 3-3)*

4. Make a line plot of the data. *(Lesson 3-4)*

11	19	16
19	17	20
15	13	16
20	18	19
15	14	18
17	18	15
18	17	14

5. Find the mode, median, and mean. *(Lesson 3-5)*

6. Which number best describes the data, mode, median, or mean? Explain. *(Lesson 3-5)*

3-5B Are You Average?

A Follow-Up of Lesson 3-5

Objective

Use mean, median, and mode to describe students in your school.

Materials

pencil
paper
markers
ruler

How would you describe an "average" student in your school? The average student in your school may vary quite a bit from the average student in another. In this lab, you will find out what the average student in your math class is like.

Try this!

Work together as a class.

- List at least ten questions you would like to ask to find out what the average student is like. For example:

   What is your height in inches?
   What is your age in months?
   How many children are in your family?
   What is your favorite TV program?

- Prepare a survey with your ten questions. Each student in the class should complete the survey.

Work in groups of three.

- Each group should take at least two of the questions and compile the data in a frequency table or on a line plot.
- Find the mode, median, and mean. Decide which one best describes the data.

Work together as a class.

- Compile the results of all the groups.
- Prepare a poster that describes the "average" student in the classroom.

What do you think?

1. Which did you use for your data: mode, median, or mean? Explain why.
2. If you surveyed students in another class, would you expect the same results? Why or why not?

Extension

3. Run a contest in your school. Find the student that best meets the description of the "average" student.

3-6 Stem-and-Leaf Plots

Objective

Construct stem-and-leaf plots.

Words to Learn

stem-and-leaf
  plot
stem
leaf

DID YOU KNOW

Theodore Roosevelt was William McKinley's vice president. When McKinley was killed in 1901, Roosevelt assumed the presidency. At age 42, he was the youngest person to serve as president.
John F. Kennedy, however, was the youngest person to be *elected* president. He was 43 when he took office.

Did you know that Theodore Roosevelt was the youngest United States president to be sworn into office? He was 42 years old. He served as president from 1901–1909.

The oldest president to be sworn into office was Ronald Reagan. He was 69 years old. He served as president from 1981–1989.

As of 1991, there had been 41 presidents. The ages of each United States president when sworn into office are listed at the right.

57	61	57	57	58	57	61	54
68	51	49	64	50	48	65	52
56	46	54	49	50	47	55	55
54	42	51	56	55	51	54	51
60	62	43	55	56	61	52	69	64

When there are many numbers in a set of data, one way to show the data and make it easy to read is to construct a **stem-and-leaf plot.**

- Find the least number and the greatest number in the data set above. The greatest number, 69, has a 6 in the tens place and the least number, 42, has a 4 in the tens place.

- Draw a vertical line and write the digits in the tens places from 4 through 6 to the left of the line. The tens digits form the **stems.**

```
4 |
5 |
6 |
```

- The units digits are written to the right of the line.

```
4 | 9869723
5 | 7778741026405541655141562
6 | 1184502194
```

The units digits form the **leaves.**

- Rewrite the units digits in each row from the least to greatest.

```
4 | 2367899
5 | 00111122444455556667778
6 | 0111244589
```

- Include an explanation.   4 | 2 means 42.

It is easy to see now that most of the presidents were between 50 and 58 years when they were sworn into office. What other things can you find out quickly from the stem-and-leaf plot?

Example 1

Low temperatures on a March day for 19 midwestern cities are 34°, 57°, 56°, 27°, 58°, 46°, 53°, 28°, 34°, 9°, 56°, 57°, 45°, 34°, 29°, 52°, 57°, 57°, 8°. In which interval (0°–9°, 10°–19°, and so on) do most of the temperatures fall?

Make a stem-and-leaf plot.

The tens digit of a single-digit number is 0. So the stems start with 0 and end with 5. Include 1 as a stem even though none of the numbers have a 1 in the tens place.

0	89
1	
2	789
3	444
4	56
5	236677778

5 | 2 *means* 52°.

Most of the temperatures fall in the 50°–59° interval.

The mode, median, and any outliers can easily be found using a stem-and-leaf plot. For the data above, the mode is 57°, the median is 46°, and the outliers are 8° and 9°.

Checking for Understanding

Communicating Mathematics

Read and study the lesson to answer each question.

1. **Tell** whether you would make a stem-and-leaf plot if you had five numbers in your set of data. Explain.

2. **Discuss**, in a small group, some conclusions you can draw from the stem-and-leaf plot of the presidents' ages on page 109. Do you think that the next president will be over 70 years old when he or she is sworn into office? Why or why not?

Guided Practice

Determine the stems for each set of data.

3. 28, 22, 41, 39, 26, 33, 17, 14, 22, 56, 44

4. 42, 33, 9, 21, 7, 11, 14, 6, 40, 5, 9, 5

5. 100, 132, 135, 92, 86, 136, 106, 89, 128, 112

6. The ages of eleven people randomly surveyed in the Paramount Theater are 23, 67, 45, 35, 16, 25, 29, 35, 41, 36, and 32.
   a. Make a stem-and-leaf plot of this data.
   b. How many people are under 30 years old?
   c. In which age group do most of the people fall?

Exercises

Independent Practice
Make a stem-and-leaf plot for each set of data.

7. 58, 27, 34, 53, 24, 36, 38, 20, 43, 45, 54, 78, 35, 36, 47, 58

8. $0.83, $0.94, $0.54, $0.92, $0.85, $0.54, $0.96, $0.89, $0.75, $1.17
   Hint: Think of $0.54 as 54¢.

The stem-and-leaf plot at the right shows the high temperatures for the first 14 days of January in Buffalo, New York.

```
0 | 6
1 | 1145899
2 | 3556
3 | 36
```
1 | 1 means 11°.

9. What were the highest and lowest temperatures?

10. In which interval do most of the temperatures lie?

Mixed Review
11. **Algebra**   Evaluate $3a - 2b + 6(a - b)$ if $a = 14$ and $b = 2$.   *(Lesson 1–8)*

12. Find the mean, median, and mode for this set of data. $10, $18, $15, $6, $13, $12, $10   *(Lesson 3–5)*

Problem Solving and Applications
13. **Sports**   The table at the right shows the points scored by each player in a high school basketball game.
    a. Make a stem-and-leaf plot of this data.
    b. How many players scored over 10 points?
    c. Find the mode, median, and mean of the points scored. Which one best describes the data?

Player	Points
Bill	6
Alonso	4
Gary	15
Rich	3
Leroy	11
Javier	22
Ed	2
Mike	9
Doug	4
Andy	2

14. **Collect Data**   Ask 15 of your friends how much change they have in their wallets or pockets.
    a. Make a stem-and-leaf plot of this data.
    b. How many of your friends have more than 50¢?
    c. Write a sentence that describes the shape of the stem-and-leaf plot.

15. **Critical Thinking**   How is a stem-and-leaf plot like a bar graph?

3-7A How Much Is a Handful?

A Preview of Lesson 3-7

Objective

Use, mode, median, and mean to make predictions.

Materials

popcorn
pencil
paper

Jeremy and his older brother ask their mother if they can have some popcorn. She says, "Only one handful apiece." How much is a handful? Will Jeremy and his older brother get the same amount of popcorn?

In this lab, we will predict how many popped kernels are in a handful.

Try this!

- Each student should grab a handful of popped popcorn.
- Count the number of popped kernels in each handful and have one person write the numbers on the board.

Work in groups of three.

- Find the mode, median, and mean of the data.
- Choose one of the following ways to present the data: frequency table, bar graph, line graph, line plot, or stem-and-leaf plot.
- Display your data presentation to the rest of the class. Explain.

What do you think?

1. Which number (mode, median, or mean) best describes the data?
2. Which data presentation seemed most effective? Why?
3. Predict how many popped kernels would be in a handful for a student your age who is not in your class.

Extension

4. Randomly choose 10 students that are not in your math class and have them grab a handful of popcorn. Compare their handfuls with your prediction.
5. Do you think you could use your findings to predict a handful for an adult? Randomly choose 10 adults and have them grab a handful of popcorn. Compare their handfuls with your data.

3-7 Making Predictions

Objective

Make predictions from graphs.

At Disney World, in Epcot's Future World, visitors were shown a short video and asked to answer questions about space travel. A computer then counted the responses and displayed them on a video screen. The responses of one set of visitors are shown at the right.

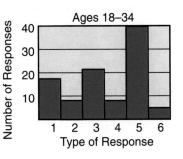

Where would you like to visit? (Responses are out of 100)				
	AGES			
	18-34	35-49	50-64	65+
1. A floating space station	17	27	32	33
2. A lunar base	8	9	9	9
3. A Martian settlement	22	16	17	14
4. Jupiter or Saturn	8	4	5	5
5. Another solar system	40	36	29	21
6. No response	5	8	8	18

A bar graph of the responses for ages 18–34 is shown below.

You can review bar graphs and line graphs on page 90.

Ages 18–34

(bar graph: Number of Responses vs. Type of Response)

If 100 others in the same age group were asked the same questions, how do you think they would respond? What if 200 others were asked the same questions?

You can use graphs to make predictions about how other people, similar to the ones in the survey, would respond to a given question.

You can use line graphs to predict future events.

Example 1 *Problem Solving*

Health   Yoko has been walking a mile each morning before school. She has recorded her times on a line graph. If she continues to walk each day, what do you think her time will be in 5 more days? Use the graph to make your prediction.

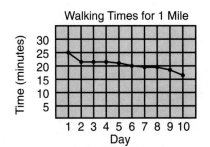

Walking Times for 1 Mile

(line graph: Time (minutes) vs. Day)

Because the line graph is sloping downward, you could predict that her times will probably continue to decrease each day. A good prediction would be 15 minutes. Do you think her time will continue to decrease indefinitely? Why or why not?

Line graphs can also be used to make decisions.

Example 2 *Problem Solving*

Business The manager of a McDonald's® restaurant wants to determine when extra help is needed. She makes a line graph of the number of orders placed each hour.

If extra help is needed whenever the number of orders in one hour exceeds 60, during which hours would the manager need extra help?

Orders Per Hour

The line graph goes above 60 between 11:00 A.M. and 2:00 P.M. and between 5:00 P.M. and 9:00 P.M. These are the times when extra help would be needed.

Checking for Understanding

Communicating Mathematics

Read and study the lesson to answer each question.

1. **Tell** three ways graphs can be used to make predictions.

2. **Draw** a bar graph of the responses of the 35–49 age group in the Disney World example on page 113.

Guided Practice

3. On the day before class elections, 100 students were asked who they would choose for class president. The graph at the right shows the results. Who do you think will win?

4. The line graph at the right shows the number of TV stations an average U.S. household receives. How many TV stations per household would you predict in 1992?

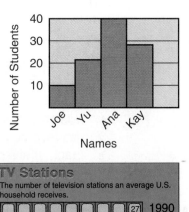

TV Stations
The number of television stations an average U.S. household receives.

27 1990
25 1988
19 1986
17 1984

= 3 stations

Exercises

5. A shoe store needs to know how many shoes of each color to order. The graph at the right shows how many of each color were sold last month. Based on the graph, for which color should they order the most?

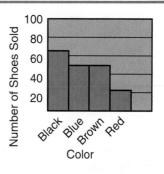

6. George has been practicing basketball free throws. What do you predict his percent will be on day 14?

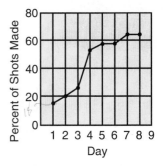

Mixed Review

7. Order 25.9, 21.6, 23, 30.2, and 27.4 from least to greatest. *(Lesson 2–1)*

8. **Final Exams** An American History class gets the following grades on their final exam: 83, 76, 91, 88, 72, 60, 75, 86, 94, 81. Construct a stem-and-leaf plot for this set of data. *(Lesson 3–6)*

Problem Solving and Applications

9. **Sports** The graph at the right shows the attendance at professional baseball games from 1977 to 1990.

   a. Based on the graph, what do you predict the attendance to be in 1995?

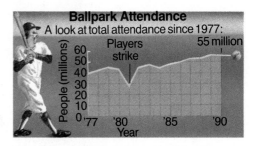

   b. What other factors, besides the graph, would you want to take into consideration when making your prediction?

10. **Critical Thinking** If you were the restaurant manager in Example 2, would you use the graph shown to predict the extra help needed on a holiday? Why or why not?

11. **Journal Entry** Collect at least three graphs from magazines and newspapers from which you could make a prediction. Include these in your journal. Write a statement about your predictions for each graph.

Lesson 3-7 Making Predictions   **115**

3-8 Misleading Statistics

Objective

Recognize when statistics and graphs are misleading.

Everywhere you look it seems as if people are dieting. People are very weight-conscious these days, sometimes to an extreme.

As a result of this increasing weight-consciousness, weight-loss programs have become big business.

One weight-loss program wants to convince potential customers that they will lose weight if they follow their program. The table at the right was prepared to show the pounds lost for each of 8 customers.

Customer	Number of Pounds Lost
Bob	25
Karen	34
Patty	29
Rosa	21
Mike	35
Betty	28
Chun	32
Linda	24

Did the company only give the weight lost by the 8 people who lost the most weight? How long did it take these people to lose the weight? How much did they weigh when they started the program? What else might be *misleading* about this data?

When you are given insufficient background information or an incomplete picture of the data, the data may be misleading.

Example 1 *Problem Solving*

Business Both line graphs below show monthly CD (compact discs) sales for one year at Barnie's Music Barn. Which graph could be misleading?

LOOKBACK

You can review scales on page 96.

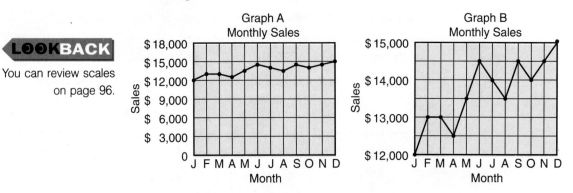

Both graphs show the same data, but they look very different. Graph B could be misleading because the vertical scale on the graph does not begin with zero.

Example 2 *Problem Solving*

Human Resources A company is interviewing people for a job opening. Mr. Eckert, a prospective employee, is told that the mean annual salary is $47,050. Study the table of salaries below. Do you think Mr. Eckert would make $47,050 as a beginning employee?

Position	Number of Employees	Annual Salary
President	1	$400,000
Vice President	2	$100,000
Sales Person	10	$ 25,000
Clerical Person	7	$ 13,000

Mr. Eckert would probably not make $47,050. The very high salaries of the president and vice presidents make this statistic misleading.

 **LOOKBACK**

You can review outliers on page 101.

Whenever there are outliers in the data, the mean is not a good way to describe the data. Which number would be a better way to describe the salaries above?

Checking for Understanding

Communicating Mathematics

Read and study the lesson to answer each question.

1. **Tell**, in your own words, three ways data can be misleading.

2. **Tell** which graph in Example 1 the manager of Barnie's Music Barn might prefer to use to show the owner of the store that sales are increasing?

Guided Practice

Rosio wants to show her parents that her math test scores have really improved.

3. Which graph below would she probably show them? Why?

4. How is the graph at the left misleading?

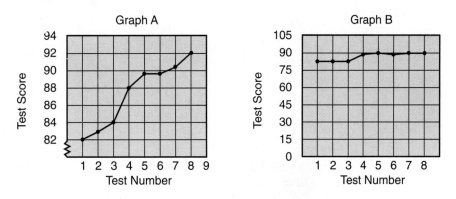

Graph A

Graph B

5. A school newspaper reporter randomly stops 50 students leaving gym class and asks them if they thought there was enough time between classes. Forty-five students said that there was not enough time. Do you think this survey reflects the opinions of all students? Why or why not?

Exercises

Independent Practice

6. Tei's history test scores, in order, are 89, 87, 90, 92, 95, 97, 99, and 100.

   a. Draw two line graphs of this data, one with a vertical scale of 0 to 100 and the other with a vertical scale of 80 to 100.

   b. What conclusion might you make from the first graph?

   c. What conclusion might you make from the second graph?

7. One hundred teenagers were asked how much spending money they get from their parents each week.

Amount of Money	Number of Teenagers
$25	8
$20	16
$15	10
$10	21
$ 5	45

   a. Find the mode, median, and mean of the data.

   b. Which number is misleading?

   c. Which would most accurately describe the data?

Mixed Review

8. Evaluate 5^4. *(Lesson 1–9)*

9. The bar graph shows responses to the question, "What is your favorite type of music?" Of which type of music will a music store probably sell the most? *(Lesson 3–8)*

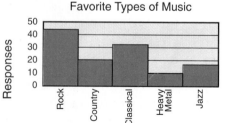

Problem Solving and Applications

10. **Entertainment** A movie advertisement says "Thousands of people rate this movie as the BEST MOVIE OF THE YEAR." What kind of information would you want to know before believing that this movie really is the best one of the year?

11. **Advertising** Super Sandwiches, a new fast-food restaurant, wants to promote its products. Super Burgers have 540 calories and 25 grams of fat. Super Dogs have 235 calories and 43 grams of fat.

Select data from the chart at the right to compare with the Super Burgers and Super Dogs. Draw bar graphs that will promote Super Burgers and Super Dogs as healthier food in each of the following areas.

Item	Calories	Fat (gm)
HAMBURGERS		
Burger King Whopper	660	41
Jack-in-the-Box Jumbo Jack	538	28
McDonald's Big Mac	591	33
Wendy's Old Fashioned	413	22
SANDWICHES		
Roy Rogers Roast Beef	356	12
Burger King Chopped-Beef Steak	445	13
Hardee's Roast Beef	251	17
Arby's Roast Beef	370	15

a. Super Burgers: calories   b. Super Burgers: fat

c. Super Dogs: calories   d. Super Dogs: fat

12. **Critical Thinking** Do outliers affect the *median* of a set of data? Give several examples to support your answer.

13. **Journal Entry** Find at least four newspaper or magazine ads. Describe how each one might be misleading.

Save Planet Earth

Garbage in Schools The amount of trash created in schools has grown steadily over the last 10 years as cafeterias have switched to paper and throwaway plastic instead of conventional tableware, glass, or reusable plastic. In schools around the country, napkins, plates, cups, forks, knives, and spoons are tossed into the garbage thousands of times per day.

How You Can Help

• Take lunch to school in reusable containers instead of bags, plastic wrap, and waxed paper.

• Work with your student council or parent-teacher organization to reduce the amount of garbage your school produces.

   a. Stop using throwaway dishes for regular school meals.

   b. Put uneaten food in a separate garbage can. The waste can be composted and used as fertilizer.

3 Study Guide and Review

Communicating Mathematics

State whether each sentence is *true* or *false.* If false, replace the underlined word to make the sentence true.

1. The <u>scale</u> of a set of data is the difference between the greatest number and the least number.

2. It is possible to have more than one choice of <u>interval</u> for a set of data.

3. An <u>outlier</u> is a number that differs greatly from the rest of the data.

4. The <u>median</u> of a set of data is the number or item that appears most often.

5. In a stem-and-leaf plot, the units digits form the <u>stems</u>.

6. In your own words, describe a situation for which the mean would not be the best way to describe the average of a set of data.

Skills and Concepts

Objectives and Examples

Upon completing this chapter, you should be able to:

Review Exercises

Use these exercises to review and prepare for the chapter test.

- choose appropriate scales and intervals for graphs *(Lesson 3-3)*

  Find the range and appropriate scale and interval for the data 11, 24, 2, and 26. Draw a number line.

  range = 26 − 2 = 24
  scale = 0 to 30, interval of 5

Find the range for each set of data. Choose an appropriate scale and interval. Then draw a number line to show the scale and intervals.

7. 3, 12, 1, 43, 25, 16
8. 75, 150, 100, 400, 550
9. 2.3, 11.9, 7.6, 1.3, 4.8
10. 35,000, 24,500, 53,000, 18,500

- construct line plots *(Lesson 3-4)*

  Make a line plot and circle any outliers for the data 5, 7, 3, 8, 3, 2, 8, 4, 15.

Make a line plot for each set of data. Circle any outliers.

11. 10, 12, 10, 8, 13, 10, 8, 22
12. 7.9, 8.3, 10.2, 8.7, 8.3, 3.8
13. 1,550, 1,700, 1,625, 1,590, 1,655, 1,780, 1,760, 1,575

Objectives and Examples

- find the mean, median, and mode for a set of data *(Lesson 3-5)*

  21, 18, 19, 20, 23, 19, 20, 19

  $$\text{mean} = \frac{21+18+19+20+23+19+20+19}{8}$$

  $$= 19.875$$

  median = 19.5   mode = 19

- construct stem-and-leaf plots *(Lesson 3-6)*

  49, 58, 42, 63, 55, 42, 59, 44

  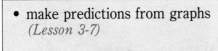

  ```
  4 | 2249
  5 | 589      4 | 2 means 42.
  6 | 3
  ```

- make predictions from graphs *(Lesson 3-7)*

  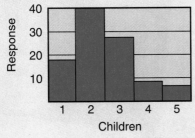

  Children

  The most common response to the question "How many children are in your family?" will be 2.

- recognize when statistics and graphs are misleading *(Lesson 3-6)*

  When you are given insufficient background information or an incomplete picture of the data, the data may be misleading.

Review Exercises

Find the mode(s), median, and mean for each set of data.

14. 2, 3, 4, 3, 4, 3, 8, 7, 2
15. 24, 26, 18, 23, 31
16. 89, 76, 93, 100, 72, 86, 74
17. 54,000, 49,000, 112,000, 89,000, 76,000, 65,000

Make a stem-and-leaf plot for each set of data.

18. 75, 61, 83, 99, 78, 85, 87, 92, 77, 78, 60, 53, 87, 89, 91, 90
19. $0.29, $0.54, $0.31, $0.26, $0.38, $0.46, $0.23, $0.21, $0.32, $0.37

Predict the answer in the following situation.

20. Isabel displayed her salary for the last five years in the line graph below. Predict her salary after three more years.

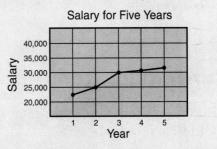

Salary for Five Years

Identify the part of the following situation which could be considered misleading.

21. The mean score on a math exam is 81.6. The set of test scores is 89, 92, 87, 86, 95, 93, 29.

Applications and Problem Solving

22. Use the graph on page 107. The United States Bureau of the Census projects that the median age of the population will be 41.8 years in the year 2030. How much higher is that than in 1980? *(Lesson 3-1)*

23. Use the graph on page 107. Was the median age of the U.S. population greater or less than 25 in 1940? *(Lesson 3-1)*

24. Use the frequency table on page 94. What was the most common height of junior high students? *(Lesson 3-2)*

25. **After-School Jobs** Eight students were asked their hourly wage at their after-school jobs. The following data resulted. $3.65, $4.15, $3.90, $4.25, $3.90, $4.00, $4.50, $3.80 Find the mode, median, and the mean. *(Lesson 3-5)*

26. **Television** A survey of the number of minutes a junior high student watches TV on a school day produced the following data.

    95, 85, 69, 75, 90, 45,
    92, 65, 50, 40, 75

    Make a stem-and-leaf plot for the data. *(Lesson 3-6)*

Curriculum Connection Projects

- **History** Write a short paper on mathematician René Descartes and the Cartesian coordinate system.

- **Current Events** Read and interpret today's graph from the front page of the *USA Today* newspaper.

Read More About It

Martin, Susan. *I Sailed With Columbus.*
Belton, John and Cramblit, Joella. *Dice Games.*
Kohn, Bernice. *Secret Codes and Ciphers.*

1. Use the graph on page 107. Based on the data presented in the line graph, predict what the graph will show by the year 3000.

2. Use the frequency table on page 94. How many more students were taller than 155 centimeters?

Find the range for each set of data. Choose an appropriate scale and interval. Then draw a number line to show the scale and intervals.

3. 2, 7, 26, 19

4. 8.9, 6.3, 10.3

5. Ann's grades on six French tests are 95, 76, 82, 90, 71, and 80. Find an appropriate scale and interval. Draw a number line to show them.

Make a line plot for each set of data. Circle any outliers.

6. 72, 76, 75, 72, 80, 55

7. 3.2, 7.2, 3.7, 3.3

8. **Family Tree** Mrs. Hanna has 5 children whose ages are 32, 20, 19, 17, and 22. Make a line plot and make a conclusion about the differences in ages.

Find the mode(s), median, and mean for each set of data.

9. 4, 6, 11, 7, 4, 11, 4

10. 12.4, 17.9, 16.5, 10.2

Make a stem-and-leaf plot for each set of data.

11. 37, 59, 26, 42, 57, 53, 31, 58

12. $0.46, $0.59, $0.42, $0.69, $0.55, $0.48, $0.66, $0.43

13. **Employment** The line graph at the right shows the percent of women holding jobs outside the home from 1975–1990. In which year were the most number of women working outside the home?

Women in Jobs

14. Using the graph at the right, predict the percent of women who will hold jobs outside the home in the year 2000.

15. The prices of 50 pairs of shoes at a local shoe store are recorded in the table at the right.
   a. Find the mode, median, and mean.
   b. Which average would the manager prefer to quote to a cost-conscious customer?

Price	Pairs
$10	6
$20	17
$30	15
$40	9
$50	3

Bonus What effect does an outlier have on the range of a set of data?

3 Academic Skills Test

Directions: Choose the best answer. Write A, B, C, or D.

1. Three tablecloths are sewn together end-to-end to make one long tablecloth. The tablecloths are about 48 inches, 64 inches, and 54 inches long. What is the combined length?

   A 102 in.   B 112 in.
   C 118 in.   D 166 in.

2. 182×6 is about

   A 6,000   B 1,200
   C 1,080   D 600

3. $2{,}396 \div 42$ is about

   A 600   B 60
   C 80   D 5

4. To mentally add 106 and 255, you can

   A add 6 to each addend.
   B add 100 and 250.
   C subtract 6 from each addend.
   D add 100 and 261.

5. What is the value of $x + y + 5$ if $x = 6$ and $y = 15$?

   A 21   B 25
   C 26   D none of these

6. Which is equivalent to 3^6?

   A 36
   B 18
   C $6 \cdot 6 \cdot 6$
   D $3 \cdot 3 \cdot 3 \cdot 3 \cdot 3 \cdot 3$

7. What is the value of $3[2(23 - 11) - (3 + 9)]$?

   A 12   B 36
   C 60   D 133

8. Which is a true statement?

   A $7.2 > 0.72$
   B $3.91 < 3.9$
   C $0.489 > 4.81$
   D $0.35 < 0.202$

9. Which is correct for rounding to the nearest tenth?

   A 0.56 rounds to 0.5
   B 0.95 rounds to 1.0
   C 1.205 rounds to 1.3
   D 0.4173 rounds to 0.42

10. $3.1 \times 2.5 =$

    A 77.5   B 7.75
    C 7.5   D 0.75

11. Earth is about 93,000,000 miles from the sun. How is this written in scientific notation?

    A 93×10^6
    B 9.3×10^7
    C 9.3×10^8
    D 0.93×10^8

12. Tomato juice is priced at three cans for $2.39. To the nearest cent, what is the cost of one can?

    A $0.08   B $0.79
    C $0.80   D $1.20

13. Suppose you need 0.65 liters of water for a science experiment but the container is measured in milliliters. How much water do you need?

A 6,500 milliliters
B 650 milliliters
C 65 milliliters
D 0.00065 milliliters

14.   Average Height for Adolescents

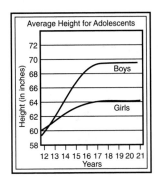

At age 16, about how much taller is the average boy than the average girl?

A 4 in.
C 63.5 in.
B 6 in.
D 67.5 in.

15. About what is the range of boys' heights shown in the graph in question 14?

A 9 years
C 21 years
B 10 in.
D 69.5 in.

16. The line plot shows how far in kilometers some students live from school. How many students are represented in the plot?

```
          X
          X  X
    X     X  X      X  X        X
  ──┼──┼──┼──┼──┼──┼──┼──┼──┼──┼──┼──
    0  1  2  3  4  5  6  7  8  9  10
```

A 5
C 9
B 7
D 10

17. Use the data in question 16. What is the median distance?

A about 3 km
C about 5 km
B about 4 km
D about 9 km

18. In a stem-and-leaf plot of the data below, what numbers would be used for the stems?

Height in Inches

52	75	49	51	57	66	69
58	59	62	61	61	73	68
76	74	78	49			

A 0–9
C 4–7
B 1–9
D none of these

19. Use the graph in question 14. About how much can a 13-year-old girl expect to grow in the next two years?

A 63 in
C 4 in.
B 61.5 in.
D 1.5 in.

20. A business had weekly profits of $5,000, $3,000, $2,000, $2,500, and $5,000. Which "average" might be misleading?

A mean
C mode
B median
D none of these

Patterns and Number Sense

Spotlight on Forestland

Have You Ever Wondered. . .

- Where most of the forestland in the United States is located?

- How many forest fires each year are caused by people?

United States Forestland

Region	Acres (Thousands)
North	**169,790**
Northeast	85,251
North Central	80,310
Great Plains	4,229
South	**199,943**
Southeast	87,744
SouthCentral	112,199
West	**358,189**
Pacific Northwest	178,956
Pacific Southwest	41,129
Rocky Mountains	138,104

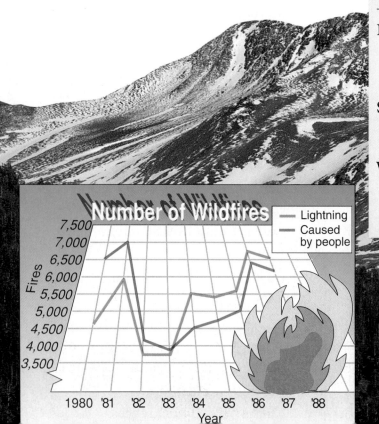

Number of Wildfires

Fires

7,500
7,000
6,500
6,000
5,500
5,000
4,500
4,000
3,500

1980 '81 '82 '83 '84 '85 '86 '87 '88
Year

Lightning
Caused by people

Chapter Project

Forestland
Work in a group.

1. Choose a tree that you see every day in your neighborhood or on your way to school.

2. Record all the characteristics of your tree. Draw a diagram of your tree.

3. Observe the tree at least three times a week for the next month. Make a list of changes that occurred on the tree.

4. Display your data on a timeline. Then compare your timeline with timelines made by your classmates.

Looking Ahead

In this chapter, you will see how patterns play an important role in mathematics.

The major objectives of the chapter are to:

- find prime factorization of a composite number

- solve problems by making an organized list

- find the greatest common factor or least common multiple of two or more numbers

- compare and order fractions and decimals

Calvin and Hobbes

by Bill Watterson

I WAS READING ABOUT HOW COUNTLESS SPECIES ARE BEING PUSHED TOWARD EXTINCTION BY MAN'S DESTRUCTION OF FORESTS.

© 1989 Universal Press Syndicate

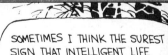

SOMETIMES I THINK THE SUREST SIGN THAT INTELLIGENT LIFE EXISTS ELSEWHERE IN THE UNIVERSE IS THAT NONE OF IT HAS TRIED TO CONTACT US.

4-1A Exploring Factors

A Preview of Lesson 4-1

Objective

Discover the factors of the whole numbers 1–30.

Materials

Thirty pieces of tagboard numbered individually from 1 to 30.

Try this!

- In order around the classroom, give thirty students each a number card from 1 to 30.
- **Step 1** Have each of these students stand up and write the number 1 on the back of his or her card.
- **Step 2** Have every second student, beginning with the student holding the "2" card, sit down and write the number 2 on the back of his or her card.
- **Step 3** Have every third student, beginning with the student holding the "3" card, stand up or sit down (depending on whether the student is already sitting or standing) and write the number 3 on the back of his or her card.
- Continue this process for each of the remaining numbers, until the thirtieth student has stood up or sat down and written the number 30 on the back of his or her card.

What do you think?

1. Make a conjecture about the numbers written on the back of each number card.

2. Use your conjecture to predict the numbers that would be written on the back of number cards from 31 to 35.

Applications

3. Which number cards have exactly two numbers on the back?

4. Which card has the fewest numbers on the back?

5. Which cards have the most numbers on the back?

6. What number cards are held by those students standing at the end of the activity?

Extension

7. Suppose there were 100 students holding number cards. Predict which number cards would be held by students standing at the end of the activity and write a statement explaining why.

4-1 Divisibility Patterns

Objective

Use divisibility rules.

Words to Learn

divisible
factor

Suppose you are baby-sitting for your neighbor's three children. To avoid arguments while the children color pictures, you decide to divide up the box of 48 crayons. Can you divide the crayons evenly among the three children?

$48 \div 3 = 16$ Since the quotient is a whole number, we say 48 is **divisible** by 3 or 3 is a **factor** of 48.

Thus, the crayons can be evenly divided with each child receiving 16 crayons.

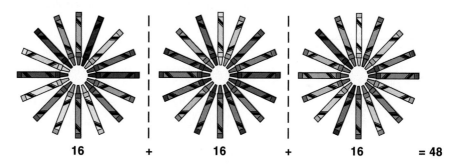

16   **+**   **16**   **+**   **16**   **= 48**

Mental Math Hint

●●●●●●●●●●●●●●

The divisibility rule for 2 can be used to determine whether an even number is also divisible by 4. Mentally divide your even number by 2. If the quotient is also even, your number is divisible by 4.

48 is also divisible by 1, 2, 3, 4, 6, 8, 12, 16, 24, and 48.

In order to quickly check the divisibility of large numbers, you can use the following rules of divisibility.

A number is divisible by:

2   if the digit in the ones place is even.

3   if the sum of the digits is divisible by 3.

4   if the number formed by the last two digits is divisible by 4.

5   if the digit in the ones place is 5 or 0.

6   if the number is divisible by both 2 and 3.

9   if the sum of the digits is divisible by 9.

10   if the digit in the ones place is 0.

1 Determine whether 126 is divisible by 2, 3, 4, 5, 6, 9, or 10.

2: The ones digit, 6, is even, so 126 is divisible by 2.

3: The sum of the digits, 9, is divisible by 3, so 126 is divisible by 3.

4: The number formed by the last two digits, 26, is not divisible by 4, so 126 is not divisible by 4.

5: The ones digit is not 5 or 0, so 126 is not divisible by 5.

6: The number is divisible by both 2 and 3, so 126 is divisible by 6.

9: The sum of the digits, 9, is divisible by 9, so 126 is divisible by 9.

10: The ones digit, 6, is not 0, so 126 is not divisible by 10.

2 Use your calculator to determine whether 397 is divisible by 7.

$$397 \; \boxed{\div} \; 7 \; \boxed{=} \; 56.714286$$

Since the quotient is not a whole number, 397 is not divisible by 7.

3 Find a number that is divisible by 3, 9, 5, and 10.

The ones digit must be 0 in order for the number to be divisible by 10 (which means the number will also be divisible by 5), and the sum of the digits must be divisible by 9 (which means the sum is also divisible by 3).

The numbers 1,260, 9,990, 333,000, and 123,210 are just a few of the numbers that meet these requirements.

Checking for Understanding

Communicating Mathematics

Read and study the lesson to answer each question.

1. **Tell** why every number that is divisible by 10 is also divisible by 5.

2. **Draw** pictures showing how 48 dots can be equally divided into 4 rows, 6 rows, or 8 rows.

3. **Tell** what the figure at the right suggests about the divisibility of the number 15.

4. **Tell** how to pick a number that is divisible by 3 but not divisible by 9. Then give two examples of such a number.

Using the divisibility rules or your calculator, determine whether the first number is divisible by the second number.

5. 113; 3   6. 2,357,890; 10   7. 81,726,354; 9   8. 7,538; 4

Using the divisibility rules, determine whether each number is divisible by 2, 3, 4, 5, 6, 9, or 10.

9. 180   10. 364   11. 95,455
12. 1,260   13. 31,212   14. 1,837

Exercises

Determine whether the first number is divisible by the second number.

15. 813; 3   16. 5,112; 6   17. 2,115; 6
18. 7,770; 10   19. 3,308; 4   20. 2,927; 9

Determine whether each number is divisible by 2, 3, 4, 5, 6, 9, or 10.

21. 510   22. 1,455   23. 101
24. 6,600   25. 775   26. 3,770

Name two numbers that are divisible by both of the given numbers.

27. 2, 5   28. 4, 9   29. 3, 10   30. 4, 5   31. 6, 5

32. Estimate $365 \div 38$. *(Lesson 1-3)*

33. Round 0.006 to the hundredths place. *(Lesson 2-2)*

34. Choose an appropriate scale and interval for the following data. Then draw a number line to display the scale and interval. 35, 42, 18, 25, 32, 47, 34 *(Lesson 3-3)*

35. **Statistics** Dave Smith reports that the average income for the five people who work in his department is $29,080. The five individual salaries are $26,700, $23,500, $24,800, $28,300, and $42,100. Why could the average quote be considered misleading? *(Lesson 3-8)*

36. **History** Thomas Jefferson was elected president of the United States in 1800. The United States holds a presidential election every 4 years. Which presidential election years since 1800 were divisible by 10?

37. **School** Mr. Variety has arranged the desks in his classroom in different ways, but he is always careful to have the same number of desks in each row. He can arrange the desks in 3 rows, 4 rows, and 6 rows. What is the fewest number of desks in his classroom?

38. **Critical Thinking** Are all numbers that are divisible by 9 also divisible by 3? Are all numbers that are divisible by 3 also divisible by 9? Give an argument or a counterexample to support your answer.

4-2 Prime Factorization

Objective

Find the prime factorization of a composite number.

Words to Learn

prime number
composite numbers
prime factorization
factor tree

Did you know that on average a hen lays 300 eggs per year? That is 25 dozen eggs. The numbers 25 and 12 are factors of 300 because $25 \times 12 = 300$. Just as 300 can be written as 25×12, the numbers 25 and 12 can also be written as the products of a pair of factors: $5 \times 5 = 25$ and $3 \times 4 = 12$. We could also have chosen $2 \times 6 = 12$.

For some numbers, the only product that can be written is 1 times the number itself. These numbers are called **prime numbers.**

A prime number is a whole number greater than 1 that has exactly two factors, 1 and itself. 2, 5, 11, and 17 are examples of prime numbers.

Notice that 0 and 1 are neither prime nor composite numbers.

Numbers that have more than two factors, such as 12 and 25, are called **composite numbers.** A composite number is a whole number greater than 1 that has more than two factors.

The Fundamental Theorem of Arithmetic states that every composite number can be written as the product of prime numbers in exactly one way if you ignore the order of the factors. This product is called the **prime factorization** of the number. Looking again at the number 300, let's write its prime factorization.

The figure formed by the steps of the factorization of 300 is called a factor tree.

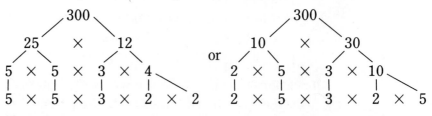

Since 2, 3, and 5 are prime numbers, then $2 \times 2 \times 3 \times 5 \times 5$ or $2^2 \times 3 \times 5^2$ is the prime factorization of 300.

Example 1

Use a factor tree to find the prime factorization of 630.

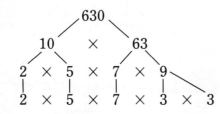

630 is divisible by 10.

The prime factorization of 630 is $2 \times 3^2 \times 5 \times 7$.

Example 2

Use your calculator to find the prime factors of 132.

$$132 \; \boxed{\div} \; 2 \; \boxed{=} \; 66 \; \boxed{\div} \; 2 \; \boxed{=} \; 33 \; \boxed{\div} \; 3 \; \boxed{=} \; 11$$

The prime factorization of 132 is $2^2 \times 3 \times 11$.
Thus, the prime factors are 2, 3, and 11.

Calculator Hint

• • • • • • • • • • • • •

Start with the divisibility rules and divide by the least prime number until the quotient is prime.

Example 3 *Connection*

Algebra Evaluate the expression $n^2 + n + 11$ for $n = 0, 1, 2,$ and 3 to find four prime numbers.

$n = 0:$	$n = 1:$	$n = 2:$	$n = 3:$
$0^2 + 0 + 11$	$1^2 + 1 + 11$	$2^2 + 2 + 11$	$3^2 + 3 + 11$
$= 0 + 0 + 11$	$= 1 + 1 + 11$	$= 4 + 2 + 11$	$= 9 + 3 + 11$
$= 11$	$= 13$	$= 17$	$= 23$

The numbers 11, 13, 17, and 23 are prime. *See Exercise 35 for more about the expression $n^2 + n + 11$.*

Checking for Understanding

Communicating Mathematics

Read and study the lesson to answer each question.

1. **Tell** why 1 is neither a prime number nor a composite number.

2. **Tell** why 0 is neither a prime number nor a composite number.

3. **Tell** why $2^4 \times 15$ is not the prime factorization of 240.

4. **Draw** a figure showing 21 dots in a rectangular array. (In a rectangular array, each row has the same number of dots and each column has the same number of dots.) What does this show?

5. **Draw** a figure showing 11 dots in a rectangular array. What do you discover?

Guided Practice

Determine whether each number is composite or prime.

6. 15　　　　　7. 29　　　　　8. 333　　　　　9. 552

Use a factor tree to find the prime factorization of each number.

10. 252　　　11. 66　　　12. 880　　　13. 270　　　14. 144

Use your calculator to find the prime factors of each number. Then write the prime factorization of each number.

15. 88　　　16. 146　　　17. 221　　　18. 250　　　19. 300

Exercises

Determine whether each number is composite or prime.

20. 75   21. 17   22. 6,453   23. 10,101

Write the prime factorization of each number.

24. 72   25. 1,260   26. 625   27. 1,300   28. 221

Use your calculator to find the prime factors of each number. Then write the prime factorization of each number.

29. 90   30. 121   31. 175   32. 236   33. 320

34. Find the missing factor: $2^3 \times \blacksquare \times 3^2 = 360$.

35. **Algebra** Find the least whole number n for which the expression $n^2 + n + 11$ is not prime.

36. **Algebra** Is the value of $2a + 5b$ prime or composite if $a = 3$ and $b = 5$?

37. Evaluate $2m + 4(m - n) + mn$ if $m = 10$ and $n = 3$. *(Lesson 1-8)*

38. Compute mentally $0.045 \times 1,000$. *(Lesson 2-5)*

39. Divide 1.08 by 1.2. *(Lesson 2-7)*

40. **Statistics** During the first two weeks of the new semester, Mrs. Jing keeps track of the attendance in her history class. Construct a stem-and-leaf plot for the following attendance data. 27, 31, 25, 19, 41, 32, 34, 36, 33, 31. *(Lesson 3-6)*

41. Use the divisibility rules to determine whether 135 is divisible by 2, 3, 4, 5, 6, 9, or 10. *(Lesson 4-1)*

42. **Business** Andre is designing cartons to ship cans of gourmet pecans. The height, width, and length of the cartons must be measured in whole cans. For example, a carton could be 3 cans high, 4 cans wide, and 2 cans long. Such a carton would contain $2 \times 3 \times 4$, or 24 cans of pecans.

    a. List two other ways to arrange 24 cans.

    b. List three ways to arrange 36 cans in a carton.

    c. In how many different ways could you arrange 30 cans?

    d. How would you arrange 29 cans?

    e. From his research, Andre knows that shipping from 30 to 35 cans is cost-effective. What number and arrangement of cans would you recommend be shipped in each carton?

43. a. **Number Sense** Determine the prime factorization for each of the following square numbers: 4, 9, 16, 25, 36, 49, 64.

b. What pattern do you notice?

c. Test your hypothesis on 81, 100, and 121.

44. **Number Sense** Will the sum of any two prime numbers be prime? Justify your answer.

45. **Critical Thinking** Numbers that can be represented by a triangular arrangement of dots, such as those below, are called **triangular** numbers.

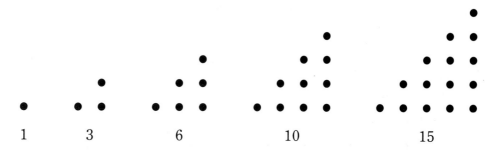

Determine the prime factorization for the first ten triangular numbers and describe the pattern.

46. Using the square at the right, multiply the middle number by 9. What does this product represent? Explain.

4	5	6
7	8	9
10	11	12

47. Jenny found a pair of prime numbers, 5 and 7, that differed by 2. These numbers are called **twin primes.** Find all the twin primes that are less than 100.

48. **Number Sense** Find the least 3-digit number that has 2, 3, 5, and another prime number as its prime factors.

49. **Number Sense** Name two numbers that have exactly twelve factors, including 1 and the number itself, and are divisible by both 5 and 7.

50. **Journal Entry** Explain why any composite number of dots can be arranged in a rectangular array that has at least two rows.

COMPUTER

CONNECTION

51. **Computer Connection** Many formulas have been tried to see whether they generate prime numbers. The computer program below will evaluate the expression $n^2 + n + 17$ for $n = 0, 1, 2, \ldots, 20$. Run the program and identify each number from the output as prime or composite.

```
10  FOR N = 0 TO 20
20  P = N^2 + N + 17
30  PRINT N, P
40  NEXT N
```

4-3 Sequences

Objective

Recognize and extend a pattern for sequences.

Words to Learn

sequence
terms
arithmetic
  sequence
geometric
  sequence

The bamboo plant is the fastest-growing plant in the world. One bamboo plant was observed to have grown 36 inches in 24 hours. If the plant grew at a constant rate during the 24 hours, how many inches would the bamboo plant have grown after 8 hours?

Making a table or organized list can help you discover a pattern.

Since the plant grew 36 inches in 24 hours and it is assumed to have grown at a constant rate, it was growing $36 \div 24$, or 1.5 inches per hour.

DID YOU KNOW

A Sequoia tree starts out as a seed weighing $\frac{1}{100}$ of an ounce. It increases its weight 100,000,000,000 times and can grow as tall as 340 feet.

Hours	1	2	3	4
Growth in inches during each hour	1.5	1.5	1.5	1.5
Total growth in inches after the hour	1.5	3.0	4.5	6.0

By studying the sequence of numbers in the last row of the table, you can identify a pattern and then use the pattern to find the growth after 8 hours. A **sequence** of numbers is a list in a specific order. The numbers in the sequence are called **terms.**

Since the growth each hour is a constant 1.5 inches, you can find the growth at the end of any hour by multiplying the number of hours by 1.5.

$$8 \times 1.5 = 12$$

The bamboo plant grew 12 inches in 8 hours.

If you can always find the next term in the sequence by adding the same number to the previous term, the sequence is called an **arithmetic sequence.**

$$3, \quad 6, \quad 9, \quad 12, \quad 15, \ldots$$
$$+3 \quad +3 \quad +3 \quad +3$$

Example 1

Identify the pattern in the sequence 3, 8, 13, 18, 23, . . . and describe how the terms are created. Then find the next three terms.

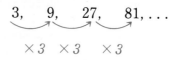

$$3, \quad 8, \quad 13, \quad 18, \quad 23, \ldots$$

$$+5 \quad +5 \quad +5 \quad +5$$

This is an arithmetic sequence in which each term after the first is created by adding 5 to the previous term.

$$23 + 5 = 28 \qquad 28 + 5 = 33 \qquad 33 + 5 = 38$$

The next three terms are 28, 33, and 38.

If you can always find the next term in the sequence by multiplying the previous term by the same number, the sequence is called a **geometric sequence.**

$$3, \quad 9, \quad 27, \quad 81, \ldots$$

$$\times 3 \quad \times 3 \quad \times 3$$

Example 2

Identify the pattern in the sequence 64, 32, 16, 8, . . . and describe how the terms are created. Then find the next three terms.

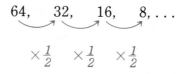

$$64, \quad 32, \quad 16, \quad 8, \ldots$$

$$\times \tfrac{1}{2} \quad \times \tfrac{1}{2} \quad \times \tfrac{1}{2}$$

This is a geometric sequence in which each term after the first is created by multiplying the previous term by $\frac{1}{2}$.

$$8 \times \tfrac{1}{2} = 4 \qquad 4 \times \tfrac{1}{2} = 2 \qquad 2 \times \tfrac{1}{2} = 1$$

The next three terms are 4, 2, and 1.

There are many sequences that are neither arithmetic nor geometric.

Example 3

Identify the pattern in the sequence 1, 4, 9, 16, 25, ... and describe how the terms are created. Then find the next three terms.

The terms are squares of the numbers 1, 2, 3, 4, 5, ... That is,

1, 4, 9, 16, 25, ... can be written as $1^2, 2^2, 3^2, 4^2, 5^2, \ldots$.

This is neither an arithmetic nor geometric sequence.

The next three terms are $6^2, 7^2, 8^2$, or 36, 49, 64.

Checking for Understanding

Communicating Mathematics

Read and study the lesson to answer each question.

1. **Tell** why the sequence in Example 3 is neither arithmetic nor geometric.
2. **Draw** a model using dots for each number in the sequence 1, 3, 5, 7, 9, ... to show that the sequence is arithmetic.
3. **Tell** how to find the next term in the sequence 0, 3, 8, 15, 24, 35, ...
4. **Write** a rule for generating a sequence of your own. Then have a classmate try to create the sequence using only the first number and your rule.

Guided Practice

Describe the pattern in each sequence. Identify the sequence as arithmetic, geometric, or neither. Then find the next three terms.

5. 7, 14, 21, 28, 35, 42, ...
6. 2, 6, 18, 54, ...
7. 0, 1, 3, 6, 10, 15, ...
8. 15, 30, 45, 60, ...
9. 1, 6, 4, 9, 7, 12, 10, ...
10. 9, 3, 1, $\frac{1}{3}$, $\frac{1}{9}$, ...

Create a sequence using each of the given rules. Provide at least four terms for each sequence beginning with a number of your choice. State whether the sequence is arithmetic, geometric, or neither.

11. Add 3 to each term.
12. Multiply each term by $\frac{1}{3}$.
13. Add 0.1 to the 1st term, add 0.2 to the 2nd term, add 0.3 to the 3rd term, and so on.

Exercises

Independent Practice

Identify each sequence as arithmetic, geometric, or neither. Then find the next three terms in each sequence.

14. 0.1, 0.3, 0.5, 0.7, ...
15. 2, 1, 0.5, 0.25, ...
16. 12, 17, 22, 27, ...
17. 1, 2.1, 3.2, 4.3, ...
18. 100, 101, 103, 106, 110, ...
19. 4, 1, $\frac{1}{4}$, $\frac{1}{16}$, ...
20. 0, 17, 34, 51, ...
21. 11, 22, 33, 44, ...
22. 1, 2, 2, 3, 3, 3, 4, 4, 4, 4, ...
23. 1, 8, 27, 64, ...

Create a sequence using each of the given rules. Provide at least four terms for each sequence beginning with the given number. State whether the sequence is arithmetic, geometric, or neither.

24. Add 0.6 to each term; 10.

25. Square consecutive odd integers; 1.

26. Add $\frac{1}{2}$ times to each term; 12

Mixed Review 27. Solve mentally $3y = 63$. *(Lesson 1-10)*

28. **Science Experiment** Juan's science experiment requires 2.3 mL of sodium. How many liters of sodium are required? *(Lesson 2-9)*

29. **Statistics** Find the mean, median, and mode for the following set of data. 3, 5, 2, 3, 3, 4, 3, 2, 2. *(Lesson 3-5)*

30. Use a factor tree to find the prime factorization of 630. *(Lesson 4-2)*

Problem Solving and Applications 31. **Health** Sara has decided to start an exercise program and she knows that it is important to begin gradually. She plans to begin by working out for 5 minutes and then double her exercise time each day for 1 week. Write a sequence showing the length of time she exercises each day. Is her plan reasonable? Why or why not?

32. See Exercise 45 on page 135. Write a sequence formed by the first eight triangular numbers. Write a rule for generating the sequence.

33. **Critical Thinking** A magic square is a number square in which the rows, columns, and diagonals all have the same sum. Make a magic square by using the sequence 2.3, 3.2, 4.1, 5, ... Find the pattern and write the first nine terms. Then place the terms in the appropriate squares.

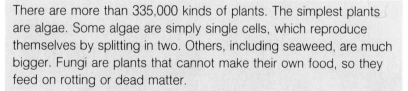

34. **Mathematics and Science** Read the following paragraph.

There are more than 335,000 kinds of plants. The simplest plants are algae. Some algae are simply single cells, which reproduce themselves by splitting in two. Others, including seaweed, are much bigger. Fungi are plants that cannot make their own food, so they feed on rotting or dead matter.

The giant kelp seaweed is found in the Pacific Ocean. One plant grows 3 feet the first two days. If it continues to grow at the same rate, what would be the length of the seaweed at the end of 80 days?

4-3B Exploring Geometric and Arithmetic Sequences

A Follow-Up of Lesson 4-3

Objective

Recognize the amount of change in arithmetic and geometric sequences.

Materials

calculator
notebook paper
graph paper

In business, science, sports, and our daily lives, we can see patterns of growth. Two common patterns are arithmetic sequences and geometric sequences.

Activity One

- Fold a piece of notebook paper in half and record the number of layers of paper. (See the table below.)
- Shade one side of the folded paper.
- Open the piece of paper and record the fractional part of the paper that is *not* shaded. Refold the piece of paper.
- Fold your paper in half again so the unshaded side is on the outside and record the number of layers of paper.
- Shade one side of the folded paper.
- Open the piece of paper and record the fractional part of the paper that is *not* shaded. Completely refold the paper.
- Continue folding, shading, and recording until you can no longer fold your paper (at least five folds).

Number of Folds	Layers of Paper	Fraction of Paper Unshaded
1	2	$\frac{1}{2}$
2	4	$\frac{1}{4}$
3	8	
4		
⋮		

What do you think?

1. Examine the sequence of numbers in the "Layers" column of your table. Is this sequence arithmetic or geometric?

2. Study the sequence in the "Fraction" column. Is this sequence arithmetic or geometric?

Activity Two

Imagine continuing the paper-folding process from Activity One forever. Assuming that your unfolded sheet of notebook paper is 0.002 inches thick, make a table similar to the one below for the first five folds.

Number of Folds	Thickness of the Folded Paper	
	In Layers	In Inches
1	2	0.004
2	4	0.008
3	8	
⋮		

What do you think?

3. How many folds would it take until the paper is as tall as you?

4. One mile is 5,280 feet. How many folds would it take until the folded paper is one mile tall?

5. How would your data change if you were folding the paper into thirds instead of halves?

Extension

6. a. Write the first ten terms of the geometric sequence 3, 9, 27, . . . , which is created by multiplying the previous number by 3.

   b. Write the sequence of numbers formed by ones digits of successive terms in the sequence.

   c. What pattern do you see?

   d. Find the sum of the digits for each term in the sequence created in part a.

   e. Which sums are divisible by 3?

   f. Which sums are divisible by 9? Describe the pattern.

Mathematics Lab 4-3B Exploring Geometric and Arithmetic Sequences   **141**

4-4 Make a List

Objective

Solve problems by making an organized list.

Leonardo Fibonacci, a mathematical genius during the Middle Ages, is best known for a sequence of numbers. Fibonacci introduced his sequence by making up a story. The story begins on January 1 with one pair of rabbits named A that live in a pen.

On February 1, a pair of bunnies named B is born. *Now there are 2 pairs of rabbits: 1 adult pair and 1 baby pair.*

By March 1, the original parents have had another pair of bunnies and the original bunnies have grown into adult rabbits. *Now there are 3 pairs of rabbits: 2 adult pairs and 1 baby pair.*

The pattern continues. Find the pattern.

Explore

What do you know?
You know that:
On January 1, there is 1 pair of rabbits in a pen.
On February 1, there are 1 pair of rabbits and 1 pair of bunnies.
On March 1, there are 2 pairs of rabbits and 1 pair of bunnies.
On April 1, the pattern continued.

What are you trying to find?
You are trying to find the pattern.

Plan

Make an organized list. Use A for adult pairs of rabbits and B for pairs of bunnies. Organize a list of the number of As, Bs, and total pairs. Use this to find a pattern.

Solve

Date													Number of As	Number of Bs	Total
1/1						A							1	0	1
2/1					B		A						1	1	2
3/1				A		B		A					2	1	3
4/1			B		A		A		B		A		3	2	5
5/1		A	B	A	B	A	A	B	A				5	3	8
6/1	B	A	A	B	A	A	B	A	B	A	A	B A	8	5	13

You can see a pattern in each of the columns. Look at the numbers in column A.

1, 1, 2, 3, 5, 8, . . . is the **Fibonacci sequence.**

The terms are called Fibonacci numbers. For discussion purposes, we will label each term as F1, F2, and so on.

F1 = 1, F2 = 1, F3 = 2, F4 = 3, F5 = 5, F6 = 8, . . .

By looking at each of the columns, you can see that
F1 + F2 = F3, F2 + F3 = F4, F3 + F4 = F5,
F4 + F5 = F6, and so on.

Examine   After the first two terms, 1 and 1, of the Fibonacci sequence, the sum of each two consecutive terms gives the next term.

$$1 + 1 = 2 \qquad 1 + 2 = 3 \qquad 2 + 3 = 5 \qquad 3 + 5 = 8$$

Checking for Understanding

Communicating Mathematics

Read and study the lesson to answer each question.

1. **Tell** the next three numbers on page 142 for columns A, B, and Total.

2. a. **Write** in your own words how to extend the list on page 142 through July 1.
   b. **Show** the list on page 142 completed through July 1.

Guided Practice

Solve by making a list.

3. List the first 20 Fibonacci numbers.

4. Refer to the list on page 142. What is the relationship between the numbers under *Total* and the numbers under *Number of As?*

Problem Solving

Practice

Solve using any strategy.

5. Explore the Fibonacci sequence using your calculator to complete the table.
   a. Write the first 10 terms of the Fibonacci sequence. Divide each term by the succeeding term. Round to the nearest thousandth.
   b. In your own words, describe the pattern of quotients.

Term	Quotient
1	—
1	$1 \div 1 = 1$
2	$1 \div 2 = 0.5$
3	$2 \div 3 = 0.667$
⋮	⋮

Strategies
••••••••••
Look for a pattern.
Solve a singular pattern.
Act it out.
Guess and check.
Draw a diagram.
Make a chart.
Work backwards.

6. **Geography** Mount Everest, the tallest mountain on Earth, is 8,872 meters tall. The tallest mountain on Mars is Olympus Mons, which is 23,775 meters tall. How much taller is Olympus Mons than Mount Everest?

7. Jo sees that every third Fibonacci number is even (F3 = 2, F6 = 8, F9 = 34). Find F30.

8. John took a bag of cookies to the school play rehearsal. Half were given to the actors and five to the director of the play. That left John with 15 cookies. How many cookies did he take to rehearsal?

9. Find two numbers so that, when added together, they equal 56 and, when multiplied, they equal 783.

10. **Science** Telephone calls travel through optical fibers at the speed of light, which is 186,000 miles per second. A millisecond is 0.001 of a second. How far can your voice travel over an optical line in 1 millisecond?

11. Are there any perfect squares (for example: $4 \times 4 = 16$) or perfect cubes (for example: $4 \times 4 \times 4 = 64$) in the first 12 terms of the Fibonacci sequence? If so, name them.

12. Sarah is experimenting with the Fibonacci numbers. She has these numbers on her paper.

    a. Describe the pattern in the second column.

    b. Tell what she did to get the pattern in the third column.

1	1	
1	1	2
2	4	5
3	9	13
5	25	34
8	64	89
13	169	...

13. **Measurement** There are about 2.54 centimeters in 1 inch. Suppose a field mouse is about 2.5 inches long. How long is it in centimeters?

14. **Aircraft** A Boeing 747 Jumbo Jet, the largest capacity jetliner, is 70.51 meters long. The Stits Skybaby, the smallest fully functional aircraft, is only 2.794 meters long. How many Skybabies, set end-to-end, would it take to equal the length of the 747?

Greatest Common Factor

Objective

Find the greatest common factor of two or more numbers.

Words to Learn

greatest common factor (GCF)

Eratosthenes, a Greek mathematician, lived during the 3rd century B.C. One of his many contributions is a device, now generally known as the *sieve of Eratosthenes,* which can be used to find prime numbers.

By using the sieve to find prime numbers, you can discover the **greatest common factor** of two or more numbers. The greatest common factor (GCF) of two or more numbers is the greatest number that is a factor of each number.

Mini-Lab

Work with a partner.
Materials: 4 colored pencils

- Copy the array of numbers shown at the right.
- Use a different colored pencil for each step below.

```
    2  3  4  5  6  7  8  9 10
11 12 13 14 15 16 17 18 19 20
21 22 23 24 25 26 27 28 29 30
31 32 33 34 35 36 37 38 39 40
41 42 43 44 45 46 47 48 49 50
```

Follow this "sieve" process.

Step 1 Circle 2, the first prime number. Cross out every second number after 2.

Step 2 Circle 3, the second prime number. Cross out every third number after 3.

Step 3 Circle 5, the third prime number. Cross out every fifth number after 5.

Step 4 Circle 7, the fourth prime number. Cross out every seventh number after 7.

Talk About It

a. In the process, 30 was crossed off when using which primes?
b. In the process, 42 was crossed off when using which primes?
c. What prime factors do 30 and 42 have in common? What composite factor do they have in common?
d. What is the greatest common factor of 30 and 42?

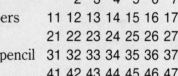

When am I ever going to use this?

Health Care Professional

There are many career options in this field from home health care aides to highly trained nurses and physicians.

In addition to specific training, health care workers need to have estimation and problem solving skills.

For more information contact:
National Health Council
350 Fifth Avenue,
   Suite 1118
New York, NY 10118

One of the following two methods is usually used to find the greatest common factor (GCF) of two or more numbers.

Method 1: List the factors of each number. Then identify the common factors and choose the greatest of these common factors, the GCF.

You can review prime factorization on page 132.

Method 2: Write the prime factorization of each number. Then identify all common prime factors and find their product, the GCF.

Examples

Problem-Solving Hint

• • • • • • • • • • • • •

Make a list.

1 Find the GCF of 45 and 54 by listing the factors of each number.

factors of 45: 1, 3, 5, 9, 15, 45
factors of 54: 1, 2, 3, 6, 9, 18, 27, 54

common factors: 1, 3, 9

Thus, the GCF of 45 and 54 is 9.

2 Find the GCF of 180 and 675 by writing the prime factorization of each number.

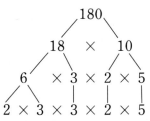
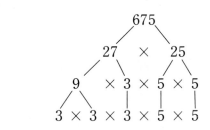

common prime factors: 3, 3, 5

Thus, the GCF of 180 and 675 is $3 \times 3 \times 5$ or 45.

3 Find the GCF of 510, 714, and 306 by writing the prime factorization of each number.

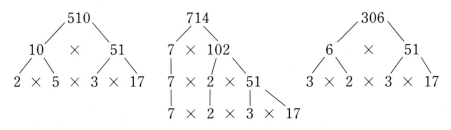

common prime factors: 2, 3, 17

Thus, the GCF of 510, 714, and 306 is $2 \times 3 \times 17$ or 102.

Checking for Understanding

Communicating Mathematics

Read and study the lesson to answer each question.

1. **Write** all the common factors of 12 and 20. Which is the greatest common factor?
2. **Draw** factor trees for 210 and 150, and then circle the common factors. What is the GCF of 210 and 150?
3. **Tell** why the GCF of 56 and 140 is *not* 14, even though $56 = 14 \times 4$ and $140 = 14 \times 10$.

Guided Practice

Find the GCF of each pair of numbers by listing the factors of each number.

4. 16, 36   5. 12, 30   6. 33, 121   7. 28, 84

List the common prime factors for each pair of numbers. Then write the GCF.

8. $60 = 2^2 \times 3 \times 5$ 9. $36 = 2^2 \times 3^2$
   $105 = 3 \times 5 \times 7$ $27 = 3^3$

Find the GCF of each pair of numbers by writing the prime factorization of each number.

10. 45, 75   11. 100, 30   12. 12, 78   13. 39, 91

Exercises

Independent Practice

Find the GCF of each set of numbers.

14. 360, 540   15. 132, 108   16. 120, 72   17. 14, 33
18. 18, 54   19. 20, 30   20. 16, 28   21. 8, 9
22. 6, 8, 12   23. 10, 15, 20   24. 18, 42, 60   25. 54, 90, 126

26. Name two different pairs of numbers whose GCF is 28.

Mixed Review

27. Jamie Hyatt has 12 gallons of gasoline in her car. She needs to drive 465 miles to a business meeting. Her car averages 37 miles per gallon. Can Jamie make the trip without stopping for gasoline? *(Lesson 1-1)*
28. Estimate 23.69 divided by 4.05. *(Lesson 2-3)*
29. **Statistics** Construct a line plot for 76, 65, 82, 93, 75, 72, 81, and 90. *(Lesson 3-4)*
30. Describe the pattern and find the next three terms for the sequence 13, 26, 52, 104, . . . *(Lesson 4-3)*

Problem Solving and Applications

31. In the sequence 15, 30, 45, 60, 75, . . . , what is the GCF of all the numbers in the sequence?
32. **School** A band director chooses to have the band march on and off the field in a rectangular array. She likes to have at least 3 rows. On a piece of graph paper, sketch all the possible rectangular arrays for 48 musicians.
33. **Number Sense** Can the GCF of a set of numbers be greater than any one of the numbers? Explain.
34. **Critical Thinking** A set of numbers whose GCF is 1 are *relatively prime*. Find the two least composite numbers that are relatively prime.

DECISION MAKING

Sponsoring a Retirement Center

Situation

Suppose your student body voted unanimously to sponsor the Crown Retirement Center this year. For most of the year, you scheduled in-house activities with music, writers and readers, and parties. There are two trips scheduled for outside the retirement center, but some of the patients can't leave. What activities will you plan for them to make February and August special? Since the student body has $650 in their treasury, Meg suggests that they order fruit from the Big Bounty Catalogue.

Hidden Data

There are shipping and handling charges when ordering from a catalogue. There may be reduced prices or club prices. You may need to watch the quantity on membership prices.

To avoid long distance charges, see if there is an 800 number for toll-free calling.

Big Bounty Catalogue's
Fruit of the Month Club

January Crisp Valley Apples	$19.99
February Sumptuous Grapefruit	19.99
March Lucious Large Oranges	20.99
April Exotic Pineapple	22.99
May Maypole Apples	21.99
June Plumet Plums	20.99
July Golden Tangerines	22.99
August Natural Nectarines	21.99
September Russet Pears	20.99
October Blue Giant Grapes	21.99
November Best Bosc Pears	19.99
December Japanette Apples	23.99

3 BOX CLUB
(Jan., Feb., Mar.)................$53.99

5 BOX CLUB
(3 box plus June, Aug.).........85.99

8 BOX CLUB
(5 box plus May, July , Oct.).145.99

12 BOX CLUB
(All 12 boxes shown)...........210.99

Each box is approximately 5 pounds.

Deluxe Clubs
Each box is approximately 7 pounds.

3 BOX CLUB **$82.99**
5 BOX CLUB **111.99**
8 BOX CLUB **162.99**
12 BOX CLUB **229.99**

Giant Party Drums -
Giant pears and apples, cookies, candies, corn puffs, cheeses, and nuts: ORDER ANY TIME.

$7\frac{1}{2}$ gal. drum, about 10 lb **$60.00**
4 gal. drum, about $5\frac{1}{2}$ lb **$40.00**

Standard Delivery Charge	
Merchandise Total	Add
Up to $25.00	$3.95
$25.01-$50.00	$5.95
$50.01-$100.00	$9.95
$100.01-$150.00	$14.95
$150.01-$200.00	$19.95
$200.01-$250.00	$24.95
$250.01-$400.01	$29.95

We will compute all applicable state and local taxes.

Analyzing the Data

1. What is the standard delivery charge for a 3 BOX CLUB?
2. What is the standard delivery charge for a DELUXE 12 BOX CLUB?
3. What other items can be ordered besides the BOX CLUB?

Making a Decision

4. **Can you** find a club item for February and August?
5. **Would you** buy extra items to have the club selections?
6. **Is there** a charge for state and local taxes?
7. **Can you** order gifts for less than one per person?
8. **Can you** order one large gift that all patients can share?
9. **Did you** check with the nursing home staff for patient diet information?

Making Decisions in the Real World

10. **Request** special prices for your school project from local shops in return for advertising possibilities.
11. **Find** the cost of buying separate items and packaging them at school.

4-6 Fractions in Simplest Form

Objective

Express fraction in simplest form.

Words to Learn

simplest form

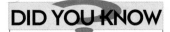

DID YOU KNOW

The Appalachian dulcimer is an American variety of the zither. Various forms of the zither were played in Europe during the 18th century. The Swedish *hummel*, the Norwegian *langleik*, and the French *épinette des Vosges* are but a few of the regional forms of the zither.

A dulcimer is a stringed musical instrument that produces a soft, melodic sound. The most common dulcimer in the United States is the Appalachian dulcimer. It is used to play a variety of folk music. Music is produced by strumming or hammering the strings. When struck, a string vibrates and produces a specific tone. For example, when middle C is played, the string vibrates at a frequency of 256 vibrations per second, producing the tone for middle C. The tone for the next higher C is produced by a frequency of 512 vibrations per second.

Can you express the fractions $\frac{256}{512}$ in simplest form? A fraction is in **simplest form** when the GCF of the numerator and denominator is 1.

To express a fraction in simplest form:

- find the greatest common factor (GCF) of the numerator and denominator,

- divide the numerator and the denominator by the GCF, and

- write the resulting fraction.

To simplify $\frac{256}{512}$, first find the GCF of 256 and 512 by listing the factors of each number.

factors of 256: 1, 2, 4, 8, 16, 32, 64, 128, **256**
factors of 512: 1, 2, 4, 8, 16, 32, 64, 128, **256**, 512

The GCF of 256 and 512 is 256.

$$\frac{256}{512} = \frac{256 \div 256}{512 \div 256} = \frac{1}{2}$$

If the numerator and denominator do not have any common factors other than 1, then the fraction is in simplest form.

Examples

Express each fraction in simplest form.

1 $\dfrac{36}{54}$

Since $36 = 2 \times 2 \times 3 \times 3$ and $54 = 2 \times 3 \times 3 \times 3$, the GCF of 36 and 54 is $2 \times 3 \times 3$ or 18.

Now divide the numerator and denominator by using the GCF.

$$\frac{36}{54} = \frac{36 \div 18}{54 \div 18} = \frac{2}{3}$$

2 $\dfrac{45}{75}$

Since $45 = 3 \times 3 \times 5$ and $75 = 3 \times 5 \times 5$, the GCF of 45 and 75 is 3×5 or 15.

$$\frac{45}{75} = \frac{45 \div 15}{75 \div 15} = \frac{3}{5}$$

3 $\dfrac{17}{35}$

Since $35 = 5 \times 7$ and 17 is prime, 17 and 35 have no common factors greater than 1. Therefore, $\frac{17}{35}$ is in simplest form.

Checking for Understanding

Communicating Mathematics

Read and study the lesson to answer each question.

1. **Tell** how to determine if a fraction is in simplest form.
2. **Tell** why $\frac{13}{17}$ is in simplest form, but $\frac{2}{4}$ is not in simplest form.
3. **Tell** why the method shown below does not produce an equivalent fraction in simplest form.

$$\frac{35}{45} = \frac{3\cancel{5}}{4\cancel{5}} = \frac{3}{4}$$

Guided Practice

Express each fraction in simplest form.

4. $\dfrac{15}{35}$

5. $\dfrac{150}{350}$

6. $\dfrac{25}{45}$

7. $\dfrac{99}{66}$

8. $\dfrac{81}{90}$

9. $\dfrac{44}{160}$

10. $\dfrac{64}{80}$

11. $\dfrac{300}{400}$

12. Write two different fractions that can be expressed in simplest form as $\frac{5}{7}$.

Exercises

Express each fraction in simplest form.

13. $\frac{16}{32}$ 14. $\frac{50}{75}$ 15. $\frac{250}{450}$ 16. $\frac{42}{72}$

17. $\frac{27}{33}$ 18. $\frac{120}{100}$ 19. $\frac{49}{63}$ 20. $\frac{9}{123}$

21. $\frac{56}{128}$ 22. $\frac{125}{625}$ 23. $\frac{22}{33}$ 24. $\frac{20}{36}$

Write three different fractions that can be expressed in simplest form as each of the following.

25. $\frac{2}{3}$ 26. $\frac{3}{10}$ 27. $\frac{3}{4}$ 28. $\frac{5}{6}$

29. Compute $234 - 86$ mentally. *(Lesson 1-4)*

30. Divide 0.81 by 0.3 *(Lesson 2-7)*

31. **Statistics**   Refer to the graph in Exercise 5 on page 115. Which color of shoe would you consider to be the least popular? *(Lesson 3-7)*

32. Find the greatest common factor of 72 and 270. *(Lesson 4-5)*

33. **School**   Clark Middle School has 525 girls and 600 boys. Express the number of girls as a fraction of the total student population. Write the fraction in simplest form.

34. **Sports**   Kara has won 24 of her 30 tennis matches during the tennis season.
    a. In simplest form, what fraction of her matches has she won?
    b. In simplest form, what fraction of her matches has she lost?

35. **Daily Living**   Jacob made the following table showing how he spends his day. Find the fraction of the day he spends on each activity.
    a. Express all fractions in simplest form.

Activity	Number of Hours
Sleeping	8
Eating	2
Dressing & bathing	1
School	7
Studying	3
Swim practice	2
Miscellaneous	1

    b. Make a table for your day. Include at least six activities.

36. **Critical Thinking** Ling read 125 pages of his 300-page book.
   a. What fraction of his book has he read?
   b. If Ashley has read the same fraction of her 240-page book, how many pages has she read?

Solve. Use the chart at the right.

37. Suppose a female, age 13, took in 2,000 calories in one day. What fraction, in simplest form, of the recommended daily calories did she take in?

38. Suppose a male, age 25, took in 3,600 calories in one day. What fraction, in simplest form, represents the amount he took in compared to the recommended daily calories?

39. **Journal Entry** Could a fraction have more than one simplest form? Explain your answer.

Recommended Daily Calories (moderately active person)		
	Females	**Males**
Age	**Calories**	**Calories**
11–14	2,200	2,700
15–18	2,100	2,800
19–22	2,100	2,900
23–50	2,000	2,700
51+	1,800	2,400

4 Mid-Chapter Review

Determine whether the first number is divisible by the second number. *(Lesson 4-1)*

1. 811; 5
2. 392; 4
3. 5,739; 3
4. 3,025; 9

Write the prime factorization for each number. *(Lesson 4-2)*

5. 28
6. 57
7. 72
8. 108

Identify each sequence as arithmetic, geometric, or neither. Then find the next three terms in each sequence. *(Lesson 4-3)*

9. 12, 19, 26, 33, . . .
10. 3, 15, 75, 375, . . .
11. 16, 2, $\frac{1}{4}$, $\frac{1}{32}$, . . .

Solve by making a list. *(Lesson 4-4)*

12. Show that the sum of the first ten Fibonacci numbers is divisible by 11. Then show that the sum of the second through the eleventh terms is also divisible by 11. *Recall the sequence: 1, 1, 2, 3, 5, 8, 13, 21, ...*

Find the GCF of each set of numbers. *(Lesson 4-5)*

13. 12, 56
14. 110, 215
15. 88, 383
16. 35, 72

Express each fraction in simplest form. *(Lesson 4-6)*

17. $\frac{18}{63}$
18. $\frac{7}{98}$
19. $\frac{228}{336}$
20. $\frac{120}{35}$

4-7 Fractions and Decimals

Objective

Express terminating decimals as fractions and express fractions as decimals.

Words to Learn

terminating decimal
repeating decimal

Are you a "lefty?" President Bush, Paul McCartney, and Martina Navratilova are all left-handed. Actually, about 3 out of every 25 people are left-handed. The fraction $\frac{3}{25}$ can be expressed as a decimal.

There are two methods for expressing a fraction as a decimal.

Method 1: Find an equivalent fraction with a denominator of 10, 100, or any other power of 10. This equivalent fraction can easily be expressed as a decimal.

$$\overset{\curvearrowright \times 4 \searrow}{\frac{3}{25}} = \frac{12}{100} = 0.12$$
$$\underset{\times 4 \nearrow}{}$$

Method 2: A fraction indicates division. Divide the numerator of the fraction by the denominator.

$$\frac{3}{25} \rightarrow \overset{0.12}{25\overline{)3.00}} \quad \text{or} \quad 3 \boxed{\div} 25 \boxed{=} \text{0.12}$$

This method may be more efficient than Method 1.

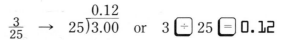

Examples

1 Express $\frac{33}{500}$ as a decimal by finding an equivalent fraction with a denominator of 10, 100, or 1,000.

$$\overset{\curvearrowright \times 2 \searrow}{\frac{33}{500}} = \frac{66}{1,000} = 0.066$$
$$\underset{\times 2 \nearrow}{}$$

2 Express $\frac{7}{8}$ as a decimal by dividing the numerator of the fraction by the denominator.

$$\frac{7}{8} \rightarrow \overset{0.875}{8\overline{)7.000}} \quad \text{or} \quad 7 \boxed{\div} 8 \boxed{=} \text{0.875}$$

In both of these examples, the result is a **terminating decimal.** Every terminating decimal can be written as a fraction with a denominator of 10, 100, 1,000, and so on. Examples 3–5 involve fractions whose decimal equivalent is a **repeating decimal.** A repeating decimal is a decimal whose digits repeat forever.

Express each fraction as a decimal using division.

3 $\frac{2}{3}$

Pencil and paper

$$
\begin{array}{r}
0.666... \\
3\overline{)2.000} \\
-18 \\
\hline
20 \\
-18 \\
\hline
2
\end{array}
$$

We will continue to get a remainder of 2.

Calculator

2 ÷ 3 = 0.6666667

A calculator can only show a certain number of digits. Even though the digit 6 will repeat forever, the calculator may round the final digit to 7.

Use a bar to indicate that the digit 6 repeats. So, $\frac{2}{3} = 0.\overline{6}$.

4 $\frac{5}{6}$

5 ÷ 6 = 0.8333333

$\frac{5}{6} = 0.8\overline{3}$

5 $\frac{23}{99}$

23 ÷ 99 = 0.2323232

$\frac{23}{99} = 0.\overline{23}$.

You can also express decimals as fractions. Express the decimal as a fraction with a denominator of the power of 10 indicated by the place value of the final digit of the decimal. Simplify the fraction.

$$0.12 = \frac{12}{100} = \frac{3}{25}$$ *The GCF of 12 and 100 is 4.*

Examples

Express each decimal as a fraction.

6 0.45

Write as a fraction. $0.45 = \frac{45}{100}$

Simplify. $= \frac{9}{20}$ *The GCF of 45 and 100 is 5.*

7 0.8

$0.8 = \frac{8}{10}$ or $\frac{4}{5}$

8 0.125

$0.125 = \frac{125}{1,000}$ or $\frac{1}{8}$

Checking for Understanding

Communicating Mathematics

Read and study the lesson to answer each question.

1. **Tell** how you know that $\frac{1}{5}$, $\frac{2}{5}$, $\frac{3}{5}$, and $\frac{4}{5}$ will all be terminating decimals.

2. **Tell** why 0.45 and 0.450 can be represented by the same fraction. Can 0.450 and 0.045 be represented by the same fraction? Why or why not?

Express each fraction as a decimal by finding an equivalent fraction with a denominator of 10, 100, or 1,000.

3. $\frac{4}{5}$ 4. $\frac{17}{50}$ 5. $\frac{6}{25}$ 6. $\frac{7}{250}$ 7. $\frac{56}{50}$

Express each fraction as a decimal by dividing. Use bar notation if necessary.

8. $\frac{14}{25}$ 9. $\frac{7}{40}$ 10. $1\frac{3}{8}$ 11. $\frac{7}{9}$ 12. $\frac{4}{11}$

Express each decimal as a fraction in simplest form.

13. 0.85   14. 0.625   15. 0.50   16. 0.83   17. 0.075

Exercises

Express each fraction as a decimal. Use bar notation if necessary.

18. $\frac{8}{25}$ 19. $\frac{11}{20}$ 20. $\frac{8}{250}$ 21. $\frac{6}{50}$ 22. $\frac{12}{200}$

23. $\frac{14}{16}$ 24. $\frac{7}{12}$ 25. $\frac{3}{8}$ 26. $\frac{17}{25}$ 27. $\frac{8}{11}$

28. $\frac{2}{3}$ 29. $\frac{20}{30}$ 30. $\frac{34}{125}$ 31. $\frac{7}{8}$ 32. $\frac{15}{9}$

Express each decimal as a fraction in simplest form.

33. 0.09   34. 0.64   35. 0.375   36. 8.407   37. 2.5

38. 0.10   39. 0.48   40. 0.540   41. 12.205   42. 21.48

43. Evaluate 3^4. *(Lesson 1-9)*

44. Order from least to greatest: 2.3, 0.23, 23, 1.6, 2.29, 0.22. *(Lesson 2-1)*

45. **Statistics** Angela collects the following heights (in inches) from the girls in her gym class: 62, 58, 68, 63, 64, 56, 65, 61, 70. Find the mean and the median height. *(Lesson 3-5)*

46. Use a factor tree to find the prime factorization of 245. *(Lesson 4-2)*

47. Express $\frac{39}{81}$ in simplest form. *(Lesson 4-6)*

48. **Taxes** A mill is a unit of money that is used in assessing taxes. One mill is $\frac{1}{1,000}$ of a U.S. dollar or $\frac{1}{10}$ of a cent. Use these facts to copy and complete the following table.

Mills	1,000	750	375	100	?	?	1
U.S. Dollars	1	?	?	?	1.50	?	?
Cents	100	?	?	?	?	12.5	?

49. **Critical Thinking** Which of the fractions, $\frac{1}{2}, \frac{1}{3}, \frac{1}{4}, \frac{1}{5}, \frac{1}{6}, \frac{1}{7}, \frac{1}{8}$, and $\frac{1}{9}$, are terminating decimals and which are repeating?

 50. **Journal Entry** Use long division to express $\frac{1}{9}$ and $\frac{2}{9}$ as decimals. Then, predict the decimal equivalents of $\frac{3}{9}, \frac{4}{9}, \frac{5}{9}, \frac{6}{9}, \frac{7}{8}$, and $\frac{8}{9}$. Use your calculator to check your predictions. Write a paragraph comparing your predictions to the decimals computed using your calculator.

4-8 **Simple Events**

Objective

Find the probability of a simple event.

Words to Learn

event
probability
random

Your school is raffling off a bicycle to raise money for new gym equipment. A total of 500 raffle tickets were sold. If your family bought 10 of the tickets, what is the probability that your family will win the bicycle?

The **event,** or specific outcome, we are interested in is your family winning the bicycle.

Probability	**In words:**	The probability of an event is the ratio of the number of ways an event can occur to the number of possible outcomes.
	In symbols:	$P(\text{event}) = \dfrac{\text{number of ways event occurs}}{\text{number of possible outcomes}}$

Outcomes occur at **random** if each outcome is equally likely to occur. Since the winning raffle ticket will be drawn at random,

$$P(\text{your family winning}) = \frac{\text{number of ways your family can win}}{\text{number of ways one ticket can be drawn}}$$

$$= \frac{10}{500} \qquad \textit{Your family has 10 tickets. 500 tickets were sold.}$$

$$= \frac{1}{50} \qquad \textit{Express the fraction in simplest form.}$$

Thus, your family has a 1-in-50 chance of winning the bicycle. This probability can also be expressed as the decimal 0.02.

Example 1

LOOKBACK

You can review prime numbers on page 132.

Find the probability of drawing a card with a prime number on it from a deck of cards numbered 1 to 24.

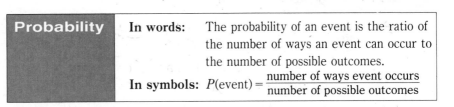

$$P(\text{prime}) = \frac{\text{number of ways of drawing a card with a prime number}}{\text{number of ways a card can be drawn}}$$

$$= \frac{9}{24} \qquad \textit{The primes are 2, 3, 5, 7, 11, 13, 17, 19, and 23. There are 24 cards.}$$

$$= \frac{3}{8} \qquad \textit{The GCF of 9 and 24 is 3.}$$

The probability of drawing a card with a prime number is $\frac{3}{8}$ or 0.375.

Example 2 *Problem Solving*

Business Burger House is having a contest. Each time you visit, you can place your name in a barrel. Each week they will draw one winner who will receive a $20 gift certificate. Burger House serves about 450 customers per day every day of the week. Lisa has put her name in the barrel twice this week. What is the probability that she will win? Assume half of the customers put their name in the barrel.

First find the total number of cards in the barrel for one week. Then divide by the number of cards that Lisa placed in the barrel.

number of cards: 450 ⌈÷⌉ 2 ⌈×⌉ 7 ⌈=⌉ **1,575**
number of cards with Lisa's name on them: 2

$$P(\text{Lisa}) = \frac{\text{number of cards with Lisa's name}}{\text{number of cards in the barrel}}$$

$$= \frac{2}{1,575}$$

$$= \frac{2}{1,575} \qquad 2 \;⌈÷⌉\; 1575 \;⌈=⌉\; \mathbf{0.0012698}$$

The probability of Lisa's name being drawn is $\frac{2}{1,575}$ or about 0.0013.

Mini-Lab

Work with a partner.

Materials: two number cubes

- Roll the number cubes 100 times.
- Make a frequency table and record the sum of the numbers on the two cubes.
- Copy and complete the table at the right. The table shows the sum of the numbers rolled on two cubes.

Number Cube A

Number Cube B	1	2	3	4	5	6
1	2	3	4	5	6	7
2	3	4	5	6		
3	4	5				
4						
5						
6						

Talk About It

a. What patterns do you see in the table?
b. What is the most frequent sum in the table?
c. What sum(s) did you roll most frequently?
d. What is the least frequent sum in the table?
e. What sum(s) did you roll least frequently?
f. There are 36 entries in the table and the number 6 is shown in five of them. This means that the probability of rolling a sum of 6 using two number cubes is $\frac{5}{36}$ or $0.13\overline{8}$. Find the probability for rolling each sum from 2 to 12.
g. Describe how your actual results of rolling the number cubes compare to the probabilities found in part f.

Checking for Understanding

Communicating Mathematics

Read and study the lesson to answer each question.

1. **Tell** how you could make sure that the winning ticket in a raffle is drawn at random.

2. **Make a model** showing all the ways a dime and a penny could land after they are tossed in the air.

3. **Tell** how you could figure the probability of being chosen at random to give your oral report on the first day if all 25 students in your history class have prepared oral reports, but only 5 students will report the first day.

Guided Practice

The spinner shown at the right is equally likely to stop on each of the regions numbered 1 to 12. Find the probability that the spinner will stop on each of the following.

4. an odd number

5. a factor of 12

6. a number less than 4

7. a prime number

A package of erasers contains 4 red, 3 orange, 6 green, and 5 yellow erasers. If you reach in the package and choose one eraser at random, what is the probability that you will select each of the following? Express each ratio as both a fraction and a decimal.

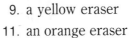

8. a red eraser

9. a yellow eraser

10. a green eraser

11. an orange eraser

Exercises

Independent Practice

A certain spinner is equally likely to stop on each of its regions numbered 1 to 24. Find the probability that the spinner will stop on each of the following.

12. an even number

13. a factor of 12

14. a factor of 24

15. a number less than 24

16. a composite number

17. the GCF of 9 and 15

A bag of marbles contains 15 red, 10 clear, 12 black, 16 blue, and 7 yellow marbles. If you reach in the bag and draw one marble at random, what is the probability that you will draw each of the following? Express each ratio as both a fraction and a decimal.

18. a yellow marble

19. a red marble

20. a blue marble

21. either a red or a black marble

22. a red, clear, or yellow marble

23. Evaluate $2(3+5) \div 4 - 2$. *(Lesson 1-7)*

24. **Smart Shopping** Suzanne buys 3.5 yards of fabric for $7.52. Find the price per yard rounded to the nearest cent. *(Lesson 2-8)*

25. Compute mentally 314×0.001. *(Lesson 2-5)*

26. Express $\frac{4}{9}$ as a decimal. *(Lesson 4-7)*

27. All the factors of 36 are written on separate cards. If you randomly choose a card, what is the probability of choosing a prime number?

28. **School** The students and faculty at South High School are selling 1,000 raffle tickets for $5 each. The prize is a new television set.

    a. If you spend $20 on tickets, what is the probability that you will win?

    b. How many tickets would you have to buy so that your probability of winning would be greater than your probability of losing? How much would it cost?

29. **Number Sense** Describe an event that has a probability of 1.

30. **Number Sense** Describe an event that has a probability of 0.

31. **Critical Thinking** If you know the probability of an event is 0.32, how can you determine the probability of the event *not* occurring? Explain.

Save Planet Earth

Test the Water The investigation of the White House drinking water in 1991 has led many Americans to worry about the safety of their drinking water. A 1990 EPA study estimated that at least one of six Americans drinks water containing an unacceptable amount of lead.

This may be a problem in school buildings as well. The solder that connects pipes in most plumbing systems is about 50% lead. In addition, many drinking fountains contain lead-lined cooling tanks or pipes. The longer the water sits unused in the lead fountains or pipes, the more lead may dissolve into it.

How You Can Help

- Let the water run before taking a drink.
- Ask school officials to check for lead in the drinking water and change the piping as necessary.
- Contact the EPA's toll-free drinking water hot line at 1-800-426-4791 for more information on water-testing.

4-9 Least Common Multiple

Objective

Find the least common multiple of two or more numbers.

Words to Learn

multiple
least common
   multiple (LCM)

Presidents of the United States are elected to four-year terms. United States Senators are elected to six-year terms. In 1988, a presidential election year, Robert Byrd of West Virginia was reelected to the United States Senate. If he continues to run for reelection each time his term expires, in what year will he again campaign during a presidential election year?

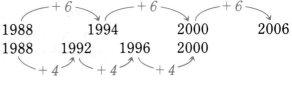

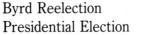

Byrd Reelection   1988   1994   2000   2006
Presidential Election   1988   1992   1996   2000

The next presidential election year that is also a possible election year for Senator Byrd is the year 2000.

When you multiply a number by the whole numbers 0, 1, 2, 3, 4, and so on, you get **multiples** of the number. The **least common multiple (LCM)** of two or more numbers is the least of the common positive multiples of all the numbers. The least common multiple of 4 years and 6 years is 12 years.

One of the following two methods is usually used to find the least common multiple of two or more numbers.

Method 1: List several multiples of each number. Then identify the common multiples and choose the least of these common multiples, the LCM.

Method 2: Write the prime factorization of each number. Identify all common prime factors. Then find the product of the prime factors using each common prime factor only once and any remaining factors. The product is the LCM.

Example 1

Problem-Solving Hint

• • • • • • • • • • • •

Make a list.

Find the LCM of 8 and 12 by listing the multiples of each number.

multiples of 8: 8, 16, **24**, 32, . . .   *Zero is a multiple of every number.*
multiples of 12: 12, **24**, 36, 48, . . .

The LCM of 8 and 12 is 24.

Notice that a factor need only appear in one prime factorization for it to be used to compute the LCM.

2 Find the LCM of 10, 12, and 15 by writing the prime factorization of each number.

$$10 = 2 \times 5$$
$$12 = 2 \times 2 \times 3 = 2^2 \times 3$$
$$15 = 3 \times 5$$

The LCM of 10, 12, and 15 is $2^2 \times 3 \times 5$ or 60.

3 Mentally compute the LCM of 5, 6, and 10 by listing multiples.

Think: The positive multiples of 10 are 10, 20, 30, 40, ...

Ask: What is the least of these multiples that is divisible by both 5 and 6? It's 30.

So, the LCM of 5, 6, and 10 is 30.

4 Use your calculator to find the LCM of 63 and 14.

Write the multiples of the greater number, 63: 63, 126, 189, . . .
Find the least multiple that is divisible by 14:

$$63 \div 14 = 4.5 \qquad 126 \div 14 = 9 \quad \checkmark$$

Thus, the LCM of 63 and 14 is 126.

Checking for Understanding

Communicating Mathematics

Read and study the lesson to answer each question.

1. **Tell** why the LCM of 24 (which is $2^3 \times 3$) and 54 (which is 2×3^3) must have factors of 2^3 and 3^3. What is the LCM of 24 and 54?

2. **Write** a description of the types of problems where you could easily find the LCM mentally.

Guided Practice

Find the LCM for each set of numbers by listing the multiples of each number.

3. 60, 12   4. 30, 15   5. 2, 3, 5   6. 20, 30, 50

Find the LCM of each set of numbers by writing the factorization of each number.

7. 6, 12, 18   8. 17, 6, 34   9. 22, 11, 4   10. 35, 25, 49

Exercises

Independent Practice

Find the LCM of each set of numbers.

11. 3, 15   12. 16, 176   13. 4, 10, 9   14. 55, 44, 33

15. 24, 12, 6   16. 42, 16, 7   17. 300, 18   18. 625, 30

19. 60, 80   20. 12, 15   21. 6, 9, 12   22. 10, 12, 15

Mixed Review

23. Estimate the sum of 125 and 2,347. *(Lesson 1-2)*

24. Multiply 2.6 by 3.15. *(Lesson 2-3)*

25. **Statistics** Construct a stem-and-leaf plot for the following data: 12, 17, 23, 5, 9, 25, 13, 16, 2, 25, 31, 10. *(Lesson 3-6)*

26. **Probability** Ryu purchases 5 tickets in support of a raffle to raise money for intramural sports at his high school. A total of 500 tickets are sold. One ticket is to be selected to win the grand prize, which is a season pass to the Washington Redskins football games. What is the probability that Ryu will win the season pass? *(Lesson 4-8)*

Problem Solving and Applications

27. **Health** Robbie visits the dentist every 6 months. He sees his optometrist every 18 months and gets a physical for football every August 1. If he schedules all his appointments on August 1 this year, in how many years will they all fall on August 1 again?

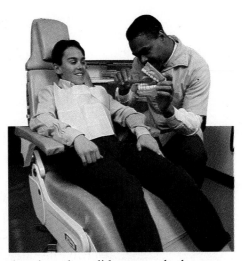

28. **Scheduling** Sam, Lilly, and Tom were hired to work a regular schedule in the evenings at the library, which is open every day. They each started work on the same day, but they did not work the same evening again until 30 days later. Sam and Lilly worked the same evening every 6 days, and Sam and Tom worked the same evening every 10 days. If Sam worked every second day, what schedules did Lilly and Tom have?

29. **Number Sense** When will the LCM of two numbers be one of the numbers?

30. **Number Sense** When will the LCM of two numbers be the product of the two numbers?

31. **Critical Thinking** Write a set of three numbers whose LCM is the product of the numbers.

32. **Critical Thinking** Write two numbers whose LCM is 225 and whose GCF is 15.

4-10 Comparing and Ordering Fractions and Decimals

Objective

Compare and order fractions by first writing them as equivalent fractions with a common denominator.

Words to Learn

common denominator
least common denominator (LCD)

In Ms. Mapp's geography class, Laura has earned 30 points out of a possible 35 points on tests, projects, and oral reports. In English class she worked hard writing her short story and book report, earning 42 out of a possible 48 points. In which class has Laura earned a greater portion of the possible points?

That is, which fraction is greater, $\frac{30}{35}$ or $\frac{42}{48}$?

One way to compare these two fractions is to first write each fraction in simplest form.

$$\frac{30}{35} \overset{\div 5}{\underset{\div 5}{=}} \frac{6}{7} \qquad \frac{42}{48} \overset{\div 6}{\underset{\div 6}{=}} \frac{7}{8}$$

To compare $\frac{6}{7}$ and $\frac{7}{8}$, rewrite each fraction using the same denominator. Then you need only compare the numerators.

A **common denominator** is a common multiple of the denominators of two or more fractions. The **least common denominator (LCD)** is the least common multiple (LCM) of the denominators of two or more fractions.

To rewrite $\frac{6}{7}$ and $\frac{7}{8}$ with the same denominators, first find the LCD by listing the multiples of each denominator.

multiples of 7: 7, 14, 21, 28, 35, 42, 49, **56**, ...
multiples of 8: 8, 16, 24, 32, 40, 48, **56**, ...

The LCD of the fractions $\frac{6}{7}$ and $\frac{7}{8}$ is 56, since 56 is the LCM of 7 and 8. So, rewrite each fraction using a denominator of 56.

$$\frac{6}{7} = \frac{6 \times 8}{7 \times 8} = \frac{48}{56} \qquad \frac{7}{8} = \frac{7 \times 7}{8 \times 7} = \frac{49}{56}$$

Now, compare $\frac{49}{56}$ and $\frac{48}{56}$. Since $49 > 48$, then $\frac{49}{56} > \frac{48}{56}$, and Laura has earned a greater portion of the possible points in English than in geography.

DID YOU KNOW

One-fourth of U.S. schools include geography in their curriculum.

Three strategies can be used to compare fractions. The first strategy was used on the previous page.

1. Express each fraction in simplest form. Then write equivalent fractions using the LCD.
2. Consider each fraction in relationship to the nearest whole number.
3. Express each fraction as a decimal.

Examples

1 Compare $\frac{7}{15}$ and $\frac{4}{10}$ by writing equivalent fractions using the least common denominator.

The LCM of 15 and 10 is 30.

$$\frac{7}{15} = \frac{7 \times 2}{15 \times 2} = \frac{14}{30} \qquad \frac{4}{10} = \frac{4 \times 3}{10 \times 3} = \frac{12}{30}$$

Since $\frac{12}{30} < \frac{14}{30}$, then $\frac{4}{10} < \frac{7}{15}$.

2 The diameter of an Oreo® cookie is $1\frac{3}{4}$ inches long. The diameter of an Archway® gingersnap cookie is $1\frac{3}{16}$ inches long. Which cookie has the longer diameter?

Since both diameters are at least one inch long, we will compare only the fractions. Let's use strategy 2.

$\frac{3}{16}$ is nearest to 0. $\frac{3}{4}$ is nearest to 1.

So, $\frac{3}{4} > \frac{3}{16}$ and $1\frac{3}{4} > 1\frac{3}{16}$.

The Oreo® cookie has the longer diameter.

Example 3   *Problem Solving*

Sports Kiesha was proud that her soccer team had won 7 of their 10 matches. Her cousin wrote her a letter announcing that his team had won 11 of their 16 matches. Whose team won a greater fraction of their matches?

Since $\frac{7}{10}$ and $\frac{11}{16}$ are already in simplest form and yet are still difficult to compare, express each fraction as a decimal and then compare (strategy 3).

$$7 \boxed{\div} 10 \boxed{=} 0.7 \qquad\qquad 11 \boxed{\div} 16 \boxed{=} 0.6875$$

Since $0.7 > 0.6875$, then $\frac{7}{10} > \frac{11}{16}$ and Kiesha's team won a greater fraction of their matches.

Work with a partner.

The figure below is called a **Venn diagram.** The circle at the lower left contains all fractions greater than $\frac{1}{3}$. The circle at the lower right contains all fractions less than 1. Thus the region labeled G, where only these two circles overlap, contains all fractions that are greater than $\frac{1}{3}$ *and* less than 1.

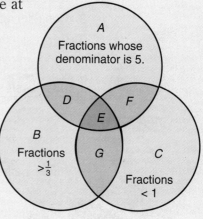

Talk About It

a. Write a description of the fractions contained in each of regions A through F.

b. For each of regions A through F, name a fraction that would be located in that region.

c. Identify the region where each of the following fractions would be located.

$$\frac{1}{5}, \frac{2}{9}, \frac{8}{5}, \frac{7}{8}, \frac{12}{7}, \frac{4}{5}, \frac{1}{3}, \frac{3}{3}$$

d. Which regions contain an infinite number of fractions? a finite number of fractions? no fractions?

Checking for Understanding

Communicating Mathematics

Read and study the lesson to answer each question.

1. **Write** a problem that involves comparing fractions and state the strategy that would be easiest to use to make the comparison.

2. **Tell** how you can use a calculator to compare fractions.

Guided Practice

Find the LCD for each pair of fractions.

3. $\frac{5}{9}, \frac{7}{12}$ 4. $\frac{7}{5}, \frac{14}{11}$ 5. $\frac{3}{13}, \frac{4}{26}$ 6. $\frac{6}{7}, \frac{13}{16}$

Replace each ⬤ with < or > to make a true statement.

7. $\frac{5}{8}$ ⬤ $\frac{2}{3}$ 8. $\frac{9}{13}$ ⬤ $\frac{14}{20}$ 9. $\frac{5}{9}$ ⬤ $\frac{8}{15}$ 10. $\frac{3}{4}$ ⬤ $\frac{7}{8}$

Exercises

Independent Practice

Find the LCD for each pair of fractions.

11. $\frac{2}{5}, \frac{2}{6}$ 12. $\frac{4}{5}, \frac{8}{9}$ 13. $\frac{5}{4}, \frac{9}{8}$ 14. $\frac{11}{16}, \frac{3}{4}$

15. $\frac{3}{10}, \frac{1}{3}$ 16. $\frac{7}{8}, \frac{23}{24}$ 17. $\frac{13}{17}, \frac{3}{4}$ 18. $\frac{2}{3}, \frac{19}{27}$

19. $\frac{1}{6}, \frac{2}{15}$ 20. $\frac{3}{10}, \frac{1}{12}$ 21. $\frac{6}{7}, \frac{15}{21}$ 22. $\frac{4}{5}, \frac{6}{8}$

Replace each ● with $<$, $>$, or $=$ to make a true statement.

23. $\frac{8}{13}$ ● $\frac{8}{17}$ 24. $\frac{17}{20}$ ● $\frac{36}{50}$ 25. $\frac{45}{90}$ ● $\frac{15}{30}$ 26. $\frac{4}{3}$ ● $\frac{8}{7}$

27. $\frac{4}{5}$ ● $\frac{6}{7}$ 28. $\frac{1}{3}$ ● $\frac{3}{5}$ 29. $\frac{5}{6}$ ● $\frac{7}{8}$ 30. $\frac{5}{8}$ ● $\frac{4}{7}$

31. $\frac{3}{10}$ ● $\frac{1}{5}$ 32. $\frac{4}{7}$ ● $\frac{5}{8}$ 33. $\frac{5}{12}$ ● $\frac{3}{8}$ 34. $\frac{2}{9}$ ● $\frac{5}{14}$

Mixed Review

35. Evaluate $15(xy) - (x + y)$ if $x = 4$ and $y = 1$. *(Lesson 1-8)*

36. **Sports** During a basketball game, it is announced that the attendance for the game is 10,300. Express the attendance in scientific notation. *(Lesson 2-6)*

37. **Statistics** Choose an appropriate scale and interval for the following set of data. Then construct a number line displaying the scale and interval. 1.2, 3.4, 2.7, 4.3, 1.9, 2.5, 3.7, 1.8 *(Lesson 3-3)*

38. Find the prime factorization of 255. *(Lesson 4-2)*

39. Find the least common multiple of 27 and 30. *(Lesson 4-9)*

Problem Solving and Applications

Statistics Jolie's math class uses cooperative learning groups to study mathematics. Each group keeps score based on points earned out of number of points attempted. Find the median score for each group by ranking the scores from least to greatest and locating the middle score.

40. **Group 1:** $\frac{3}{7}, \frac{4}{7}, \frac{4}{8}, \frac{5}{8}, \frac{5}{6}$ 41. **Group 2:** $\frac{7}{10}, \frac{8}{9}, \frac{4}{6}, \frac{9}{11}, \frac{7}{9}$

42. **Civics** Amy and Jose are handing out pamphlets urging people to register to vote. Amy has delivered 72 pamphlets of the 108 pamphlets assigned to her. Jose has delivered 84 pamphlets of the 126 pamphlets assigned to him. Which student has completed more of the assignment?

43. **Critical Thinking** When is the least common denominator of two fractions equal to one of the denominators? Give two examples.

44. **Number Sense** Which fraction is nearest to 2? Explain why. $1\frac{15}{16}, \frac{63}{32}, \frac{17}{8}$

45. **School** If the brass section makes up $\frac{4}{9}$ of the marching band and the woodwind section makes up $\frac{2}{7}$, which section is larger?

46. **Data Search** Refer to pages 126 and 127.
   a. What fraction of the U.S. forestland is in the West? North? South?
   b. Is there a greater portion of forestland in the North or in the South?

4 Study Guide and Review

Communicating Mathematics

Choose the correct term or number to complete each sentence.

1. The number 72 is said to be divisible by 8 because the quotient is a(n) __?__ of 72.

2. A(n) __?__ number is a whole number greater than 1 that has exactly two factors, 1 and itself.

3. Every composite number can be written as the __?__ of at least two prime numbers.

4. In a geometric sequence, you can always find the next term by __?__ the previous term by the same number.

5. A fraction is in simplest form when the GCF of the numerator and the denominator is __?__ .

6. The __?__ is the least common multiple of the denominators of two or more fractions.

7. In your own words, explain the difference between a terminating decimal and a repeating decimal.

GCF
adding
factor
multiple
10
whole number
prime
sum
LCM
1
composite
LCD
product
multiplying

Skills and Concepts

Objectives and Examples

Upon completing this chapter, you should be able to:

Review Exercises

Use these exercises to review and prepare for the chapter test.

- use divisibility rules *(Lesson 4-1)*
  Determine whether 336 is divisible by 2, 3, 4, 5, or 6.

  2: The ones digit, 6, is even, so 336 is divisible by 2.

  3: The sum of the digits, 12, is divisible by 3, so 336 is divisible by 3.

  4: The number formed by the last two digits, 36, is divisible by 4, so 336 is divisible by 4.

  5: The ones digit is not 5 or 0, so 336 is not divisible by 5.

  6: The number is divisible by both 2 and 3, so 336 is divisible by 6.

Determine whether each number is divisible by 2, 3, 4, 5, 6, 9, or 10.

8. 221	9. 1,225
10. 630	11. 1,300
12. 828	13. 707
14. 452	15. 594
16. 255	17. 93

Objectives and Examples

- find the prime factorization of a composite number *(Lesson 4-2)*

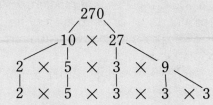

The prime factorization of 270 is $2 \times 3^3 \times 5$.

- recognize and extend sequences *(Lesson 4-3)*

32, 16, 8, 4, 2, . . . is a geometric sequence created by multiplying the previous term by $\frac{1}{2}$. The next three terms are $1, \frac{1}{2}, \frac{1}{4}$.

- find the greatest common factor of two or more numbers *(Lesson 4-5)*

$45 = \boxed{3} \times 3 \times \boxed{5}$
$75 = \boxed{3} \times 5 \times \boxed{5}$

The GCF of 45 and 75 is 3×5 or 15.

- express fractions in simplest form *(Lesson 4-6)*

Write $\frac{45}{81}$ in simplest form.

$\frac{45}{81} = \frac{45 \div 9}{81 \div 9} = \frac{5}{9}$ *The GCF of 45 and 81 is 9.*

- express terminating decimals as fractions and express fractions as decimals *(Lesson 4-7)*

$\frac{6}{20} = \frac{30}{100} = 0.30$

$\frac{1}{6} \rightarrow 6)\overline{1.000} \rightarrow 0.1\overline{6}$
$\phantom{\frac{1}{6} \rightarrow 6)}\underline{-6}$
$\phantom{\frac{1}{6} \rightarrow 6)}40$
$\phantom{\frac{1}{6} \rightarrow 6)}\underline{-36}$
$\phantom{\frac{1}{6} \rightarrow 6)}40$

$0.32 = \frac{32}{100} = \frac{8}{25}$ *The GCF of 32 and 100 is 4.*

Review Exercises

Write the prime factorization of each number.

18. 1,000
19. 144
20. 950
21. 77
22. 96
23. 300
24. 2,800
25. 1,450

Identify each sequence as arithmetic, geometric, or neither. Then find the next three terms.

26. 16, 21, 26, 31, 36, . . .
27. 1, 4, 16, 64, 256, . . .
28. 0, 2, 6, 12, 20, 30, . . .
29. 10, 100, 1,000, 10,000, . . .

Find the GCF of each set of numbers.

30. 36, 81
31. 40, 65
32. 252, 336
33. 57, 240
34. 56, 280, 400

Express each fraction in simplest form.

35. $\frac{56}{70}$
36. $\frac{250}{750}$
37. $\frac{26}{39}$
38. $\frac{60}{18}$
39. $\frac{77}{121}$
40. $\frac{57}{95}$

Express each fraction as a decimal and express each decimal as a fraction in simplest form.

41. $\frac{10}{25}$
42. 0.36
43. $\frac{3}{8}$
44. 1.25
45. $\frac{5}{9}$
46. $\frac{33}{300}$

Objectives and Examples	Review Exercises

- find the probability of a simple event *(Lesson 4-8)*

  When a die is rolled,

  $P(\text{odd}) = \frac{3}{6}$

  $\qquad = \frac{1}{2}$ or 0.5

A bag contains 6 red, 3 pink, and 3 white bows. If you draw a bow at random, what is the probability of drawing each of the following? Express each ratio as both a fraction and a decimal.

47. red
48. either red or white

- find the least common multiple of two or more numbers *(Lesson 4-9)*

  $4 = 2 \times 2 = 2^2$

  $18 = 2 \times 3 \times 3 = 2 \times 3^2$

  The LCM of 4 and 18 is $2^2 \times 3^2$ or 36.

Find the LCM of each set of numbers.

49. 6, 15
50. 42, 56
51. 16, 40
52. 15, 125, 600
53. 21, 81, 147
54. 48, 81, 270

- compare and order fractions *(Lesson 4-10)*

  $\frac{5}{9} = \frac{10}{18}$ $\frac{4}{6} = \frac{12}{18}$

  Since $10 < 12$, then $\frac{5}{9} < \frac{4}{6}$.

Replace each ● with $<$ or $>$ to make a true statement.

55. $\frac{2}{3}$ ● $\frac{3}{4}$
56. $\frac{11}{12}$ ● $\frac{8}{9}$
57. $\frac{3}{8}$ ● $\frac{5}{12}$
58. $\frac{7}{10}$ ● $\frac{13}{25}$

Applications and Problem Solving

59. Find the greatest common divisor of the twelfth and fifteenth terms of Fibonacci sequence, 1, 1, 2, 3, 5, 8, 13, 21, ... *(Lesson 4-4)*

60. **School**   Beth scored 21 out of 25 on her spelling test. Ted scored 37 out of 40 on his test. Who scored higher? *(Lesson 4-10)*

Curriculum Connection Projects

- **Communications**   Open a telephone book ten times at random and write down the page numbers. Find the prime factors for each page number listed.

- **Automotive**   Check auto repair manuals to find the cylinder firing sequences for three different 6-cylinder cars.

Read More About It

Cooper, Clare. *Ashar of Qarius.*
Pluckrose, Henry. *Know About Patterns.*
Charosh, Mannis. *Mathematical Games for One or Two.*

4 Test

Determine whether each number is divisible by 2, 3, 4, 5, 6, 9, or 10.

1. 639   2. 2,350

Write the prime factorization of each number.

3. 250   4. 1,296   5. 2,400

6. Find two terms in the Fibonacci sequence that are divisible by the seventh term.
   1, 1, 2, 3, 5, 8, 13, . . .

Identify each sequence as arithmetic, geometric, or neither. Then find the next three terms.

7. 9, 15, 21, 27, 33, . . .   8. 2, 6, 18, 54, 162, . . .

9. **Savings** Elena plans to open a savings account with $50 from her January paycheck and then increase the amount she deposits by $5 each month. How much will Elena deposit from her April check?

Find the GCF of each set of numbers.

10. 52, 100   11. 95, 150, 345

Express each fraction in simplest form.

12. $\frac{33}{55}$ 13. $\frac{24}{64}$ 14. $\frac{60}{135}$

15. **Vacation** Mina spent 8 of her 14 vacation days in Florida. Express this fraction of her vacation in simplest form.

Express each fraction as a decimal and each decimal as a fraction in simplest form.

16. $\frac{28}{70}$ 17. 0.32   18. $\frac{4}{20}$

The spinner shown at the right is equally likely to stop on each of the regions. Find the probability that the spinner will stop on each of the following.

19. a prime number   20. a factor of 24

Find the LCM of each set of numbers.

21. 8, 28   22. 14, 21, 27

Replace each ● with < or > to make a true statement.

23. $\frac{5}{8}$ ● $\frac{12}{20}$ 24. $\frac{12}{15}$ ● $\frac{9}{12}$

25. **Civics** On election day in a small town, 175 of the 200 registered Republicans voted, and 160 of the 200 registered Democrats voted. Which party had the better turnout?

Bonus Why is 2 the only even prime number?

Applications with Fractions

Spotlight on Business

Have You Ever Wondered. . .

- What it means when a news reporter says that IBM closed at $94\frac{3}{8}$, down $3\frac{1}{2}$?

- How newspapers, radio stations, and movie theaters can be different divisions of the same industry?

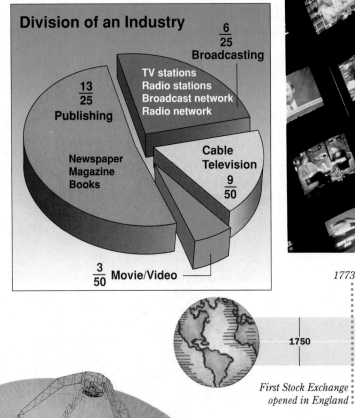

Division of an Industry

$\frac{6}{25}$ **Broadcasting**

TV stations
Radio stations
Broadcast network
Radio network

$\frac{13}{25}$
Publishing

Newspaper
Magazine
Books

Cable Television

$\frac{9}{50}$

$\frac{3}{50}$ **Movie/Video**

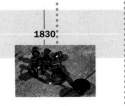

1773 *1792* *1835* *1867*

1750 **1790** **1830** **1870**

*First Stock Exchange
opened in England*

*New York Stock
Exchange organized*

*Stock tickers, which print
on ticker tape, introduced*

Chapter Project

Business

Work in a group.

1. Use the financial pages of a newspaper to choose four stocks. In a table, keep track of your stocks for one month on a daily basis.

2. Note which stocks increased and decreased overall.

3. Pretend that you bought 100 shares of each of the four stocks. If you sold them all at the end of the month, would you have gained or lost money?

4. To calculate price, multiply the number of shares by the closing price. Compare the price on the first day with that on the last day.

Looking Ahead

In this chapter, you will see how mathematics can be used to answer the questions about business and the stock market. The major objectives of the chapter are to:

- express mixed numbers as improper fractions and vice versa

- add and subtract fractions and mixed numbers

- find perimeters and circumferences

- multiply and divide fractions and mixed numbers

- find expected values of outcomes

New York Exchange

	Sales (In 100s)	Close	Chng
		54 5/8	-1/2
Amer.Cyan.	3508	36 1/2	-1/2
AT&T	11089	40 1/2	+1/8
Avon	3338	28 7/8	-1/4
BankNY	246	58 3/4	-3/8
BectnD	1222	21 5/8	-1/2
ChemBank	3692	23 5/8	—
Champion	1949	11 3/4	
1/2Chrysler	4454	10 5/8	+1/2
Citicorp	16045	24 7/8	-3/4
FedPaper	346	37 5/8	+3 1/2
Gannett	1464	94 3/8	+ 1/4
IBM	10014	67 3/8	+1/2
IntPaper	2482	4 1/8	+1/2
LoneStar	219	75 3/8	-1/8
Nynex	812	37 3/4	+1/2
O&R	44	3 7/8	-3/4
ParPharm	122	37 7/8	-1
Para Comm	1110	36	-1 1/8
Sears	2406	41	—
Sequa	183	1 1/2	-1/2
USG	479	14 3/8	
U. Water	90	61 5/8	
Xerox	1600		

New York Times.

STOCKS COLLAPSE IN 16,410,030-SHARE DAY,
BUT RALLY AT CLOSE CHEERS BROKERS;
BANKERS OPTIMISTIC TO CONTINUE AID

1897			1967	1987	
	1929		1977		
1910		**1950**		**1990**	

Stock market crashes 508 points in one day

Oil begins flowing through the Alaska Pipeline

First American subway completed in Boston, MA

First electronic hand-held calculator, developed at Texas Instruments in 1967

5-1 Mixed Numbers and Improper Fractions

Objective

Change mixed numbers to improper fractions and vice versa.

Words to Learn

mixed number
proper fraction
improper fraction

How much sleep do you get at night? Doctors recommend that we get 8 to $8\frac{1}{2}$ hours of sleep. The average American, however, gets only $7\frac{1}{5}$ hours.

Mixed numbers such as $8\frac{1}{2}$ and $7\frac{1}{5}$ indicate the sum of a whole number and a fraction.

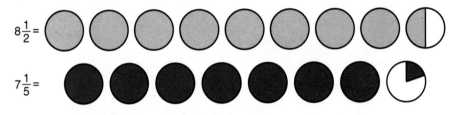

$8\frac{1}{2} =$

$7\frac{1}{5} =$

A fraction that has a numerator less than the denominator like each of those given below, is a **proper fraction.**

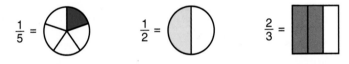

$\frac{1}{5} =$ $\frac{1}{2} =$ $\frac{2}{3} =$

A fraction that has a numerator that is greater than or equal to the denominator is an **improper fraction.** Here are three examples.

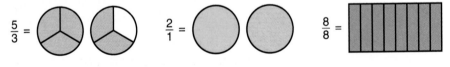

$\frac{5}{3} =$ $\frac{2}{1} =$ $\frac{8}{8} =$

Mixed numbers can be written as improper fractions, as shown below.

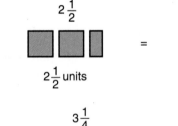

$2\frac{1}{2}$ $\frac{5}{2}$

$2\frac{1}{2}$ units   =   5 halves

$3\frac{1}{4}$ $\frac{13}{4}$

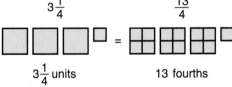

$3\frac{1}{4}$ units   13 fourths

To change a mixed number to an improper fraction, multiply the whole number and the denominator. Then add the numerator. Write the sum over the denominator.

Change each mixed number to an improper fraction.

1 $2\frac{1}{2}$

$$2\frac{1}{2} = \frac{(2 \times 2) + 1}{2}$$
$$= \frac{5}{2}$$

2 $3\frac{2}{5}$

$$3\frac{2}{5} = \frac{(5 \times 3) + 2}{5}$$
$$= \frac{17}{5}$$

A whole number can be changed to an improper fraction.

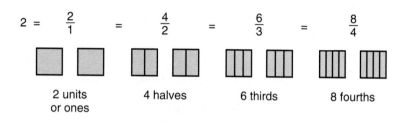

$$2 = \quad \frac{2}{1} \quad = \quad \frac{4}{2} \quad = \quad \frac{6}{3} \quad = \quad \frac{8}{4}$$

| 2 units or ones | 4 halves | 6 thirds | 8 fourths |

An improper fraction can be changed to either a whole number or a mixed number.

$$\frac{7}{2} \qquad\qquad 3\frac{1}{2}$$

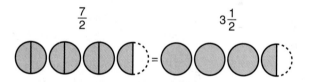

To change an improper fraction to a whole number or a mixed number, divide the numerator by the denominator.

Examples

Change each improper fraction to a mixed number in simplest form or a whole number.

3 $\frac{23}{4}$

$$\begin{array}{r} 5 \\ 4\overline{)23} \\ \underline{20} \\ 3 \end{array} \quad \rightarrow \quad 5\frac{3}{4}$$

So, $\frac{23}{4} = 5\frac{3}{4}$.

4 $\frac{21}{7}$

$$\begin{array}{r} 3 \\ 7\overline{)21} \\ \underline{21} \\ 0 \end{array} \quad \rightarrow \quad 3$$

So, $\frac{21}{7} = 3$.

Checking for Understanding

Read and study the lesson to answer each question.

1. **Make a model** to show the meaning of $\frac{7}{4}$.

2. **Write**, in your own words, how to change a mixed number to an improper fraction.

3. **Tell** what number this model represents.

Guided Practice

Identify each number as a proper fraction, an improper fraction, or a mixed number.

4. $\frac{7}{3}$ 5. $2\frac{1}{3}$ 6. $\frac{3}{8}$ 7. $\frac{4}{4}$ 8. $\frac{15}{11}$

Complete the diagram to change each number to an improper fraction or a mixed number.

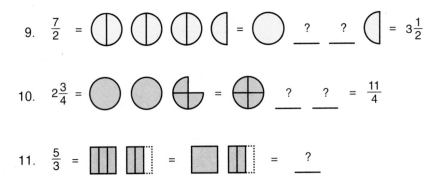

9. $\frac{7}{2}$ =   =   ?   ?   = $3\frac{1}{2}$

10. $2\frac{3}{4}$ =   =   ?   ?   = $\frac{11}{4}$

11. $\frac{5}{3}$ =   =   =   ?

Change each improper fraction to a mixed number in simplest form or a whole number.

12. $\frac{9}{2}$ 13. $\frac{12}{4}$ 14. $\frac{14}{6}$ 15. $\frac{7}{7}$ 16. $\frac{45}{10}$

Change each mixed number or whole number to an improper fraction.

17. $1\frac{7}{8}$ 18. $2\frac{3}{4}$ 19. 3   20. $4\frac{5}{9}$ 21. $13\frac{2}{3}$

Exercises

Identify each number as a proper fraction, an improper fraction, or a mixed number.

22. $\frac{9}{9}$ 23. $2\frac{6}{7}$ 24. $\frac{8}{9}$ 25. $\frac{5}{3}$ 26. $6\frac{1}{3}$

Change each improper fraction to a mixed number in simplest form or a whole number.

27. $\frac{9}{7}$ 28. $\frac{7}{4}$ 29. $\frac{12}{5}$ 30. $\frac{8}{8}$ 31. $\frac{20}{8}$

32. $\frac{17}{5}$ 33. $\frac{6}{2}$ 34. $\frac{21}{9}$ 35. $\frac{26}{6}$ 36. $\frac{23}{7}$

Change each mixed number or whole number to an improper fraction.

37. $3\frac{4}{5}$ 38. $2\frac{2}{3}$ 39. 5   40. $6\frac{1}{4}$ 41. $3\frac{7}{8}$

42. $2\frac{2}{9}$ 43. $4\frac{3}{8}$ 44. $3\frac{6}{7}$ 45. $2\frac{7}{10}$ 46. $3\frac{5}{12}$

47. Change $5\frac{4}{7}$ to an improper fraction.

48. Change $\frac{27}{6}$ to a mixed number in simplest form.

Mixed Review 49. Use front-end estimation to estimate $12,768 − $3,428. *(Lesson 1-2)*

50. **Pets**   Tom's dog Echo will stay for 7.5 minutes after Tom issues the command "stay." Kathy's dog Sam will stay for 2.3 times longer than Echo. How long will Sam stay? *(Lesson 2-4)*

51. Divide $0.0081 ÷ 0.09$. *(Lesson 2-7)*

52. **Statistics**   Draw a number line to show the scale and interval for test scores of 76, 85, 99, 45, 82, 70, and 94. *(Lesson 3-3)*

53. Find the next three terms in the following sequence. 2, 5, 12.5, 31.25, . . . *(Lesson 4-5)*

54. Order $\frac{1}{2}$, $\frac{7}{8}$, $\frac{1}{16}$, $\frac{5}{6}$, and $\frac{2}{3}$ from least to greatest. *(Lesson 4-10)*

Problem Solving and Applications 55. **Business**   Ruth's Cafe sold 28 pieces of apple pie today. If each piece was an eighth of a pie, how many apple pies did they sell?

56. **Music**   How many whole notes are there in eight quarter notes? *Hint: quarter means $\frac{1}{4}$.*

57. **Journalism**   Ren sold ten quarter-page ads for the school newspaper. How many pages of advertisement did she sell?

58. **Coin Collection**   Elise has $10 worth of quarters and Juan has $10 in half-dollar coins. Which one has the greatest number of coins?

59. **Critical Thinking**   If $5 + \frac{1}{4}$ can be written as the improper fraction $\frac{21}{4}$, how would you write $c + \frac{1}{4}$ as an improper fraction?

60. **Journal Entry**   Use circle models to show that $1\frac{3}{4} = \frac{7}{4}$.

Estimating with Fractions

Objective

Estimate sums, differences, products, and quotients of fractions and mixed numbers.

Marie plans to make three loaves of banana bread. The recipe for a single loaf calls for $1\frac{3}{4}$ cups of flour. To make sure she has enough flour before starting, she estimates how much flour will be needed for all three loaves. Marie estimates the product of 3 and $1\frac{3}{4}$.

To estimate the sum, difference, or product of mixed numbers, round each mixed number to the nearest whole number.

$$3 \times 1\frac{3}{4} \rightarrow 3 \times 2 = 6$$

Marie will need *about* six cups of flour.

Example 1

Estimate the sum of $6\frac{1}{2}$ and $2\frac{1}{3}$.

$$6\frac{1}{2} + 2\frac{1}{3} \rightarrow 7 + 2 = 9$$ *When a mixed number contains $\frac{1}{2}$, the number is rounded up.*

$6\frac{1}{2} + 2\frac{1}{3}$ is *about* 9.

Estimation Hint

• • • • • • • • • • • • • •

A fraction is close to 1 when the numerator and denominator are close in value. A fraction is close to $\frac{1}{2}$ when the numerator is about half of the denominator. A fraction is close to 0 when the numerator is much smaller than the denominator.

To estimate the sum or difference of proper fractions, round each fraction to 0, $\frac{1}{2}$, or 1, whichever is closest.

Examples

2 Estimate $\frac{3}{8} + \frac{6}{7}$.

$$\frac{3}{8} + \frac{6}{7} \rightarrow \frac{1}{2} + 1 = 1\frac{1}{2}$$

$\frac{3}{8} + \frac{6}{7}$ is *about* $1\frac{1}{2}$.

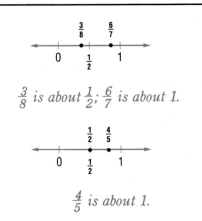

$\frac{3}{8}$ *is about* $\frac{1}{2}$; $\frac{6}{7}$ *is about 1.*

3 Estimate $\frac{4}{5} - \frac{1}{2}$.

$$\frac{4}{5} - \frac{1}{2} \rightarrow 1 - \frac{1}{2} = \frac{1}{2}$$

$\frac{4}{5} - \frac{1}{2}$ is *about* $\frac{1}{2}$.

$\frac{4}{5}$ *is about 1.*

You can also use compatible numbers to estimate with fractions.

You can review compatible numbers on page 11.

Examples

Estimate each product or quotient.

4 $\frac{1}{3} \times 14$ *$\frac{1}{3} \times 14$ means $\frac{1}{3}$ of 14.*

$\frac{1}{3}$ of 14 $\rightarrow$ $\frac{1}{3}$ of 15 or 5   *3 and 15 are compatible numbers.*

$\frac{1}{3}$ of 14 is *about* 5.

5 $34\frac{3}{4} \div 5\frac{1}{2}$

First, round $5\frac{1}{2}$ to 6. Then replace the dividend with a compatible number.

$34\frac{3}{4} \div 5\frac{1}{2}$ $\rightarrow$ $34\frac{3}{4} \div 6$

$\rightarrow$ $36 \div 6$ or 6   *6 and 36 are compatible numbers.*

$34\frac{3}{4} \div 5\frac{1}{2}$ is *about* 6.

Checking for Understanding

Communicating Mathematics

Read and study the lesson to answer each question.

1. **Tell** whether each point marked on the number line is closest to 0, $\frac{1}{2}$, or 1.

$$\frac{1}{7} \quad \frac{4}{9} \quad \frac{5}{7} \quad \frac{7}{8}$$
$$0 \qquad \frac{1}{2} \qquad 1$$

2. **Write** how you would use rounding to estimate $\frac{1}{2} + \frac{5}{6}$.

3. **Tell** how you would use compatible numbers to estimate $\frac{1}{4} \times 21$.

Guided Practice

Round each fraction to 0, $\frac{1}{2}$, or 1.

4. $\frac{11}{12}$ 5. $\frac{1}{6}$ 6. $\frac{3}{5}$ 7. $\frac{3}{4}$ 8. $\frac{2}{3}$

Round to the nearest whole number.

9. $5\frac{3}{4}$ 10. $9\frac{1}{6}$ 11. $2\frac{1}{2}$ 12. $4\frac{1}{8}$ 13. $6\frac{3}{4}$

Estimate. Tell which strategy you used.

14. $\frac{1}{2} + \frac{7}{8}$ 15. $\frac{3}{8} - \frac{1}{10}$ 16. $5\frac{1}{3} - 2\frac{3}{4}$

17. $\frac{3}{4} \times 11$ 18. $\frac{1}{8} \times \frac{3}{4}$ 19. $\frac{4}{5} \div \frac{7}{8}$

Exercises

Round each fraction to 0, $\frac{1}{2}$, or 1.

20. $\frac{1}{8}$

21. $\frac{4}{5}$

22. $\frac{2}{5}$

23. $\frac{5}{6}$

24. $\frac{1}{4}$

25. $\frac{1}{7}$

26. $\frac{11}{12}$

27. $\frac{3}{10}$

28. $\frac{5}{12}$

29. $\frac{9}{10}$

Round to the nearest whole number.

30. $7\frac{1}{3}$

31. $4\frac{1}{10}$

32. $6\frac{7}{8}$

33. $3\frac{1}{2}$

34. $1\frac{3}{8}$

35. $5\frac{3}{4}$

36. $10\frac{4}{9}$

37. $8\frac{7}{12}$

38. $11\frac{2}{3}$

39. $7\frac{5}{12}$

Estimate.

40. $\frac{1}{3} + \frac{1}{8}$

41. $\frac{5}{8} - \frac{1}{12}$

42. $\frac{3}{4} - \frac{2}{5}$

43. $4\frac{1}{4} + 3\frac{4}{5}$

44. $9\frac{7}{8} - 2\frac{3}{4}$

45. $3\frac{1}{3} \times 2\frac{2}{3}$

46. $5\frac{5}{7} \times 8\frac{2}{3}$

47. $21\frac{1}{2} \div 1\frac{3}{4}$

48. $13\frac{1}{6} \div 4\frac{1}{8}$

49. $\frac{1}{2} \times 17$

50. $18 \div \frac{3}{8}$

51. $\frac{7}{8} \times \frac{3}{5}$

52. Estimate the product of $\frac{1}{4}$ and $\frac{2}{9}$.

53. Estimate the quotient of $2\frac{4}{5}$ and $\frac{7}{8}$.

54. Estimate the quotient of $\frac{3}{5}$ and $\frac{7}{12}$.

55. Compute the sum of 65 and 59 mentally to see if they are equal to 136. *(Lesson 1-4)*

56. Evaluate $5(6 + 3) \div (3 + 2)$. *(Lesson 1-7)*

57. Estimate the difference of 14.82 and 5.13. *(Lesson 2-3)*

58. Change 36 millimeters to liters. *(Lesson 2-9)*

59. **Attendance** The number of students who missed school during the past three weeks at Lincoln Junior High are 3, 15, 24, 0, 31, 14, 7, 9, 10, 13, 4, 9, 3, 8, and 1. Make a stem-and-leaf plot for the data. *(Lesson 3-6)*

60. Write the prime factorization for 36. *(Lesson 4-2)*

61. Write $3\frac{5}{8}$ as an improper fraction. *(Lesson 5-1)*

62. **Smart Shopping** Kenny bought a tomato for a chef's salad he is making. The tomato he chose weighed 7 ounces. If tomatoes cost 90¢ a pound, *about* how much did Kenny pay for the tomato? *Hint: 1 pound = 16 ounces.*

63. **Critical Thinking**   If the dividend in a division problem is rounded up and the divisor is rounded down, what is the effect on the quotient?

64. **Construction**   For a new home, a carpenter is building a built-in bookcase with eight shelves that are $3\frac{1}{2}$ feet long each. *About* how many shelves can he cut from a 12-foot board?

65. **History**   In 1986, *Voyager* became the first plane to fly nonstop around the world without refueling in midair. *Voyager* weighed 2,000 pounds, but at take-off it carried *about* $3\frac{1}{2}$ times its weight in fuel. *About* how many pounds of fuel did *Voyager* carry at take-off?

66. **Travel**   Tim Brody traveled from New York to Chicago to visit his cousin. His car's odometer read 48,297 when he left for Chicago and 50,000 when he returned home. *About* how many miles did he travel?

67. **Baking**   Suppose a cookie recipe called for $2\frac{1}{2}$ cups of flour and $1\frac{2}{3}$ cups of sugar. *About* how many cups of dry ingredients are in the recipe?

DATA SEARCH

68. **Data Search**   Refer to pages 172 and 173. What fraction of the media industry is made up of publishing and cable television?

CULTURAL KALEIDOSCOPE

Fannie Merrit Farmer

The printing press revolutionized cooking by making cookbooks widely available. The first known cookbook was printed in 1485. It was produced by an Italian who recorded recipes for marzipans and other sweets.

One of the most successful and popular cookbooks of all time was produced in the United States in 1896, when Fannie Farmer took on the editorship of *The Boston Cooking-School Cook Book*. She was the first to standardize the methods and measurements of her recipes. Before then, most recipes were written with vague directions such as "a pinch of salt," "a handful of flour," and "a dash of pepper."

5-3 Adding and Subtracting Fractions

Objective

Add and subtract fractions.

You can help prevent the greenhouse effect by using public transportation and avoid using hair spray, deodorant, and air freshener in aerosol cans.

The greenhouse effect is a concern of many people today. Some scientists claim that industrial gases emitted into the atmosphere are slowly causing the Earth's temperature to rise. This temperature increase may lead to unusual weather patterns which could eventually threaten crops, wildlife, and our very existence.

Carbon dioxide gas is said to be responsible for $\frac{1}{2}$ of the greenhouse effect. Chlorofluorocarbons are said to account for another $\frac{1}{6}$ of it. Together, how much are these two gases responsible for the greenhouse effect? You need to find the sum of $\frac{1}{2}$ and $\frac{1}{6}$.

To find the sum or difference of numbers, the units of measure must be the same.

Mini-Lab

Work with a partner.

Materials: fraction models

- To add 3 oranges and 2 apples, the common unit of measure is fruit.

  3 oranges + 2 apples $\rightarrow$ 3 fruits + 2 fruits = 5 fruits

- To find the difference of 1 yard and 2 feet, the common unit of measure is feet.

$$1 \text{ yd } - 2 \text{ ft } = 3 \text{ ft } - 2 \text{ ft } = 1 \text{ ft}$$

- To add $\frac{1}{2}$ and $\frac{1}{4}$, the common unit of measure is fourths.

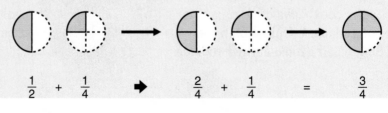

$$\frac{1}{2} + \frac{1}{4} \quad \Rightarrow \quad \frac{2}{4} + \frac{1}{4} = \frac{3}{4}$$

Talk About It

a. Name a unit of measure that would allow 4 cats and 3 dogs to be added. Find the sum.

b. Write a subtraction problem with two different units of measure so that the difference is 10 inches.

c. Name a unit of measure that would allow $\frac{1}{2}$ and $\frac{1}{3}$ to be subtracted. What is the difference?

d. What conclusion can you draw about units of measures for fractions that are to be added or subtracted?

Now solve the problem at the beginning of this lesson.

$$\frac{1}{2} + \frac{1}{6} \rightarrow \frac{3}{6} + \frac{1}{6} = \frac{4}{6} \text{ or } \frac{2}{3}$$

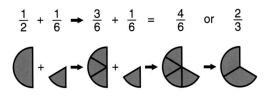

The denominator names the units to be added. The common unit is sixths.

Carbon dioxide gas and chlorofluorocarbons are responsible for about $\frac{2}{3}$ of the greenhouse effect.

Adding and Subtracting Fractions with Unlike Denominators	To add or subtract fractions: 1. Rename the fractions with a common denominator as necessary. 2. Add or subtract the numerators. 3. Simplify.

The least common multiple (LCM) can be used to rename fractions for addition and subtraction.

LOOKBACK

You can review LCM on page 161.

Example 1

Find $\frac{5}{8} - \frac{1}{6}$. Write the difference in simplest form.

Estimate: $\frac{1}{2} - 0 = \frac{1}{2}$

$$\frac{5}{8} \atop -\frac{1}{6} \quad \rightarrow \quad \begin{matrix} 8 = 2 \times 2 \times 2 \text{ and } 6 = 2 \times 3 \\ \textit{The LCM of 8 and 6 is} \\ 2 \times 2 \times 2 \times 3 \text{ or } 24. \\ \frac{5 \times 3}{8 \times 3} = \frac{15}{24} \text{ and } \frac{1 \times 4}{6 \times 4} = \frac{4}{24} \end{matrix} \quad \rightarrow \quad \begin{matrix} \frac{15}{24} \\ -\frac{4}{24} \\ \hline \frac{11}{24} \end{matrix}$$

So, $\frac{5}{8} - \frac{1}{6} = \frac{11}{24}$.

Example 2

Find $\frac{7}{9} + \frac{5}{12}$. Write the sum in simplest form.

Estimate $1 + \frac{1}{2} = 1\frac{1}{2}$

$$\begin{array}{ccccc}
\frac{7}{9} & & 9 = 3 \times 3 \text{ and } 12 = 2 \times 2 \times 3 & & \frac{28}{36} \\
& \rightarrow & \text{The LCM of 9 and 12 is} & \rightarrow & \\
& & 2 \times 2 \times 3 \times 3 \text{ or } 36. & & \\
+\frac{5}{12} & & \frac{7 \times 4}{9 \times 4} = \frac{28}{36} \text{ and } \frac{5 \times 3}{12 \times 3} = \frac{15}{36} & & +\frac{15}{36} \\
& & & & \frac{43}{36}
\end{array}$$

Rename $\frac{43}{36}$ as $1\frac{7}{36}$. So, $\frac{7}{9} + \frac{5}{12} = 1\frac{7}{36}$.

Example 3 *Problem Solving*

Movies   For the movie *The Wizard of Oz*, the wardrobe department had to make emerald green costumes for the residents of Emerald City. One-third of the shoes and stockings dyed for the movie were for the gentlemen and their wives, and $\frac{4}{15}$ were for the shopkeepers and their wives. What part of the shoes and stockings dyed for the movie were for the gentlemen and their wives and for the shopkeepers and their wives?

We need to add $\frac{1}{3}$ and $\frac{4}{15}$.

Estimate: $\frac{1}{2} + 0 = \frac{1}{2}$

$$\begin{array}{ccccc}
\frac{1}{3} & & 15 = 5 \times 3 & & \frac{5}{15} \\
& \rightarrow & \text{The LCM of 3 and 15} & \rightarrow & \\
+\frac{4}{15} & & \text{is } 3 \times 5 \text{ or } 15. & & +\frac{4}{15} \\
& & \frac{1 \times 5}{3 \times 5} = \frac{5}{15} & & \frac{9}{15}
\end{array}$$

Rename $\frac{9}{15}$ as $\frac{3}{5}$.

Three-fifths of the dyed shoes and stockings were for the gentlemen and their wives and the shopkeepers and their wives.

Calculator Hint
● ● ● ● ● ● ● ● ● ● ● ● ● ●
You can add and subtract fractions using a calculator. To find $\frac{1}{3} + \frac{4}{15}$, enter:

1 ⌒/⌒ 3 ⊞ 4 ⌒/⌒
15 ⊟ **9/15.**

To get the answer in simplest form, press

⌊Simp⌋ ⌊=⌋ until

N/D → n/d no longer appears on the screen.

Checking for Understanding

Communicating
Mathematics

Read and study the lesson to answer each question.

1. **Tell** why you must have a common denominator to add or subtract fractions.

2. **Tell** what is a common unit of measure that can be used to add 1 yard and 8 inches.

3. **Draw** a circle diagram that shows $\frac{1}{2} + \frac{1}{3} = \frac{5}{6}$.

Add or subtract. Write each sum or difference in simplest form.

4. $\dfrac{1}{5}$
   $+\dfrac{2}{5}$
   _____

5. $\dfrac{2}{6}$
   $-\dfrac{1}{6}$
   _____

6. $\dfrac{3}{8}$
   $-\dfrac{1}{12}$
   _____

7. $\dfrac{3}{4} + \dfrac{7}{20}$

8. $\dfrac{9}{10} - \dfrac{1}{6}$

9. $\dfrac{7}{15} + \dfrac{5}{9}$

Exercises

Add or subtract. Write each sum or difference in simplest form.

10. $\dfrac{2}{9}$
    $+\dfrac{4}{9}$
    _____

11. $\dfrac{2}{4}$
    $-\dfrac{1}{4}$
    _____

12. $\dfrac{5}{8}$
    $-\dfrac{1}{2}$
    _____

13. $\dfrac{3}{5}$
    $+\dfrac{1}{15}$
    _____

14. $\dfrac{4}{5}$
    $-\dfrac{1}{6}$
    _____

15. $\dfrac{3}{7}$
    $+\dfrac{4}{5}$
    _____

16. $\dfrac{5}{8} - \dfrac{5}{12}$

17. $\dfrac{5}{9} + \dfrac{5}{6}$

18. $\dfrac{3}{7} + \dfrac{9}{14}$

19. $\dfrac{4}{15} + \dfrac{9}{10}$

20. $\dfrac{7}{11} - \dfrac{1}{4}$

21. $\dfrac{19}{24} - \dfrac{1}{4}$

22. Find the sum of $\dfrac{8}{9}$ and $\dfrac{7}{15}$.

23. Find the difference of $\dfrac{5}{8}$ and $\dfrac{5}{36}$.

24. Find the sum of $\dfrac{11}{12}$ and $\dfrac{9}{20}$.

25. **Statistics** Find the mean, median, and mode for 12, 18, 25, 38, 44, and 49. *(Lesson 1-5)*

26. **Statistics** Make a line plot for $150, $1,200, $475, $235, $895, $1,075, and $390. Circle any outliers. *(Lesson 3-4)*

27. Find the least common multiple for 16 and 20. *(Lesson 4-9)*

28. Estimate $1\dfrac{7}{12} \times 12\dfrac{1}{5}$. *(Lesson 5-2)*

29. **Energy** Cars use about $\dfrac{4}{9}$ of the energy consumed by the transportation industry. Buses and trains use $\dfrac{1}{6}$. How much more energy is used by cars than by buses and trains?

30. **Home Economics** Mrs. Keaton used $\dfrac{1}{4}$ pound of cheddar and $\dfrac{1}{3}$ pound of monterey jack cheese to make nachos. How much cheese did she use in all?

31. **Critical Thinking** Does $\dfrac{3}{4} + \dfrac{7}{8} - \dfrac{5}{6} = \dfrac{7}{8} + \dfrac{5}{6} - \dfrac{3}{4}$? Explain.

32. **Journal Entry** Make up a problem that can be solved using the addition expression $\dfrac{1}{3} + \dfrac{3}{4}$.

5-4 Adding and Subtracting Mixed Numbers

Objectives

Add and subtract mixed numbers with unlike denominators.

Did you know that you can make your own window-washing solution that is environmentally safe? Just mix the four ingredients at the right in a large bucket.

$1\frac{1}{3}$ cups ammonia

$\frac{1}{4}$ cup baking soda

$1\frac{1}{2}$ cups vinegar

2 gallons water

What is the total amount of ammonia and vinegar used? Find the sum of $1\frac{1}{3}$ and $1\frac{1}{2}$.

Adding and Subtracting Mixed Numbers	To add or subtract mixed numbers: 1. Add or subtract the fractions. If necessary, rename the fractions first. 2. Add or subtract the whole numbers. 3. Rename and simplify.

Calculator Hint

• • • • • • • • • • • • •

You can add and subtract mixed numbers using a calculator. To find $18\frac{3}{4} + 13\frac{5}{6}$, enter:

18 [Unit] 3 [/] 4

[+] 13 [Unit] 5

[/] 6 [=] [Ab/c]

32 7/12.

To get the answers in simplest form, press [Simp] [=] until N/D → n/d no longer appears on the screen.

To find the total amount of ammonia and vinegar used, add $1\frac{1}{3}$ and $1\frac{1}{2}$.

Estimate: 1 + 2 = 3

$$1\frac{1}{3} \quad \rightarrow \quad \begin{array}{l} \textit{Use the LCM of 2} \\ \textit{and 3 to rename } \frac{1}{3} \\ \textit{as } \frac{2}{6} \textit{ and } \frac{1}{2} \textit{ as } \frac{3}{6}. \end{array} \quad \rightarrow \quad 1\frac{2}{6}$$

$$+1\frac{1}{2} \qquad\qquad\qquad\qquad\qquad +1\frac{3}{6}$$

$$\overline{\phantom{+1\frac{1}{2}}} \qquad\qquad\qquad\qquad\qquad \overline{2\frac{5}{6}}$$

The total amount of ammonia and vinegar used is $2\frac{5}{6}$ cups.

Examples

Add or subtract. Write each sum or difference in simplest form.

1 $8\frac{3}{4} - 2\frac{5}{12}$

Estimate: 9 − 2 = 7

$$8\frac{3}{4} \quad \rightarrow \quad 8\frac{9}{12}$$

$$-2\frac{5}{12} \qquad\qquad -2\frac{5}{12}$$

$$\overline{\phantom{-2\frac{5}{12}}} \qquad\qquad \overline{6\frac{4}{12} \text{ or } 6\frac{1}{3}}$$

2 $18\frac{3}{10} + 13\frac{5}{6}$

Estimate: 18 + 14 = 32

$$18\frac{3}{10} \quad \rightarrow \quad 18\frac{9}{30}$$

$$+13\frac{5}{6} \qquad\qquad +13\frac{25}{30}$$

$$\overline{\phantom{+13\frac{5}{6}}} \qquad\qquad \overline{31\frac{34}{30}}$$

$$31\frac{34}{30} = 31 + \frac{34}{30}$$
$$= 31 + 1\frac{4}{30}$$
$$= 32\frac{4}{30} \text{ or } 32\frac{2}{15}$$

When you subtract two mixed numbers where the fraction in the first mixed number is less than the fraction in the second mixed number, you need to rename the first mixed number before subtracting.

Example 3

Find $3\frac{1}{3} - 1\frac{2}{3}$.

$\frac{1}{3}$ is less than $\frac{2}{3}$, so you need to rename $3\frac{1}{3}$.

Think: $3\frac{1}{3} = 2\frac{\square}{3}$

Use circle diagrams.

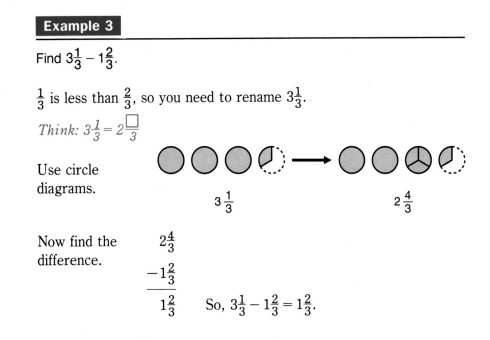

$3\frac{1}{3}$ $2\frac{4}{3}$

Now find the difference.

$$\begin{array}{r} 2\frac{4}{3} \\ -1\frac{2}{3} \\ \hline 1\frac{2}{3} \end{array}$$ So, $3\frac{1}{3} - 1\frac{2}{3} = 1\frac{2}{3}$.

Checking for Understanding

Communicating Mathematics

Read and study the lesson to answer each question.

1. **Show** that $2\frac{1}{4} = 1\frac{5}{4}$ using circle models.

2. **Show** that $\frac{10}{3} = 3\frac{1}{3}$ using circle models.

Guided Practice

Complete.

3. $5\frac{1}{6} = 4\frac{\square}{6}$

4. $3\frac{10}{7} = 4\frac{\square}{7}$

5. $8\frac{3}{5} = 7\frac{\square}{5}$

6. $2\frac{12}{9} = \square\frac{1}{3}$

Add or subtract. Write each sum or difference in simplest form.

7. $5\frac{1}{8} + 3\frac{3}{8}$

8. $6\frac{5}{6} - 2\frac{1}{3}$

9. $3\frac{1}{2} - 1\frac{3}{4}$

10. $7\frac{5}{6} + 9\frac{3}{8}$

11. $4 - 2\frac{3}{15}$

12. $13\frac{7}{8} + 15\frac{7}{10}$

Exercises

Independent Practice

Complete.

13. $6\frac{1}{2} = 5\frac{\square}{2}$ 14. $3\frac{6}{4} = 4\frac{\square}{2}$ 15. $9\frac{9}{8} = 10\frac{\square}{8}$ 16. $6\frac{1}{2} = 5\frac{\square}{2}$

17. $4\frac{5}{6} = 3\frac{\square}{6}$ 18. $7\frac{14}{10} = 8\frac{\square}{10}$ 19. $12\frac{9}{5} = \square\frac{4}{5}$ 20. $9\frac{2}{3} = \square\frac{5}{3}$

Add or subtract. Write each sum or difference in simplest form.

21. $3\frac{1}{6} + 5\frac{1}{6}$ 22. $8\frac{7}{9} - 3\frac{1}{9}$ 23. $7\frac{3}{8} + 4\frac{7}{8}$

24. $9\frac{4}{5} - 2\frac{3}{10}$ 25. $5\frac{5}{6} - 3\frac{2}{3}$ 26. $3\frac{7}{12} + 8\frac{3}{4}$

27. $6\frac{13}{15} - 2\frac{3}{5}$ 28. $7\frac{3}{8} + 9\frac{1}{6}$ 29. $8\frac{3}{4} - 1\frac{7}{10}$

30. $4\frac{3}{10} - 1\frac{3}{4}$ 31. $13\frac{1}{8} - 1\frac{7}{10}$ 32. $3\frac{1}{6} + 5\frac{1}{2} + 2\frac{7}{8}$

33. Find the difference of $7\frac{1}{3}$ and $3\frac{5}{9}$.   34. Find the sum of $5\frac{7}{12}$ and $6\frac{5}{8}$.

Mixed Review

35. **Measurement** Find the number of feet in 3 yards. *(Lesson 1-9)*

36. Multiply 0.00003×10^6. *(Lesson 2-6)*

37. Find the next three terms in the following sequence: 22.5, 25, 27.5, 30, ... *(Lesson 4-4)*

38. **Travel** Don drove to an out-of-town business meeting. It took $11\frac{4}{5}$ hours. About how many hours did he drive? *(Lesson 5-1)*

39. Add $\frac{5}{6}$ and $\frac{2}{3}$. *(Lesson 5-3)*

Problem Solving and Applications

40. **Television** A video tape will record 6 hours in the EP mode. Mike has recorded $2\frac{5}{6}$ hours of a baseball game. He wants to record $3\frac{1}{2}$ hours more on the same tape. Can he do this? Explain your answer.

41. **Sports** Sergei Bubka, a pole vaulter from the former Soviet Union, has set several world records in pole vaulting. Use the graph at the right to find the difference between his world record vault in 1988 and his world record vault in August 1991.

Bubka's Pole-Vault Records

20 ft $\frac{1}{4}$ in. (1991)

19 ft 10 $\frac{1}{2}$ in. (1988)

19 ft 6 in. (1985-86)

19 ft (1984)

42. **Critical Thinking** A string is cut in half and one half is used to bundle newspapers. Then one fifth of the remaining string is cut off and used to tie a balloon. The piece left is 8 feet long. How long was the string originally?

5-5A Multiplying Fractions and Mixed Numbers

A Preview of Lesson 5-5

Objective

Find the product of fractions by using models.

Materials

fraction circles
sheets of paper

As in multiplication of whole numbers, multiplication of fractions and mixed numbers represents repeated addition.

The product $a \times b$ means a sets of size b. You can show this by making a model. Consider 3×2.

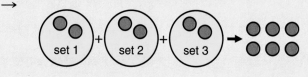

$$
\begin{array}{ccc}
3 & \times & 2 \\
\text{\textit{number}} & & \text{\textit{size of}} \\
\text{\textit{of sets}} & & \text{\textit{each set}}
\end{array}
$$

Therefore, $3 \times 2 = 6$.

You can make a model to show $3 \times \frac{1}{2}$ by using fraction circles. Use 3 sheets of paper to represent 3 sets. Put a half circle in each set. Find the product.

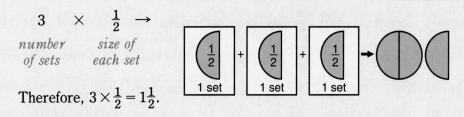

$$
\begin{array}{ccc}
3 & \times & \frac{1}{2} \\
\text{\textit{number}} & & \text{\textit{size of}} \\
\text{\textit{of sets}} & & \text{\textit{each set}}
\end{array}
$$

Therefore, $3 \times \frac{1}{2} = 1\frac{1}{2}$.

Activity One

Work in groups of three.

- Make a model to show the product of $4 \times \frac{1}{3}$ by using fraction circles and sheets of paper.

What do you think?

1. How many sheets of paper were needed?
2. What part of a circle did each sheet of paper contain?
3. What is the product?
4. How is this model similar to or different from the model for whole numbers?
5. Repeat Activity One for $2 \times \frac{3}{4}$, $5 \times \frac{2}{3}$, and $6 \times \frac{3}{8}$.

Activity Two

- Make a model to show the product of $2\frac{1}{2} \times \frac{1}{2}$. Use $2\frac{1}{2}$ sheets of paper to represent $2\frac{1}{2}$ sets. Put a half circle on each full sheet of paper to represent $\frac{1}{2}$. Since a full sheet contains $\frac{1}{2}$, put $\frac{1}{4}$ on the half sheet of paper. Then find the product.

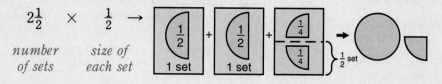

Therefore, $2\frac{1}{2} \times \frac{1}{2} = 1\frac{1}{4}$.

- Make a model to show the product of $1\frac{1}{2} \times \frac{2}{3}$.

What do you think?

6. a. What part of a circle should the full sheet of paper contain?
   b. What part of a circle should the half sheet of paper contain?
7. What is the product?
8. Repeat this activity for $2\frac{1}{2} \times \frac{1}{4}$.

Activity Three

- Make a model to show the product of $\frac{1}{3} \times \frac{3}{4}$. Use a full sheet of paper to represent one set. Since a full sheet contains $\frac{3}{4}$, put $\frac{1}{4}$ on the one-third sheet. Now find the product.

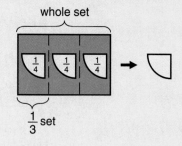

Therefore, $\frac{1}{3} \times \frac{3}{4} = \frac{1}{4}$.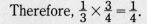

- Make a model to show $\frac{1}{3} \times \frac{3}{8}$.

What do you think?

9. a. What part of a sheet of paper is needed?
   b. What part of a circle does the one-third sheet of paper contain?
10. What is the product?
11. Repeat this activity for $\frac{1}{2} \times \frac{2}{4}$ and $\frac{1}{3} \times \frac{6}{8}$.

Multiplying Fractions and Mixed Numbers

Objective

Multiply fractions and mixed numbers.

MegaCorp purchased land that is $\frac{2}{3}$ mile long and $\frac{1}{2}$ mile wide. How many square miles is the land?

Draw a model of a square mile. Divide the length in thirds and the width in halves. If the land is $\frac{2}{3}$ mile long and $\frac{1}{2}$ mile wide, then its area is $\frac{2}{6}$ or $\frac{1}{3}$ of a square mile.

You also get $\frac{1}{3}$ if you multiply the fractions.

$\frac{2}{3} \times \frac{1}{2} = \frac{2 \times 1}{3 \times 2}$

$= \frac{2}{6}$ or $\frac{1}{3}$

The area of the land will be $\frac{1}{3}$ square mile.

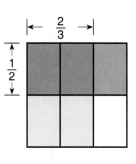

Multiplying Fractions	**In Words:** To multiply fractions, multiply the numerators and then multiply the denominators.
	Arithmetic $\qquad$ **Algebra**
	$\frac{1}{4} \times \frac{1}{2} = \frac{1}{8}$ $\qquad$ $\frac{a}{b} \times \frac{c}{d} = \frac{ac}{bd}$

Example 1

Find $\frac{2}{3} \times \frac{3}{4}$.

2×3

$\frac{2}{3} \times \frac{3}{4} = \frac{6}{12}$

3×4

$= \frac{1}{2}$

$\frac{2}{3}$ $\qquad \times \qquad$ $\frac{3}{4}$ $\qquad \rightarrow$

number of sets $\qquad$ *size of set*

whole set

$\frac{2}{3}$ set

When the numerator and denominator of either fraction have a common factor, you can simplify before you multiply.

Examples

Multiply.

2 $\frac{3}{4} \times \frac{5}{6}$

Estimate: $1 \times 1 = 1$

$\frac{3}{4} \times \frac{5}{6} = \frac{\overset{1}{\cancel{3}}}{4} \times \frac{5}{\underset{2}{\cancel{6}}}$

$= \frac{5}{8}$

The GCF of 3 and 6 is 3.

Divide 3 and 6 by 3.

3 $\frac{6}{25} \times \frac{5}{8}$

Estimate: $0 \times \frac{1}{2} = 0$

$\frac{6}{25} \times \frac{5}{8} = \frac{\overset{3}{\cancel{6}}}{\underset{5}{\cancel{25}}} \times \frac{\overset{1}{\cancel{5}}}{\underset{4}{\cancel{8}}}$

$= \frac{3}{20}$

The GCF of 6 and 8 is 2.

The GCF of 25 and 5 is 5.

Estimation Hint

Estimate $\frac{2}{3} \times \frac{3}{4}$.

Round $\frac{2}{3}$ to $\frac{1}{2}$ and $\frac{3}{4}$ to 1.

$\frac{1}{2} \times 1 = \frac{1}{2}$

LOOKBACK

You can review GCF on page 145.

Multiplying Mixed Numbers	To multiply mixed numbers, rename each mixed number as an improper fraction. Multiply the fractions.

Example 4

Find $2 \times 2\frac{1}{4}$.

Estimate: $2 \times 2 = 4$

$$2 \times 2\frac{1}{4} = 2 \times \frac{9}{4}$$
$$= \frac{2}{1} \times \frac{9}{4}$$
$$= \frac{9}{2} \text{ or } 4\frac{1}{2}$$

Checking for Understanding

Communicating Mathematics

Read and study the lesson to answer each question.

1. **Draw** a model using sets to show what $2\frac{1}{2} \times \frac{2}{3}$ means.

2. **Tell** how to multiply fractions.

Guided Practice

Multiply. Write each product in simplest form.

3. $\frac{3}{5} \times \frac{1}{2}$　　　　4. $\frac{2}{3} \times \frac{5}{6}$　　　　5. $\frac{2}{3} \times \frac{3}{8}$

6. $2 \times \frac{3}{4}$　　　　7. $2\frac{1}{2} \times 2\frac{2}{3}$　　　　8. $1\frac{1}{6} \times \frac{3}{7} \times \frac{1}{3}$

9. $\frac{4}{5} \times \frac{1}{8}$　　　　10. $4 \times \frac{2}{5}$　　　　11. $3\frac{1}{4} \times 2\frac{2}{3}$

Exercises

Independent Practice

Multiply. Write each product in simplest form.

12. $\frac{1}{8} \times \frac{3}{4}$　　　　13. $\frac{1}{5} \times \frac{1}{2}$　　　　14. $\frac{1}{4} \times \frac{4}{5}$

15. $\frac{3}{8} \times \frac{4}{5}$　　　　16. $\frac{3}{7} \times \frac{2}{3}$　　　　17. $\frac{4}{5} \times \frac{1}{8}$

18. $\frac{5}{6} \times \frac{3}{5}$　　　　19. $\frac{3}{5} \times \frac{10}{21}$　　　　20. $\frac{4}{9} \times \frac{2}{3}$

21. $\frac{1}{2} \times \frac{5}{8}$　　　　22. $\frac{3}{7} \times \frac{5}{6}$　　　　23. $\frac{5}{9} \times \frac{9}{10}$

24. $3\frac{2}{3} \times 9$　　　　25. $1\frac{4}{7} \times 4\frac{2}{3}$　　　　26. $5\frac{1}{3} \times \frac{4}{5}$

27. Find the product of $4\frac{1}{2}$ and $1\frac{1}{3}$.

28. Find the product of 3 and $2\frac{1}{7}$.

29. Evaluate $3m + 4(p + m) - 2mnp$ if $m = 2$, $n = 3$, and $p = 1$. *(Lesson 1-8)*

30. **Income** The mean income for a group of accountants was $26,266.67. The incomes were $17,500, $26,100, $19,800, $23,400, $21,300, and $49,500. In what way is the mean misleading? *(Lesson 3-8)*

31. Tell whether 240 is divisible by 2, 3, 4, 5, 6, 9, or 10. *(Lesson 4-1)*

32. Find the sum of $6\frac{3}{4}$ and $9\frac{7}{8}$. *(Lesson 5-4)*

33. **Landscaping** A brick is about $2\frac{1}{4}$ inches thick, and the mortar joint is about $\frac{1}{2}$ inch thick. If six layers of brick are to be laid around a new flower bed, about how high will the wall be?

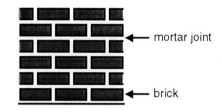

mortar joint

brick

34. **Coin Collecting** Susan B. Anthony was a leader in the women's suffrage movement. In 1979, the United States honored her efforts by minting a special one-dollar coin with her portrait on the front. The coin is $\frac{3}{4}$ copper and $\frac{1}{4}$ nickel. It weighs $8\frac{1}{2}$ grams. How many grams of copper are in each coin?

35. **Critical Thinking** Observe that $3 \times \frac{1}{3} = 1$, $4 \times \frac{1}{4} = 1$, and $9 \times \frac{1}{9} = 1$. What number times $1\frac{1}{2}$ equals one? times $2\frac{1}{2}$?

Mid-Chapter Review

Change each mixed number to an improper fraction. *(Lesson 5-1)*

1. $2\frac{1}{2}$ 2. $4\frac{2}{3}$ 3. $3\frac{1}{3}$ 4. $5\frac{4}{5}$

Estimate. *(Lesson 5-2)*

5. $\frac{3}{4} + \frac{1}{8}$ 6. $\frac{5}{6} - \frac{1}{5}$ 7. $4\frac{1}{6} \times 5\frac{1}{3}$

8. **Machinery** A steel rod has a diameter of $\frac{3}{4}$ inch. It must be made into a $\frac{9}{16}$-inch rod to fit on a tractor. How much must the diameter be reduced? *(Lesson 5-3)*

Add or subtract. Write each sum or difference in simplest form. *(Lesson 5-4)*

9. $8\frac{3}{8} + 6\frac{5}{6}$ 10. $2\frac{1}{4} - 1\frac{5}{8}$ 11. $8\frac{7}{9} + 1\frac{1}{3}$

Multiply. Write each product in simplest form. *(Lesson 5-5)*

12. $\frac{3}{8} \times \frac{2}{5}$ 13. $\frac{5}{9} \times \frac{3}{7}$ 14. $\frac{4}{5} \times \frac{5}{6}$

5-6 Perimeter

Objective

Find perimeter using fractional measurements.

Words to Learn

perimeter

Mr. Kirby has a pond he uses for irrigation and fishing. He wishes to fence the field around the pond to protect small children that have moved to the neighborhood. How much fencing does he need?

Mr. Kirby needs to know the distance around his field. The distance around a geometric figure is called its **perimeter.**

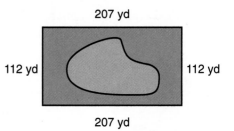

To find the perimeter, P, of the field, Mr. Kirby adds the measures of the sides.

$P = 207 + 112 + 207 + 112$

$\quad = 638$ Mr. Kirby needs 638 yards of fencing.

Perimeter of a Rectangle	**In Words:** The perimeter of a rectangle is the sum of the measures of the sides. **In symbols:** $P = \ell + w + \ell + w$ $P = 2\ell + 2w$

 **When am I ever going to use this?**

Suppose you wanted to frame a 6-inch × 7-inch cross-stitch design with a 2-inch mat around the needlework. What is the perimeter of the frame you will need?

Examples

1 Find the perimeter of a rectangle with a length of 7 feet and a width of 4 feet.

$P = 2\ell + 2w$

$\quad = 2(7) + 2(4)$ *Replace ℓ with 7 and w with 4.*

$\quad = 14 + 8$

$\quad = 22$ The perimeter is 22 feet.

2 Find the perimeter of the figure at the right.

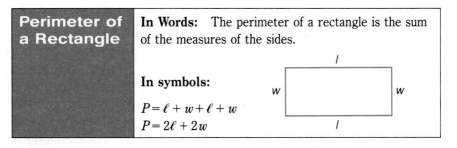

Estimate:

5 + 10 + 2 + 4 + 3 + 6 = 30

$P = 4\frac{1}{2} + 10\frac{1}{3} + 1\frac{1}{2} + 4 + 3 + 6\frac{1}{3}$

$\quad = 29\frac{2}{3}$ The perimeter is $29\frac{2}{3}$ yards.

Example 3   *Connection*

Measurement   Find the perimeter of the rectangle. Measure to the nearest eighth inch.

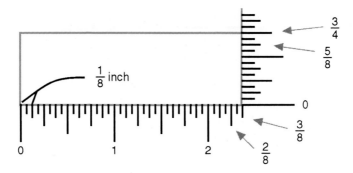

To the nearest eighth inch, the width is $\frac{6}{8}$ or $\frac{3}{4}$ inch.

To the nearest eighth inch, the length is $2\frac{3}{8}$ inches.

Estimate: $2 + 1 + 2 + 1 = 6$

$$P = 2\ell + 2w$$

$$= (2 \times 2\tfrac{3}{8}) + (2 \times \tfrac{3}{4}) \qquad \textit{Replace } \ell \textit{ with } 2\tfrac{3}{8} \textit{ and } w \textit{ with } \tfrac{3}{4}.$$

$$= (2 \times \tfrac{19}{8}) + (2 \times \tfrac{3}{4}) \qquad \textit{Change } 2\tfrac{3}{8} \textit{ to } \tfrac{19}{8}.$$

$$= \tfrac{38}{8} + \tfrac{6}{4} \qquad\qquad \textit{Multiply within each set of parentheses.}$$

$$= \tfrac{38}{8} + \tfrac{12}{8} \qquad\qquad \textit{Change } \tfrac{6}{4} \textit{ to } \tfrac{12}{8}.$$

$$= \tfrac{50}{8} \qquad\qquad\qquad \textit{Add.}$$

$$= 6\tfrac{2}{8} \text{ or } 6\tfrac{1}{4} \qquad\quad \text{The perimeter is } 6\tfrac{1}{4} \text{ inches.}$$

Checking for Understanding

Communicating Mathematics   Read and study the lesson to answer each question.

1. **Show** where $3\frac{7}{8}$ inches is located on a ruler.

2. **Write**, in your own words, how to find the perimeter of a rectangular figure.

Guided Practice   Find the perimeter of each figure shown or described below.

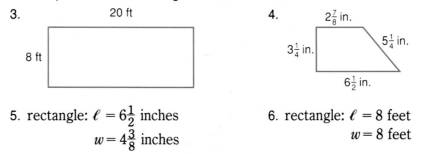

3.   20 ft   8 ft

4. $2\frac{7}{8}$ in. $3\frac{1}{4}$ in. $5\frac{1}{4}$ in. $6\frac{1}{2}$ in.

5. rectangle: $\ell = 6\frac{1}{2}$ inches
   $w = 4\frac{3}{8}$ inches

6. rectangle: $\ell = 8$ feet
   $w = 8$ feet

Find the perimeter of each figure. Measure to the nearest eighth inch.

7.

8.

Exercises

Independent Practice

Find the perimeter of each figure shown or described below.

9. $12\frac{1}{2}$ mi

10. 20 ft / 16 ft / 20 ft

11. 13 in. / 7 in. / 9 in. / 10 in. / 15 in.

12. rectangle: $\ell = 5$ yards
    $w = 2$ yards

13. rectangle: $\ell = 3.5$ miles
    $w = 1.7$ miles

14. Find the perimeter of a square with side 15 yards.

15. Find the perimeter of a rectangle with length $13\frac{1}{2}$ feet and width $7\frac{3}{4}$ feet.

Find the perimeter of each figure. Measure to the nearest eighth inch.

16.

17.

Mixed Review

18. **Money** Jill has a total of $76.89 in pennies. How many pennies does she have? *(Lesson 2-5)*

19. Express $\frac{18}{24}$ in simplest form. *(Lesson 4-6)*

20. Find $\frac{5}{9} \times \frac{3}{4}$. *(Lesson 5-5)*

Problem Solving and Applications

21. **Landscaping** Mrs. Knowles plans to plant azalea bushes across the back and down two sides of her yard. Her lot is 100 feet wide and 160 feet deep.
    a. Draw and label a diagram of Mrs. Knowles' yard.
    b. How many bushes will she need to buy if she plants them 4 feet apart?

22. **Critical Thinking** Khoa has 36 feet of fencing for a rectangular dog pen. He plans to use 22 feet of the garage wall for one side of the pen.
    a. Draw and label a diagram of the pen.
    b. Find the width of the pen.

23. **Journal Entry** Using one or two sentences, describe a situation where you needed to find the perimeter of an object.

5-7 Circles and Circumference

Objective

Find the circumference of circles.

Words to Learn

circle
center
diameter
radius
circumference

Today, you don't need a watchdog to watch your house. A robot can do that for you! A recently developed robot, 2 feet high and shaped like a dome, can roam around your house and "watch" for intruders. If anything moves within a 30-foot radius of the robot, the robot will detect this motion and send a silent alarm to the local police. What is the circumference of the circle the robot guards? *This question will be answered on page 198.*

Let's study some terms and properties related to circles. A **circle** is the set of all points in a plane that are the same distance from a given point called the **center.**
The **diameter (d)** of a circle is the distance across the circle through its center. The **radius (r)** of a circle is the distance from the center to any point on the circle. The **circumference (C)** of a circle is the distance around the circle.

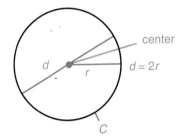

Mini-Lab

Work with a partner.
Materials: metric ruler, string, circular objects of various sizes

- Find a circular object.
- Use a metric ruler to measure the diameter of your circular object. Record your finding.
- Wrap a string around the circular object once. Mark the string where it meets itself.
- Lay the string out straight and measure the length of the string with your ruler. Record your finding. This is the circumference of the circle.
- Divide the circumference by the diameter. Record your answer.
- Repeat this activity with circular objects of various sizes.

Talk About It

a. Compare your results. What do you find?

b. How is the circumference related to the diameter?

Calculator Hint

• • • • • • • • • • • • •

π is such an important number in mathematics that it usually has its own key on a calculator. What is displayed on your calculator when you press $\boxed{\pi}$?

The Greek letter π is used to represent the circumference divided by the diameter $\left(\frac{C}{d}\right)$. Often used approximations for π are 3.14 and $\frac{22}{7}$.

Formulas for Circumference of a Circle	**In words:** The circumference of a circle is equal to π times its diameter or π times twice the radius. (The diameter is twice the radius.)
	In symbols: $C = \pi d$ $C = 2\pi r$

Now let's solve the problem given at the beginning of the lesson. Find the circumference of the circle the robot guards. Since you know the radius, use the formula $C = 2\pi r$.

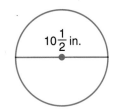

$C = 2\pi r$ ≈ *means "approximately equal to."*
$C \approx 2 \times 3.14 \times 30$ *Use 3.14 for π. Replace r with 30.*
$C \approx 188.4$

The robot guards a circle with a circumference of about 188.4 feet.

Examples

1 Find the circumference of a circle with a diameter of $10\frac{1}{2}$ inches. Use $\frac{22}{7}$ for π.

$C = \pi d$

$C \approx \frac{22}{7}\left(10\frac{1}{2}\right)$ *Estimate: $3 \times 11 = 33$*

$C \approx \overset{11}{\cancel{\frac{22}{7}}} \times \overset{3}{\underset{1}{\cancel{\frac{21}{2}}}}$

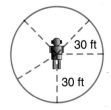

$C \approx 33$ The circumference is about 33 inches.

2 Find the circumference of a circle with a radius of 8.5 meters.

$C = 2\pi r$ *Estimate: $2 \times 3 \times 9 = 54$*
$C \approx 2\pi(8.5)$
$2 \boxed{\times} \boxed{\pi} \boxed{\times} 8.5 \boxed{=} 53.407075$
$C \approx 53.4$ The circumference is about 53.4 meters.

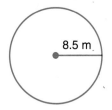

Checking for Understanding

Read and study the lesson to answer each question.

1. **Write** the steps taken in finding the circumference of a circle whose diameter is 25 feet.

2. **Write** *an estimate* for the circumference of a circle with a diameter of 5 inches.

3. **Tell** two numbers that are approximations of π.

4. **Tell** how to find the circumference of a circle.

Guided Practice   Find the circumference of each circle.

5. 8 ft

6. 3.5 m

7. $4\frac{1}{2}$ yd

8. 12 cm

9. $d = 14$ in.   10. $r = 3.2$ m   11. $r = 10\frac{1}{2}$ in.   12. $d = 6.5$ in.

13. What is the radius of a circle whose diameter is 6 meters long?

14. What is the diameter of a circle whose radius is 7.5 meters long?

Exercises

Independent Practice   Find the circumference of each circle.

15. $1\frac{3}{4}$ ft

16. 0.5 mi

17. 9.5 km

18. 12.2 m

19. $d = 17\frac{1}{2}$ mi   20. $r = \frac{3}{4}$ in.   21. $r = 1.5$ km   22. $d = 6$ cm

23. $r = 6.2$ cm   24. $d = 4.5$ yd   25. $d = 8\frac{3}{4}$ ft   26. $r = 1\frac{1}{3}$ yd

27. What is the diameter of a circle whose radius is 13 feet?

28. Find the circumference of a circle whose radius is 6.5 inches.

Mixed Review   29. **Packaging**   If 2,365 light bulbs are packaged in cases of 12, *about* how many cases of bulbs are there? *(Lesson 1-3)*

30. Order 3.4, 2.6, 3.8, 1.9, 2.3, and 3.6 from least to greatest. *(Lesson 2-1)*

Lesson 5-7 Geometry Connection: Circles and Circumference   **199**

31. Express $\frac{7}{8}$ as a decimal. *(Lesson 4-7)*

32. **Carpentry**   The deck on Jack's house is $25\frac{4}{5}$ feet long and $12\frac{1}{2}$ feet wide. One length of the deck is against the house. How many feet of wood does Jack need to buy to build a railing around the deck. *(Lesson 5-6)*

Problem Solving and Applications

33. **Recreation**   The pedals on a bicycle of the 1870s were on the front wheel. For speed, the front wheel was large, with a diameter of about 64 inches. The back wheel was about 12 inches. How far did a cyclist travel each time the pedals made a complete turn?

34. **Critical Thinking**   In Exercises 33, how many times does the back wheel go around for each complete turn of the front wheel?

COMPUTER

CONNECTION

35. **Computer Connection**   The BASIC program below will compute the circumference of a circle with a given diameter.

```
10 PRINT "WHAT IS THE DIAMETER?"
20 INPUT D
30 C = 3.14 * D
40 PRINT C
```

a. Run the program for diameters of 2, 4, 8, 16, 32, and 64 units.

b. Write a sentence that describes how the circumference changes when the diameter is doubled.

36. **Recreation**   The swimming pool at the community center is a circle having a radius of 50 feet. Find the circumference of the pool. Use $\pi = \frac{22}{7}$.

37. **Mathematics and Cycling**   Read the following paragraphs.

The Tour de France is considered to be the world's most important bicycle race. It was established in 1903 by Henri Desgrange, a French cyclist and journalist.

The annual race involves 120 or more professional contestants and covers about 2,500 miles of flat and mountainous country, mostly in France and Belgium. The bicycle and the rider with the lowest total time for all stages at the completion of the race is the winner.

a. The longest race ever held was in 1926. It covered 3,562 miles and took 3 weeks to complete. If the cyclists rode one third of the race in the first week, how many miles did they travel?

b. The diameter of a bicycle wheel is 28 inches. How far will you travel after 1 complete turn of the wheel?

5-8 Expected Value

Objective

Find expected value of outcomes.

Words to Learn

expected value

The Boy Scouts are having a fund-raiser for camp. They are selling 400 tickets for $1 each. The prizes are one $80-bicycle, two $20-Walkman stereos, and ten $4-theater passes.

Dawn wants to know whether buying a fund-raiser ticket is wise. To do so, she calculates the expected value of her ticket. The **expected value** is the long-term average of what she could expect to win by repeatedly buying a ticket.

LOOKBACK

You can review probability on page 157.

To find the expected value, she first needs to find the probability of winning each prize.

$P(\text{bicycle}) = \frac{1}{400}$ *number of bicycle-winning tickets*
total number of tickets

$P(\text{Walkman}) = \frac{2}{400}$ *number of Walkman winning-tickets*
total number of tickets

$= \frac{1}{200}$

$P(\text{theater passes}) = \frac{10}{400}$ *number of theater-winning tickets*
total number of tickets

$= \frac{1}{40}$

Next she needs to find the probability of *not* winning a prize.

$P(\text{not winning}) = \frac{387}{400}$ *number of nonwinning tickets*
total number of tickets

Using these probabilities, she can find the expected value of a fund-raiser ticket. The expected value is found by multiplying each probability by the value of that prize and then adding these products together.

Calculator Hint

To change $\frac{80}{400}$ to a decimal, press

80 ÷ 400 =
0.20

$$80\left(\tfrac{1}{400}\right) + 20\left(\tfrac{1}{200}\right) + 4\left(\tfrac{1}{40}\right) + 0\left(\tfrac{387}{400}\right) = \tfrac{80}{400} + \tfrac{20}{200} + \tfrac{4}{40} + 0$$
$$= 0.20 + 0.10 + 0.10 + 0$$
$$= 0.40$$

The expected value of a ticket is 40¢. That means that if Dawn buys one ticket after another, she could expect to win an average of 40¢ in prizes per ticket. Since this is less than the cost of a ticket, Dawn believes the ticket is *not* a good buy. But she is happy to help the Boy Scouts anyway because they will make a profit of $1.00 − $0.40 or $0.60 per ticket.

Games Barry's hometown of Bucyrus, Ohio, holds a bratwurst festival each year. One of the games at the festival is called Spin-O-Rama. Each spin costs $1. The number on the spinner tells you how much you win. Is this game a good buy?

First find the probability of each event.

$P(\text{spinning a } 0) = \frac{1}{2}$

$P(\text{spinning a } 1) = \frac{1}{4}$

$P(\text{spinning a } 2) = \frac{1}{8}$

$P(\text{spinning a } 3) = \frac{1}{8}$

Now find the expected value of a spin.

$$0\left(\frac{1}{2}\right) + 1\left(\frac{1}{4}\right) + 2\left(\frac{1}{8}\right) + 3\left(\frac{1}{8}\right) = \frac{1}{4} + \frac{1}{4} + \frac{3}{8}$$

$$= \frac{7}{8}$$

Calculator Hint

• • • • • • • • • • • • •

To change $\frac{7}{8}$ to a decimal, press

7 $\boxed{\div}$ 8 $\boxed{=}$ 0.875

The expected value of a spin is $\frac{7}{8}$ of a dollar or about 88¢. That means that if you play the game over and over, you could expect to win an average of 88¢ per spin. Since this is less than the cost of a spin, the game is *not* a good buy. However, many people will play anyway to try to beat "the odds."

Checking for Understanding

Communicating Mathematics

Read and study the lesson to answer each question.

1. **Write**, in your own words, how you would find the expected value of a raffle ticket.

2. **Tell** how knowing the expected value of an event helps you to make a good buy.

Guided Practice

Use Spinner A to find each of the following.

3. $P(\text{spinning a } 0)$

4. $P(\text{spinning a } 1)$

5. $P(\text{spinning a } 2)$

Spinner A

6. If the number on the spinner tells the number of dollars you win, find the expected value of a spin.

7. Over the long-term, would you expect to win, lose, or break even if each spin costs 50¢?

Exercises

Independent Practice

Use Spinner B to answer each of the following.

Spinner B

8. If the number on the spinner tells the number of dollars you win, find the expected value of a spin.

9. Over the long term, would you expect to win, lose, or break even if each spin cost $2?

10. What should each spin cost to be a fair game (you have an equal chance of winning or losing)?

Use a die to answer each of the following.

11. If the number on the die tells the number of dollars you win, find the expected value of a roll.

12. Would you play the game if each roll costs $2? Why or why not?

Mixed Review

13. **Statistics**   The bar graph at the right displays the responses to the question "What is your favorite soft drink?" Which soft drinks should the manager of the local grocery always keep in stock? *(Lesson 3-7)*

Favorite Soft Drinks

14. **Geometry**   Find the circumference of a circle having a radius of $4\frac{7}{8}$ inches. *(Lesson 5-7)*

Problem Solving and Applications

15. **Science**   Use the table below to determine the expected number of girls in a family having three children. Leave your answer as a mixed number.

number of girls	0	1	2	3
probability	$\frac{1}{8}$	$\frac{3}{8}$	$\frac{3}{8}$	$\frac{1}{8}$

16. **Critical Thinking**   If the numbers were rearranged on Spinner C as shown at the right, would the expected value in Exercise 8 increase, decrease, or remain the same? Explain your answer.

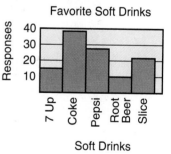

Spinner C

5-9 Properties

Objective

Use addition, multiplication, and distributive properties to solve problems mentally.

Words to Learn

commutative
associative
identity
inverse
multiplicative
  inverse
reciprocal
distributive

Don't ever try to outrun a cheetah. The cheetah is the fastest land animal in the world. Its speed is $2\frac{1}{2}$ times that of the fastest human's speed. If the fastest recorded speed for a human is 28 miles per hour, how fast can the cheetah run? You can use fraction properties of multiplication to compute the speed mentally. *You will solve this problem in Example 1.*

Addition and multiplication of fractions have the same properties you learned for addition and multiplication of whole numbers. These properties are important when you are computing sums and products. The properties are summarized in the chart below.

Property	Arithmetic	Algebra
Commutative	$\frac{1}{5} + \frac{2}{5} = \frac{2}{5} + \frac{1}{5}$	$a + b = b + a$
	$\frac{1}{2} \times \frac{3}{8} = \frac{3}{8} \times \frac{1}{2}$	$a \times b = b \times a$
Associative	$\left(\frac{1}{8} + \frac{3}{8}\right) + \frac{5}{8} = \frac{1}{8} + \left(\frac{3}{8} + \frac{5}{8}\right)$	$(a + b) + c = a + (b + c)$
	$\left(\frac{1}{2} \times \frac{2}{3}\right) \times \frac{1}{4} = \frac{1}{2} \times \left(\frac{2}{3} \times \frac{1}{4}\right)$	$(a \times b) \times c = a \times (b \times c)$
Identity	$\frac{1}{2} + 0 = \frac{1}{2}$	$a + 0 = a$
	$\frac{1}{2} \times 1 = \frac{1}{2}$	$a \times 1 = a$

There is a property that applies to multiplication. It is called the **inverse** property of multiplication. Two numbers whose product is 1 are **multiplicative inverses,** or **reciprocals,** of each other.

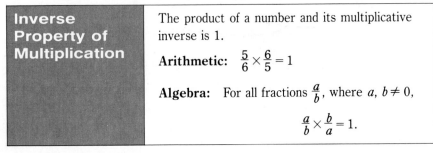

Inverse Property of Multiplication	The product of a number and its multiplicative inverse is 1.
	Arithmetic: $\frac{5}{6} \times \frac{6}{5} = 1$
	Algebra: For all fractions $\frac{a}{b}$, where $a, b \neq 0$,
	$$\frac{a}{b} \times \frac{b}{a} = 1.$$

The **distributive** property involves two operations.

Distributive Property of Multiplication over Addition	The sum of two addends multiplied by a number is the sum of the product of each addend and the number.
	Arithmetic: $\frac{1}{2} \times \left(\frac{2}{5} + \frac{1}{3}\right) = \frac{1}{2} \times \frac{2}{5} + \frac{1}{2} \times \frac{1}{3}$
	Algebra: $a \times (b + c) = a \times b + a \times c$

Examples

1 Find the speed of a cheetah, using the information given in the lesson opener.

$$2\frac{1}{2} \times 28 = \left(2 + \frac{1}{2}\right) \times 28$$

$$= 2 \cdot 28 + \frac{1}{2} \cdot 28 \qquad \textit{Distributive property}$$

$$= 56 + 14$$

$$= 70$$

The cheetah can run 70 miles per hour.

2 Name the multiplicative inverse of $2\frac{3}{4}$.

$$2\frac{3}{4} = \frac{11}{4} \qquad \textit{Rename the mixed number as an improper fraction.}$$

$$\frac{11}{4} \times \square = 1 \qquad \textit{What number can you mutiply by } \frac{11}{4} \textit{ to get 1?}$$

$$\frac{11}{4} \times \frac{4}{11} = 1$$

The multiplicative inverse of $2\frac{3}{4}$ is $\frac{4}{11}$.

Mental Math Hint
• • • • • • • • • • • •
The distributive property allows you to "break apart" one of the factors into a sum. You can then add the two products mentally.

3 Compute $\frac{1}{4} \times 8\frac{4}{5}$ mentally.

$$\frac{1}{4} \times 8\frac{4}{5} = \frac{1}{4}\left(8 + \frac{4}{5}\right)$$

$$= \frac{1}{4} \times 8 + \frac{1}{4} \times \frac{4}{5} \qquad \textit{Think: } \frac{1}{4} \textit{ of 8 is 2.}$$

$$= 2 + \frac{1}{5} \textit{ or } 2\frac{1}{5} \qquad\qquad\qquad \frac{1}{4} \textit{ of } \frac{4}{5} \textit{ is } \frac{1}{5}.$$

Checking for Understanding

Communicating Mathematics

Read and study the lesson to answer each question.

1. **Write** an arithmetic sentence that shows the associative property of addition.

   Tell how the identity property of addition is similar to the identity property of multiplication.

Name the property shown by each statement.

3. $\frac{7}{8} + 0 = \frac{7}{8}$ 4. $\frac{1}{8} + \frac{3}{4} = \frac{3}{4} + \frac{1}{8}$

State which pairs of numbers are multiplicative inverses. Write *yes* or *no*.

5. $\frac{2}{3}, 3$ 6. $\frac{4}{5}, 1\frac{1}{4}$ 7. $7, \frac{1}{7}$ 8. $\frac{3}{10}, \frac{10}{3}$

Compute mentally.

9. $2 \times 1\frac{1}{6}$ 10. $\frac{1}{2} \times 4\frac{2}{5}$ 11. $8\frac{1}{2} \times \frac{1}{4}$

Exercises

Independent Practice

Name the multiplicative inverse of each number.

12. $\frac{9}{10}$ 13. $\frac{7}{8}$ 14. $1\frac{2}{3}$ 15. 3

Name the property shown by each statement.

16. $\left(\frac{1}{3} + \frac{1}{2}\right) + 7 = \frac{1}{3} + \left(\frac{1}{2} + 7\right)$ 17. $\frac{3}{5} \times 1\frac{2}{3} = 1$

18. $\frac{3}{10} \times 1 = \frac{3}{10}$ 19. $\frac{2}{3} \times \left(\frac{1}{2} + \frac{3}{4}\right) = \frac{2}{3} \times \frac{1}{2} + \frac{2}{3} \times \frac{3}{4}$

Compute mentally.

20. $4 \times 6\frac{1}{8}$ 21. $2 \times 5\frac{1}{2}$ 22. $1\frac{1}{3} \times 6$ 23. $\frac{1}{3} \times 9\frac{1}{2}$

Mixed Review

24. A typist types 1,950 words in 30 minutes. What is her typing rate in words per minute? *(Lesson 1-1)*

25. Divide 12.36 by 3.5. Round to the nearest tenth. *(Lesson 2-8)*

26. **Probability**   There are 36 ways for two dice to land. What is the probability that you will roll an 8? *(Lesson 4-8)*

Problem Solving and Applications

27. **Critical Thinking**   Use the properties of addition and mutliplication to compute $50\left(3\frac{3}{8}\right) + 50\left(6\frac{5}{8}\right)$ mentally.

28. **Cooking**   The shortcake recipe at the right is for 6 servings. Write the recipe that will make 12 servings.

> $2\frac{1}{3}$ cups of biscuit mix
>
> $\frac{1}{2}$ cup milk
>
> 3 tablespoons of sugar
>
> 3 tablespoons of margarine

29. **Geometry**   Find the area of the rectangle at the right.

$\frac{1}{2}$ in.

$2\frac{1}{3}$ in.

5-10 Dividing Fractions and Mixed Numbers

Objective

Divide fractions and mixed numbers.

DID YOU KNOW

The four Hs in 4-H stand for head, heart, hands, and health.

Ellen is in charge of buying groceries for her 4-H overnight camping trip. She figures that each person will drink an average of $1\frac{1}{3}$ cups of orange juice for breakfast. If she buys one quart of orange juice for five people, will there be enough juice? *Recall that 1 quart = 4 cups.*

To solve this problem, we need to find how many $1\frac{1}{3}$ cups are in 4 cups. Divide 4 by $1\frac{1}{3}$.

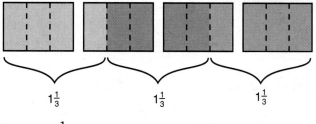

$$1\frac{1}{3} \qquad 1\frac{1}{3} \qquad 1\frac{1}{3}$$

So, $4 \div 1\frac{1}{3} = 3$.

We can also divide by a fraction or mixed number. To do this multiply by its multiplicative inverse.

$4 \div 1\frac{1}{3} = \frac{4}{1} \div \frac{4}{3}$ *Rename 4 as $\frac{4}{1}$ and $1\frac{1}{3}$ as $\frac{4}{3}$.*

$\qquad\quad = \frac{4}{1} \times \frac{3}{4}$ *Dividing by $\frac{4}{3}$ is the same as*

$\qquad\quad = \frac{3}{1}$ or 3   *multiplying by $\frac{3}{4}$.*

One quart of orange juice will be enough for 3 people, *not* 5 people.

Division of Fractions and Mixed Numbers	To divide by a fraction, multiply by its multiplicative inverse.
	Arithmetic: $\frac{4}{5} \div \frac{2}{3} = \frac{4}{5} \cdot \frac{3}{2}$
	Algebra: $\frac{a}{b} \div \frac{c}{d} = \frac{a}{b} \cdot \frac{d}{c}$, where b, c, and $d \neq 0$.

Divide. Write each quotient in simplest form.

1 $\frac{1}{2} \div \frac{2}{3}$ *Estimate:* $\frac{1}{2} \div 1 = \frac{1}{2}$

$\frac{1}{2} \div \frac{2}{3} = \frac{1}{2} \times \frac{3}{2}$ *Dividing by* $\frac{2}{3}$ *is the same as multiplying by* $\frac{3}{2}$.

$\qquad = \frac{3}{4}$

2 $\frac{3}{4} \div 4\frac{1}{2}$ *Estimate:* $1 \div 5 = \frac{1}{5}$

$\frac{3}{4} \div 4\frac{1}{2} = \frac{3}{4} \div \frac{9}{2}$ *Rename* $4\frac{1}{2}$ *as* $\frac{9}{2}$.

$\qquad = \frac{\overset{1}{\cancel{3}}}{\underset{2}{\cancel{4}}} \times \frac{\overset{1}{\cancel{2}}}{\underset{3}{\cancel{9}}}$ *Dividing by* $\frac{9}{2}$ *is the same as multiplying by* $\frac{2}{9}$.

$\qquad = \frac{1}{6}$

3 $2\frac{2}{3} \div 1\frac{1}{2}$ *Estimate:* $3 \div 2 = 1\frac{1}{2}$

$2\frac{2}{3} \div 1\frac{1}{2} = \frac{8}{3} \div \frac{3}{2}$ *Rename* $2\frac{2}{3}$ *as* $\frac{8}{3}$ *and* $1\frac{1}{2}$ *as* $\frac{3}{2}$.

$\qquad = \frac{8}{3} \times \frac{2}{3}$ *Dividing by* $\frac{3}{2}$ *is the same as multiplying by* $\frac{2}{3}$.

$\qquad = \frac{16}{9}$

$\qquad = 1\frac{7}{9}$ *Rename as a mixed number.*

Example 4 *Problem Solving*

Food Mrs. Lazo had $\frac{2}{3}$ of a pie left for dinner. She divided it into 4 equivalent parts for her family. What part of a pie will each family member get?

$\frac{2}{3} \div 4 = \frac{2}{3} \div \frac{4}{1}$ *Rename 4 as* $\frac{4}{1}$.

$\qquad = \frac{\overset{1}{\cancel{2}}}{3} \times \frac{1}{\underset{2}{\cancel{4}}}$ *Dividing by* $\frac{4}{1}$ *is the same as multiplying by* $\frac{1}{4}$.

$\qquad = \frac{1}{6}$ Each serving is $\frac{1}{6}$ of a pie.

Checking for Understanding

Read and study the lesson to answer each question.

1. **Draw a model** to show what $4\frac{1}{2} \div \frac{3}{4}$ means.

2. **Write** the multiplicative inverse of $2\frac{1}{4}$.

Name the multiplicative inverse of each number.

3. $\frac{3}{5}$ 4. 2   5. $\frac{1}{3}$ 6. $4\frac{1}{2}$

Divide. Write each quotient in simplest form.

7. $\frac{3}{4} \div \frac{1}{2}$ 8. $3 \div \frac{6}{7}$ 9. $2\frac{2}{3} \div 4$ 10. $1\frac{1}{4} \div 3\frac{1}{2}$

11. $\frac{1}{8} \div \frac{1}{3}$ 12. $4\frac{2}{3} \div \frac{7}{8}$ 13. $\frac{9}{10} \div 2\frac{1}{4}$ 14. $\frac{3}{4} \div \frac{1}{2}$

Exercises

Name the multiplicative inverse of each number.

15. $\frac{5}{6}$ 16. 3   17. $\frac{4}{5}$ 18. $2\frac{9}{10}$ 19. $3\frac{3}{5}$

Divide. Write each quotient in simplest form.

20. $\frac{2}{3} \div \frac{1}{2}$ 21. $\frac{3}{5} \div \frac{1}{4}$ 22. $\frac{5}{6} \div \frac{2}{3}$ 23. $\frac{1}{6} \div \frac{1}{4}$

24. $6 \div \frac{1}{2}$ 25. $\frac{3}{8} \div \frac{6}{7}$ 26. $\frac{4}{9} \div 2$ 27. $\frac{5}{9} \div \frac{5}{6}$

28. $\frac{3}{4} \div \frac{3}{8}$ 29. $\frac{2}{3} \div 2\frac{1}{2}$ 30. $5 \div 1\frac{1}{3}$ 31. $2\frac{1}{4} \div \frac{2}{3}$

32. $2\frac{2}{3} \div 5\frac{1}{3}$ 33. $1\frac{1}{9} \div 1\frac{2}{3}$ 34. $5\frac{1}{4} \div 3$ 35. $4\frac{1}{2} \div 6\frac{3}{4}$

36. Find the quotient of $\frac{1}{8}$ and $\frac{1}{9}$.

37. Solve the equation $x = \left(\frac{1}{5} + \frac{1}{12}\right) \div 3\frac{1}{2}$.

38. Solve mentally $\frac{t}{5} = 15$. *(Lesson 1-10)*

39. Round 10.2573 to the nearest tenth. *(Lesson 2-2)*

40. Add $\frac{1}{3}$ and $\frac{1}{4}$. *(Lesson 5-3)*

41. **Health** Tim jogs on a circular track that has a diameter of 77 feet. How far does Tim jog each time he goes around the track? *(Lesson 5-7)*

42. Compute mentally $\frac{5}{6} \times 3\frac{3}{7}$. *(Lesson 5-9)*

43. **Housing** A contractor is going to develop land for single-family homes near the Eastland Mall. If she buys 12 acres of land, how many $\frac{3}{4}$-acre lots can she sell?

44. **Food** Teresa and two friends bought $\frac{1}{2}$ pound of trail mix to eat as an after school snack. How much mix will each girl get?

45. **Critical Thinking** Will the quotient $5\frac{1}{8} \div 3\frac{1}{4}$ be a proper fraction or a mixed number? Why?

5-10B Fraction Patterns

A Follow-Up of Lesson 5-10

Objective

Find unit price.

Words to Learn

unit price

Materials

play money
sheets of paper

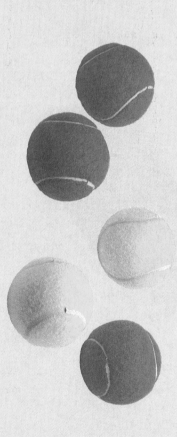

Anna wants to find the **unit price** for each item she bought at the grocery store. To find the unit price means to find the cost per pound, yard, gallon, square foot, or some other unit of measure.

Grocery Item	Cost
5 pounds of potatoes	$3.75
$3\frac{1}{2}$-ounce box of ground cinnamon	$2.10
half-pound box of turkey stuffing mix	$1.09

Activity One

Help Anna find the unit price of potatoes by using 5 sheets of paper to represent 5 pounds of potatoes. Using the play money, distribute $3.75 evenly on the 5 sheets of paper to find the price of 1 pound.

| 1 lb | 1 lb | 1 lb | 1 lb | 1 lb |

Find $3.75 ÷ 5. What is the price per pound?

What do you think?

A can of three tennis balls costs $2.40. What is the cost per tennis ball?
1. How many sheets of paper are needed?
2. What is the cost per tennis ball?
3. Find $2.40 ÷ 3 to check your answer.

Extension

4. Doughnuts are $1.50 for a half dozen. What is the cost per donut?

Activity Two

To help Anna find the unit price of ground cinnamon, use $3\frac{1}{2}$ sheets of paper to represent $3\frac{1}{2}$ ounces. Using the play money, distribute $2.10 so that the half sheet of paper has half the money as each full sheet of paper.

1 oz   1 oz   1 oz $\frac{1}{2}$ oz

Find $2.10 ÷ $3\frac{1}{2}$. What is the price per ounce?

What do you think?

Mrs. Grant paid $3.75 for $2\frac{1}{2}$ yards of dress material. What is the cost per yard?

5. How many sheets of paper are needed?

6. How much money should a half sheet of paper contain?

7. What is the cost of dress material per yard?

8. Find $3.75 ÷ $2\frac{1}{2}$ to check your answer.

Extension

9. Mr. O'Grady bought $1\frac{1}{2}$ pounds of nails for 90¢. What was the price per pound?

Activity Three

To help Anna find the unit price of the turkey stuffing, use a half sheet of paper to represent $\frac{1}{2}$ pound. Put $1.09 on the half sheet of paper. A whole pound would be represented by two half sheets of paper. Put $1.09 on a second half sheet of paper to find the price per pound.

$\frac{1}{2}$ lb $\frac{1}{2}$ lb

Find $1.09 ÷ $\frac{1}{2}$. What is the price per pound?

What do you think?

A $\frac{1}{4}$-pound box of chocolates sells for 95¢. How much would a pound of chocolates cost?

10. What is the cost of one pound of chocolates?

11. Find 95¢ ÷ $\frac{1}{4}$ to check your answer.

Extension

12. One-third of an acre of residential property costs $5,000. What is the cost per acre?

5-11 Eliminate Possibilities

Objective

Solve problems by using estimation to eliminate possibilities.

Today the Camera Club had a show of its best work. After the show, the six winners shook hands. Each winner shook hands once with the other five. How many handshakes were there in all? Choose the best estimate from the list below.

36   30   15   6

Explore

What do you know?
There were six winners. Each winner shook hands one time with the other five winners.

What are you trying to find?
You are trying to find the best estimate of how many handshakes there were.

Plan

Look at the choices of estimates to see which ones are impossible and eliminate those. Then concentrate on the other estimates to choose the best one.

Solve

Handshakes	Interpret	Eliminate?
6 shakes	One person shakes the hands of the other five. Then there is only one shake left. So 6 is too few handshakes.	yes
36 shakes	Six times six means that six persons each shake hands with 6 different people. But no person shakes his own hand, so 36 is too many handshakes.	yes
30 shakes	Six times five means that six persons each shake hands with 5 different people. But once the first person shakes hands with the second person, those two don't shake again. So 30 is too many handshakes.	yes

The best estimate must be 15 shakes.

Examine   The first person shakes hands with the other five people. Then the next person only has 4 handshakes left. The third has 3, the fourth has 2, and the final shake is that of the fifth and sixth winners.

$$5 + 4 + 3 + 2 + 1 = 15$$

The total is 15 handshakes.

Checking for Understanding

Communicating Mathematics   Read and study the lesson to answer each question.

1. **Write** a problem with multiple-choice answers.
2. **Tell** how you use the strategy of eliminating possibilities to solve a problem.

Guided Practice   Solve by eliminating the possibilities.

3. Jessica's math test scores were 80, 78, 87, 70, and 81. Choose the best average for her test scores.

    75   80   83   86

4. Jeff bought felt-tipped pens, 3 for $1.39; pencils, 2 for $0.49; and an eraser for $0.29. Choose the best estimate for the amount of change he will get from $5.

    $0.30 $1.80 $2.70 $3.30

Problem Solving

Practice   Solve using any stragegy.

Strategies
● ● ● ● ● ● ● ● ● ●
Look for a pattern.
Solve a simpler problem.
Act it out.
Guess and check.
Draw a diagram.
Make a chart.
Work backward.

5. A taxi charges $1.15 for the first $\frac{1}{5}$ mile and $0.50 for each additional $\frac{1}{5}$ mile. Choose the best estimate for the cost of a 4-mile taxi ride.

    $13.65 $12.65 $11.65 $10.65

6. Meagan has 4 less than 3 times as many compact discs as Jason. Meagan has 92 compact discs. How many CDs does Jason have?

7. Luis bought a VCR for $280. He could have paid for it in 12 monthly installments of $26.50 each. If he paid cash, choose the best estimate for the amount of money he saved.

    $80 $54 $38 $28

5 Study Guide and Review

Communicating Mathematics

Choose the letter that best matches each phrase.

1. a fraction that has a numerator that is less than the denominator
2. the sum of a whole number and a fraction
3. the least common denominator of $\frac{1}{8}$ and $\frac{1}{6}$
4. the distance around a rectangle
5. the distance across a circle through the center
6. the property that states $(a + b) + c = a + (b + c)$
7. In your own words, explain how the distributive property can be used to compute the product of a mixed number and a whole number.

a. mixed number
b. perimeter
c. 24
d. circumference
e. commutative
f. proper fraction
g. 48
h. radius
i. expected value
j. diameter
k. associative
l. distributive

Skills and Concepts

Objectives and Examples

Upon completing this chapter, you should be able to:

Review Exercises

Use these exercises to review and prepare for the chapter test.

- change mixed numbers to improper fractions and vice versa *(Lesson 5-1)*

$$3\frac{4}{9} = \frac{(3 \times 9) + 4}{9} = \frac{31}{9}$$

$$\frac{12}{8} \rightarrow 8\overline{)12} \overset{1}{\underset{4}{\underset{-8}{}}} \rightarrow 1\frac{4}{8} \text{ or } 1\frac{1}{2}$$

Change each mixed number to an improper fraction.

8. $5\frac{2}{5}$ 9. $3\frac{4}{7}$ 10. $2\frac{9}{10}$

Change each improper fraction to a mixed number in simplest form or a whole number.

11. $\frac{63}{4}$ 12. $\frac{12}{5}$ 13. $\frac{21}{8}$

- estimate sums, differences, products, and quotients of fractions and mixed numbers *(Lesson 5-2)*

Estimate $14\frac{5}{9} + 2\frac{1}{6}$.

$15 + 2 = 17$

Estimate.

14. $\frac{2}{3} + \frac{7}{9}$ 15. $6\frac{5}{4} + 11\frac{2}{7}$

16. $\frac{1}{3} \times \frac{5}{8}$ 17. $\frac{2}{5} \div \frac{3}{4}$

18. $\frac{14}{15} - \frac{1}{8}$ 19. $99\frac{9}{10} - 9\frac{9}{10}$

20. $1\frac{21}{25} \times 16\frac{4}{13}$ 21. $6\frac{1}{8} \div 1\frac{5}{6}$

Objectives and Examples

- add and subtract fractions *(Lesson 5-3)*

$$\frac{11}{14} - \frac{1}{6} = \frac{33}{42} - \frac{7}{42}$$

$$= \frac{26}{42} \text{ or } \frac{13}{21}$$

- add and subtract mixed numbers with unlike denominators *(Lesson 5-4)*

$$6\frac{3}{4} + 9\frac{2}{3} = 6\frac{9}{12} + 9\frac{8}{12}$$

$$= 15\frac{17}{12} \text{ or } 16\frac{5}{12}$$

- multiply fractions and mixed numbers *(Lesson 5-5)*

$$\frac{2}{5} \times \frac{1}{3} = \frac{2}{15}$$

$$4\frac{1}{4} \times 2\frac{2}{3} = \frac{17}{\underset{1}{4}} \times \frac{\overset{2}{8}}{3}$$

$$= \frac{34}{3} \text{ or } 11\frac{1}{3}$$

- find perimeter using fractional measurements *(Lesson 5-6)*

Find the perimeter of a rectangle $4\frac{3}{4}$ feet long and $2\frac{1}{3}$ feet wide.

$$P = 4\frac{3}{4} + 4\frac{3}{4} + 2\frac{1}{3} + 2\frac{1}{3}$$

$$= 14\frac{2}{12} \text{ or } 14\frac{1}{6} \text{ feet}$$

- find the circumference of circles *(Lesson 5-7)*

If the diameter is $3\frac{1}{2}$ feet, then find the circumference.

$$C = \pi d$$

$$= \frac{22}{7} \times 3\frac{1}{2}$$

$$= \frac{22}{7} \times \frac{7}{2}$$

$$= 11 \quad \text{The circumference is 11 feet.}$$

Review Exercises

Add or subtract. Write each sum or difference in simplest form.

22. $\begin{array}{r} \frac{3}{5} \\ + \frac{2}{7} \\ \hline \end{array}$

23. $\begin{array}{r} \frac{3}{4} \\ - \frac{1}{2} \\ \hline \end{array}$

24. $\frac{7}{33} + \frac{1}{3}$

25. $\frac{5}{8} - \frac{3}{20}$

Add or subtract. Write each sum or difference in simplest form.

26. $9 + 5\frac{4}{9}$

27. $4\frac{5}{12} + 3\frac{5}{6}$

28. $2\frac{5}{8} - \frac{1}{6}$

29. $16\frac{1}{7} - 2\frac{1}{2}$

Multiply. Write each product in simplest form.

30. $\frac{4}{9} \times \frac{2}{5}$

31. $\frac{21}{5} \times 4\frac{1}{3}$

32. $5 \times 3\frac{1}{8}$

33. $6\frac{2}{7} \times 5\frac{1}{2}$

Find the perimeter of each figure.

34. rectangle: $\ell = 4\frac{1}{2}$ yd

$$w = 3\frac{3}{4} \text{ yd}$$

35. rectangle: $\ell = 25\frac{1}{7}$ feet

$$w = 12\frac{2}{3} \text{ feet}$$

Find the circumference of each circle. Use $\pi = \frac{22}{7}$.

36. $d = 6$ in.

37. $r = 3.7$ yd

38. $r = \frac{7}{9}$ ft

39. $d = 6\frac{3}{5}$ ft

Objectives and Examples

- find expected value of outcomes *(Lesson 5-8)*

Find the expected winnings.

Amount Won	$1	$3	$5
Probability	$\frac{2}{5}$	$\frac{3}{10}$	$\frac{3}{10}$

$$1\left(\frac{2}{5}\right) + 3\left(\frac{3}{10}\right) + 5\left(\frac{3}{10}\right) = \$2.80$$

- use addition and multiplication properties to solve problems mentally *(Lesson 5-9)*

$$\frac{1}{5} \times 10\frac{3}{4} = \frac{1}{5}\left(10 + \frac{3}{4}\right)$$
$$= 2 + \frac{3}{20} \text{ or } 2\frac{3}{20}$$

- divide fractions and mixed numbers *(Lesson 5-10)*

$$1\frac{2}{5} \div \frac{7}{10} = \frac{\overset{1}{\cancel{7}}}{\cancel{5}} \times \frac{\overset{2}{\cancel{10}}}{\cancel{7}} = \frac{2}{1} \text{ or } 2$$

Review Exercises

40. Find the expected winnings.

Amount Won	$0	$1	$5	$10
Probability	$\frac{17}{20}$	$\frac{1}{10}$	$\frac{1}{50}$	$\frac{1}{100}$

Compute mentally.

41. $6 \times 2\frac{1}{8}$ 42. $\frac{5}{9} \times 3\frac{1}{5}$

43. $\frac{2}{3} \times 6\frac{1}{12}$ 44. $4\frac{1}{6} \times 3$

Divide. Write each quotient in simplest form.

45. $\frac{3}{5} \div \frac{1}{6}$ 46. $1\frac{7}{8} \div 2\frac{4}{9}$

Applications and Problem Solving

47. **Cooking** A recipe calls for $2\frac{3}{4}$ cups of flour, $1\frac{3}{8}$ cups of sugar, and $1\frac{1}{3}$ cups of brown sugar. How many cups are in the mixture? *(Lesson 5-4)*

48. Lou parked at a parking lot from 8:30 A.M. until 2:45 P.M. It cost $2.25 for the first hour and $0.90 for each additional hour or part of an hour. How much did Lou pay when he left the lot? *(Lesson 5-11)*

Curriculum Connection Projects

- **Physical Education** Find how far you travel with each complete turn of your bicycle wheel.
- **Science** How would you find the perimeter of one side of the school building without using a ladder? Try it!

Read More About It

Barron, T. A., *Heartlight.*
Sitomer, Mindel and Harry. *Circles.*

Buchman, Dian Dincin, and Seli Groves. *What If? Fifty Discoveries that Changed the World.*

Express each improper fraction as a mixed number or vice versa.

1. $\frac{124}{12}$

2. $6\frac{5}{9}$

3. **Music** Philip's rock band recorded 13 songs during an afternoon recording session. Each song took $\frac{1}{6}$ of a side of a tape to record. How many sides were used during the session?

Estimate.

4. $1\frac{1}{3} + \frac{7}{8}$

5. $11\frac{4}{5} - 6\frac{1}{8}$

6. $16\frac{1}{8} \div \frac{5}{6}$

Add, subtract, multiply, or divide. Write each sum, difference, product, or quotient in simplest form.

7. $\frac{3}{5} + \frac{4}{9}$

8. $\frac{5}{6} - \frac{13}{27}$

9. $\frac{5}{12} \times \frac{3}{10}$

10. $2\frac{5}{12} \div \frac{3}{8}$

Add or subtract. Write each sum or difference in simplest form.

11. $4\frac{1}{3} + 3\frac{1}{4}$

12. $16\frac{5}{9} - 3\frac{1}{12}$

13. $14\frac{2}{3} - 8\frac{7}{8}$

14. Find the perimeter of a rectangle with a length of $5\frac{3}{4}$ yards and a width of 3 yards.

Find the circumference for each circle. Use $\pi = \frac{22}{7}$.

15. $d = 5\frac{3}{4}$ in.

16. $r = \frac{2}{3}$ yd

17. The spinner to the right tells the number of dollars you win. Over the long run, would you expect to win, lose, or break even if each spin costs $2.

Compute mentally.

18. $6 \times 3\frac{1}{5}$

19. $\frac{4}{5} \times 1\frac{1}{4}$

20. Kate found a $\frac{1}{3}$-off sale. About how much did she pay for a $39.95 golf jacket and a set of $18.65 sweats?

$9.07 $14.20 $20.00 $39.07

Bonus The wheel of a car is $2\frac{6}{11}$ feet in diameter. If the car travels 80 feet, how many times does the wheel rotate?

6

An Introduction to Algebra

Spotlight on Temperature

Have You Ever Wondered. . .

- Why it seems cooler than it actually is when there is less humidity?

- What the highest recorded temperatures are in the United States?

Apparent Temperature for Values of Room Temperature and Relative Humidity										

Relative Humidity (%)

Room Temperature (°F)	0	10	20	30	40	50	60	70	80	90	100
75	68	69	71	72	74	75	76	76	77	78	79
74	66	68	69	71	72	73	74	75	76	77	78
73	65	67	68	70	71	72	73	74	75	76	77
72	64	65	67	68	70	71	72	73	74	75	76
71	63	64	66	67	68	70	71	72	73	74	75
70	63	64	65	66	67	68	69	70	71	72	73
69	62	63	64	65	66	67	68	69	70	71	72
68	61	62	63	64	65	66	67	68	69	70	71
67	60	61	62	63	64	65	66	67	68	68	69

RATS! ANOTHER "D-MINUS"!

LIFE HAS ITS SUNSHINE AND ITS RAIN, SIR ..ITS DAYS AND ITS NIGHTS..ITS PEAKS AND ITS VALLEYS...

IT'S RAINING TONIGHT IN MY VALLEY!

© 1988 United Feature Syndicate, Inc.

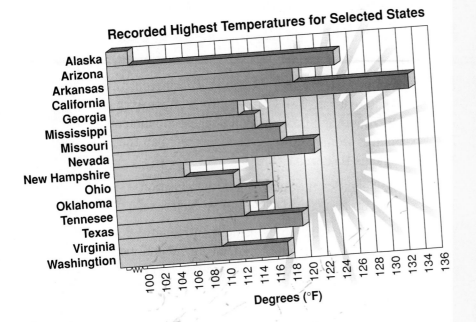

Recorded Highest Temperatures for Selected States

Alaska
Arizona
Arkansas
California
Georgia
Mississippi
Missouri
Nevada
New Hampshire
Ohio
Oklahoma
Tennesee
Texas
Virginia
Washingtion

100 102 104 106 108 110 112 114 116 118 120 122 124 126 128 130 132 134 136

Degrees (°F)

Looking Ahead

In this chapter, you
will see how
mathematics can be
used to answer the
questions about
temperature and
humidity. The major
objectives of the
chapter are to:

- write simple
  algebraic
  expressions

- solve equations by
  using mental
  math, models, and
  inverse operations

- solve addition,
  subtraction,
  multiplication, and
  division equations

- find the area of
  rectangles and
  parallelograms

Chapter Project

Temperature

Work in a group.

1. Keep a chart of the high
   temperature in your
   community each day for
   one month.

2. Be creative in showing
   what the temperature was
   on each day.

3. Show how the temperature
   on each day of the month
   varied (higher or lower)
   from the temperature on
   the first day of the project.
   Include any other
   descriptions about the
   weather that would help
   make your presentation
   clearer.

6-1 Solving Equations Using Inverse Operations

Objective

Solve equations using inverse operations.

Words to Learn

inverse operation

The Washington Monument is located in Washington, D.C., and stands 555 feet tall. The Gateway to the West Arch is located in St. Louis and is our nation's tallest monument. The Washington Monument is 75 feet shorter than the Arch. How tall is the Arch?

Let a be the height of the Arch. Then the equation $555 = a - 75$ can be used to solve the problem.

You could solve this equation by using mental math. But, as the numbers get larger or more difficult for mental math, you need another strategy. One such strategy is described below.

You can review variables on page 28.

$$555 = a - 75$$
$$555 + 75 = a \qquad \textit{Undo the subtraction by adding.}$$
$$630 = a \qquad \textit{Check: } 555 = 630 - 75$$

The height of the Arch is 630 feet.

You can undo the operation because addition and subtraction are **inverse operations.** *Related* addition and subtraction sentences are shown below.

$$11 + 4 = 15 \quad \rightarrow \quad 15 - 11 = 4 \qquad 18 + d = 24 \quad \rightarrow \quad 24 - 18 = d$$
$$\rightarrow \quad 15 - 4 = 11 \qquad\qquad\qquad\qquad \rightarrow \quad 24 - d = 18$$

Examples

Solve each equation by using inverse operations. This can be shown by writing a related sentence.

1 $t - 5 = 11$
$t = 11 + 5 \qquad \textit{Write a related addition sentence.}$
$t = 16 \qquad\qquad \textit{Check: } 16 - 5 = 11$

2 $b + 3\frac{1}{5} = 12\frac{3}{5}$
$b = 12\frac{3}{5} - 3\frac{1}{5} \qquad \textit{Write a related subtraction sentence.}$
$b = 9\frac{2}{5}$

Estimation Hint

In Example 2, think $b + 4 = 13$. The solution is *about* 9.

Similarly, multiplication and division are *inverse operations. Related* multiplication and division sentences are shown below.

$$7 \cdot 8 = 56 \quad \rightarrow \quad 56 \div 7 = 8 \qquad\qquad 7m = 35 \quad \rightarrow \quad 35 \div 7 = m$$
$$\rightarrow \quad 56 \div 8 = 7 \qquad\qquad\qquad\qquad \rightarrow \quad 35 \div m = 7$$

3 $\frac{m}{2} = 7$

$m = 7 \cdot 2$ *Write a related multiplication sentence.*

$m = 14$

4 $2.3n = 75$

$n = 75 \div 2.3$ *Write a related division sentence.*

$75 \boxed{\div} 2.3 \boxed{=} \mathsf{32.608695}$

$n \approx 32.6$ Round to the nearest tenth.   *≈ means "is approximately equal to"*

Checking for Understanding

Communicating Mathematics

Read and study the lesson to answer each question.

1. **Tell** what the inverse of a temperature increase of three degrees would be.
2. **Tell** the operation you would use to undo subtraction.
3. **Write** an open sentence using multiplication. Then write two related division sentences.

Guided Practice

For each sentence, write a related sentence using the inverse operation.

4. $3 + 5 = 8$ 5. $9 - 5 = 4$ 6. $m + 7 = 19$

7. $n - 12 = 21$ 8. $6 \cdot 5 = 30$ 9. $\frac{54}{9} = 6$

10. $\frac{a}{2} = 13$ 11. $c = 3 \cdot 14$ 12. $4e = 48$

Exercises

Independent Practice

Solve each equation by using the inverse operation. Round to the nearest tenth.

13. $f + 7 = 15$ 14. $l + 4 = 13$ 15. $9 + n = 16$

16. $t - 5 = 17$ 17. $p - 9 = 22$ 18. $63 = r - 8$

19. $\frac{a}{6} = 19$ 20. $\frac{b}{7} = 25$ 21. $60 = \frac{c}{5}$

22. $4d = 64$ 23. $9e = 104$ 24. $102 = 6g$

25. $n + 3.8 = 17.2$ 26. $m + 14.1 = 26.5$

27. $1\frac{3}{4} = r + \frac{1}{2}$ 28. $s - 5.6 = 8.9$

29. $t - 18.8 = 3.2$ 30. $5\frac{3}{8} = d - 1\frac{1}{4}$

31. $\frac{x}{2.3} = 4$ 32. $\frac{y}{4.7} = 3.9$

33. $5.6 = \frac{z}{2.4}$ 34. $1.8a = 9.72$

35. What is the solution to $2.6b = 2.08$?

36. Solve the equation $0.79 = 0.6c$.

37. Estimate the quotient $\$4,724 \div 6$. *(Lesson 1-3)*

38. **Smart Shopping**   Pilar bought 3.75 pounds of beef that is priced at $2.29 a pound. Find the total amount she spent on the beef. *(Lesson 2-4)*

39. **Statistics**   Construct a stem-and-leaf plot for 12, 7, 23, 9, 10, 20, 0, 4, 19, 13, 5, 7, 2, 13, 18, and 2. *(Lesson 3-6)*

40. Write $\frac{18}{24}$ in simplest form. *(Lesson 4-6)*

41. Divide $4\frac{2}{5}$ by $\frac{1}{2}$. *(Lesson 5-10)*

Problem Solving and Applications   42. **Geometry**   Suppose that the sum of the measures of two angles is $180°$. One angle measures $96°$. Solve the equation $a + 96 = 180$ to find a, the measure of the second angle.

43. **Statistics**   The range of a set of scores is found by subtracting the lowest score from the highest. The range of a set of scores is 54 and the lowest score is 44. Solve the equation $s - 44 = 54$ to find s, the highest score.

44. **Sports**   A skateboarder travels 100 meters in 16 seconds. Solve the equation $100 = 16r$ to find r, the average speed in meters per second.

45. **Sports**   Michael Jordan has been averaging 33.6 points per game. In last night's game, he scored 38 points. Solve the equation $33.6 + p = 38$ to find p, how many points he scored over his average.

46. **Critical Thinking**   Think about the division sentence $\frac{3}{0} = h$ and its related multiplication sentence, $0 \cdot h = 3$. Use these sentences to explain why division by 0 is not defined.

Save Planet Earth

Animal Testing   Consumers have pressured many companies to stop using animals for testing their products, and now some are putting labels on their products that say so. More than 70 firms use one of the two labels shown below. One is used by the Humane Society and the other by People for Ethical Treatment of Animals. Some companies, such as Avon and Revlon, don't use animals for testing, but don't use a label on their products.

How You Can Help
For a list of 350 animal-friendly companies, send $2.95 to PETA, P.O. Box 42516, Washington, D.C. 20015. Ask for the current Shopping Guide for Caring Consumers.

6-2A Solving Equations Using Models

A Preview of Lesson 6-2

Objective

Solve equations using models.

Materials

cups
counters
mats

In the preceding lesson, you learned to solve an equation by using inverse operations. In this lab, you will use models to discover another method for solving equations.

Activity One

Work with a partner.

In this lab, a counter represents 1 and a cup represents the variable.

- Find the solution to the equation $x + 3 = 5$ by using models.

- On the first mat, place a cup and 3 counters. On the second mat, place 5 counters.

- To find the value of x, remove (subtract) 3 counters from the first mat and 3 counters from the second mat.

- The solution is $x = 2$.

What do you think?

1. Using the models, how could you check your solution?

2. Study the model. In your own words, what did you do to find the solution?

Mathematics Lab 6-2A Solving Equations Using Models   **223**

3. Repeat Activity 1 for the equations $w + 7 = 12$ and $8 + b = 14$. Explain to your group how you arrived at your solutions.

4. Draw a picture of the steps taken to find the solution to the equation $m + 5 = 11$.

Activity Two

- Find the solution to the equation $4x = 12$ by using models.

- Place 4 cups on the first mat. On the second mat, place 12 counters.

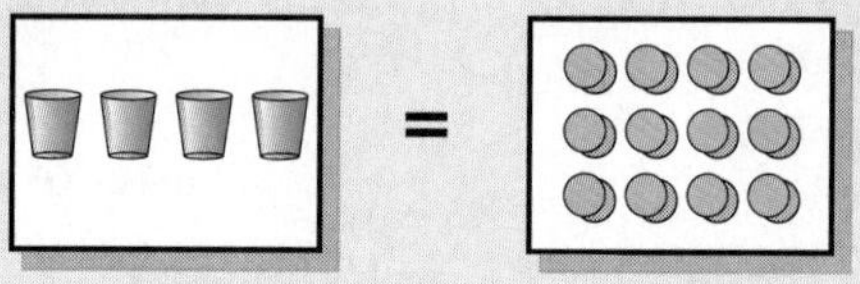

- Each cup must contain the same number of counters.

What do you think?

5. How would you complete the model shown above to find the value of x? *(Hint: Separate each side into the same number of groups.)*

6. What is the solution?

7. How could you check your solution?

8. In your own words, what did you do to find the solution?

9. Repeat the activity for the equations $5e = 20$ and $2t = 24$. Explain to your group how you arrived at your solutions.

10. Draw a picture of the steps taken to find the solution to the equation $3n = 18$.

11. Write two equations that are related to $7q = 63$. How are they related to each other?

Extension

12. In the equation $x + 9 = 15$, 9 was added to x. To solve it, you subtracted 9. Suppose you were solving the equation $2x - 3 = 13$. What operations would you use with 2 and 3 to find x?

6-2 Solving Addition and Subtraction Equations

Objective
Solve equations using the addition and subtraction properties of equality.

Words to Learn
equivalent equations
addition property
  of equality
subtraction
  property of equality

Orville and Wilbur Wright repaired bicycles for a living, but dreamed of inventing a flying machine. On December 17, 1903, in Kitty Hawk, North Carolina, they flew their airplane called the *Flyer I* for the first time. Wilbur was the first to fly the plane, but his flight ended up being 244 feet shorter than Orville's. If Wilbur flew a total of 120 feet, how long was Orville's flight?

Let f represent the number of feet Orville flew the plane. Then, use the equation $f - 244 = 120$ to solve the problem.

In the previous Mathematics Lab, you learned that an equation can be solved by keeping the equation balanced. If you add or subtract the same quantity to or from each side of the equation, the result is an equation that has the same solution. Such equations are called **equivalent equations.**

Solve $f - 244 = 120$.

$$f - 244 = 120$$
$$f - 244 + \underline{244} = 120 + \underline{244} \qquad \textit{Add 244 to each side.}$$
$$f = 364 \quad \text{The solution is 364. Orville flew 364 feet.}$$

The property that was used to add 244 to both sides of the equation is called the **addition property of equality.**

Addition Property of Equality	**In words:** If you add the same number to each side of an equation, then the two sides remain equal.
	<table><tr><td>**Arithmetic**</td><td>**Algebra**</td></tr><tr><td>$4 = 4$</td><td>$a = b$</td></tr><tr><td>$4 + 3 = 4 + 3$</td><td>$a + c = b + c$</td></tr><tr><td>$7 = 7$</td><td></td></tr></table>

There is a similar property when subtraction is used.

Subtraction Property of Equality	**In words:** If you subtract the same number from each side of an equation, then the two sides remain equal.
	<table><tr><td>**Arithmetic**</td><td>**Algebra**</td></tr><tr><td>$4 = 4$</td><td>$a = b$</td></tr><tr><td>$4 - 3 = 4 - 3$</td><td>$a - c = b - c$</td></tr><tr><td>$1 = 1$</td><td></td></tr></table>

Examples

Solve each equation. Check your solution.

1
$$x + 3.6 = 12.4$$
$$x + 3.6 - 3.6 = 12.4 - 3.6 \quad \textit{Subtract 3.6 from each side.}$$
$$12.4 \boxed{-} 3.6 \boxed{=} \text{8.8}$$
$$x = 8.8$$

Check: $x + 3.6 = 12.4$
$$8.8 \boxed{+} 3.6 \boxed{=} \text{12.4} \quad \textit{Replace x with 8.8.}$$
$$12.4 = 12.4 \checkmark \quad \text{The solution is 8.8.}$$

You can review adding
and subtracting
fractions on page 185.

2
$$8\tfrac{2}{3} + m = 15\tfrac{1}{6}$$
$$8\tfrac{2}{3} - 8\tfrac{2}{3} + m = 15\tfrac{1}{6} - 8\tfrac{2}{3} \quad \textit{Subtract } 8\tfrac{2}{3} \textit{ from each side.}$$
$$m = 14\tfrac{7}{6} - 8\tfrac{4}{6} \quad 15\tfrac{1}{6} = 14\tfrac{7}{6},\ 8\tfrac{2}{3} = 8\tfrac{4}{6}$$
$$m = 6\tfrac{3}{6} \text{ or } 6\tfrac{1}{2}$$

Check: $8\tfrac{2}{3} + m = 15\tfrac{1}{6}$
$$8\tfrac{2}{3} + 6\tfrac{1}{2} \stackrel{?}{=} 15\tfrac{1}{6} \quad \textit{Replace m with } 6\tfrac{1}{2}.$$
$$8\tfrac{4}{6} + 6\tfrac{3}{6} \stackrel{?}{=} 15\tfrac{1}{6} \quad 8\tfrac{2}{3} = 8\tfrac{4}{6},\ 6\tfrac{1}{2} = 6\tfrac{3}{6}$$
$$14\tfrac{7}{6} \stackrel{?}{=} 15\tfrac{1}{6}$$
$$15\tfrac{1}{6} = 15\tfrac{1}{6} \checkmark \quad \text{The solution is } 6\tfrac{1}{2}.$$

Checking for Understanding

Communicating Mathematics

Read and study the lesson to answer each question.

1. **Tell** how to check your solution to an equation.

2. **Model** the equation $y + 3 = 9$. Use the model to solve the equation.

3. **Tell** whether $m + 5 = 14$ and $m = 9$ are equivalent equations. Explain why or why not.

4. **Write** a sentence or two explaining how to solve $d - 8 = 12$.

Guided Practice

Complete the solution of each equation.

5. $$p + 21 = 46$$
$$p + 21 - 21 = 46 - 21$$
$$p = 25$$

6. $$t - 14 = 22$$
$$t - 14 + 14 = 22 + 14$$
$$t = 36$$

7. $$58 = s + 19$$
$$58 - 19 = s + 19 - 19$$
$$39 = s$$

Solve each equation. Check your solution.

8. $a + 3 = 12$

9. $k - 7.2 = 4.5$

10. $m - 8 = 13$

11. $2\tfrac{1}{4} + p = 6\tfrac{1}{2}$

12. $27 = 18 + g$

13. $b - 4\tfrac{1}{3} = 3\tfrac{5}{12}$

Exercises

Independent
Practice
Solve each equation. Check your solution.

14. $32 + c = 56$ 24

15. $m + 18 = 34$

16. $x + 27 = 39$

17. $y + 43 = 68$

18. $17\frac{3}{4} + a = 51\frac{1}{8}$ $33\frac{3}{8}$

19. $35 + n = 73$

20. $b - 63 = 14$

21. $e - 56 = 17$

22. $h - 13 = 47$

23. $42\frac{1}{4} = k - 5\frac{1}{6}$

24. $102 = x - 15$ 117

25. $44 = f - 83$

26. $a + 3.9 = 5.6$

27. $e + 11.8 = 13.1$

28. $p + 4.7 = 13.2$

29. $s - 5.9 = 4.8$

30. $e - 0.4 = 14.3$

31. $h - 28\frac{2}{5} = 47\frac{2}{3}$ $76\frac{1}{15}$

32. Solve the equation $46 + f = 98$.

33. Find the solution to $e - 6.9 = 13.3$.

Mixed Review
34. Express 9,800 in scientific notation. *(Lesson 2-6)*

35. **Statistics** Find the range and appropriate scale and interval for 45, 29, 31, 38, and 25. Then draw a number line to show the scale and interval. *(Lesson 3-3)*

36. Find the LCM for 8 and 12. *(Lesson 4-9)*

37. Solve the equation $\frac{r}{12} = 4$. *(Lesson 6-1)*

Problem Solving
and
Applications

38. **Stock Market** Cho purchased a share of Hershey's stock at $38\frac{1}{4}$. The next month it was selling for $40\frac{3}{8}$.

   a. Solve the equation $38\frac{1}{4} + m = 40\frac{3}{8}$ to find the value of m, the increase in the stock.

   b. How much would Cho make if she sold her stock?

39. **Law Enforcement** At the end of each working day, highway police must report the total amount of time spent performing certain types of tasks. During one 8-hour shift, an officer spent 30 minutes aiding vehicles in distress, $1\frac{3}{4}$ hours at the scene of an accident, and 2 hours 30 minutes writing tickets.

   a. How much time did the officer spend doing these tasks?

   b. Use the equation $a + \frac{1}{2} + 1\frac{3}{4} + 2\frac{1}{2} = 8$ to find a, the amount of time the officer spent performing other tasks.

40. **Critical Thinking** Place one of the digits 1, 2, 4, 5, 6, and 8 in each of the boxes so that a true sentence results. Use each digit exactly once.

$$\blacksquare\blacksquare - \blacksquare\blacksquare = \blacksquare\blacksquare$$

41. **Journal Entry** Write an equation in which 123 has to be added to each side of the equation in order to solve it.

6-3 Solving Multiplication and Division Equations

Objective

Solve equations using the multiplication and division properties of equality.

Words to Learn

division
  property of equality
multiplication
  property of equality

How can you tell the difference between fraternal twins and identical twins? Fraternal twins do not necessarily look alike. They are not always the same sex, so you can't always tell they are twins. Identical twins look alike and are of the same sex. In the United States, an average of 434 twin babies are born each day. How many sets of twins are born each day?

You know that twins means two. So two times the number of sets of twins is the number of twin babies. If 434 twin babies are born each day, you can solve the equation $434 = 2s$ to find s, the average number of sets of twins born each day.

Since multiplication and division are inverse operations, equations that involve multiplication can be solved by dividing each side of the equation by the same number. Solve $434 = 2s$ using this method.

$434 = 2s$
$\dfrac{434}{2} = \dfrac{2s}{2}$ *Divide each side by 2 to undo*
$217 = s$ *the multiplication by 2.*

Check: $434 = 2s$
      $434 \stackrel{?}{=} 2 \cdot 217$ *Replace s with 217.*
      $434 = 434$ ✓

The solution is 217. On the average, 217 sets of twins are born each day in the United States.

Division Property of Equality	**In words:** If each side of an equation is divided by the same nonzero number, then the two sides remain equal.
	Arithmetic **Algebra**
	$8 = 8$ $a = b$
	$\dfrac{8}{2} = \dfrac{8}{2}$ $\dfrac{a}{c} = \dfrac{b}{c}, c \neq 0$
	$4 = 4$

Equations that involve division can be solved by multiplying each side of the equation by the same number.

Example 1 *Problem Solving*

You can review fractions on page 190.

Marketing Did you know that chewing gum loses its flavor after only about 20 minutes ($\frac{1}{3}$ hour)? However, scientists have recently invented chewing gum that will keep its flavor longer, using synthetically derived polymers. If the newly-developed polymer chewing gum keeps its flavor 30 times as long, use the equation $\frac{1}{3} = \frac{h}{30}$ to find h, the number of hours it keeps its flavor.

$$\frac{1}{3} = \frac{h}{30}$$
$$\frac{1}{3} \cdot 30 = \frac{h}{30} \cdot 30 \qquad \textit{Multiply each side by 30}$$
$$\qquad\qquad\qquad\quad \textit{to undo the division by 30.}$$
$$10 = h$$

Check: $\frac{1}{3} = \frac{h}{30}$

$\frac{1}{3} \overset{?}{=} \frac{10}{30}$ *Replace h with 10.*

$\frac{1}{3} = \frac{1}{3}$ ✓

The solution is 10. The newly-developed polymer chewing gum may keep its flavor up to 10 hours.

After World War II, various waxes, plastics, and synthetic rubber virtually replaced chicle in making chewing gum. Artificially sweetened chewing gum found a wide market in the U.S. in the late 20th century, with mint being the favorite flavor.

Multiplication Property of Equality	**In words:** If each side of an equation is multiplied by the same number, then the two sides remain equal.
	Arithmetic **Algebra**
	$4 = 4$ $a = b$
	$4 \cdot 2 = 4 \cdot 2$ $ac = bc$
	$8 = 8$

Example 2

Estimation Hint

• • • • • • • • • • • • •

In Example 2, think $360 \div 2 = 180$. The solution is about 180.

Solve $368 = 2.3b$. Check your solution.

$$368 = 2.3b$$
$$\frac{368}{2.3} = \frac{2.3b}{2.3} \qquad \textit{Divide each side by 2.3.}$$
$$368 \;\boxed{÷}\; 2.3 \;\boxed{=}\; \text{160}$$
$$160 = b$$

Check: $368 = 2.3b$
$$368 \overset{?}{=} 2.3 \cdot 160 \qquad \textit{Replace b with 160.}$$
$$368 = 368 ✓$$

The solution is 160.

Checking for Understanding

Communicating Mathematics

Read and study the lesson to answer each question.

1. **Write** an equation in the form of $\frac{x}{a} = b$. Then explain why a cannot be 0.

2. **Tell** if 3 is a solution of $\frac{y}{3} = 12$. Explain why or why not.

3. **Write** the equation shown by the model at the right. Then find the solution.

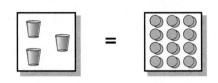

Guided Practice

Complete the solution of each equation.

4. $4m = 20$
$\frac{4m}{4} = \frac{20}{?}$
$m = \underline{\quad}$

5. $\frac{n}{12} = 3$
$(12)\frac{n}{12} = (?)3$
$n = \underline{\quad}$

6. $42 = \frac{r}{7}$
$42(?) = \frac{r}{7}(?)$
$\underline{\quad} = r$

Solve each equation. Check your solution.

7. $7c = 49$

8. $\frac{a}{3.1} = 7.75$

9. $9e = 54$

10. $\frac{4}{5} = \frac{1}{2}f$

11. $72 = \frac{x}{12}$

12. $\frac{y}{4} = 24$

Exercises

Independent Practice

Solve each equation. Check your solution.

13. $3c = 21$ 7

14. $\frac{1}{2}f = \frac{2}{5}$

15. $12x = 156$

16. $34 = 2g$ 17

17. $54 = 3p$

18. $182 = 13s$

19. $\frac{a}{3} = 17$

20. $x \div \frac{1}{8} = \frac{1}{2}$

21. $\frac{m}{4} = 11$

22. $28 = \frac{e}{4}$

23. $96 = \frac{n}{8}$ 768

24. $13 = \frac{t}{5}$

25. $\frac{m}{5} = 1.2$

26. $\frac{p}{3.6} = 0.8$

27. $\frac{t}{2.4} = 13.5$

28. Find the solution of the equation $1.2x = 2.4$.

29. Solve the equation $0.4m = 16$.

30. **Earning Money** If Max Stahler receives 26 paychecks a year and each check is for $763.50, what is his yearly salary? To solve, use the equation $s \div 26 = 763.50$.

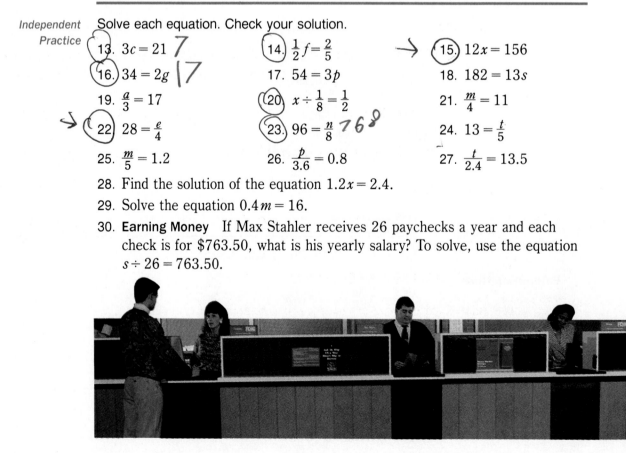

230 **Chapter 6** An Introduction to Algebra

31. Evaluate b^5 if $b = 3$. *(Lesson 1-9)*

32. **Statistics.** Zina has seven brothers and sisters. Their ages are 5, 12, 8, 17, 14, 20, and 22. Find the mean age and median age. *(Lesson 3-5)*

33. Write the prime factorization of 24. *(Lesson 4-2)*

34. Subtract $1\frac{1}{3}$ from $4\frac{3}{4}$. *(Lesson 5-4)*

35. Solve the equation $p - 25.55 = 74.45$. *(Lesson 6-2)*

36. **Engineering** In designing gasoline storage tanks, engineers multiply the government-required minimum thickness by a factor of 2.5 for added safety. Use the equation $2.5m = 1.625$, where m is the minimum thickness.
    a. Solve for m.
    b. What is the required minimum thickness?

37. **Energy** Hydrogen is being considered as a safe alternative fuel. It may also be more economical. Over long distances, the amount of hydrogen equivalent to a kilowatt-hour of electricity can be transported via pipeline for about $\frac{1}{8}$ the cost of sending the electricity through transmission lines. Use the equation $400 = \frac{1}{8}e$ to find the cost of transmitting electricity to a site where it costs only $400 to transmit hydrogen.

38. **Critical Thinking** What is wrong with the equation $m \div 0 = n$? Explain.

39. **Computers** Computer modems transmit data at different speeds. One type of modem transmits at 9,600 bits per second. This is four times faster than a second modem.
    a. Write an equation that when solved will give the speed of the second modem.
    b. Solve the equation and give the speed of the second modem.

6 Mid-Chapter Review

Solve each equation by using the inverse operation. *(Lesson 6-1)*

1. $23 + p = 71$ 2. $1.8n = 2.16$ 3. $68 = g - 89$

Solve each equation. Check your solution. *(Lessons 6-2, 6-3)*

4. $41 + w = 71$ 5. $s - 33 = 35$ 6. $y + 19 = 24$

7. $\frac{c}{1.5} = 0.3$ 8. $11b = 121$ 9. $\frac{k}{9} = 34$

Cooperative Learning

6-3B Solving Two-Step Equations

A Follow-Up of Lesson 6-3

Objective

Solve two-step equations using models.

Materials

cups
counters
mats

In this lab, you will use what you know about solving one-step equations to solve two-step equations like $2x + 3 = 7$.

Try this!

Work with a partner.

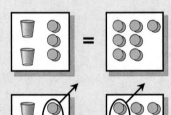

- First, let's build the equation $2x + 3 = 7$ using models. On the left side, we need two cups and three counters. On the right side, we need seven counters.
- Remember, the goal is to get the cup(s) by itself on one side of the mat. Remove three counters from each side of the mat. Now the model shows the equation $2x = 4$.
- Each cup must contain the same number of counters. Therefore, $x = 2$.

What do you think?

1. Why is an equation like $2x + 3 = 7$ called a two-step equation?
2. Write the equation that is shown at the right.

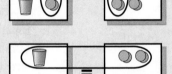

Application

Make a model of each equation. Solve the equation.

3. $3x + 1 = 7$ 4. $2y + 4 = 12$ 5. $5x + 1 = 11$

6. $9 = 4m + 1$ 7. $3x + 5 = 14$ 8. $3 = 2a + 3$

Extension

9. Make a model of the equation $2x + 5 = 10$.
   a. Solve the equation.
   b. Write a paragraph that describes the process you used. In your paragraph, describe any difference between the solution of this equation and the equations you solved previously.

232   **Chapter 6** An Introduction to Algebra

6-4 Writing Algebraic Expressions

Objective
Write simple algebraic expressons from verbal phrases.

Suppose you are attending Space Camp in Huntsville, Alabama, and you take a simulated trip to the moon. When you get to the moon, you find that you weigh more on Earth than you do on the moon. In fact, you weigh 6 times more on Earth than you do on the moon. The chart at the right shows weights that are equivalent on the moon and on Earth.

Weight (In Pounds)	
On Moon	**On Earth**
8	6 · 8 or 48
20	6 · 20 or 120
40	6 · 40 or 240
n	6 · n

You can review variables and expressions on page 28.

You can review variables and expressions on page 28.

In this lesson, you will learn how to translate verbal phrases such as *6 times a number* into an algebraic expression.

Words and phrases often suggest addition, subtraction, multiplication, and division. Here are some examples.

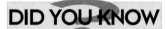

+	−	×	÷
plus	minus	times	divided
sum	difference	product	quotient
more than	less than	multiplied	
increased by	subtract		
total	decreased by		

DID YOU KNOW

Although your weight is less on the moon, your mass is exactly the same.

You can write algebraic expressions to represent verbal phrases.

Examples

Translate each phrase into an algebraic expression.

1 7 more runs scored than the Bruins

Let *b* represent the number of runs scored by the Bruins. The words *more than* suggest addition.

The algebraic expression is $b + 7$.

2 John's test score decreased by 5

Let *s* represent John's test score. The words *decreased by* suggest subtraction.

The algebraic expression is $s - 5$.

Lesson 6-4 Writing Algebraic Expressions **233**

3 the number of shirts divided among four teams

Let s represent the number of shirts. The words *divided by* suggest division.

The algebraic expression is $\frac{s}{4}$.

4 Write verbal phrases for the algebraic expression $r - 6$.

Several phrases can represent $r - 6$.
- a number decreased by 6
- 6 points less than Ben's score
- subtract 6 from a number
- a number of trees minus 6

Checking for Understanding

Communicating Mathematics

Read and study the lesson to answer each question.

1. **Tell** which operation the phrase *increased by* suggests.
2. **Write** two verbal phrases for the algebraic expression $\frac{x}{3}$.
3. **Tell** what the expression $n - 2$ could represent if n is the number of games the Dodgers won.
4. **Tell** how the expressions $n - 3$ and $3 - n$ are different.

Guided Practice

Translate each phrase into an algebraic expression.

5. seven more than t
6. the sum of r and 2
7. eight less than p
8. the difference of g and 4
9. twelve plus s
10. eighteen minus y
11. c divided by 4
12. three times a
13. the quotient of b and 2
14. the product of a number and 7

Write two verbal phrases for each algebraic expression.

15. $t - 10$ 16. $4 \div d$ 17. $10n$ 18. $14 + h$

Exercises

Independent Practice

Translate each phrase into an algebraic expression.

19. five more hits than the Yankees
20. ten fewer points than the Bulls
21. twice as many calories as a slice of pizza
22. five less n
23. your age divided by 3
24. nine increased by x
25. the quotient of a and 6
26. seventeen less than p
27. nineteen less r
28. the product of b and 4
29. Jan's salary plus $1,110
30. five years older than Paul
31. the difference of 8 and d
32. four times as many bees
33. six divided by k
34. l increased by 4
35. Sue's score decreased by 8
36. 19 divided into n
37. x decreased by 15

An airplane is at an altitude of t feet. Write a related situation for each expression.

38. $t - 1,000$ 39. $2t$ 40. $t + 6,500$

Mixed Review 41. Solve $32t = 8$ mentally. *(Lesson 1-10)*

42. **Measurement** Bill is 5.875 feet tall and Diego is 5.785 feet tall. Who is taller? *(Lesson 2-1)*

43. Find the GCF of 36 and 48. *(Lesson 4-5)*

44. Subtract $\frac{24}{25} - \frac{2}{3}$. *(Lesson 5-3)*

45. Dr. Green charges $45.50 for a half-hour office visit. Solve the equation $\frac{d}{3} = 45.50$ to find the charge for an office visit that lasts $1\frac{1}{2}$ hours. *(Lesson 6-3)*

Problem Solving and Applications 46. **Critical Thinking** If x is an odd number, how would you represent the odd number immediately following it? preceding it?

47. **Statistics** Use the information in the graph at the right to answer each question.

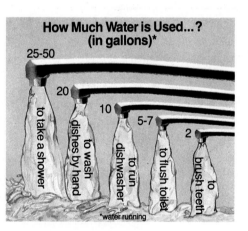

How Much Water is Used...?
(in gallons)*

25-50 to take a shower
20 to wash dishes by hand
10 to run dishwasher
5-7 to flush toilet
2 to brush teeth
*water running

a. Suppose n represents the most water used by the average American to take a shower. Which activity can be expressed by $n - 48$?

b. If t represents the amount of water used to wash dishes by hand, write an expression to represent how much water can be saved by running a dishwasher.

48. **Collect Data** Find yesterday's high and low temperature. Let t represent the low temperature. Write an expression for the high temperature.

49. **Biology** A blue whale gains an average of 2.3 tons a month for the first year of life. Copy and complete the table below.

Current weight (tons)	15	23.7	38	?	45	53.9	?
Weight in 1 month (tons)	?	?	?	44.1	?	?	63

50. **Journal Entry** Write a verbal phrase for the algebraic expression $5 \div n$.

6-5 Use an Equation

Objective

Solve problems by using an equation.

The Jacobson's sold their house for $135,000. This price is four times the amount they originally paid for it 15 years ago. How much did they originally pay for the house?

Real Estate Agent

A real estate agent helps people to buy and sell properties. Being a real estate agent requires patience, a knack for selling, and consumer math skills to compute down payment amounts, interest rates, closing costs, taxes, and insurance. Most agents work on commission.

For more information, contact the Department of Regulatory Agencies, Real Estate Commission, in your state.

Explore   What do you know?
The Jacobson's sold their house for $135,000. The price is four times the amount they paid for it.

What do you need to find?
You need to find how much they originally paid for the house.

Plan   Choose a variable and decide what unknown number it will represent. Write an equation for the problem. Then solve the equation.

Solve   Let p represent the price they originally paid for the house.

The price is 4 times the amount they paid for it.   *Selling Price*

$$4p = \$135{,}000$$

$$\frac{4p}{4} = \frac{\$135{,}000}{4}$$

$$135000 \; \boxed{\div} \; 4 \; \boxed{=} \; \textsf{33750}$$

$$p = \$33{,}750$$

The Jacobson's originally paid $33,750 for their house.

Examine   You can check the answer by replacing the p in $4p = 135{,}000$ with 33,750.

$$4 \; \boxed{\times} \; 33750 \; \boxed{=} \; \textsf{135000}$$

Checking for Understanding

Communicating Mathematics

Read and study the lesson to answer each question.

1. **Tell** what a variable is.

2. **Write** an alternative strategy for solving the problem in the example.

Guided Practice

Solve by using an equation.

3. The number of receivers on a football team is three times the number of quarterbacks. If there are nine receivers on a team, how many quarterbacks are there?

4. Twelve is subtracted from a number. If the difference is 98, what was the original number?

Problem Solving

Practice

Solve using any strategy.

Strategies
••••••••••
Look for a pattern.
Solve a simpler problem.
Act it out.
Guess and check.
Draw a diagram.
Make a chart.
Work backwards.

5. A linen manufacturer sends 250 tablecloths to several discount stores. Each store was sent the same number of tablecloths. How many did each store receive?

6. The greater of two numbers is 25 more than the other number. If the greater number is 82, find the lesser number.

7. There are 10 red, 5 white, and 3 blue marbles in a bag. What is the probability of picking a red marble from the bag without looking?

8. **Mathematics and Sociology** Read the following paragraph.

> Although homeless people have become a common sight in many communities, there is very little reliable information available about who they are, and how many of them live in shelters or on the street. The various federal government programs to assist the homeless are grouped under the Stewart B. McKinney Homeless Assistance Act. In 1990, Congress appropriated $1.1 billion for these programs.

A 1988 U.S. Department of Education survey estimated that there were 220,000 homeless school-age children in this country. There are 90,000 more homeless children that go to school than those that do not. How many homeless children go to school and how many do not?

6-6 Changing Units in the Customary System

Objective

Changing units in the customary system

Words to Learn

ounce
pound
ton
cup
pint
quart
gallon

A newborn hooded seal pup weighs about 45 pounds at birth. In the first four days of life, the pup gains about 56 ounces a day. How many pounds will the average pup gain in one day?

Customary units of weight are **ounce, pound,** and **ton.** The following table gives the relationships among these units.

> 1 pound (lb) = 16 ounces (oz)
> 1 ton = 2,000 pounds (lb)

DID YOU KNOW

The hooded seal got its name because of the bright red hood on the male seal's head. When he is angry, he blows air into it and it expands like a balloon. It is used to frighten attackers.

In the seal pup problem above, you need to change 56 ounces to pounds. When you change from a smaller unit to a larger unit, divide.

56 oz = _?_ lb

56 ⌷÷⌷ 16 ⌷=⌷ **3.5**   *Since 16 oz = 1 lb, divide by 16.*

The average pup will gain 3.5 pounds in one day.

Sometimes you will need to convert from a larger unit to a smaller unit. In this case, multiply.

Example 1   *Problem Solving*

Highway Safety   Have you ever crossed over a small country bridge? Normally a sign will be posted just before the entrance to the bridge warning drivers of the weight limit. Suppose you cross a small bridge with a $3\frac{1}{2}$ ton weight limit. How many pounds can the bridge hold?

$3\frac{1}{2}$ ton = _?_ lb

$3\frac{1}{2} \cdot 2,000 = 7,000$ *Since 2,000 lb = 1 ton, multiply by 2,000.*

The bridge will hold 7,000 pounds.

Customary units of liquid capacity are **cup, pint, quart,** and **gallon.** The following table shows the relationships among these units.

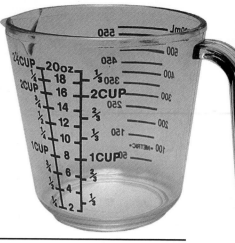

> 1 cup (c) = 8 fluid ounces (fl oz)
> 1 pint (pt) = 2 cups (c)
> 1 quart (qt) = 2 pints (pt)
> 1 gallon (gal) = 4 quarts (qt)

Example 2

How many fluid ounces are in 3 cups?

$3 \text{ c} = \underline{?} \text{ fl oz}$

$3 \cdot 8 = 24$ *Multiply by 8 since there are 8 fl oz in a cup.*

$3 \text{ c} = 24 \text{ fl oz}$

There are 24 fluid ounces in 3 cups.

Remember that dividing by any number is the same as multiplying by its multiplicative inverse. For example, dividing by 4 is the same as multiplying by $\frac{1}{4}$. You can use this fact when you change from smaller units to larger units.

Example 3

$19 \text{ qt} = \underline{?} \text{ gal}$

- Method 1: Divide by 4.

  $19 \boxed{\div} 4 \boxed{=} 4.75$

  $19 \text{ qt} = 4.75 \text{ gal}$

- Method 2: Multiply by $\frac{1}{4}$.

  $19 \text{ qt} = \underline{?} \text{ gal}$

  $19 \times \frac{1}{4} = 4\frac{3}{4}$

  $19 \text{ qt} = 4\frac{3}{4} \text{ gal}$

There are 4.75 or $4\frac{3}{4}$ gallons in 19 quarts.

Checking for Understanding

Communicating Mathematics

Read and study the lesson to answer each question.

1. **Tell** which operation is needed to convert from pints to gallons. Explain how you know.

2. **Tell** how to change 8 quarts to gallons.

Guided Practice

Complete.

3. $3 \text{ pt} = \underline{?} \text{ c}$

4. $2 \text{ tons} = \underline{?} \text{ lb}$

5. $5 \text{ lb} = \underline{?} \text{ oz}$

6. $2 \text{ gal} = \underline{?} \text{ qt}$

7. $6,000 \text{ lb} = \underline{?} \text{ tons}$

8. $12 \text{ qt} = \underline{?} \text{ gal}$

9. $15 \text{ pt} = \underline{?} \text{ qt}$

10. $12 \text{ c} = \underline{?} \text{ pt}$

11. $128 \text{ oz} = \underline{?} \text{ lb}$

12. How many tons are in 3,600 pounds?

13. A customer orders 1,500 pounds of rock. The supply truck holds $\frac{3}{4}$ ton. Will the truck be able to deliver the order with one load?

Exercises

Complete.

14. 3 lb = _?_ oz   15. 5 tons = _?_ lb   16. 3 c = _?_ fl oz

17. 5 pt = _?_ c   18. 4,000 lb = _?_ tons   19. 48 oz = _?_ lb

20. 16 qt = _?_ gal   21. 8 pt = _?_ qt   22. 2.5 lb = _?_ oz

23. 2.5 qt = _?_ pt   24. 0.5 gal = _?_ qt   25. 4.5 pt = _?_ c

26. 1 pt = _?_ qt   27. 2 fl oz = _?_ c   28. 5 c = _?_ pt

29. How many pints are in $4\frac{1}{2}$ quarts?

30. Change 11 quarts to gallons.

31. Convert 4 ounces to pounds.

32. Evaluate the expression $3 \cdot 0.50 + 4 \cdot 0.75$. *(Lesson 1-7)*

33. **Earning Money** Ginny receives a paycheck for 12 days of work in the amount of $732.25. How much did she earn per day? Round to the nearest cent. *(Lesson 2-8)*

34. **Statistics** Construct a line plot for 5, 2, 0, 3, 2, 5, 15, 4, 3, and 0. Circle any outliers. *(Lesson 3-4)*

35. Multiply $5\frac{3}{8}$ and $\frac{2}{3}$. *(Lesson 5-5)*

36. Translate the phrase *12 more than d* into an algebraic expression. *(Lesson 6-4)*

37. **Critical Thinking** The owner's manual of Car A states that it has a capacity of 13.2 gallons of gasoline. Car B has a capacity of 13 gallons 1 quart.
    a. Which car has the greater gasoline capacity?
    b. How many more quarts will it hold?

38. **Biology** About 190 million years ago, a giant lizard-like dinosaur called a brontosaurus was roaming Earth. It weighed about 30 tons. Today the largest living land animal is the African elephant. It weighs about 16,500 pounds. How many more tons did the brontosaurus weigh than today's elephant?

39. **Data Search** Refer to pages 218 and 219. What is the apparent temperature when the room temperature is 73°F with a relative humidity of 20%? What is the apparent temperature when the room temperature is 73°F, but the relative humidity is 90%?

DATA SEARCH

Cooperative Learning

6-7A Preview of Geometry: Area

A Preview of Lesson 6-7

Objective
Use models to find the area of rectangles and parallelograms.

Words to Learn
area
parallelogram
base
height

Materials
grid paper
scissors
pencil

In this lab, you will investigate the areas of rectangles and parallelograms by using models.

Activity One

- On grid paper, draw a rectangle with a length of 6 units and a width of 4 units as shown at the right.
- The **area** of a geometric figure is the number of square units needed to cover the surface within the figure.

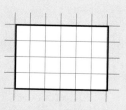

What do you think?

1. How many squares are found within this rectangle?
2. How does the area relate to the length and width of the rectangle?

A **parallelogram** is a four-sided figure whose opposite sides are parallel. One of its sides may be identified as its **base**. The distance from the base to the opposite side is called the **height**.

Activity Two

- On grid paper, draw a parallelogram with a base of 6 units and a height of 4 units as shown at the right.
- Draw a line to represent the height as shown.
- Use your scissors to cut out the parallelogram.
- Then cut the parallelogram along the line for the height as shown below.

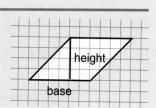

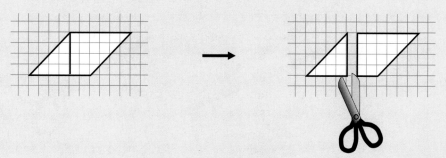

- Move the triangle to the opposite end of the parallelogram to form a rectangle.

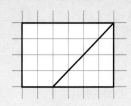

Talk About It

3. What is the area of the newly-formed rectangle?
4. How are the area of the original parallelogram and the area of the newly-formed rectangle related?
5. How does the area of the parallelogram relate to its base and height?
6. How does this compare to the number of square units in the rectangle you drew?

Applications

Use your grid paper to draw the following figures. Cut them out and find their areas by counting squares.

7. rectangle: ℓ, 5; w, 3
8. parallelogram: b, 5; h, 3
9. rectangle: ℓ, 4; w, 2
10. parallelogram: b, 4; h, 2
11. What is the height of a parallelogram if its area is 36 square units and its base is 9 units long?
12. Think about the length of the base and the height of a parallelogram if its area is 16 square units.
    a. Can other parallelograms be made with the same area, but different base and height?
    b. If so, what is the base and height of your other parallelograms? Draw some examples.

Extension

13. If you double the length and width of a rectangle, how does its area change? Explain your reasoning in words and by drawing diagrams.

6-7 Area

Objective

Find the area of rectangles and parallelograms.

Basketball was invented in 1891 by Dr. James Naismith of Springfield, Massachusetts. He wanted a game that could be played indoors during the winter and in the evening. Dr. Naismith used two wooden peach baskets nailed to the balcony of the school gym.

Today we use net baskets and play on a court shaped like a rectangle. The length of a regulation size court for professional and college basketball is 94 feet and the width is 50 feet. What is the area of the court?

One way to find the area is to count the number of square feet by marking off a grid on the court. This method is not practical, however. In the Mathematics Lab, you discovered that the area of a rectangle can be found as follows.

Area of Rectangles	**In words:** The area of a rectangle equals the product of its length (ℓ) and width (w).
	In symbols: $A = \ell w$

Examples

1 Find the area of a regulation size basketball court with a length of 94 feet and a width of 50 feet.

$A = \ell w$ *Write the formula for area.*

$A = 94 \cdot 50$ *Replace l with 94 and w with 50.*

94 ⊠ 50 ⊜ **4700**

$A = 4,700$ The area of a regulation size basketball court is 4,700 square feet (ft²).

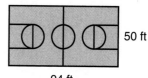
50 ft
94 ft

Problem-Solving Hint

● ● ● ● ● ● ● ● ● ● ● ● ●

For problems like the one in Example 2, it may be helpful to make a drawing.

2 Find the width of a rectangle with an area of 260.4 square inches and a length of 16.8 inches.

$A = \ell w$ *Write the formula for area.*

$260.4 = 16.8w$ *Replace A with 260.4 and l with 16.8.*

$\dfrac{260.4}{16.8} = \dfrac{16.8w}{16.8}$ *Divide each side by 16.8.*

260.4 ⊡ 16.8 ⊜ **15.5**

$15.5 = w$

Check: $260.4 = 16.8w$

$260.4 \overset{?}{=} 16.8 \cdot 15.5$ *Replace with 15.5.*

$260.4 = 260.4$ ✓

The width is 15.5 inches.

You also discovered in the Mathematics Lab that the area of a parallelogram is closely related to the area of a rectangle.

Area of Parallelograms	**In words:** The area of a parallelogram equals the product of its base (*b*) and its height (*h*).
	In symbols: $A = bh$

Example 3

Find the area of the parallelogram at the right.

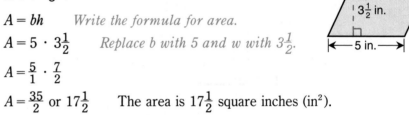

$A = bh$ *Write the formula for area.*

$A = 5 \cdot 3\frac{1}{2}$ *Replace b with 5 and w with $3\frac{1}{2}$.*

$A = \frac{5}{1} \cdot \frac{7}{2}$

$A = \frac{35}{2}$ or $17\frac{1}{2}$ The area is $17\frac{1}{2}$ square inches (in²).

Checking for Understanding

Communicating Mathematics   Read and study the lesson to answer each question.

1. **Model** a parallelogram with a base of 8 units and a height of 5 units using grid paper. What is its area?

2. **Tell** how the formula $A = bh$ is also appropriate for finding the area of a rectangle.

3. **Write** two sentences comparing the similarities and differences of rectangles and parallelograms.

Guided Practice   Find the area of each figure shown or described below.

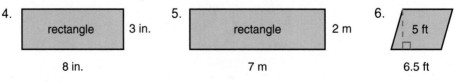

4. rectangle — 8 in., 3 in.

5. rectangle — 7 m, 2 m

6. 6.5 ft, 5 ft

7. rectangle: ℓ, 1 in.; w, 6 in.

8. rectangle: ℓ, 18 cm; w, 12 cm

Exercises

Independent Practice

Find the area of each figure shown or described below.

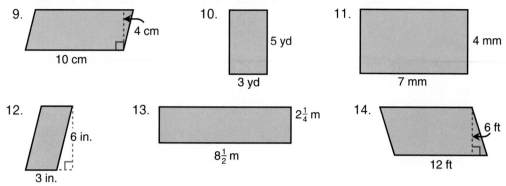

9. 4 cm, 10 cm

10. 5 yd, 3 yd

11. 4 mm, 7 mm

12. 6 in., 3 in.

13. $2\frac{1}{4}$ m, $8\frac{1}{2}$ m

14. 6 ft, 12 ft

15. parallelogram: b, 23 ft; *h*, 18 ft

16. rectangle: ℓ, 8.5 mm; *w*, 7.5 mm

17. parallelogram: *b*, 12.6 ft; *h*, 11.3 ft

18. rectangle: ℓ, $6\frac{1}{3}$ yd; *w*, 5 yd

19. Find the area of a parallelogram with a base of 4.5 centimeters and a height of 4 centimeters.

20. What is the length of a rectangle with an area of $31\frac{2}{3}$ square yards and a width of 5 yards?

Mixed Review

21. Use front-end estimation to find the difference of 365 and 151. *(Lesson 1-2)*

22. Multiply 100 and 30. *(Lesson 2-5)*

23. How many $\frac{1}{2}$-cup servings of ice cream are there in a gallon of chocolate ice cream? *(Lesson 6-6)*

Problem Solving and Applications

24. **Critical Thinking** If you cut a cardboard tube, you will find that it is a rectangle. How much cardboard is used in a tube if its length is $4\frac{1}{2}$ inches and its circumference is 5 inches?

25. **Construction** Philip Frazier wants to build a deck with an area of at least 120 square feet. He has space for a length of up to 14 feet, but no more than 9 feet for the width.
    a. Will he be able to build a deck as large as he wants?
    b. If so, what will be the area of the largest deck possible?

26. **Journal Entry** Collect data by measuring the playing surface of a court or playing field at your school. Make and label a scale drawing of it. What is its area?

6 Study Guide and Review

Communicating Mathematics

State whether each sentence is *true* or *false*. If false, replace the underlined word to make a true sentence.

1. Equations that have the same solution are called <u>equivalent</u> equations.
2. The multiplication property of equality states that if you multiply each side of an equation by the same number, the two sides will then be <u>unequal</u>.
3. The words "more than" suggest <u>multiplication</u> in algebraic terms.
4. To change from pounds to tons, the number of pounds should be <u>multiplied</u> by 2,000 since 1 ton equals 2,000 pounds.
5. To change 6 cups to fluid ounces, 6 should be <u>multiplied</u> by 8.
6. The formula for the area of a <u>parallelogram</u> is $A = \ell w$.
7. In your own words, explain the addition property of equality.

Skills and Concepts

Objectives and Examples	*Review Exercises*
Upon completing this chapter, you should be able to:	*Use these exercises to review and prepare for the chapter test.*

• solve equations using inverse operations *(Lesson 6-1)*

Solve $p + 3 = 5$ *Write the related subtraction sentence.*
$$p = 5 - 3$$
$$p = 2$$

Solve each equation by using the inverse equation.

8. $x - 14 = 15$ 9. $m + \frac{7}{2} = 5$

10. $t - 3.6 = 10.1$ 11. $26 + p = 26$

12. $5q = 6$ 13. $\frac{2}{3}y = \frac{1}{4}$

14. $\frac{s}{2} = 13$ 15. $\frac{b}{4.6} = 3.1$

• solve equations using the addition and subtraction properties of equality *(Lesson 6-2)*

Solve $c - 32 = 112$
$$c - 32 + 32 = 112 + 32$$
$$c = 144$$

Solve each equation. Check your solution.

16. $w + 13 = 25$ 17. $f - 3\frac{1}{2} = 3$

18. $10.9 + r = 11$ 19. $54 = m - 9$

20. $3\frac{2}{5} = 1\frac{1}{3} + x$

21. $s + 3.75 = 5.25$

22. $\frac{100}{9} = \frac{100}{9} + k$

Objectives and Examples

- solve equations using the multiplication and division property of equality *(Lesson 6-3)*

Solve $33y = 132$

$$\frac{33y}{33} = \frac{132}{33}$$ *Divide each side by 33.*

$$y = 4$$

- write algebraic expressions from verbal phrases *(Lesson 6-4)*

Translate "4 times the price" into an algebraic expression.

Let p represent the price. The word *times* suggests multiplication.
The algebraic expression is $4p$.

- convert within the customary system *(Lesson 6-6)*

Complete: 12 lb = _?_ oz

Since 1 lb = 16 oz, multiply by 16.
$12 \cdot 16 = 192$
12 lb = 192 oz

- find the area of rectangles and parallelograms *(Lesson 6-7)*

Find the area of a parallelogram with a base of 6 inches and a height of 3 inches.

$A = bh$
$A = 6 \cdot 3$ *Replace b with 6 and*
$A = 18$ *h with 3.*

The area is 18 square inches.

Review Exercises

Solve each equation. Check your solution.

23. $4b = 32$

24. $\frac{p}{6} = 4.5$

25. $\frac{3}{2}m = \frac{5}{4}$

26. $5.9r = 0.59$

27. $64 = 16a$

28. $\frac{g}{15} = \frac{3}{5}$

29. $\frac{g}{3.6} = 10$

30. $1.333t = 0$

Translate each phrase into an algebraic expression.
31. the sum of x and 5
32. 13 less than s
33. 13 less r
34. e multiplied by 2
35. 9 increased by q
36. the quotient of z and 15
37. 23 fewer than b
38. t times 3.6
39. c divided by 100

Complete.

40. 3 qt = _?_ pt
41. 5 gal = _?_ qt
42. 6,000 lb = _?_ tons
43. 54 oz = _?_ lb
44. 24 fl oz = _?_ c
45. $5\frac{3}{4}$ qt = _?_ pt

Find the area of each figure described below.
46. rectangle: ℓ, 12 in.; w, 4 in.
47. parallelogram: b, 15 yd; h, 3 yd
48. rectangle: ℓ, $5\frac{1}{2}$ ft; w, $\frac{3}{4}$ ft

49. parallelogram: b, 13.6 in.; h, 2.7 in.
50. rectangle: ℓ, 12.5 ft; w, 1.6 ft
51. parallelogram: b, $3\frac{5}{8}$ in.; h, $1\frac{2}{3}$ in.

Applications and Problem Solving

52. **Fund Raiser** In 1991, Central Junior High School raised $335 more during their annual fund-raising drive than they had during the previous year's drive. Write an algebraic expression for the amount raised in 1991. Let r represent the amount raised in 1990. *(Lesson 6-4)*

53. **Highway Safety** Ben Johnson is driving a semitrailer-tractor truck that weighs in at 13,596 pounds when fully loaded. He approaches a bridge that is marked with a sign stating the maximum weight allowed is 6 tons. If Bob's semitrailer-tractor truck is fully loaded, should he attempt to cross the bridge? *(Lesson 6-6)*

54. Patrick earned $65 in January shoveling snow. The total was 4 times more than his January earnings last winter. How much did he earn last January? *(Lesson 6-5)*

55. Ellen is making invitations for a surprise party to celebrate her mother's birthday. The invitations are designed in the shape of a parallelogram with a base of 5 inches and a height of $2\frac{1}{2}$ inches. How much paper will Ellen need to make 20 invitations? *(Lesson 6-7)*

Curriculum Connection Projects

- **Health** Have a friend measure your pulse rate while resting and again after running in place for a short time. Write a formula for finding the difference in the two pulse rates.

- **Art** List and measure rectangles and parallelograms that you find in patterns on clothing, quilts, wallpaper, wall hangings, carpets, curtains, and so on. Then find the area of each item you have listed.

Read More About It

Manes, Stephen. *Chocolate Covered Ants.*
Angell, Judie. *Leave the Cooking to Me.*
Bjork, Christina and Anderson, Lena. *Elliot's Extraordinary Cookbook.*

Test

Solve each equation by using the inverse operations.

1. $h + 6 = 19$
2. $3.5d = 14.7$
3. $7\frac{3}{8} = x - \frac{15}{2}$

4. **Chemistry** During a chemical reaction, 3 mL of the original chemical evaporates, leaving 2.6 mL in the test tube. Solve the equation $a - 3 = 2.6$ to find a, the amount of chemical in the test tube before the reaction occurs.

Solve each equation. Check your solution.

5. $p + 21 = 35$
6. $\frac{3}{2} + r = \frac{8}{3}$
7. $12e = 120$

8. $\frac{3}{2}f = 3$
9. $\frac{7}{6} = b - \frac{1}{8}$
10. $0.01m = 50$

11. $s - 5.9 = 12.1$
12. $\frac{y}{3} = 36$
13. $0.997 + t = 1$

14. Ann prices a sweater in two stores and finds it is $29 in the first store and $34 in the second. Solve the equation $29 + p = 34$ to find p, the price difference between the two stores.

15. **Inflation** Economists say that, on average, the 1991 price of a gallon of gas was 1.7 times what it was in 1980. The average 1991 price of a gallon of gas was $1.19. Solve the equation $1.7g = 1.19$ to find g, the average price in 1980.

Translate each phrase into an algebraic expression.

16. 26 less x
17. t increased by 23
18. 5 divided into w

19. Debbie's aunt is 60 years old. Her age is six years more than Debbie's age. How old is Debbie?

Complete.

20. $24 \text{ qt} = \underline{\;?\;} \text{ gal}$
21. $3\frac{1}{4} \text{ lb} = \underline{\;?\;} \text{ oz}$
22. $4{,}500 \text{ lb} = \underline{\;?\;} \text{ tons}$

23. **Sports** The attendance at the last football game of the season was 97 fewer than the attendance at the first game. Let y represent the attendance at the first game. Write an expression for the last game's attendance.

24. A typical soft drink can holds 12 fluid ounces. How many cups is this?

25. **Geometry** Theresa argues that the area of a parallelogram with a base of 6 inches and a height of 3 inches is the same as the area of a rectangle with a length of 10 inches and width of 1.8 inches. Is she correct?

Bonus Write an expression for the difference of two consecutive even numbers.

6 Academic Skills Test

Directions: Choose the best answer. Write A, B, C, or D.

1. Mike is reading a 258-page novel. If he reads 8 pages an hour, *about* how long will it take him to read the entire book?

   A 8 hours   B 16 hours
   C 24 hours   D 30 hours

2. Sandy is knitting scarves for her 3 sisters. She needs to buy 9 skeins of yarn. Each skein costs $2.59, including tax. Which information do you need to find the cost of the yarn?

   A 3 sisters, 9 skeins, $2.59 each
   B 3 sisters, $2.59 each
   C 3 sisters, 9 skeins
   D 9 skeins, $2.59 each

3. What is the value of x^5 if $x = 3$?

   A 243   B 125
   C 15   D 3

4. Which is the best estimate for the total cost?

$1.89
2.08
2.00
1.95
1.88
2.10

   A $8.00
   B $10.00
   C $12.00
   D $13.00

5. 2.36×100

   A 0.236   B 2.36
   C 236   D 2,360

6. George earned $18.20 for babysitting 6.5 hours. How much was he paid per hour?

   A $6.50   B $3.80
   C $3.00   D $2.80

7. The frequency table below contains data about students' test scores on a 20-point test.

Score	14	15	16	17	18	19	20
No. of Students	2	1	5	4	6	5	2

   How many students had a score greater than 16?

   A 4   B 5
   C 17   D 22

8. Use the data in Exercise 7. What is the mode of the scores?

   A 6   B 17
   C 17.36   D 18

9. The stem-and-leaf plot shows the Wildcats' basketball scores for this season.

   ```
   1 | 8 9
   2 | 0 2 3 3 6 8 8
   3 | 0 1 4 4 5 6 8 9
   4 | 0 1 2
   ```
   1 | 8 = 18 points

   In how many games did they score at least 30 points?

   A 8   B 9
   C 11   D 20

10. Which number comes next in this pattern?

$$0.5, 2, 3.5, 5, \ldots$$

A 8 B 7.5

C 7 D 6.5

11. Which fraction is in simplest form?

A $\frac{4}{15}$ B $\frac{6}{21}$

C $\frac{8}{10}$ D $\frac{9}{27}$

12. What is the least common multiple of 25 and 45?

A 45 B 225

C 1,125 D none of these

13. Change $3\frac{5}{8}$ to an improper fraction.

A $\frac{35}{8}$ B $\frac{20}{8}$

C $\frac{15}{8}$ D $\frac{29}{8}$

14. $\frac{3}{4} + \frac{1}{2}$

A $\frac{1}{4}$ B $\frac{3}{8}$

C $\frac{4}{6}$ D $1\frac{1}{4}$

15. $\frac{1}{4} \times 2\frac{1}{2}$

A 10 B 1

C $\frac{5}{8}$ D $\frac{3}{8}$

16. To the nearest whole number, what is the circumference of the circle?

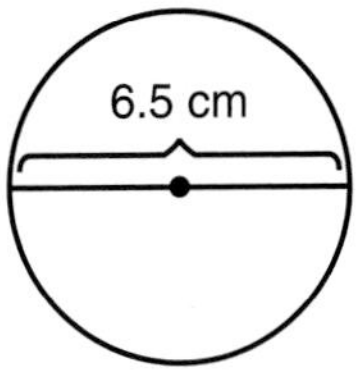

6.5 cm

A 10 cm B 20 cm

C 32 cm D 41 cm

17. If $n + 3.9 = 4.2$, what is the value of n?

A 1.7 B 3.9

C 8.1 D none of these

18. If $15a = 20$, what is the value of a?

A $\frac{3}{4}$ B $1\frac{1}{3}$

C 5 D 100

19. Which expression represents *five more than a number*?

A $5 - n$ B $n - 5$

C $5n$ D $n + 5$

20. A coffee can contains 1 pound 10 ounces of coffee. What is this equivalent to?

A 18 oz B 20 oz

C 26 oz D 42 oz

Academic Skills Test

Integers

Spotlight on Wind Storms

Have You Ever Wondered. . .

- How fast wind can actually blow?
- At what time of the year most tropical storms and hurricanes occur?

Beaufort Scale

Sailors commonly use the Beaufort Scale to describe wind strengths. The scale classifies strong winds as storms or hurricanes. When using a number to describe a wind strength, the word force is used. A force 9 means the wind is a strong gale.

Beaufort Number	Type of Wind	Wind Speed	Description and Effect
0	Calm	less than 1 mi/h	still; smoke rises vertically
1	Light Air	1-5 mi/h	wind direction shown by smoke drift; weather vanes inactive
2	Light breeze	6-11 mi/h	wind felt on face; leaves rustle; weather vanes active
3	Gentle breeze	12-19 mi/h	leaves and small twigs move constantly; wind extends lightweight flags
4	Moderate breeze	20-28 mi/h	raises dust and loose paper; moves twigs and thin branches
5	Fresh breeze	29-38 mi/h	small trees in leaf begin to sway
6	Strong breeze	39-49 mi/h	large branches move; telephone wires whistle; umbrella difficult to control
7	Moderate gale	50-61 mi/h	whole trees sway; somewhat difficult to walk
8	Fresh gale	62-74 mi/h	twigs broken off trees; walking against wind very difficult
9	Strong gale	75-88 mi/h	slight damage to buildings, shingles blown off roof
10	Whole gale	89-102 mi/h	trees uprooted; considerable damage to buildings
11	Storm	103-117 mi/h	widespread damage, rarely occurs inland
12-17	Hurricane	greater than 117 mi/h	extreme destruction

Chapter Project

Wind Storms

Work in a group.

1. Make a chart showing wind speed and its direction in your area for one month.

2. Include a brief description of the weather for each day.

3. Explain how the wind changed throughout the month and how it is related to the weather by presenting a TV weather show.

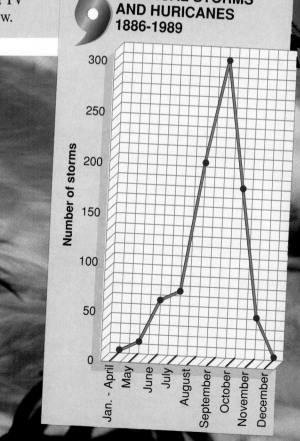

TROPICAL STORMS AND HURICANES 1886-1989

Number of storms

Jan. - April, May, June, July, August, September, October, November, December

Looking Ahead

In this chapter, you will see how mathematics can be used to answer the questions about wind storms.

The major objectives of the chapter are to:

- identify positive and negative integers

- compare and order integers

- add, subtract, multiply, and divide integers

- express decimal powers of ten using negative integer exponents

253

7-1 Integers

Objectives

Read and write integers.
Find the opposite and
absolute value of
an integer.

Words to Learn

integer
positive integer
negative integer
opposite
absolute value

Do you like cold weather? If so, you might want to move to Alaska.
The average high temperature in Anchorage, Alaska, for January
is 7 degrees below zero. You can express this temperature using
the negative number −7. This number is a member of the
set of **integers.**

Integers	An integer is any number from the set $\{ \ldots, -4, -3, -2, -1, 0, +1, +2, +3, +4, \ldots \}$.

Integers greater than 0 are
positive integers. Integers
less than zero are **negative
integers.** Zero itself is
neither positive nor
negative. Positive integers
usually are written without
the + sign, so +4 and 4 are
the same. Use negative and
positive counters to
represent integers.

Calculator Hint

● ● ● ● ● ● ● ● ● ● ● ● ●

To enter a negative
integer on a
calculator, use the
[+/−] key. For
example, to enter
−7, press 7 [+/−].

You can also represent integers as points on a number line. On a
horizontal number line, positive integers are represented as points to
the right of 0, and negative integers as points to the left of 0. The
integers −4 and +5 are graphed on the number line below.

Two numbers are **opposites** of one another if they are represented
by points that are the same distance from zero, but on opposite sides
of zero. The number line below shows that −4 and 4 are opposites.

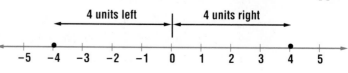

The **absolute value** of an integer tells how far the point that represents it is from the point for 0.

Absolute Value	The absolute value of a number is the number of units its graph is from the graph of 0 on a number line.

The *absolute value of n* is written as $|n|$. So, $|-4| = 4$ and $|4| = 4$.

Example

Find the opposite and the absolute value of -7.

On the number line, -7 is at the point 7 units to the left of 0. The opposite of -7 would be at the point 7 units to the right of 0.

7 units left | 7 units right

$-7\ -6\ -5\ -4\ -3\ -2\ -1\ \ 0\ \ 1\ \ 2\ \ 3\ \ 4\ \ 5\ \ 6\ \ 7$

So, the opposite of -7 is 7.

The point that represents -7 is 7 units from 0, so the absolute value of -7 is 7. Write $|-7| = 7$.

Checking for Understanding

Communicating Mathematics

Read and study the lesson to answer each question.

1. **Show** -7 by using counters and also on a number line.
2. **Draw** a number line that shows 5 and its opposite.
3. **Tell** how to find the absolute value of -8.
4. **Write** about an everyday situation in which negative integers are used.

Guided Practice

Write an integer for each situation.

5. a profit of $4
7. 12 yards gained
9. a loss of 6 points

6. a withdrawal of $5
8. a gain of 6 pounds
10. 10°F above zero

Write the integer represented by the point for each letter. Then find its opposite and its absolute value.

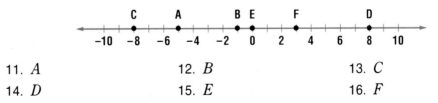

11. *A*
14. *D*

12. *B*
15. *E*

13. *C*
16. *F*

17. Name the least positive integer.
18. **Weather** In 12 hours on December 24, 1924, the temperature in Fairfield, Montana, fell from 63°F above zero to 21°F below zero. Write these temperatures as integers.

Exercises

Write an integer for each situation.

19. a deposit of $6

20. a loss of $10

21. a gain of 2 yards

22. 4 seconds before liftoff

23. 25 points lost

24. 10°F above 0

Write the integer represented by the point for each letter.
Then find its opposite and its absolute value.

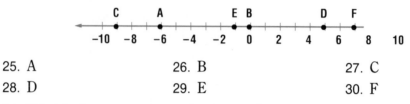

25. A   26. B   27. C

28. D   29. E   30. F

31. What is the greatest negative integer?

32. Find the absolute value of the opposite of −8.

33. Compute 196 − 18 mentally. *(Lesson 1–4)*

34. **Statistics** Construct a line plot for the following data: 18, 21, 19, 18, 18, 17, 18, 22, 19. *(Lesson 3–4)*

35. Express $\frac{100}{9}$ as a mixed number. *(Lesson 5–1)*

36. **Real Estate** A house advertised in the real estate section of the newspaper claims to have a rectangular lot with a length of 250 feet and a width of 120 feet. What is the area of the lot? *(Lesson 6–7)*

37. **Geography** The highest point in the United States is Mount McKinley, Alaska, which rises 20,320 feet above sea level. The lowest point is Death Valley, California, which is 282 feet below sea level. Write integers to represent these elevations.

38. **Calculators** If you press 6 [+/−], you get −6.

   a. What is the result when you press 6 [+/−] [+/−]?

   b. What is the result when you press 6 [+/−] [+/−] [+/−]?

   c. What can you conclude about the number of times you press [+/−] and the result?

39. **Critical Thinking** Complete each sentence using either the word *positive* or *negative*.

   a. If the absolute value of an integer is equal to the integer itself, then the integer is either __?__ or zero.

   b. If the absolute value of an integer is equal to the opposite of the integer, then the integer is __?__.

40. **Journal Entry** Write a sentence about each of three different real-world situations that might be represented by the integer −5.

256   **Chapter 7** Integers

7-2 Comparing and Ordering Integers

Objective

Compare and order integers.

John and Barry were playing Jeopardy®. In Jeopardy®, a player can end up with negative scores. John's final score was -800, and Barry's final score was -200. Whose score was greater?

You can use a number line to answer this question. On a number line, values increase as you go right and decrease as you go left.

John   Barry

-800 -600 -400 -200 0 200 400 600 800

LOOK BACK

You can review inequalities on page 48.

Since Barry's score is to the right of John's score on the number line, Barry's score is greater than John's. You can write $-200 > -800$ or $-800 < -200$.

Examples

1 Replace ● with $<$, $>$, or $=$ in -5 ● -1 to make a true sentence. Draw the graph of each integer on a number line.

-7 -6 -5 -4 -3 -2 -1 0 1 2 3 4 5 6 7

Since -5 is to the left of -1 on the number line, $-5 < -1$.

Mental Math Hint

• • • • • • • • • • • • •

On a number line, any positive integer is to the right of a negative integer. So, when comparing a positive and negative integer, the positive integer will always be greater.

$1 > -1{,}000{,}000$

$-1{,}000{,}000 < 1$

2 Order the integers -4, 5, 3, -2, and 0 from least to greatest. Draw the graph of each integer on a number line.

-7 -6 -5 -4 -3 -2 -1 0 1 2 3 4 5 6 7

Order the integers by reading from left to right.

$$-4, -2, 0, 3, 5$$

Example 3 *Connection*

LOOK BACK

You can review median on page 102.

Statistics Find the median of the temperatures in the chart.

Monthly Record Low Temperatures for Juneau, Alaska (°F)

J	F	M	A	M	J	J	A	S	O	N	D
-22	-22	-15	6	25	31	36	27	23	11	-5	-21

List the temperatures in order from least to greatest.

$$-22, -22, -21, -15, -5, \textbf{6, 11}, 23, 25, 27, 31, 36$$

There are two middle numbers, 6 and 11. So the median is $\frac{6+11}{2}$ or 8.5. The median of the temperatures is $8.5°$F.

Checking for Understanding

Communicating
Mathematics

Read and study the lesson to answer each question.

1. **Show** that $-3 < 4$ using a number line.

2. **Tell** how to determine when one integer is greater than another integer.

Guided Practice

Replace each ● with $<$, $>$, or $=$ to make a true sentence.

3. -9 ● -19 4. -19 ● -9 5. -6 ● 0

6. -12 ● -121 7. -87 ● -78 8. 0 ● -5

9. Order 56, -1, 31, -98, 14, and -76 from least to greatest.

Exercises

Independent
Practice

Replace each ● with $>$, $<$, or $=$ to make a true sentence.

10. 1 ● -9 11. 5 ● -98 12. -2 ● -9

13. -98 ● -987 14. -1 ● 1 15. -8 ● -5

Order the integers from least to greatest.

16. 3, -4, 0, -9, 29, -76 17. 2, -9, -76, -91, 32, 18, -6

18. Which is greater, -11 or 9?

19. Order 13, 5, -7, 0, -1, and -14 from least to greatest.

Mixed Review

20. **Income Tax**   A U.S. citizen pays an average of $\$2,245.48$ in income tax annually. How much is this to the nearest dollar? *(Lesson 2-2)*

21. Name two numbers that are divisible by both 3 and 8. *(Lesson 4-1)*

22. Find the opposite and absolute value of -21. *(Lesson 7-1)*

Problem Solving and Applications

23. **Weather**   The table below gives the monthly record low temperatures (°F) in Wilmington, Delaware. Find the median of the temperatures.

J	F	M	A	M	J	J	A	S	O	N	D
-14	-6	2	18	30	41	48	43	36	24	14	-7

COMPUTER

24. **Critical Thinking**   Points *A*, *B*, *C*, and *D* are different points on a number line. Using the following clues, order the integers for *A*, *B*, *C*, and *D*.
    - *D* is the least of the integers.
    - Point *C* is a positive integer.
    - Points *A* and *D* are the same distance from 0.
    - Point *C* is closer to *D* than it is to *B*.

CONNECTION

25. **Computer Connection**   In the BASIC computer language, the INT(X) function finds the greatest integer that is *not* greater than X. For example, INT(3.5) = 3 because 3 is not greater than 3.5. Find INT(-2.1).

26. **Journal Entry**   Make up a puzzle like the one in Exercise 24. Show one possible solution.

7-3 The Coordinate System

Objective

Graph points on a coordinate plane.

Words to Learn

coordinate
 system
x-axis
y-axis
origin
quadrant
ordered pair
x-coordinate
y-coordinate

DID YOU KNOW

Most volcanoes in the world are in the "Ring of Fire." The "Ring of Fire" is a circle of volcanic activity surrounding the Pacific Ocean.

One way that geologists keep track of volcanic activity is by graphing volcanoes on a world map. The longitudinal (vertical) and latitudinal (horizontal) lines help you to locate specific volcanoes. Mt. St. Helens, for example, is at 122°W longitude and 48°N latitude.

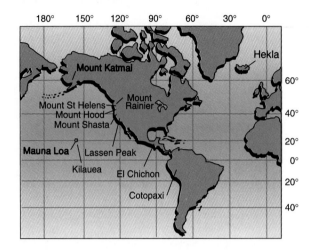

In mathematics, a **coordinate system,** or coordinate plane, is used to graph points in a plane. It is made up of a horizontal number line and a vertical number line that intersect. On the vertical number line, positive integers are represented as points above 0 and negative integers as points below 0.

The horizontal line is called the **x-axis,** and the vertical line is called the **y-axis.** They intersect at their zero points. This is called the **origin.** Together, they make up a coordinate system that separates the plane into four **quadrants.**

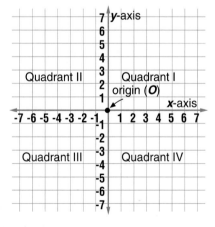

Points graphed on a coordinate system are identified using **ordered pairs.** The first number in an ordered pair is the **x-coordinate,** and the second number is the **y-coordinate.**

An ordered pair is written in this form.
    (x-coordinate, y-coordinate)

1 Name the ordered pair for point *A* and identify its quadrant.

Start at the origin, *O*. Locate point *A* by moving right 3 units along the *x*-axis. The *x*-coordinate is +3.

Now move down 5 units along the *y*-axis. The *y*-coordinate is −5.

The ordered pair is (3, −5). Point *A* is in quadrant IV.

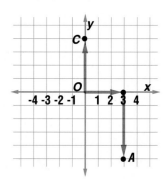

2 Name the ordered pair for point *C* and identify its quadrant.

The ordered pair is (0, 4). Point *C* is not in a quadrant because it is on an axis.

To graph a point in a coordinate system, draw a dot at the location named by its ordered pair.

Examples

3 Graph the point *D*(3, 5).

First draw a coordinate system. Start at the origin, *O*. Move 3 units to the right. Then move 5 units up to locate the point. Draw a dot and label it *D*(3, 5).

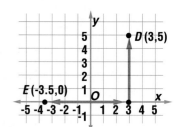

4 Graph the point *E*(−3.5, 0).

Start at *O*. Move 3.5 units to the left. Do not move up or down. Draw a dot and label it *E*(−3.5, 0).

Checking for Understanding

Communicating Mathematics

Read and study the lesson to answer each question.

1. **Write** the name of the point where the *x*-axis and the *y*-axis intersect. What is the ordered pair for this point?

2. **Show** why the point (2, 5) is different from the point (5, 2).

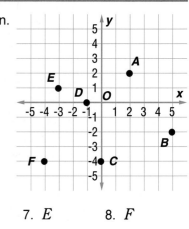

Guided Practice

Name the *x*-coordinate and *y*-coordinate for each point labeled at the right. Then tell in which quadrant the point lies.

3. *A*   4. *B*   5. *C*   6. *D*   7. *E*   8. *F*

On graph paper, draw a coordinate plane. Then graph and label each point.

9. $T(-2, 5)$
10. $N(3, -4)$
11. $P(-\frac{1}{2}, -1)$
12. $L(2.5, 7)$
13. $B(0, -2)$
14. $W(4, 0)$

Exercises

Independent Practice

Name the ordered pair for each point on the city map at the right.

15. bank
16. library
17. grocery
18. gas station
19. theater
20. city hall

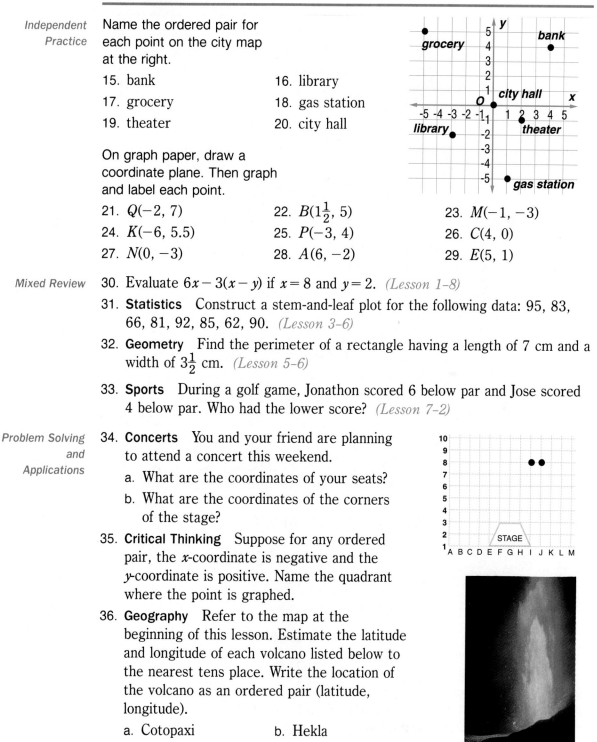

On graph paper, draw a coordinate plane. Then graph and label each point.

21. $Q(-2, 7)$
22. $B(1\frac{1}{2}, 5)$
23. $M(-1, -3)$
24. $K(-6, 5.5)$
25. $P(-3, 4)$
26. $C(4, 0)$
27. $N(0, -3)$
28. $A(6, -2)$
29. $E(5, 1)$

Mixed Review

30. Evaluate $6x - 3(x - y)$ if $x = 8$ and $y = 2$. *(Lesson 1-8)*

31. **Statistics** Construct a stem-and-leaf plot for the following data: 95, 83, 66, 81, 92, 85, 62, 90. *(Lesson 3-6)*

32. **Geometry** Find the perimeter of a rectangle having a length of 7 cm and a width of $3\frac{1}{2}$ cm. *(Lesson 5-6)*

33. **Sports** During a golf game, Jonathon scored 6 below par and Jose scored 4 below par. Who had the lower score? *(Lesson 7-2)*

Problem Solving and Applications

34. **Concerts** You and your friend are planning to attend a concert this weekend.
    a. What are the coordinates of your seats?
    b. What are the coordinates of the corners of the stage?

35. **Critical Thinking** Suppose for any ordered pair, the x-coordinate is negative and the y-coordinate is positive. Name the quadrant where the point is graphed.

36. **Geography** Refer to the map at the beginning of this lesson. Estimate the latitude and longitude of each volcano listed below to the nearest tens place. Write the location of the volcano as an ordered pair (latitude, longitude).
    a. Cotopaxi
    b. Hekla
    c. Mauna Loa
    d. Mount Katmai

7-4A Adding Integers

A Preview of Lesson 7-4

Objective

Add integers by using models.

Materials

counters of two colors

mat

In this lab, you will use counters to model addition with integers. Let one color of counter represent positive integers and another color represent negative integers.

Try this!

Work with a partner.

- Remember that $5 + 3$ means *combine a set of five items with a set of three items.* In this lab, the addition $5 + 3$ tells you to combine a set of 5 positive counters with a set of 3 positive counters.

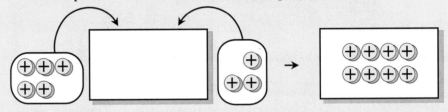

So, $5 + 3 = 8$.

- Place 5 negative counters on the mat. Place 3 more negative counters on the mat. Use your results to complete the addition sentence $-5 + (-3) = \underline{\quad ? \quad}$.

- Place 5 positive counters and 3 negative counters on the mat.

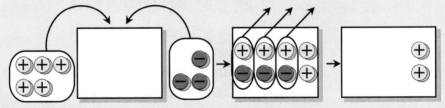

In this case, it is possible to pair a positive counter with a negative counter. This is called a *zero pair.* Remove as many zero pairs as possible. Use your result to complete this addition sentence $5 + (-3) = \underline{\quad ? \quad}$.

- Use counters to model $-5 + 3$. Then write an addition sentence.

What do you think?

1. What is the result when you add two negative integers?

2. What is the result when you add a positive and a negative integer?

Model each addition. Use your result to write an addition sentence.

3. $7 + 2$ 4. $7 + (-2)$ 5. $-7 + 2$ 6. $-7 + (-2)$

7-4 Adding Integers

Objective

Add integers.

Words to Learn

additive inverse

DID YOU KNOW

Without the weather to spread the Sun's heat around the world, the tropics would get hotter and the poles would get colder. Nothing would be able to live on Earth.

Did you know that the moon has mountains taller than Mount Everest, but has no rain or snow? In 1966, the United States' spacecraft Surveyor I landed on the moon. Its purpose was to send back to Earth television pictures and moon measurements. Changes in temperature were one such measurement.

At lunar noon, the temperature on the moon's surface was 235°F. By midnight the temperature had dropped 485 degrees. You can write the drop in temperature as -485 degrees. What was the temperature at midnight? *This question will be answered in Example 5.*

To solve the problem like the one about moon temperatures, you can add integers. One way to add integers is by using arrows on a number line. Positive integers are represented by arrows pointing *right*. Negative integers are represented by arrows pointing *left*.

$$5 + 2 = 7$$

$$-5 + (-2) = -7$$

Adding Integers with the Same Sign	The sum of two positive integers is positive. The sum of two negative integers is negative.

Examples

1 Solve $a = 60 + 15$.
  Use a number line.
  Start at 60. Since 15
  is positive, go 15 units
  to the right.

  $a = 60 + 15$
  $a = 75$

2 Solve $-3 + (-2) = t$.
  Use counters. Put in
  3 negative counters.
  Add 2 more negatives.

  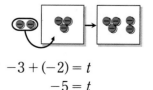

  $-3 + (-2) = t$
  $\qquad -5 = t$

What happens when you add integers with different signs? Let's first look at the sum $5 + (-2)$: $5 + (-2) = 3$.

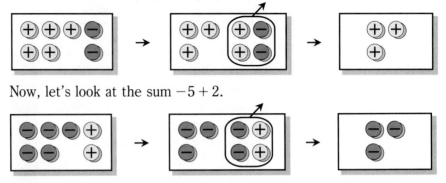

Now, let's look at the sum $-5 + 2$.

$$-5 + 2 = -3$$

The results above suggest the following rule for adding two integers with different signs.

Adding Integers with Different Signs	To add integers with different signs, subtract their absolute values. The sum is: • positive if the positive integer has the greater absolute value. • negative if the negative integer has the greater absolute value.

Examples

3 Solve $a = 23 + (-18)$.
$|23| > |-18|$, so the sum is positive.
The difference of 23 and 18 is 5.
So, $a = 5$.

4 Solve $-39 + 19 = w$.
$|-39| > |19|$, so the sum is negative.
The difference of 39 and 19 is 20.
So, $w = -20$.

Example 5 *Problem Solving*

Geology Refer to the problem in the lesson introduction. What was the temperature at midnight?

Explore   At noon, the temperature was 235°F. By midnight, the temperature dropped 485 degrees. You can write this drop as -485 degrees.

Plan   Let $t =$ the temperature at midnight.
Solve the equation $t = 235 + (-485)$.

Solve $|-485| > |235|$, so the sum is negative.
The difference of 485 and 235 is 250. So, $t = -250$.
The temperature at midnight is 250°F below zero.

Examine   Use a calculator to check.

235 $\boxed{+}$ 485 $\boxed{+/-}$ $\boxed{=}$ -250 It checks.

___

What happens when you add two integers that are opposites, like 2 and -2? In a sense, adding its opposite "undoes" the first integer, and the result is 0. For this reason, two integers that are opposites of each other are called **additive inverses.**

Additive Inverse Property	**In words:** The sum of any number and its additive inverse is zero.
	Arithmetic **Algebra**
	$2 + (-2) = 0$ $a + (-a) = 0$

Checking for Understanding

Communicating Mathematics

Read and study the lesson to answer each question.

1. **Tell** how you know the sign of the result when adding integers with different signs.

2. **Show** three examples of integers and their additive inverses.

3. **Draw a model** that shows $4 + (-4) = k$. How many zero pairs are there?
4. **Write** the addition sentence shown by each model.

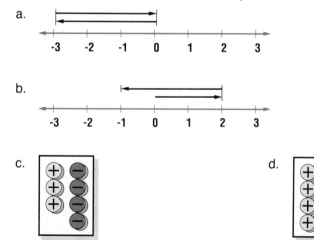

a.

b.

c.

d.

Guided Practice

Tell whether the sum is positive, negative, or zero.

5. $6 + (-6)$ 6. $-9 + (-13)$ 7. $-5 + 9$ 8. $-7 + 12$
9. $7 + (-1)$ 10. $8 + (-11)$ 11. $-12 + (-7)$ 12. $-11 + 18$

Solve each equation.

13. $y = 4 + (-9)$ 14. $-9 + (-1) = f$ 15. $y = -3 + 14$
16. $-5 + 12 = w$ 17. $-5 + 5 = r$ 18. $h = 17 + (-8)$

19. **Sports** The Bulldogs lost five yards on one play, then gained nine yards on the next play. What was the total number of yards gained or lost?

Exercises

Independent Practice

Solve each equation.

20. $3 + (-8) = u$
21. $q = 5 + (-9)$
22. $c = -10 + 12$
23. $-7 + 13 = p$
24. $-9 + -19 = u$
25. $b = 35 + (-10)$
26. $-3 + 10 = i$
27. $200 + (-100) = v$
28. $-57 + 10 = k$

Evaluate each expression if $a = -10$, $b = 10$, and $c = -5$.

29. $a + (-3)$
30. $-6 + b$
31. $c + (-1)$
32. $0 + a$
33. $a + b$
34. $c + b$

35. Evaluate $a + 9$ if $a = -23$.
36. Evaluate $b + a$ if $b = -4$ and $a = 19$.

Mixed Review

37. **Smart Shopping** Angela purchases 3.9 pounds of coffee for $23.75. Estimate the price per pound of the coffee. *(Lesson 2-3)*

38. **Statistics** Find the range for the following set of data. Choose an appropriate scale and interval. Then draw a number line to show the scale and interval. *(Lesson 3-3)*

$$57, 62, 75, 55, 59, 63, 63, 45, 83, 61$$

39. Find the next three terms in $\frac{1}{2}, \frac{1}{4}, \frac{1}{8}, \ldots$ *(Lesson 4-4)*

40. Solve using the inverse operation $8t = 64$. *(Lesson 6-1)*

41. On graph paper, draw coordinate axes. Then graph the point $B(-4, 4)$. *(Lesson 7-3)*

Problem Solving and Applications

42. **Personal Finance** Rosa opened a checking account with a balance of $150. She wrote a check for $87.
    a. Write an addition sentence to represent this situation.
    b. How much money remained in the account?

43. **Space Travel** During a space shuttle launch, a maneuver is scheduled to begin at T minus 75 seconds, which is 75 seconds before liftoff. The maneuver lasts 2 minutes. At what time will this maneuver be complete?

44. **Critical Thinking** Jack made up a game of darts using the target at the right. Each person throws three darts. The score is the sum of the numbers in the regions that the darts hit. If all the darts hit the target, list all possible scores.

7-5A Subtracting Integers

A Preview of Lesson 7-5

Objective

Subtract integers by using models.

Materials

counters of
 two colors
mat

In this lab, you will use counters to model subtraction with integers.

Try this!

Work with a partner.

● Consider $7 - 3$. Place 7 positive counters on the mat and then remove 3.

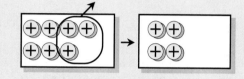

So, $7 - 3 = 4$.

● Consider $-7 - (-3)$. Place 7 negative counters on the mat. Remove 3 of them. Use your result to complete this subtraction sentence $-7 - (-3) = \underline{\ ?\ }$.

Now consider $7 - (-3)$. This means start with a set of 7 *positive* counters and remove 3 *negative* counters. Add 3 zero pairs to the set. The value of the set does not change. Now you can remove 3 negative counters.

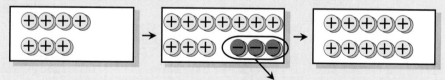

Use the result to complete this sentence $7 - (-3) = \underline{\ ?\ }$.

What do you think?

1. Model $7 - 5$ and $7 + (-5)$. How are they alike? How are they different? Describe any relationship in your own words.

2. If you subtract a positive integer from a lesser positive integer, is the difference positive or negative?

Model each subtraction. Use your result to write a subtraction sentence.

3. $8 - 2$ 4. $-8 - (-2)$ 5. $8 - (-2)$ 6. $-8 - 2$

7. $1 - 4$ 8. $1 - (-4)$ 9. $-1 - (-4)$ 10. $-1 - 4$

7-5 Subtracting Integers

Objective

Subtract integers.

Global warming, or the greenhouse effect, makes Earth liveable. However, because of atmospheric changes people are causing, scientists fear the greenhouse effect could increase too much.

The greenhouse effect occurs on other planets. The thermometer shows how it affects temperatures on Mars, Earth, and Venus. What is the difference between the actual temperature and the temperature without the greenhouse effect on each planet? *You will answer this question in Exercise 45.*

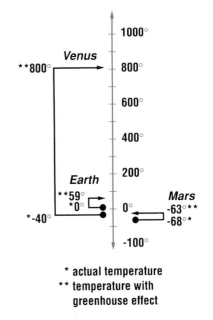

* actual temperature
** temperature with greenhouse effect

When am I ever going to use this?

Suppose you are traveling from Seattle to Sydney, Australia. How many hours ahead will you have to set your watch?

Time zones separate the world into 24 sections. The 12th zone east ($+12$) and the 12th zone west (-12) are separated by the International Date Line.

Seattle is in zone -8 and Sydney is in zone $+10$, so you would have to set your watch $-8 + (-10)$ or ahead 18 hours.

Before we can find each difference, let's see how addition and subtraction of integers are related. Let's compare the subtraction $5 - 2$ to the addition $5 + (-2)$.

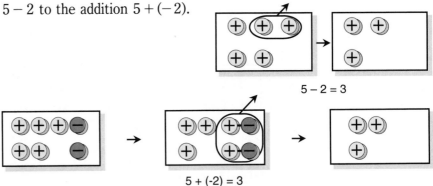

$$5 - 2 = 3$$

$$5 + (-2) = 3$$

The diagrams above show that $5 - 2 = 5 + (-2)$. This example suggests that adding the additive inverse of an integer produces the same result as subtracting it.

Subtracting Integers	**In words:** To subtract an integer, add its additive inverse.
	Arithmetic **Algebra**
	$5 - 2 = 5 + (-2)$ $a - b = a + (-b)$

Examples

1 Solve $x = -8 - 5$.

$$\begin{aligned} x &= -8 - 5 \\ &= -8 + (-5) \qquad \text{\textit{To subtract 5, add} } -5. \\ &= -13 \end{aligned}$$

2 Solve $-13 - (-6) = w$.

$$\begin{aligned} -13 - (-6) &= w \\ -13 + 6 &= \qquad \text{\textit{To subtract} } -6, \text{\textit{ add 6.}} \\ -7 &= \end{aligned}$$

3 Solve $t = 54 - (-4)$.

$$\begin{aligned} t &= 54 - (-4) \\ &= 54 + 4 \qquad \text{\textit{To subtract} } -4, \text{\textit{ add 4.}} \\ &= 58 \end{aligned}$$

Example 4   *Connection*

Algebra   Evaluate $a - b$ where $a = 3$ and $b = 12$.

$$\begin{aligned} a - b &= 3 - 12 \qquad \text{\textit{Replace a with 3 and b with 12.}} \\ &= 3 + (-12) \qquad \text{\textit{To subtract 12, add} } -12. \\ &= -9 \end{aligned}$$

Communicating Mathematics

Read and study the lesson to answer each question.

1. **Tell** how addition and subtraction of integers are related.
2. **Tell** how to find the additive inverse of an integer.
3. **Show** how to use the additive inverse to solve $6 - 8 = n$.
4. **Write** a subtraction sentence shown by each model.

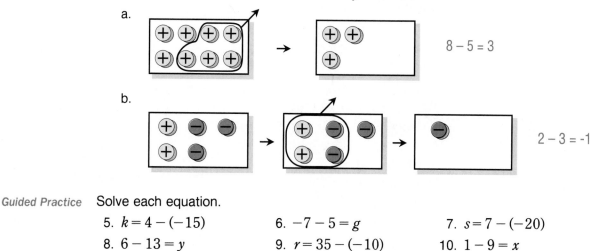

a. $8 - 5 = 3$

b. $2 - 3 = -1$

Guided Practice

Solve each equation.

5. $k = 4 - (-15)$
6. $-7 - 5 = g$
7. $s = 7 - (-20)$
8. $6 - 13 = y$
9. $r = 35 - (-10)$
10. $1 - 9 = x$

Exercises

Independent Practice

Solve each equation.

11. $9 - 12 = t$
12. $54 - (-8) = w$
13. $-12 - 15 = q$
14. $-3 - (-5) = a$
15. $y = -12 - 56$
16. $g = 6 - (-35)$
17. $10 - (-10) = q$
18. $17 - (-9) = w$
19. $-87 - (-87) = k$
20. $-50 - 45 = f$
21. $3 - 86 = h$
22. $k = 6 - (-23)$
23. $-8 - (-8) = q$
24. $42 - 100 = y$
25. $m = -8 - (-19)$

Evaluate each expression if $t = -2$, $y = 8$, and $e = 4$.

26. $t - 5$
27. $y - 7$
28. $-8 - e$
29. $-10 - y$
30. $3 - y$
31. $t - y$
32. $y - e$
33. $e - y$
34. $y - t$
35. $-e - y$
36. $t + e$
37. $-t - e$

38. Evaluate $m - n$ if $m = 5$ and $n = -6$.
39. Evaluate $-p - q$ if $p = -3$ and $q = 9$.

Mixed Review

40. Estimate 2,367 + 1,248. *(Lesson 1-2)*
41. Find the LCM of 14 and 21. *(Lesson 4-9)*
42. **Probability** Gwen buys one ticket for a raffle which has 1 grand prize of \$500 and 2 second-place prizes of \$100. A total of 500 tickets are sold. Find Gwen's expected winnings. *(Lesson 5-8)*
43. Translate the phrase *65 less than w* into an algebraic expression. *(Lesson 6-4)*
44. Solve the equation $d = -3 + (-2)$. *(Lesson 7-4)*

Problem Solving and Applications

45. **Environment** Refer to the problem in the lesson introduction. Find the difference between the actual temperature and the temperature without the greenhouse effect for Earth, Venus, and Mars.
46. **Critical Thinking** Evaluate $a - b$ if $a = -4$ and $b = 10$. Then evaluate $b - a$ for the same values of the variables. Use your results to make a conjecture about the relationship between $a - b$ and $b - a$. Does your conjecture hold true if $a = 7$ and $b = 9$?
47. **Weather** Use the graph at the right to answer each question.
    a. What does the zero point on this graph represent?
    b. Write an integer to represent the rainfall for each month shown.
    c. Can you use this graph to decide in which of these months the least rain fell? Explain.
    d. Write a sentence that describes the message this graph is meant to convey.

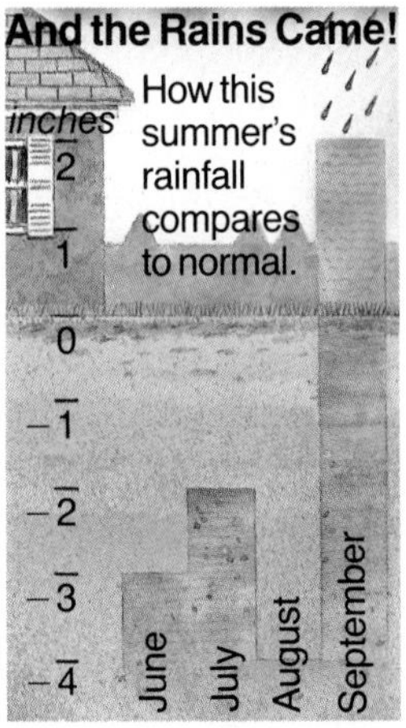

And the Rains Came!
How this summer's rainfall compares to normal.
inches
2
1
0
−1
−2
−3
−4
June
July
August
September

48. **History** Refer to the time line below.

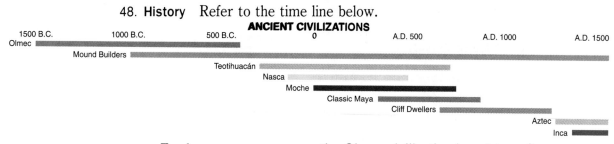

ANCIENT CIVILIZATIONS

a. For how many years was the Olmec civilization in existence?

b. How many years were there between the beginning of the Mound Builders and Columbus' arrival in America?

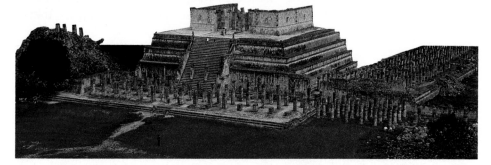

49. **Data Search** Refer to pages 252 and 253. Between the years 1886 and 1989, how many tropical storms occurred during the first five months of the year? How many tropical storms were there during August, September, and October?

Mid-Chapter Review

Find the opposite and absolute value of each integer. *(Lesson 7-1)*

1. -15 2. 23 3. 0

4. Order the integers 4, 3, -3, 5, -5, and 6 from least to greatest. *(Lesson 7-2)*

On graph paper, draw coordinate axes. Then graph and label each point. *(Lesson 7-3)*

5. $M(-3, 4)$ 6. $N(0, -5)$ 7. $P(-2, -6)$ 8. $Q(3, 0)$

Solve each equation. *(Lesson 7-4)*

9. $y = -15 + 3$ 10. $3 + (-5) = t$ 11. $-2 + (-5) = n$

Solve each equation *(Lesson 7-5)*

12. $x = 3 - 15$ 13. $3 - (-5) = p$ 14. $-2 - (-5) = r$

Planning for Good Nutrition

Situation

The challenge to your Snack Bar Committee for Sport's Night is to make $1,000 profit and still have a nutritious menu. Jack Bell, the student committee leader, has a menu from the deli at Superior Market and an agreement from the manager to supply all catered food at half price. How many of which items will you choose? How will you price them to make a profit?

Hidden Data

Will the deli furnish the paper and plastic products or is there a charge?
Does the deli have a delivery charge?
Is ice available from the deli?
What should the serving size of a salad be?
Since milk and juices are not on the deli list, will Superior Market give you a price break on these items?

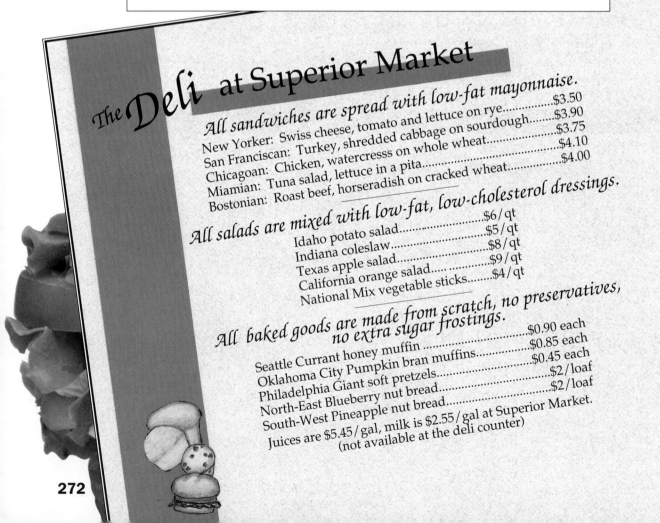

The Deli at Superior Market

All sandwiches are spread with low-fat mayonnaise.

New Yorker: Swiss cheese, tomato and lettuce on rye.............$3.50
San Franciscan: Turkey, shredded cabbage on sourdough........$3.90
Chicagoan: Chicken, watercresss on whole wheat....................$3.75
Miamian: Tuna salad, lettuce in a pita.....................................$4.10
Bostonian: Roast beef, horseradish on cracked wheat..............$4.00

All salads are mixed with low-fat, low-cholesterol dressings.

Idaho potato salad...........................$6/qt
Indiana coleslaw...............................$5/qt
Texas apple salad.............................$8/qt
California orange salad.....................$9/qt
National Mix vegetable sticks........$4/qt

All baked goods are made from scratch, no preservatives, no extra sugar frostings.

Seattle Currant honey muffin ....$0.90 each
Oklahoma City Pumpkin bran muffins................$0.85 each
Philadelphia Giant soft pretzels.........................$0.45 each
North-East Blueberry nut bread..........................$2/loaf
South-West Pineapple nut bread.........................$2/loaf
Juices are $5.45/gal, milk is $2.55/gal at Superior Market.
(not available at the deli counter)

Analyzing the Data

1. What is the total cost of a New Yorker sandwich and a Bostonian sandwich?

Making a Decision

2. **What is** the cost of 3 quarts of Texas apple salad?
3. **How much** variety is needed?
4. **Can you** feature a favorite as a "Sportsnight Sale" item?
5. **Will you** be able to return leftovers or have a sell-off?
6. **What is** the expected attendance?
7. **Will you** need an alternate place if attendance is low or if sales are low?

Making Decisions in the Real World

8. Investigate the cost of deli products from another grocery store. Ask the deli for a price cut.

7-6 **Find a Pattern**

Objective

Solve problems finding and extending a pattern.

The United States Bureau of the Census can predict the population of the world and the median age of the people. The partial chart below shows some actual and some predicted data. Use the chart to find a pattern that will help you predict the world's population in the year 2000.

The World's Population					
Year	1980	1985	1990	1995	2000
World's Total Population in millions	4,478	4,889	5,326	5,786	?
per square mile	85	93	102	110	?
Median Age	22.6	23.5	24.3	25.2	?
Industrialized Nations	1,136	1,172	1,204	1,232	?
Non-Industrialized Nations	3,343	3,717	4,122	4,554	?

Explore   What do you know?
You know the actual population for 1980 and 1990.
You know the predicted population for 1985 and 1995.

What are you trying to find?
You are trying to find the population for the year 2000.

Plan   Study the chart. You can see in the millions line that the world's population is growing. Find the pattern to predict the world's population for the year 2000.

Solve   To find how much the population is increasing, subtract each year's population from the population before it. Then subtract the differences.

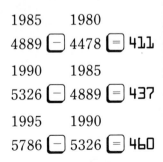

1985   1980
4889 ⊖ 4478 ⊜ 411

1990   1985
5326 ⊖ 4889 ⊜ 437

1995   1990
5786 ⊖ 5326 ⊜ 460

Since 437 − 411 is 26, and 460 − 437 is 23, you can see that the amount of growth is decreasing.

The amount of growth decreased by 3 million every 5 years. To find the amount of growth from 1995 to 2000, subtract 3 from 23. The result is 20. So, 20 + 460 is 480. This is the amount of increase from 1995 to 2000. To predict the world's population for the year 2000, add.

5,786	*predicted 1995 population*
+ 480	*amount of increase from 1995 to 2000*
6,266	*predicted population for 2000*

In the year 2000, the world's population is predicted to be 6,266,000.

Examine The answer 6,266,000 seems reasonable because the differences in population increase follow a pattern: 411, 437, 460, and 480. The differences between each of these numbers are 26, 23, and 20.

Example

Complete the pattern: 63, 48, 35, 24, __?__, __?__, __?__.

What do you add to each term to get the succeeding term?

$$48 - 63 = -15 \qquad 35 - 48 = -13 \qquad 24 - 35 = -11$$

To continue the pattern, add 2 less each time; first add -9, then -7, and then -5. So, $24 + (-9) = 15$, $15 + (-7) = 8$, and $8 + (-5) = 3$.

The next three numbers in the sequence are 15, 8, and 3.

Checking for Understanding

Communicating Mathematics

Read and study the lesson to answer each question.

1. **Tell** why the prediction for the median age in the world for the year 2000 is about 26.
2. **Write** another rule for the pattern in the example above.

Guided Practice

Solve by finding a pattern.

3. State the pattern. Then complete the sequence.

    1, 3, 1, 3, _?_, _?_, _?_

4. State two different rules for the sequence 1, 4, 9, 16, 25, ...
5. Complete the sequence 60, _?_, _?_, 42, 36.

Problem Solving

Practice

Solve. Use any strategy.

Strategies
●●●●●●●●●●

Look for a pattern.

Solve a simpler pattern.

Act it out.

Guess and check.

Draw a diagram.

Make a chart.

Work backwards.

6. Find the number of line segments determined by six points on a line.

7. Gloria made enough money with her computer graphics to buy a new printer. She told Al and Sue who each let two of their computer network friends know ten minutes later. If the news spread like this every ten minutes, how many people knew by the end of the hour?

8. Complete the pattern 100, 98, 94, __?__ , 80, __?__ .

9. You plant 10 hyacinths in exactly 5 rows. There are 4 bulbs in each row. Draw a diagram of your garden.

10. This pattern is known as Pascal's Triangle. Find the pattern and complete the 6th and 7th rows.

    1st row   1
    2nd row   1   1
    3rd row   1   2   1
    4th row   1   3   3   1
    5th row   1   4   6   4   1

11. A college student sent home this letter. If each letter stands for one digit 0-9, how much money did he ask for?

    ```
      SEND
    + MORE
     MONEY
    ```

12. Complete the chart on page 274.

CULTURAL KALEIDOSCOPE

Pat Neblett

If children are our future, Pat Neblett wants to make sure that they know and understand different cultures. With $2,000, the former real estate agent founded Tuesday's Child Books in 1988 from her Randolph, Massachusetts home. She began selling African-American books through direct mail catalogs. Now most of her business comes from sales to schools. The titles have expanded to include children's books about Asians, Hispanics, Native Americans, and minority groups in the United States. Neblett carries more than 300 titles. She read each book herself to ensure that the content positively reinforces a child's self-worth.

Sales have increased from $7,000 in her first year to a projected $25,000 in 1991. She plans to open her own retail store if her business continues to grow.

7-7A Multiplying Integers

A Preview of Lesson 7-7

Objective

Multiply integers by using models.

Materials:

counters of two colors
mat

In this lab, you will use counters to model multiplication of integers.

Try this!

Work with a partner.

- Remember that 2×4 means *two sets of four items.* Using models, 2×4 means to *place* 2 sets of 4 *positive* counters on a mat.

So, $2 \times 4 = 8$.

- Place 2 sets of 4 negative counters on the mat. Use your result to complete: $2 \times (-4) = \underline{\ ?\ }$.

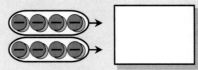

- Since -2 is the opposite of 2, -2×4 means to *remove* 2 sets of 4 *positive* counters. How can you remove 2 sets? First, put in as many zero pairs as you need. Then remove 2 sets of 4 positive counters. Use your result to complete: $-2 \times 4 = \underline{\ ?\ }$.

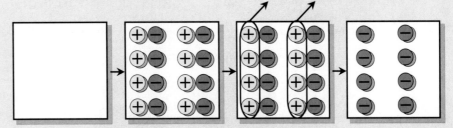

- Use counters to model $-2 \times (-4)$. Use your result to write a multiplication sentence.

What do you think?

1. How are -3×7 and $7 \times (-3)$ the same? How are they different?

Model each multiplication and write a multiplication sentence.

2. 3×5 3. $3 \times (-5)$ 4. -3×5 5. $-3 \times (-5)$

6. 5×3 7. $5 \times (-3)$ 8. -5×3 9. $-5 \times (-3)$

7-7 Multiplying Integers

Objective

Multiply integers.

The wearing away of a coastline by the action of water is called wave erosion. Wave erosion can cause a coastline to recede at a rate of a few centimeters each year.

Suppose a beach recedes 2 centimeters each year. In four years, how many centimeters will the beach recede? Let r represent the number of centimeters the beach recedes. Let -2 mean that the beach recedes 2 centimeters.

The equation $r = 4(-2)$ can be used to represent this problem. You can solve this problem by using counters or by observing a pattern.

Using Counters:

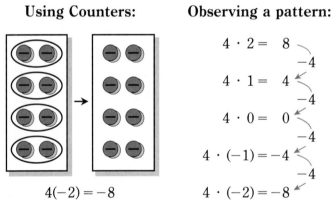

$$4(-2) = -8$$

Observing a pattern:

$$4 \cdot 2 = 8$$
$$4 \cdot 1 = 4 \qquad \searrow -4$$
$$4 \cdot 0 = 0 \qquad \searrow -4$$
$$4 \cdot (-1) = -4 \quad \searrow -4$$
$$4 \cdot (-2) = -8 \quad \searrow -4$$

So, $r = -8$. In four years, the beach will recede 8 centimeters.

The diagram and pattern above illustrate the following rule.

Multiplying Integers with Different Signs	The product of two integers with different signs is negative.

Examples

1 Solve $g = 5(-4)$.

The two integers have different signs. The product will be negative.

$$g = 5(-4)$$
$$g = -20$$

2 Solve $-6(5) = h$.

The two integers have different signs. The product will be negative.

$$-6(5) = h$$
$$-30 = h$$

How would you solve $-4(-2) = y$? You can use the same methods to multiply two integers with the same sign. When using counters, $-4(-2)$ means that you will remove 4 sets of 2 negative counters.

Using counters: **Observing a pattern:**

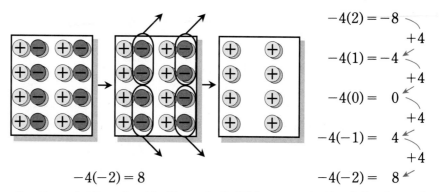

$-4(-2) = 8$ $-4(-2) = 8$

So, the solution of $-4(-2) = y$ is 8. This suggests the following rule.

Multiplying Integers with the Same Sign	The product of two integers with the same sign is positive.

Examples

3 Solve $a = -5(-4)$.
The two integers have the same sign. The product will be positive.
$$a = -5(-4)$$
$$= 20$$

4 Solve $(-7)^2 = z$.
The exponent says there are two factors of -7. The product will be positive.
$$(-7)^2 = z$$
$$(-7)(-7) =$$
$$49 =$$

Example 5 *Connection*

Algebra Evaluate abc, where $a = -2$, $b = 5$, and $c = 2$.

$abc = (-2)(5)(2)$ *Replace a with -2, b with 5, and c with 2.*
$\quad = [(-2)(5)](2)$ *Use the associative property to group the factors.*
$\quad = (-10)(2)$
$\quad = -20$

Checking for Understanding

Communicating Mathematics   Read and study the lesson to answer each question.

1. **Tell** what you can say about two integers if their product is negative.
2. **Show** a pattern to explain why the solution of $q = 4(-3)$ must be -12.
3. **Draw** a model that shows $-3(-5) = k$.
4. **Show** another way to complete the product in Example 5.

Solve each equation.

    5. $c = 21(-2)$ 6. $-4(-5) = u$ 7. $t = 6(-3)$

    8. $-9(9) = h$ 9. $-15(3) = j$ 10. $k = (-3)^2$

Evaluate each expression if $x = -9$, $y = 2$, $z = -3$, and $w = 6$.

    11. $-3w$ 12. xy 13. z^2

    14. $2xw$ 15. $-4x^2$ 16. xyz

Exercises

Solve each equation.

    17. $z = -6(-15)$ 18. $(-5)^2 = h$ 19. $-11(-11) = g$

    20. $h = -9(5)$ 21. $(-6)(-6) = p$ 22. $f = (-12)^2$

    23. $8(-12) = n$ 24. $-7(15) = k$ 25. $-34(2) = q$

    26. Find the product of -8 and -6.   27. Multiply -7 and 10.

Evaluate each expression if $a = -7$, $b = 3$, $c = -5$, and $d = 7$.

    28. $-5d$ 29. $7d$ 30. $-6bd$

    31. $-9ab$ 32. $-16c$ 33. a^2

    34. $-2d^2$ 35. $-10bc$ 36. $-2bd$

    37. $-5abc$ 38. $-acd$ 39. $-3bc$

40. Divide 2.25 by 1.5. *(Lesson 2-7)*

    41. Write $\frac{3}{8}$ as a decimal. *(Lesson 4-7)*

    42. Subtract $\frac{1}{9}$ from $\frac{5}{12}$. *(Lesson 5-3)*

    43. Solve the equation $\frac{k}{21} = 4$. *(Lesson 6-3)*

    44. **Hiking** Jay's boy scout troop is hiking on a trail that is 75 feet above sea level. They hike into a canyon that is 12 feet below sea level. How many feet has Jay's troop hiked? *(Lesson 7-5)*

45. **Earth Science** Earth's atmosphere exerts a pressure of 14.7 pounds per square inch at the ocean's surface. The pressure increases by 2.7 pounds per square inch for every 6 feet that you descend. Yung-Mi says that the pressure at -17 feet will be 132.5 pounds per square inch. Is this reasonable? Explain.

    46. **Critical Thinking** Find each product.

        a. $(-2)(3)(4)$ b. $(-2)(-3)(4)$

        c. What is the sign of the product $(2)(-3)(-4)$?

        d. Write a rule for determining the sign of the product of *three* integers.

7-8 Dividing Integers

Objective

Divide integers.

The list at the right shows some of the stocks traded on the New York Stock Exchange. Rod and Linda Conley invested money in the Heinz Company. During a 2-day period, Rod and Linda's stock price had a change of -2. What was the average change per day?

Stock	High	Low	Last	Change
Heinz	34	32	32	−2
HeleneC	7 7/8	7 3/4	7 7/8	
Hellrint	21 7/8	19 1/4	21	+3/4
HelmrP	71 3/4	67 3/4	71 1/2	+7/8
HemCap	4 1/8	4	4 1/8	+1/8
Heminc	8 1/4	8	8 1/4	
Herculs	22	20 1/2	21	−1
Hershy	25 1/2	24 3/4	25 1/2	+1/4

Let a represent the average change per day. To find a, divide -2 by 2.

$$-2 \div 2 = a$$

Division of integers is related to multiplication. The division sentence $-2 \div 2 = a$ can be written as the multiplication sentence $a \times 2 = -2$. Think: 2 times what number equals -2?

$$2(1) = 2$$
$$2(-1) = -2 \checkmark$$

So, $a = -1$. Rod and Linda's stock dropped an average of \$1 per day.

Let's see how some other division sentences are related to multiplication sentences.

$$4 \div (-2) = b \quad \rightarrow \quad -2 \times b = 4$$
$$-15 \div (-3) = c \quad \rightarrow \quad -3 \times c = -15$$
$$-63 \div 7 = d \quad \rightarrow \quad 7 \times d = -63$$

Since division is related to multiplication, you can follow the multiplication rules for signs to determine the sign of a quotient.

Dividing Integers	The quotient of two integers with the same sign is positive.
	The quotient of two integers with different signs is negative.

Examples

1 Solve $a = -20 \div (-4)$.

$a = -20 \div (-4)$ *The signs are the same.*
    $= 5$ *The quotient is positive.*

2 Solve $-10 \div 2 = x$.

$-10 \div 2 = x$ *The signs are different.*
    $-5 = x$ *The quotient is negative.*

Checking for Understanding

Read and study the lesson to answer each question.

1. **Write** a division sentence related to the multiplication sentence $6n = -12$.

2. **Tell** why the solution of $x = -50 \div 10$ is equal to the solution of $x = 50 \div (-10)$.

Guided Practice

Solve each equation.

3. $c = 45 \div (-15)$
4. $-250 \div 25 = u$
5. $a = -35 \div -7$
6. $500 \div (-25) = h$
7. $-68 \div (-4) = b$
8. $-84 \div 12 = p$

Evaluate each expression if $v = -12$, $e = 6$, and $q = 4$.

9. $\dfrac{124}{q}$
10. $\dfrac{v^2}{e}$
11. $\dfrac{-96}{v}$

Exercises

Independent Practice

Solve each equation.

12. $c = 56 \div (-2)$
13. $-88 \div (-22) = u$
14. $t = -56 \div 4$
15. $45 \div (-9) = h$
16. $-45 \div 9 = b$
17. $-33 \div 11 = p$
18. $-100 \div (-10) = k$
19. $a = 56 \div (-1)$
20. $m = -120 \div 6$

21. Divide -64 by 8.
22. Find the quotient of -28 and -7.

Evaluate each expression if $v = -24$, $e = 8$, and $q = 3$.

23. $\dfrac{v}{q}$
24. $v \div (-e)$
25. $\dfrac{v}{eq}$
26. $v^2 \div e$
27. $\dfrac{v^2}{eq}$
28. $\dfrac{eq}{v}$

Mixed Review

29. Evaluate 4^3. *(Lesson 1-9)*

30. **Statistics** A survey concerning the number of hours a junior high student spends doing homework each day produces the following data: 2, 4, 3, 2, 1, 2, 2, 4. Find the mean, median, and mode for the data. *(Lesson 3-5)*

31. Solve the equation $s = -5(12)$. *(Lesson 7-7)*

Problem Solving and Applications

32. **Engineering** The water level in a tank decreased 8 inches in 4 minutes. The tank drains at a steady rate. What is the change in the water level each minute?

33. **Statistics** Soya Gilbers recorded the following noon temperatures, in degrees Fahrenheit for one week. What is the mean?

Sun.	Mon.	Tues.	Wed.	Thurs.	Fri.	Sat.
-10	-8	-10	0	4	1	9

34. **Critical Thinking** Find all values of a and b for which $ab = \dfrac{a}{b} = \dfrac{b}{a}$.

7-9A Solving Equations

A Preview of Lesson 7-9

Objective

Solve equations by using models.

Materials

counters
cups
mats

You have used counters to solve equations in Chapter 6. These can also be used to solve equations that involve integers.

Try this!

Work in groups of three.

- Start with two empty mats. Use counters and a cup to show the equation $x + (-2) = -3$. This equation can be solved by removing 2 negative counters from each mat. So, $x = -1$.

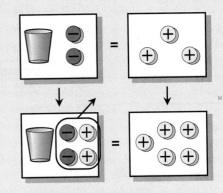

- Start again with two empty mats. Use counters to show the equation $x + (-2) = 3$. In this case, the counters on each side of the equals sign are different. Add 2 positive counters to each side of the equation. Then remove 2 zero pairs from the left side. So, $x = 5$.

What do you think?

1. Write an equation for the model shown below.

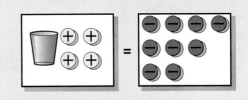

2. How do you know what type of counter to add to each side to solve an equation? How many counters need to be added?

Make a model of each equation. Then solve the equation.

3. $x + 4 = -6$	4. $c - 5 = 7$	5. $m + 6 = -20$
6. $g - (-4) = 8$	7. $b + (-1) = 0$	8. $a + (-7) = 16$
9. $y - 2 = -1$	10. $p + (-8) = 3$	11. $g + (-4) = -4$

Extension

12. Make a model of $2x = -8$. Then solve the equation.

7-9 Solving Equations

Objective

Solve equations with integer solutions.

 **LOOKBACK**

You can review addition and subtraction equations on page 225.

Chinook is an Indian word for a warm dry wind on the east side of the Rocky Mountains that causes a rapid rise in temperature. For example, a chinook once caused the temperature in Rapid City, South Dakota to increase 25°C in 15 minutes. If the final temperature was 15°C, what was the temperature before rising 25°C? *This question will be answered in Example 2.*

You solve equations involving integers in the same way you have solved equations involving whole numbers. Remember that when you perform the same addition or subtraction on each side of an equation, the two sides will remain equal.

Example 1

Solve $-5 + v = 3$.

$$-5 + v = 3$$
$$-5 + v + 5 = 3 + 5 \qquad \textit{Add 5 to each side.}$$
$$v = 8$$

Check:
$$-5 + v = 3$$
$$-5 + 8 \stackrel{?}{=} 3 \qquad \textit{Replace v with 8.}$$
$$3 = 3 \checkmark$$

Example 2 *Problem Solving*

Weather Refer to the problem in the lesson introduction. What was the temperature before it rose 25°C?

Let $t =$ the original temperature. Since the wind is increasing the temperature, the equation $t + 25 = 15$ represents the situation.

Solve $t + 25 = 15$.

$$t + 25 = 15$$
$$t + 25 - 25 = 15 - 25 \qquad \textit{Subtract 25 from each side.}$$
$$t = -10$$

Check: $t + 25 = 15$
$$-10 + 25 \stackrel{?}{=} 15 \qquad \textit{Replace t with −10.}$$
$$15 = 15 \checkmark \qquad \text{The original temperature was } -10°C.$$

You can also perform the same multiplication or division on each side of an equation.

You can review multiplication and division equations on page 228.

Examples

3 Solve $-196 = 4s$.

$$-196 = 4s$$

$$\frac{-196}{4} = \frac{4s}{4} \qquad \textit{Divide each side by 4.}$$

$$-49 = s$$

Check: Use your calculator. Replace s with -49.

4 $\boxed{\times}$ 49 $\boxed{+/-}$ $\boxed{=}$ −𝟭𝟵𝟲 ✓

4 Solve $\frac{g}{9} = -81$.

$$\frac{g}{9} = -81 \qquad \textit{$\frac{g}{9}$ means g divided by 9.}$$

$$\left(\frac{g}{9}\right)(9) = (-81)(9) \qquad \textit{Multiply each side by 9.}$$

$$g = -729$$

Check: Use your calculator. Replace g with -729.

729 $\boxed{+/-}$ $\boxed{\div}$ 9 $\boxed{=}$ −𝟴𝟭 ✓

Checking for Understanding

Communicating Mathematics

Read and study the lesson to answer each question.

1. **Tell** what operation you would use to solve the equation $17 = m + (-9)$.

2. **Write** the equation shown by each model. Then solve the equation.

a. 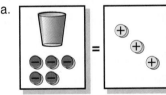 b.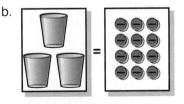

3. **Draw** a model that shows the equation in Example 2.

4. **Tell** how to check your solution to an equation.

Guided Practice

Solve each equation. Check your solution.

5. $m - 5 = -9$ 6. $5t = -140$ 7. $35 = -5m$

8. $-9 = 3 + r$ 9. $y + (-7) = 6$ 10. $18 = \frac{r}{-3}$

11. $\frac{e}{10} = -1{,}010$ 12. $24 = g - 8$ 13. $-60 = \frac{n}{-7}$

14. When a number w is multiplied by 6, the result is -48. What is the value of w?

Exercises

Independent Practice

Solve each equation. Check your solution.

15. $-168 = 3m$

16. $t + 9 = -34$

17. $y - 45 = -2$

18. $-9 + w = 12$

19. $-345 = t + 56$

20. $6 + p = 98$

21. $-570 = 3t$

22. $75 + p = -100$

23. $-8y = -368$

24. $\frac{r}{15} = -90$

25. $56 = \frac{n}{-3}$

26. $98 = \frac{a}{-4}$

Write an equation for each problem below. Then solve.

27. When a number k is multiplied by -9, the result is 45. Find k.
28. The sum of a number v and 7 is -11. Find v.
29. If you decrease a number d by -4, the result is 10. Find d.
30. The quotient when a number r is divided by -6 is 12. Find r.

Mixed Review

31. Complete: 245 mm = __?__ cm. *(Lesson 2-9)*
32. Compute $\frac{5}{8} \times 2\frac{1}{3}$ mentally. *(Lesson 5-9)*
33. Complete: 40 oz = __?__ lb. *(Lesson 6-6)*
34. **Ballooning** A hot air balloon starts a descent from 1,000 feet above the ground. After 10 minutes, the balloon is at 600 feet above the ground. Find the rate of its descent in feet per minute. *(Lesson 7-8)*

Problem Solving and Applications

Write an integer equation to represent each problem in Exercises 35–39. Then solve the problem.

35. **Diving** A diver begins an ascent from 160 feet below sea level. A few minutes later, the diver is 50 feet below sea level. How many feet did the diver ascend?
36. **Business** In her new position as manager, Jennifer earns $250 more than three times the salary she earned as a cashier. If she now earns $10,300, how much did she earn as a cashier?
37. **Geometry** Find the side of a square that has a perimeter equal to the perimeter of a quadrilateral with sides the length of 18 centimeters, 23 centimeters, 12 centimeters, and 15 centimeters.
38. **Consumer Math** Sam bought three boards that are 8 feet long, four boards that are 6 feet long, and 5 boards that are 10 feet long. How many feet of lumber did he have in the 8-foot and 10-foot lengths?
39. **Environment** Geologists calculated that a section of coastline is eroding at a steady rate of 4 centimeters per year. How many years will it take for this coastline to erode 96 cm?
40. **Critical Thinking** Describe how you would solve $2x + 5 = 4$.

7-10 Integers as Exponents

Objective
Use negative exponents.

Red blood cells are the majority of living cells that make up the blood. Their main purpose is to carry oxygen from the lungs to body tissues and to carry carbon dioxide from the body tissues to the lungs. The diameter of each red blood cell is 0.0003 inch.

 **LOOKBACK**
You can review positive exponents on page 32.

You have already seen how large numbers can be written using powers of 10. A number such as 0.0003 can also be written using powers of 10.

Study the pattern in this table.

	÷10	÷10	÷10	÷10	÷10	÷10	
10,000	1,000	100	10	1	0.1	0.01
10^4	10^3	10^2	10^1	?	?	?

The pattern indicates that, when you divide a power of 10 by 10, the exponent decreases by 1. This suggests how to define the power with a zero exponent.

$$10^0 = 1$$

The pattern also suggests how to define powers with negative exponents.

$$10^{-1} = 0.1 \qquad 10^{-2} = 0.01 \qquad 10^{-3} = 0.001$$
$$= \frac{1}{10} \text{ or } \frac{1}{10^1} \qquad = \frac{1}{100} \text{ or } \frac{1}{10^2} \qquad = \frac{1}{1,000} \text{ or } \frac{1}{10^3}$$

Calculator Hint

●●●●●●●●●●●●●
You enter numbers in scientific notation using the $\boxed{EE}$ or $\boxed{EXP}$ key. When the exponent is negative, press the $\boxed{+/-}$ key after entering the exponent. For example, use this key sequence to enter 2×10^{-9}.

2 $\boxed{EE}$ 9 $\boxed{+/-}$

Negative Exponents	**Arithmetic**	**Algebra**
	$10^{-2} = \frac{1}{10^2}$	$10^{-n} = \frac{1}{10^n}$

A number written in scientific notation with a negative exponent can be rewritten in standard form.

 **LOOKBACK**
You can review scientific notation on page 67.

Example 1

Write 4×10^{-3} in standard form.

$$4 \times 10^{-3} = 4 \times \frac{1}{10^3}$$
$$= 4 \times \frac{1}{1,000} \qquad 10^3 = 1,000$$
$$= 4 \times 0.001 \qquad \textit{Move decimal point three places to the left.}$$
$$= 0.004$$

Negative exponents can be used to write numbers between 0 and 1 in scientific notation.

Example 2 *Problem Solving*

Biology Refer to the problem in the lesson introduction. How is the diameter of a red blood cell written in scientific notation?

$$0.0003 = 3 \times 0.0001 \qquad \text{\textit{Write a 0.0003 as the product of 3}}$$
$$\textit{and a power of ten.}$$
$$= 3 \times \left(\frac{1}{10,000}\right) \qquad \textit{Rename 0.0001 as } \frac{1}{10,000}.$$
$$= 3 \times \frac{1}{10^4} \qquad\qquad 10,000 = 10^4$$
$$= 3 \times 10^{-4} \qquad\qquad \frac{1}{10^4} = 10^{-4}$$

In scientific notation, the diameter of a red blood cell is 3×10^{-4} inches.

Checking for Understanding

Communicating Mathematics

Read and study the lesson to answer each question.

1. **Tell** how many zeros are to the right of the decimal point when you write 10^{-3} as a decimal.

2. **Tell** the exponent you get when you write 0.0001 as a power of 10.

Guided Practice

Write each number in standard form.

3. 3×10^{-4} 4. 6×10^{-1} 5. 7×10^{-5}

Write each decimal in scientific notation.

6. 0.00002   7. 0.005   8. 0.0000009

9. **Computers** A nanosecond is 1×10^{-9} seconds. Write this number as a decimal.

Exercises

Independent Practice

Write each number in standard form.

10. 5×10^{-4} 11. 7×10^{-1} 12. 2×10^{-3}

13. 3×10^{-5} 14. 9×10^{-6} 15. 8×10^{-7}

Write each decimal in scientific notation.

16. 0.04   17. 0.001   18. 0.00007

19. 0.9   20. 0.0006   21. 0.0000003

22. Write 0.0005 in scientific notation.

23. Write 6×10^{-5} in standard form.

24. **Data analysis**   Refer to the graph at the top of page 114. What appears to be the slowest time of day at this particular McDonald's? *(Lesson 3-7)*

25. Use one or more strategies to compare $\frac{1}{4}$ and $\frac{2}{7}$. *(Lesson 4-10)*

26. Compare the fractions $\frac{5}{9}$ and $\frac{3}{4}$. *(Lesson 5-10)*

27. Solve the equation $\frac{g}{-16} = -2$. *(Lesson 7-9)*

28. **Physics**   The diameter of a silver atom is about 0.0000000003 meter. Write this measurement in scientific notation.

29. **Manufacturing**   The thickness of a sheet of paper is about 2×10^{-3} centimeter. Write this measurement as a decimal.

30. **Critical Thinking**   The average American male is expected to live 72.2 years. In contrast, some white blood cells live only 10 hours or 0.00114 years. Write this measurement in scientific notation.

31. **Food**   A Jelly Belly® is a gourmet jelly bean that has 4 calories and weighs 4×10^{-2} ounce. Write this weight as a decimal.

32. **Science**   The bacterium E. coli has a diameter of 0.001 millimeter. Write this number in scientific notation.

33. **Critical Thinking**   Copy and complete the table.

2^4	2^3	2^2	2^1	2^0	2^{-1}	2^{-2}
16	?	?	?	?	?	?

Write a definition for 2^{-n}, where n is a positive integer.

34. **Journal Entry**   Make a place-value chart that shows 1,000,000, 100,000, . . . , 1, 0.1, 0.01, . . . , 0.000001 and the corresponding powers of 10.

35. **Mathematics and Science**   Read the following paragraphs.

The air pressing down on Earth is called *air pressure,* or atmospheric pressure. The weight of air pressing down on each 1 square meter of Earth's surface is greater than that of a large elephant. Air pressure is greatest at ground level and decreases the higher up you go.

Barometers are used to measure air pressure. On an aneroid barometer, a needle on the dial moves as the air pressure changes. Pressure can also be measured with a mercury barometer.

A *bar* is a measure of pressure slightly less than Earth's air pressure at sea level under normal conditions, or about 29.92 inches of mercury. Earth's air pressure at sea level is 0.98 bar.

a. The atmospheric pressure at the surface of Mercury is 10^{-15} bar. Write this number in standard form.

b. Saturn's atmospheric pressure is 8,000,000 bar. Write this number in scientific notation.

7 Study Guide and Review

Communicating Mathematics

Choose the correct term or number to complete the sentence.

1. The absolute value of an integer is its (distance, direction) from 0 on a number line.
2. On a number line, values (decrease, increase) as you move to the right.
3. In an ordered pair, the second number is the (x-coordinate, y-coordinate).
4. Two integers that are opposites of each other are called (additive inverses, similar).
5. The product of two integers with (same, different) signs is negative.
6. The quotient of two integers with the same sign is (negative, positive).
7. In your own words, explain how to graph the point $A(-4, 3)$ on a coordinate system.

Skills and Concepts

Objectives and Examples	Review Exercises
Upon completing this chapter, you should be able to:	*Use these exercises to review and prepare for the chapter test.*

• read and write integers *(Lesson 7-1)*
Write 4°F below 0 as an integer.

You can write 4°F below 0 as -4°F.

Write an integer for each situation.
8. a loss of $150
9. a gain of 42 yards
10. 5°F below 0
11. a deposit of $75
12. 12 points gained
13. a loss of 5 pounds

• find the opposite and absolute value of an integer *(Lesson 7-1)*

Find the opposite and the absolute value of -7.

The opposite of -7 would be at the point 7 units to the right of 0.

The point that represents -7 is 7 units from 0, so $|-7| = 7$.

Write the integer represented by the point for each letter. Then find its opposite and its absolute value.

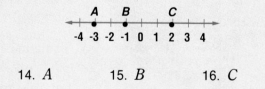

14. A 15. B 16. C

Objectives and Examples

- compare and order integers
  (*Lesson 7-2*)

  Replace the ● with <, >, or =.
  $$2 \bullet -6$$

  Since 2 is to the right of −6 on the
  number line, $2 > -6$.

- graph points on a coordinate
  plane (*Lesson 7-3*)

  Find the coordinates of point A and
  identify its quadrant.

  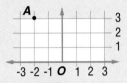

  The ordered pair is $(-2, 3)$.
  A is in the second quadrant.

- add integers (*Lesson 7-4*)

  Solve $p = 4 + (-3)$.

  $|4| > |-3|$, so the sum is positive. The
  difference of 4 and 3 is 1. So, $p = 1$.

- subtract integers (*Lesson 7-5*)

  Solve $y = -2 - 4$.

  $$\begin{aligned}
  y &= -2 - 4 \\
  &= -2 + (-4) \quad \textit{To subtract 4,} \\
  &= -6 \qquad\qquad \textit{add} -4.
  \end{aligned}$$

- multiply integers (*Lesson 7-7*)

  Solve $h = -2(5)$.

  The two integers have different signs,
  so the product will be negative.
  $$\begin{aligned}
  h &= -2(5) \\
  &= -10
  \end{aligned}$$

Review Exercises

Replace each ● with <, >, or = to make a
true sentence.

17. $-4 \bullet 3$ 18. $-18 \bullet -19$
19. $12 \bullet -12$ 20. $-100 \bullet -10$
21. $0 \bullet -8$ 22. $8 \bullet 0$

On graph paper, draw a coordinate plane.
Then graph and label each point. Identify
its quadrant.

23. $B(3, 6)$
24. $C(3, -5)$
25. $D(-1.5, -4)$
26. $E(-5, 1)$
27. $F(0, 6)$
28. $G(-2, 0)$

Solve each equation.

29. $c = 6 + (-2)$ 30. $-10 + 4 = r$
31. $-5 + 12 = m$ 32. $b = -1 + (-1)$
33. $s = 25 + (-50)$ 34. $-7 + (-6) = t$

Solve each equation.

35. $k = 5 - 7$ 36. $-13 - 4 = q$
37. $z = 6 - (-2)$ 38. $-8 - (-10) = m$
39. $a = -2 - 4$ 40. $12 - (-12) = p$

Solve each equation.

41. $w = 4(-2)$
42. $b = -6(-2)$
43. $c = (-3)^2$
44. $-8(4) = g$
45. $j = -5(5)$
46. $-100(-1) = y$

Objectives and Examples	Review Exercises
• divide integers *(Lesson 7-8)* solve $-15 \div (-3) = a$. The signs are the same, so the quotient is positive. $-15 \div (-3) = a$ $\qquad 5 = a$	Solve each equation. 47. $v = 45 \div (-9)$ 48. $s = -10 \div 10$ 49. $-12 \div (-2) = b$
• solve equations with integer solutions *(Lesson 7-9)* Solve $t + 6 = -3$. $t + 6 = -3$ $t + 6 - 6 = -3 - 6$ *Subtract 6 from* $\qquad t = -9$ *each side.*	Solve each equation. Check your solution. 50. $m - 8 = -2$ 51. $-3 + p = -6$ 52. $11s = -55$ 53. $-25c = 125$ 54. $\frac{f}{3} = -5$ 55. $240 = \frac{g}{-2}$
• use negative exponents *(Lesson 7-10)* $0.004 = \frac{4}{1,000}$ $\qquad = \frac{4}{10^3}$ $\qquad = 4 \times 10^{-3}$	Write each decimal in scientific notation. 56. 0.06 57. 0.000000005 Write each number in standard form. 58. 8×10^{-5} 59. 2×10^{-3}

Applications and Problem Solving

60. Use a pattern to find the sum of the numbers in each set. *(Lesson 7-6)*
    a. from 1 to 10   b. from 1 to 200   *Hint: Try pairing numbers.*

61. **Sports**   On the first play of the 4th quarter, the Bearcats lost 10 yards. After the second play, their position was 5 yards behind where they had started on the first play. What happened on the second play? *(Lesson 7-9)*

Curriculum Connection Projects

• **Geography**   Spin a globe and stop it with your finger. Record the latitude and longitude of the location where your finger points. Then find the number of degrees latitude and longitude of that location from your home state.

• **Recreation**   Shuffle a deck of cards. Consider black cards as positive and red cards as negative. Turn over one card at a time. Record and add each card to the one before it.

Read More About It

Buffie, Margaret. *The Haunting of Frances Rain.*
Miller, Marvin. *You Be the Jury: Courtroom 3.*
Ratner, Marilyn. *Plenty of Patches: An Introduction to Patchwork Quilting and Appliqué.*

7 Test

Write the integer represented by each letter.
Then find its opposite and its absolute value.

1. A

2. B

3. C

Replace each ● with $<$, $>$, or $=$ to make a true sentence.

4. -9 ● 6

5. 0 ● -3

6. On the same day, the thermometer registered $5°$ below zero in Cleveland, and $2°$ below zero in Cincinnati. Which city was colder?

On graph paper, draw a coordinate plane. Then graph and label each point.

7. $P(6,-3)$

8. $B(0, -9)$

9. $T(-5, 1)$

Solve each equation.

10. $g = 5 + (-3)$

11. $-9 + (-3) = m$

12. $r = -4 + 4$

13. $k = 11 - 15$

14. $-7 - (-2) = s$

15. $b = -3 - 4$

16. $c = -5(-3)$

17. $h = 12(-2)$

18. $(-7)^2 = p$

19. $q = 90 \div (-3)$

20. $(-25) \div 5 = t$

21. $m = (-72) \div (-9)$

22. $8 + x = 0$

23. $-5 = r - 3$

24. $-16f = 32$

25. $\frac{b}{15} = -3$

26. $\frac{d}{-16} = 1$

27. $-4 \cdot 5 = s$

28. Find a pattern in these statements. Then write two true statements using the pattern.

$$4 \times 4 - 4 = 3 \times 3 + 3 \qquad 70 \times 70 - 70 = 69 \times 69 + 69$$

29. **Travel** Patricia left home to drive to a friend's house. She drove 12 miles before realizing that she had gone too far. She retraced her path backward for 3 miles and arrived at her destination.

a. Write an addition sentence to represent this situation.

b. How far does Patricia live from her friend's house?

30. **Games** Byron is playing Monopoly® and has $250 left. He lands on Boardwalk and needs $200 more to pay the rent. How much is the rent on Boardwalk?

31. **Allowance** Ming is supposed to receive an allowance of $25 each month. However, for each day he forgets to take out the garbage, his allowance decreases by $2. Ming forgets to take the garbage out three times during October. Find the amount of his allowance for October.

32. Write 0.6 using negative exponents.

33. Write 3×10^{-5} as a decimal.

Bonus Evaluate $(-1)^{1357}$.

Investigations in Geometry

Spotlight on Highways and Byways

Have You Ever Wondered. . .

- How many miles there are between various U.S. cities?

- What the lengths are of some of the longest tunnels in the world?

Road Mileage Between Selected U.S. Cities

	Atlanta	Boston	Chicago	Cincinnati	Cleveland	Dallas	Denver	Detroit	Houston
Atlanta, GA	. . .	1,037	674	440	672	795	1,398	699	789
Boston, MA	1,037	. . .	963	840	628	1,748	1,949	695	1,804
Chicago, IL	674	963	. . .	287	335	917	996	266	1,067
Cincinnati, OH	440	840	287	. . .	244	920	1,164	259	1,029
Cleveland, OH	672	628	335	244	. . .	1,159	1,321	170	1,273
Dallas, TX	795	1,748	917	920	1,159	. . .	781	1,143	243
Denver, CO	1,398	1,949	996	1,164	1,321	781	. . .	1,253	1,019
Detroit, MI	699	695	266	259	170	1,143	1,253	. . .	1,265
Houston, TX	789	1,804	1,067	1,029	1,273	243	1019	1,265	. . .

Vehicular Tunnel Name	Location	Length (miles)
St.Gotthard	Alps, Switzerland	10.2
Mt. Blanc	Alps, France-Italy	7.5
Great St. Bernard	Alps, Switzerland	3.4
Lincoln	Hudson River, New York-New Jersey	2.5
Queensway Road	Mersey River, Liverpool, England	2.2
Brooklyn-Battery	East River, New York City	2.1
Holland	Hudson River, New York-New Jersey	1.7
Fort McHenry	Baltimore, Maryland	1.7
Hampton Roads	Norfolk, Virginia	1.4
Queens-Midtown	East River, New York City	1.3
Liberty Tubes	Pittsburgh, Pennsylvania	1.2
Baltimore Harbor	Baltimore, Maryland	1.2
Allegheny Tunnels	Pennsylvania Turnpike	1.2

Rail Road Tunnels

Vosges	7.0
Mont Cenis	8.5
St. Gotthard	9.3
Simpion, Switz-Italy	12.8
Seikan, Japan	33.1
Apennine, Italy	11.5
Lotschberg	9.1
New Cascade	7.8
Arlberg	6.3
Moffat	6.2
Shimuzu	6.1
Rimutaka	5.6

(in miles)

Looking Ahead

In this chapter, you will see how geometry is involved in answering questions about highways and byways.

The major objectives of the chapter are to:

- classify angles by their measure
- classify and construct polygons
- construct perpendiculars
- relate polygon shapes and angle measure to tiling and tessellations

Chapter Project

Highways and Byways

Work in a group.

1. Obtain a detailed street map of a local area.
2. List 20 major street intersections or crossroads.
3. Measure the angles formed by the various streets or the roads.
4. Make a verbal presentation of your results to the class.

ACCORDING TO THIS MAP, WE HAVE ANOTHER 800 MILES TO GO

THAT'S AWFUL!

MAYBE WE SHOULD'VE GOTTEN A SMALLER MAP

DIK BROWNE 11-4

8-1A Measuring Angles

A Preview of Lesson 8-1

Objective

Measure angles by using a protractor.

Materials

protractor
straightedge

You may need to extend the sides of your angle in order to measure it.

The instrument used by astronomers in the thirteenth century to track the movement of the stars and planets contained a semicircular unit for measuring angles. This unit is a forerunner of today's **protractor**. Protractors can be used to measure angles.

Try this!

• Draw any angle. Place the protractor on the angle so that the center is on the vertex of the angle and the 0° line lies on one side of the angle.

• There are two scales on your protractor. Use the one that begins with 0° where the side aligns with the protractor.

• Follow the scale from the 0° point to the point where the other side of the angle meets the scale. This is the angle's measure.

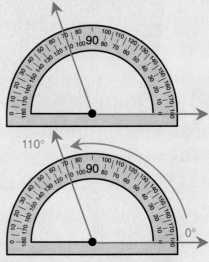

The measure of the angle is 110°

What do you think?

1. How do you know which scale to read?
2. What do you need to do if the sides of your angle do not intersect the scale of the protractor?
3. Sam says the angle above has a measure of 70°. What's wrong?

Extension

4. You can use a protractor to draw an angle of a given measure. Suppose you want to draw a 65° angle.
   a. Draw a line segment.
   b. Align your protractor on the segment with the center on one endpoint of the segment.
   c. Find the scale that starts with 0°. Go along that scale until you find 65°. Put a mark at this point.
   d. Draw a line through the endpoint of the segment and the mark.

8-1 Angles

Objective

Classify angles.

Words to Learn

vertex
degrees
right
acute
obtuse
straight
congruent
bisect

Carpenters used many tools to make sure that all the pieces of wood they cut fit together properly. One tool they use is a miter box. The miter box guides the saw so that the correct angle is cut.

When two segments or rays have a common endpoint, they form an angle. The point where they meet is called the **vertex** of the angle. An angle can be named by its vertex. To say *an angle with vertex B,* we write ∠B. Angles can also be named using a point from each side and the vertex, ∠*ABC* or ∠*CBA*. The vertex letter always goes in the middle. Another way to name an angle is to use a number inside the angle, ∠1.

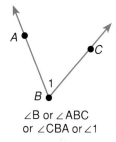

∠B or ∠ABC
or ∠CBA or ∠1

An angle is not measured by the length of its sides, but in units called **degrees.**

The corner of a picture frame is a **right** angle. Its measure is 90°. We often use the ⌐ symbol to indicate a right angle.

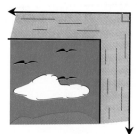

Angles that have a measure less than 90° are called **acute** angles. Those that have a measure greater than 90° but less than 180° are called **obtuse** angles. A **straight** angle has a measure of 180°.

Types of Angles and their Measures

Right Angle
exactly 90°

Straight Angle
exactly 180°

Acute Angle
acute
less than 90°

Obtuse Angle
between 90° and 180°

Lay one corner of your notebook paper on top of each angle to determine whether each angle is acute, obtuse, right, or straight.

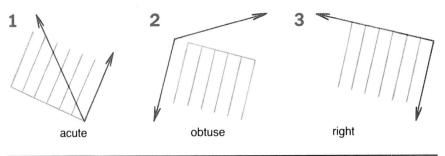

| 1 | 2 | 3 |
| acute | obtuse | right |

When two angles have the same measure, they are **congruent angles.**

$\cong$ *means is congruent to.*

| **Congruent angles** | **In words:** | If $\angle A$ has the same measure as $\angle B$, then $\angle A$ is congruent to $\angle B$. |
| | **In symbols:** | If $m \angle A = m \angle B$, then $\angle A \cong \angle B$. |

$\overrightarrow{BE}$ *is read ray BE.*

When you separate an angle into two congruent angles, you **bisect** the angle. In the figure, $\overrightarrow{BE}$ bisect $\angle ABC$. So, $m \angle 1 = m \angle 2$. This means that $\angle 1 \cong \angle 2$.

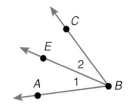

Mini-Lab

Work with a partner.
Materials: straightedge, protractor

- Use your straightedge to draw any angle.

- Fold the paper through the vertex so that the two sides match when you hold the paper up to the light.

- Unfold the paper and use your straightedge to draw a segment on the fold.

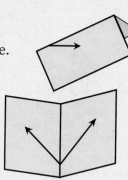

Talk About It
a. Use your protractor to measure the original angle. Then measure the two smaller angles.
b. Write a sentence to relate the measure of the smaller angles to that of the larger one.

Example 4 *Connection*

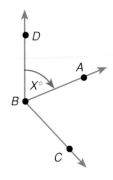

Algebra In the figure, $\overrightarrow{BA}$ bisects $\angle DBC$. Use an equation to find the value of x if the measure of $\angle DBC$ is $136°$.

$\overrightarrow{BA}$ bisects $\angle DBC$. So,
$m\angle DBA = m\angle ABC$. Since
$m\angle DBA = x$, it follows that
$m\angle ABC = x$. So, $x + x = 136$.

$$2x = 136$$
$$x = 68 \qquad \text{The value of } x \text{ is } 68.$$

Checking for Understanding

Communicating Mathematics

Read and study the lesson to answer each question.

1. **Show** how you can use the corner of your paper to classify an angle.

2. **Tell** what it means to bisect an angle.

3. **Show** whether $\overrightarrow{XZ}$ bisects $\angle WXY$ by tracing the figure at the right and folding the paper so that the two sides of the angle match.

Guided Practice

Classify each angle as acute, obtuse, right, or straight.

4.

5.

6.

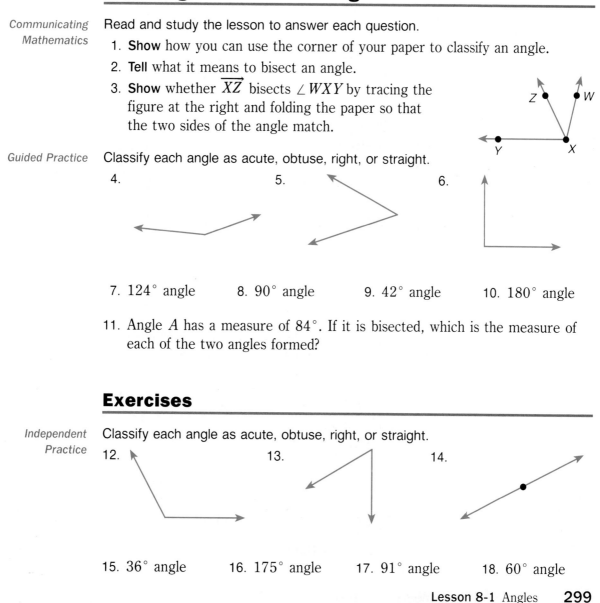

7. $124°$ angle   8. $90°$ angle   9. $42°$ angle   10. $180°$ angle

11. Angle A has a measure of $84°$. If it is bisected, which is the measure of each of the two angles formed?

Exercises

Independent Practice

Classify each angle as acute, obtuse, right, or straight.

12.   13.   14.

15. $36°$ angle   16. $175°$ angle   17. $91°$ angle   18. $60°$ angle

Tell which kind of angle is formed when each type of angle is bisected. Make a model to verify your answer to each question.

19. an acute angle

20. a right angle

21. an obtuse angle

22. a straight angle

Make a drawing for each situation. Then solve.

23. **Algebra** Angle B and angle C are congruent. If $m \angle B = 100°$ and $m \angle C = x + 80$, find the value of x.

24. **Algebra** $\overrightarrow{EF}$ bisects $\angle BED$, which is a right angle. Find the value of y, if $m \angle BEF = 5y$.

25. **Algebra** $\overrightarrow{XY}$ bisects $\angle AXE$. Suppose $m \angle AXY = 3z$ and $m \angle EXY = 4w$. If $z = 12$ and $w = 9$, find $m \angle AXE$.

Mixed Review

26. Find 2.5×0.3. *(Lesson 2-4)*

27. **Physical Fitness** Every evening, Alicia walks around a neighborhood block, which is a rectangle of length 0.75 miles and width 0.2 miles. How far does Alicia walk each evening? *(Lesson 5-6)*

28. Express 0.00006 using negative exponents. *(Lesson 7-10)*

Problem Solving and Applications

29. **Physics** Light reflects off a mirror at an outgoing angle congruent to the incoming angle. In each figure below, tell which ray is the correct outgoing ray of light.

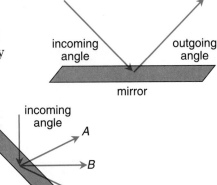

30. **Critical Thinking** Without measuring, match each angle to the appropriate measurement. Write a sentence using symbols to state your answers.

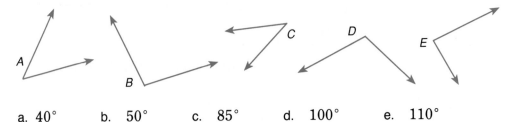

a. 40°   b. 50°   c. 85°   d. 100°   e. 110°

31. **Research** Find the name used for an angle that measures more than 180°.

32. **Journal Entry** Write a sentence telling five places where you see an angle in your classroom. Be sure to include the classification of each angle.

8-1B Perpendicular Lines

A Follow-Up of Lesson 8-1

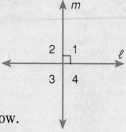

Objective

Construct a line perpendicular to another line.

Materials

compass
straightedge

Perpendicular lines are lines in the same plane that form right angles when they intersect. In the figure, line ℓ is perpendicular to line m. This can also be written as $\ell \perp m$. Two ways to construct perpendicular lines are described below.

Activity One

Construct a line perpendicular to line m through point P on m.

An arc is part of a circle.

- Draw a line and label it m. Draw a dot on the line and label it point P.

- Place the compass point on P and draw arcs to intersect line m twice. Label these points Q and R.

- Open your compass wider. Put the compass at Q and draw an arc above line m.

- With the same setting, put the compass at R and draw an arc to intersect the one you just drew. Label this intersection point S.

$\overrightarrow{PS}$ is read line PS.

- Use a straightedge to draw a line through S and P.

  By construction, $\overrightarrow{PS} \perp m$.

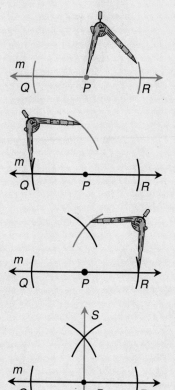

Activity Two

Construct a line perpendicular to line m through point P not on m.

- Draw a line and label it m. Draw a dot above m and label it point P.

- Open the compass to a width greater than the distance from P to m. Draw a large arc to intersect m twice. Label these points of intersection Q and R.

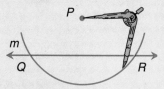

Mathematics Lab 8-1B Constructing Perpendiculars   **301**

- Put the compass at Q and draw an arc below m.

- Using the same setting, put the compass at R and draw an arc to intersect the one drawn from Q. Label the intersection point S.

- Use a straightedge to draw a line through P and S.

By construction, $\overleftrightarrow{PS} \perp m$.

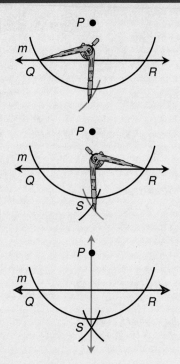

What do you think?

1. What type of angles are formed by perpendicular lines?
2. Measure the angles on your constructions. Explain why these measures may not exactly agree with your answer in Exercise 1.

Extension

3. You can construct a square by using perpendiculars.
   a. Draw line ℓ. Draw two dots on ℓ and label them points R and S. Construct a perpendicular through R.
   b. Use the compass to measure the distance from R to S. Using the same setting, place the compass at R and draw an arc on the perpendicular through R. Label this point T.
   c. Using the same setting, place the compass at T and draw an arc to the right of T. Then place the compass at S and draw an arc to intersect the one you just drew. Call this point U.
   d. Use a straightedge to draw $\overline{TU}$ and $\overline{US}$. Figure $RSUT$ is a square.

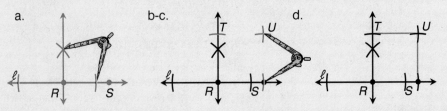

8-2 Polygons

Objective

Identify polygons.

Words to Learn

polygon
triangle
quadrilateral
pentagon
hexagon
heptagon
octagon
nonagon
decagon
undecagon
dodecagon

Vertices is the plural of vertex.

Many words in the English language have their origins in ancient Greek and Latin words. The word **polygon** comes from the prefix *poly-* meaning *many* and the suffix *-gon* meaning *angle*. So, a polygon is a many-angled figure.

Actually, a polygon is a closed figure whose sides are line segments in the same plane. Since the segments meet to form angles and angles have vertices, polygons also have vertices.

Figures A, B, and C are polygons. Figures D, E, and F are *not polygons.*

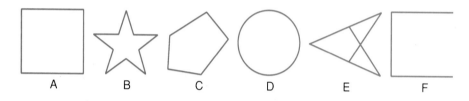

A   B   C   D   E   F

Definition of Polygon	A polygon is a closed figure in a plane that • has at least three sides, all of which are segments, • has sides that meet only at a vertex, and • has exactly two sides meeting at each vertex.

Examples

Determine which figures are polygons. If the figure is not a polygon, explain why.

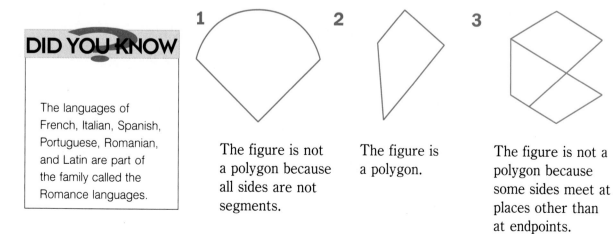

1 The figure is not a polygon because all sides are not segments.

2 The figure is a polygon.

3 The figure is not a polygon because some sides meet at places other than at endpoints.

Some polygons have special names. Triangles and quadrilaterals are polygons. The word **triangle** means *three angles*. The word **quadrilateral** means *four sides*. The names of other polygons also describe them.

Example 4   *Connection*

Language Arts   Use the meaning of each prefix to sketch each polygon.

penta-	5	*nona-*	9
hexa-	6	*deca-*	10
hepta-	7	*undeca-*	11
octa-	8	*dodeca-*	12

Each of these is only one example of the type of figure you might draw.

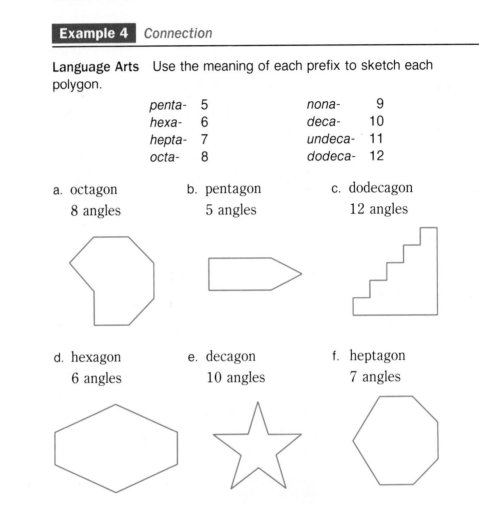

a. octagon
   8 angles

b. pentagon
   5 angles

c. dodecagon
   12 angles

d. hexagon
   6 angles

e. decagon
   10 angles

f. heptagon
   7 angles

Checking for Understanding

Communicating Mathematics

Read and study the lesson to answer each question.

1. **Tell** what conditions a figure must meet in order to be a polygon.

2. **Write** a sentence that relates the number of angles in a polygon to the number of sides in a polygon.

Guided Practice   Determine which figures are polygons. If a figure is not a polygon, explain why.

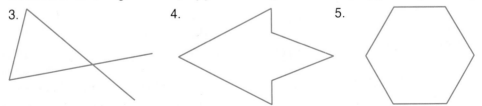

3.

4.

5.

Exercises

Independent Practice

Determine which figures are polygons. If a figure is not a polygon, explain why.

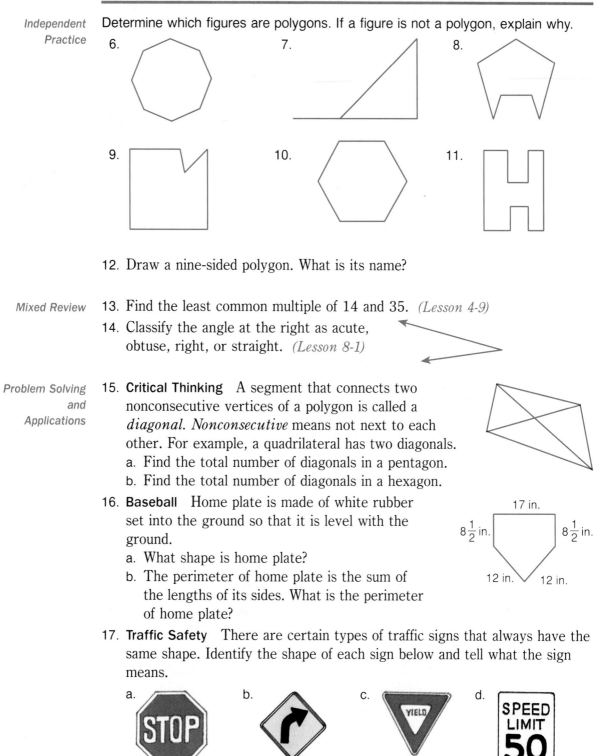

6.

7.

8.

9.

10.

11.

12. Draw a nine-sided polygon. What is its name?

Mixed Review

13. Find the least common multiple of 14 and 35. *(Lesson 4-9)*

14. Classify the angle at the right as acute, obtuse, right, or straight. *(Lesson 8-1)*

Problem Solving and Applications

15. **Critical Thinking** A segment that connects two nonconsecutive vertices of a polygon is called a *diagonal. Nonconsecutive* means not next to each other. For example, a quadrilateral has two diagonals.
    a. Find the total number of diagonals in a pentagon.
    b. Find the total number of diagonals in a hexagon.

16. **Baseball** Home plate is made of white rubber set into the ground so that it is level with the ground.
    a. What shape is home plate?
    b. The perimeter of home plate is the sum of the lengths of its sides. What is the perimeter of home plate?

17. **Traffic Safety** There are certain types of traffic signs that always have the same shape. Identify the shape of each sign below and tell what the sign means.

    a.   b.   c.   d.

18. **Journal Entry** Draw the shape of some other signs or objects you see everyday. Name each shape and identify which type of polygon it is, if any.

8-2B Sum of the Angles of a Polygon

A Follow-Up of Lesson 8-2

Objective

Discover the sum of the angle measures of any polygon.

Materials

protractor
scissors

A convex polygon is a polygon whose diagonals lie entirely within the polygon. To find the sum of the angle measures in any convex polygon, you could measure each angle and find the sum. This method may not give you an accurate sum because of human error.

Activity One

Work with a partner.

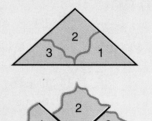

- Draw any triangle on a piece of paper and cut it out.

- Tear off the angles and arrange as shown.

What do you think?

1. What type of angle do the three angles form when put together?
2. What is the measure of this type of angle?
3. Complete this statement: The sum of the measures of the angles of a triangle is ___?___°.

Activity Two

- Draw a convex hexagon.

- Pick one vertex and draw all the diagonals possible from that vertex.

What do you think?

4. How many triangles were formed when you drew the diagonals?
5. How could you find the sum of the measures of the angles in a hexagon?
6. Find the sum of the measures of the angles of each convex polygon by using triangles.
   a. pentagon   b. heptagon   c. octagon
   d. **Algebra**   If *n* is the number of sides, write an algebraic expression that tells the sum of the measures of the angles of any polygon.

8-3 Triangles and Quadrilaterals

Objective

Classify triangles and quadrilaterals.

Words to Learn

congruent
scalene
isosceles
equilateral
rhombus
trapezoid

Early in the nineteenth century, chemists began to seek ways to classify the elements. Today, these classifications are organized into the periodic table. This table arranges elements in families according to their common characteristics.

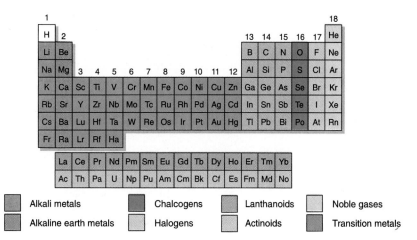

Just as with the elements, each category of polygons can be classified according to their common characteristics.

Triangles can be classified by their angle measures. Each has two acute angles. Classify using the third angle.

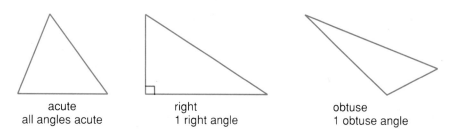

acute	right	obtuse
all angles acute	1 right angle	1 obtuse angle

They can also be classified by the number of congruent sides they have. **Congruent** sides are sides that have the same length.

Slashes show which sides are congruent.

scalene	isosceles	equilateral
no congruent sides	at least 2 congruent sides	3 congruent sides

Suppose you are building bleachers at the football field. How will you build them so they will support the weight of all the people who sit on them?

In construction, right triangles are often used to provide extra strength. In bleachers, right triangles would be used in the supports or the frame.

Classify each triangle by its angles and by its sides.

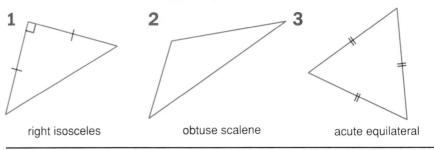

1 right isosceles

2 obtuse scalene

3 acute equilateral

You are already familiar with three types of quadrilaterals. They are squares, rectangles, and parallelograms. Two other types of quadrilaterals are the **rhombus** and the **trapezoid**. The chart below shows how the figures are related and some of their characteristics.

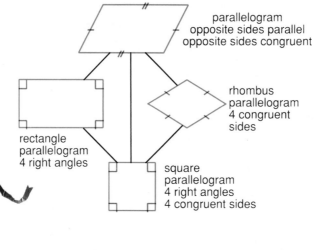

parallelogram
opposite sides parallel
opposite sides congruent

rhombus
parallelogram
4 congruent sides

rectangle
parallelogram
4 right angles

square
parallelogram
4 right angles
4 congruent sides

trapezoid
only one pair of sides parallel

Examples

Name every quadrilateral that describes each figure. Then state which name best describes the figure.

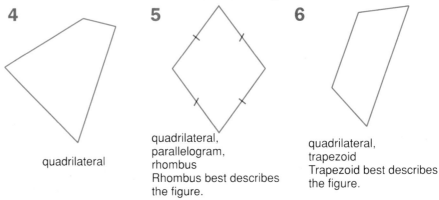

4 quadrilateral

5 quadrilateral, parallelogram, rhombus
Rhombus best describes the figure.

6 quadrilateral, trapezoid
Trapezoid best describes the figure.

Checking for Understanding

Communicating Mathematics

Read and study the lesson to answer each question.

1. **Write** a sentence that tells how a trapezoid is different from a parallelogram.

2. **Tell** why all squares are rectangles but not all rectangles are squares.

3. **Draw** an equilateral parallelogram. Name the figure you drew.

Guided Practice

Classify each triangle by its sides and by its angles.

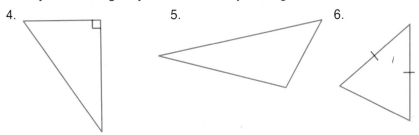

4.   5.   6.

Name every quadrilateral that describes each figure. Then underline the name that best describes the figure.

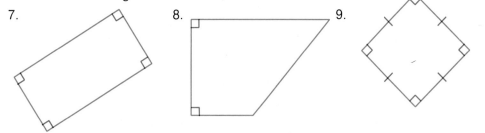

7.   8.   9.

Exercises

Independent Practice

Classify each triangle by its sides and by its angles.

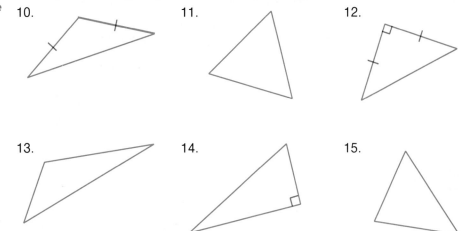

10.   11.   12.

13.   14.   15.

Name every quadrilateral that describes each figure. Then underline the name that best describes the figure.

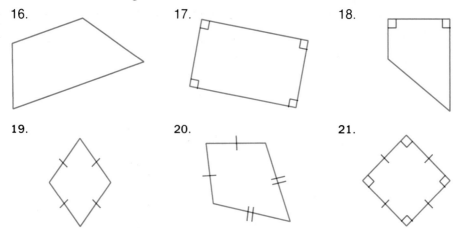

16.

17.

18.

19.

20.

21.

22. Which quadrilaterals can have four right angles?

23. A figure is a rhombus but not a rectangle. Draw the figure.

Mixed Review 24. Evaluate $3(6-2)+2+4$. *(Lesson 1-7)*

25. On his most recent math test, Ricardo scored 84 out of 100 points. Express this fraction in simplest terms. *(Lesson 4-6)*

26. Identify the polygon at the right. *(Lesson 8-2)*

Problem Solving and Applications 27. **Critical Thinking** Can a trapezoid ever have two congruent sides? Draw a figure to verify your answer. What might you call such a trapezoid?

28. **Make a Model** Begin with a large square of paper.
    a. Fold the paper to make an isosceles right triangle.
    b. Fold the triangle to make a trapezoid.
    c. Fold the triangle to make a rectangle.
    d. Fold the triangle to make a parallelogram.
    e. Fold the triangle to make a square.

29. **Mathematics and Art** Read the following paragraph.

> In abstract painting, ideas, emotions, and visual sensations are communicated through lines, shapes, colors, and textures that have no particular significance by themselves. The shapes in many of these paintings appear to rise and fall, recede and advance, or balance and float. These illusions can create moods such as joy, sadness, peace, and conflict.

Create your own abstract image using polygons, triangles, and quadrilaterals. Share you artwork with the class and discuss what it is trying to communicate.

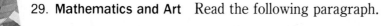

8-4A Bisecting Angles and Segments

A Preview of Lesson 8-4

Objective

Construct angle bisectors and segment bisectors.

Materials

straightedge
compass
protractor

A bicycle has two wheels. A biathlon is an event composed of two sports, cross-country skiing and rifle sharpshooting. A bicolored guinea pig is one that has two colors. The prefix *bi* means two.

To *bisect* an angle or a segment means to separate it into two congruent parts. Bisecting a segment is similar to the constructions you studied in Mathematics Lab 8-1B.

Activity One

Draw a segment and bisect it.

- Use a straightedge to draw a segment. Label the endpoints *X* and *Y*.

- Open your compass to a setting that is longer than half the length of $\overline{XY}$. Place the compass point at *X* and draw a large arc.

$\overline{XY}$ is read line segment XY.

- Using the same setting, place the compass point at *Y* and draw a large arc to intersect the first arc twice.

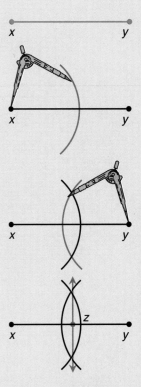

- Use a straightedge to draw a segment connecting the two intersection points. This segment intersects $\overline{XY}$. Label this point *Z*.

What do you think?

1. Use the compass to measure the distance from *X* to *Z*. Compare this to the distance from *Z* to *Y*. What do you find?

2. How is $\overline{XZ}$ *related to* $\overline{ZY}$?

3. How is the segment you drew through *Z* related to $\overline{XY}$?

Activity Two

Draw an angle and bisect it.

- Use a straightedge to draw any angle. Label the vertex P. Place the compass at the vertex and draw a large arc to intersect each side. Label the intersection points Q and R.

- Place the compass at Q and draw an arc on the inside of the angle. Using the same setting, place the compass at R and draw an arc to intersect the one you just drew. Label the intersection point W.

- Draw ray PW.

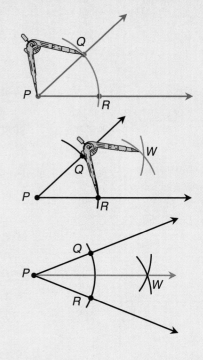

What do you think?

4. Use your protractor to measure $\angle QPW$ and $\angle WPR$. What do you find?

5. How is $\angle QPW$ related to $\angle QPR$?

Extension

6. Suppose you are given an angle and told that its measure is half that of a larger angle. How would you construct the larger angle?

   a. Draw any angle and label it $\angle A$.

   b. Draw an arc through the sides of the angle into the outside of the angle. Label the intersection points C and T.

   c. Put the compass point at T. Adjust the setting so that it measures the distance from T to C. Without removing your compass, draw an arc to intersect the large arc outside of $\angle A$. Call this point N.

   d. Draw $\overrightarrow{AN}$. $\overrightarrow{AT}$ is the bisector of $\angle CAN$.

   e. Complete: $m\angle CAT =$ ___?___ $m\angle CAN$.

8-4 Regular Polygons

Objective

Identify regular polygons.

Words to Learn

regular polygon
equiangular
exterior angle

A colony of honey bees may contain as many as 50,000 to 60,000 bees. The honeycomb, built by the worker bees, is made of cells of wax shaped like hexagons. These cells are used to raise young bees and store the food we know as honey.

These hexagons each have six congruent sides and six congruent angles. Any polygon that has all sides congruent and all angles congruent is called a **regular polygon.** So, the cells in the honeycombs are regular hexagons.

An equilateral triangle is a regular polygon. Its name means *equal sides.* An equilateral triangle can also be called an **equiangular** triangle. This means all its angles have equal measures. These words can be used to define a regular polygon.

Regular Polygon	A polygon that is both equiangular and equilateral is a regular polygon.

Examples

1 Draw a regular pentagon.

2 Draw a regular octagon.

In Mathematics Lab 8-2B, you learned to find the sum of the measures of the angles of a polygon by drawing all the diagonals from one vertex and counting the triangles formed. Then you multiplied that number by 180°. In the figure at the right, there are 3 triangles, so the sum of the measure of the angles of the pentagon is 3 · 180° or 540°.

Example 3 *Connection*

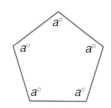

Algebra Use an equation to find the measure of each angle of a regular pentagon.

Since the pentagon is regular, all angles have the same measure.

Let a represent the measure of one angle.

The sum of the measures of the angles is 540°. So, $5a = 540$.

To solve $5a = 540$, divide each side by 5.

540 ⌷÷⌷ 5 ⌷=⌷ **108**

Each angle of a regular pentagon measures 108°.

If you extend a side of a polygon, a special angle is formed. This angle is called an **exterior angle** of the polygon. If one exterior angle is drawn at each vertex, the sum of the measures of all the exterior angles is always 360°.

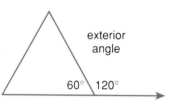

Mini-Lab

Work with a partner.
Materials: straightedge, pencil, paper

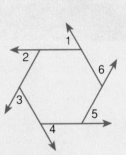

- Sketch a regular hexagon and draw one exterior angle at each vertex.

- Divide 360° by the number of exterior angles you drew. What is the measure of each exterior angle?

- The exterior angle and the angle of the hexagon form a straight angle. So the sum of the measures of the two angles is 180°. What is the measure of an angle of the regular hexagon?

Talk About It

a. Draw a regular hexagon and draw the diagonals from one vertex. Use the triangles to figure the measure of each angle of the hexagon. How does this compare with the result in the activity above?

b. How would you find the measure of each angle of a regular octagon?

Checking for Understanding

Read and study the lesson to answer each question.

1. **Tell** the difference between equiangular and equilateral.

2. **Draw** a regular quadrilateral.

3. **Draw** a figure that is equilateral, but not equiangular.

4. **Write**, in your own words, how you can use the exterior angles to find the measure of each angle in a regular octagon.

Tell whether each polygon is a regular polygon. If not, tell why.

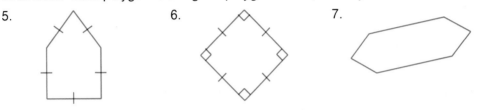

5.   6.   7.

8. Find the measure of an angle of a regular quadrilateral. Name this figure.

Exercises

Tell whether each polygon is a regular polygon. If not, tell why.

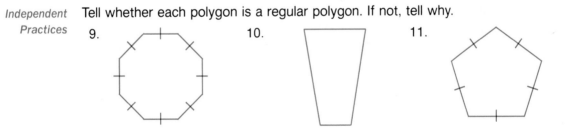

9.   10.   11.

Sketch each figure. Find the measure of an exterior angle. Then find the measure of an angle of the polygon.

12. regular heptagon   13. regular decagon   14. regular dodecagon

15. Which type of quadrilateral is an equilateral quadrilateral, but may not be an equiangular quadrilateral?

16. An equiangular triangle is also an equilateral triangle. Why doesn't this same relationship always hold true for other polygons?

17. **Statistics**   The chart at the right shows the number of births per 1,000 people in the United States. Round the birth rate for each year to the nearest whole number. *(Lesson 2-2)*

18. **Algebra**   Solve $18 = m - 5$. *(Lesson 6-2)*

Year	Birth Rate
1960	23.7
1970	18.4
1980	15.9
1989	16.2

19. **Critical Thinking** Draw a star.

   a. Which polygon describes the shape of the star?

   b. Is the star equilateral?

   c. Is the star equiangular?

20. **Carpentry** Arturo is taking an industrial technology class. He decides to make a hat rack frame by drilling three holes into six boards and attaching them with bolts. He says that making the rack in this way will form squares. His teacher says, "Not necessarily." What does his teacher mean?

21. **Computer Connection** You can use Logo software to draw regular polygons. The following program uses the exterior angle to create the regular polygon. In the program, N represents the number of sides in the polygon.

```
TO POLY :N
   REPEAT :N[FD 40 RT 360/:N]
END
```

   a. The FD command tells how long each side is. How long are the sides in this program?

   b. Use this program to draw four different regular polygons. Then use it to draw a 100-sided polygon. What do you notice?

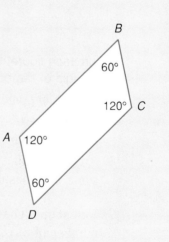

Mid-Chapter Review

8

Use the figure for Exercises 1–4.

1. Classify ∠A as acute, right, obtuse, or straight. *(Lesson 8-1)*

2. How are ∠A and ∠C related? *(Lesson 8-1)*

3. Name every quadrilateral that describes figure *ABCD.* Underline the name that best describes it. *(Lesson 8-3)*

4. Is quadrilateral *ABCD* a regular quadrilateral? Explain. *(Lesson 8-4)*

5. Draw three examples of figures that are not polygons. Tell why they are not. *(Lesson 8-2)*

8-4B Constructing Regular Polygons

A Follow-Up of Lesson 8-4

Objective

Construct a regular triangle and hexagon.

Materials

straightedge
compass
proctractor

In Mathematics Lab 8-1B, you learned to construct a square in the Extension. A square is a regular polygon. Other regular polygons can be constructed using the constructions you already have learned.

Activity One

Construct a regular triangle.

- Draw a line segment and place a point on it. Call this point *P*.

- Put the point of the compass at *P* and draw a large arc to intersect the segment. Label the intersection point *Q*.

- With the same setting, place the compass at *Q* and draw another arc to intersect the one you just drew. Label this intersection point *R*.

- Draw $\overline{PR}$ and $\overline{QR}$.

What do you think?

1. Think about the settings you used in this construction. Why do you think $\triangle PQR$ is regular?

2. Use your protractor to measure the angles in $\triangle PQR$. You may have to extend the sides. What do you find? Is the triangle regular?

Activity Two

Construct a regular hexagon.

- Use your compass to draw a circle. Put a point on the circle.

- With the same setting, place the compass on that point. Draw a small arc that intersects the circle.

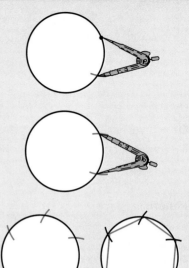

- Place the compass on the point where the arc intersects the circle. Draw another small arc to intersect the circle.

- Continue the process until you come back to the first point. Use the straightedge to connect the intersection points in order as shown.

What do you think?

3. When you finished drawing all the arcs, you may have had difficulty meeting the original point. Why do you think that happened?

4. Use your protractor to measure the angles of your hexagon. You may have to extend the sides. What do you find? Is the hexagon regular?

Extension

5. You can construct a regular dodecagon by starting with the construction for a regular hexagon.

   a. After drawing the arcs for the hexagon, draw one of the sides and bisect it.
   b. Using the setting you used to draw the circle, place the compass where the bisector meets the circle. Draw six arcs as you did before with the hexagon.
   c. Draw the 12 sides by connecting the intersection points in order.

6. You can also construct a regular octagon using a circle.

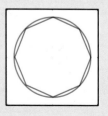

   a. Draw a circle and fold it in half.
   b. Without opening, fold it in half again and then once more.
   c. Open the circle up and connect the points where the folds intersect the circle.

8-5 Use Logical Reasoning

Objective

Solve problems by using logical reasoning.

Jason, Conrad, and Alton play safety, running back, and quarterback on a football team, but not necessarily in that order. Jason and the quarterback drove Alton to practice on Saturday. Jason does not play safety. Who is the safety?

Explore   What do you know?
Jason and the quarterback drove Alton to practice on Saturday. Jason does not play safety.

You need to find out who plays safety.

Plan   Make a chart to organize the information logically. Start with the information you already know. Then reread the problem, searching for clues. Write *no* in the chart when you use a fact to eliminate a possibility.

Solve   Begin by completing the chart with information you already know. You know that Jason does not play safety, so put a *no* in the appropriate column and row.

	Safety	Running Back	Quarterback
Jason	no		
Conrad			
Alton			

Now reread the problem searching for clues. Jason and the quarterback drove Alton to practice on Saturday. This means that neither Jason nor Alton can be the quarterback. So Conrad must be the quarterback.

	Safety	Running Back	Quarterback
Jason	no		
Conrad	no	no	yes
Alton			

Now you know that Alton is the safety.

	Safety	Running Back	Quarterback
Jason	no		
Conrad	no	no	yes
Alton	yes		

Examine You can also arrive at the answer using this reasoning: Jason and Alton are not the quarterback, so Conrad must be the quarterback. Since Jason is not the safety, then Alton must be the safety.

Checking for Understanding

Communicating Mathematics

1. Tell how you use the logical reasoning strategy to solve problems.
2. Who is the running back?

Guided Practice

Solve. Use logical reasoning.

3. Regular polygons *Q, R,* and *S* are a hexagon, a square, and an octagon but not necessarily in that order. Polygon *Q* and *S* have the same number of letters in their names. Each angle of polygon *S* measures less than 135°. Classify the polygons.

4. Use the pattern to draw the next two figures in the sequence.

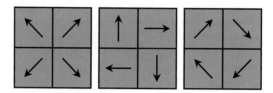

5. Jim Banker, Susan Sales, and Jessica Clerk are a banker, salesperson, and a clerk. Their occupations do not match their last names. Susan Sales is the clerk's cousin. Who is the banker?

Exercises

Independent Practice

Solve. Use any strategy.

6. Is the average of 85.6, 112.5, 90.3, 101.7, 42.2, and 66.7 about 83 or 8.3?

7. During their first possession of the ball, the Bayfield Buffalos gained 12 yards, gained 10 yards, lost 5 yards and then lost 7 yards. Was their net gain about 2 yards or 10 yards?

8. Ken, Carly, Francoise, and Kirby each have favorite sports: bowling, basketball, softball, and football. Ken's cousin's favorite sport is basketball. Francoise and Kirby do not like football. Carly's favorite sport is softball. Kirby no longer likes bowling. Which sport is Ken's favorite?

Strategies
● ● ● ● ● ● ● ● ● ●
Look for a pattern.
Solve a simpler problem.
Act it out.
Guess and check.
Draw a diagram.
Make a chart.
Work backwards.

Tessellations

Objective

Determine which regular figures can be used to form a tessellation.

Words to Learn

tiling
tessellation

Mosaic is the art of covering a surface with small squares, triangles, or other regular shapes, called *tesserae*. In Mexico, one of the most stunning mosaics is the outside structure of the Central Library of the National Autonomous University of Mexico. It was designed by Juan O'Gorman and is composed of over 7.5 million pieces.

Covering a surface with regular figures is called **tiling.** The result of tiling is called a **tessellation.** A tessellation can be made of one kind of polygon or several kinds of polygons.

 Mini-Lab

Work with a partner.
Materials: tracing paper

- Trace the equilateral triangle.

- Turn your paper and trace the triangle again so that the two triangles share a common side.
- Continue the process until you notice a pattern.

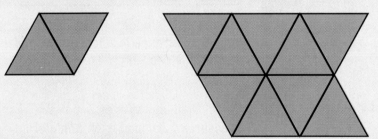

Talk About It

a. Would you be able to completely cover a large surface with equilateral triangles? Explain your answer.

b. Find a place where the vertex of several triangles meet. What is the sum of the measures of the angles whose vertices are at this point?

In the Mini-Lab, you found that where triangles meet the sum of the angle measures is 360°. The sum of the angle measures at the vertex of any tesselation must be 360°.

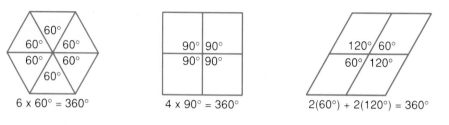

6 × 60° = 360°   4 × 90° = 360°   2(60°) + 2(120°) = 360°

Example 1 *Connection*

Algebra   The sum of the measures of the angles of an octagon is 1,080°. Can you tessellate a regular octagon by itself?

Each angle of a regular octagon has a measure of $1{,}080° \div 8$ or $135°$. To find out if a regular octagon tessellates, solve $135n = 360$, where n is the number of angles at a vertex.

To solve $135n = 360$, divide each side by 135.
360 ⌷÷⌷ 135 ⌷=⌷ **2.6666667**

The solution is not a whole number. So we cannot tessellate a regular octagon by itself.

Estimation Hint
● ● ● ● ● ● ● ● ● ● ● ● ●
Use front-end estimation to solve $135n = 360$.
$100 \times 2 = 200$
$30 \times 2 = \dfrac{60}{260}$

not close to 360

$100 \times 3 = 300$
$30 \times 3 = \dfrac{90}{390}$

over 360

As you saw in Example 1, not all regular polygons can tessellate by themselves. However, when you use a combination of polygons to form a tessellation, there is often more than one pattern possible.

Example 2 *Problem Solving*

Design   Mr. Concepción bought hexagonal stones and triangular stones to arrange in a tessellation for a patio. The sides of the stones have the same length. How many of each stone does he need at each vertex?

More than one answer is possible. Each angle of a regular hexagon measures 120° and each angle of an equilateral triangle measures 60°.

Try 1 hexagon.
total at _ one angle of
vertex   hexagon
$360° - \quad 120° \quad = 240°$
$240° = \underline{\quad ? \quad}$ triangles
$240° = 4$ triangles

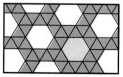

hexagon and 4 triangles
at each vertex

Try 2 hexagons.

$$\begin{array}{c}\text{total at} \_ \text{ two angles}\\ \text{vertex} \quad \text{of hexagon}\end{array}$$

$$360° - 2(120°) \quad = 120°$$
$$120° = \underline{\ ?\ } \text{ triangles}$$
$$120° = 2 \text{ triangles}$$

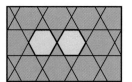

2 hexagons and 2 triangles at each vertex

Checking for Understanding

Communicating Mathematics

Read and study the lesson to answer each question.

1. **Tell** when a regular polygon can be used by itself to make a tessellation.

2. **Show** how you can make a tessellation out of squares.

Guided Practice

Assume each polygon is regular. Determine if it can be used by itself to make a tessellation.

3. hexagon   4. pentagon   5. decagon

Exercises

Independent Practice

Assume each polygon is regular. Determine if it can be used by itself to make a tessellation.

6. heptagon   7. nonagon   8. dodecagon

9. Sketch a tessellation made with equilateral triangles and squares.

The following regular polygons tessellate. Determine how many of each you need at each vertex and sketch the tessellation.

10. triangle, dodecagon   11. square, octagon

12. **Statistics** Look at the graph to answer these questions.
(Lesson 3-1)
   a. What is the estimated increase in rollerblade sales from 1989 to 1990?
   b. Which year has the greatest increase in sales?

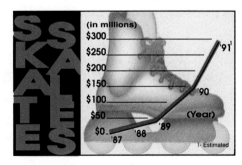

Mixed Review

13. Tell whether the polygon at the right is a regular polygon. If not, tell why.
(Lesson 8-4)

Problem Solving and Applications

14. **Critical Thinking** Can you find a non-regular triangle that tessellates? If so, sketch the tessellation.

15. **Design** Kitchen Boutique has its artists creating tile trivets for their stores. They have square, hexagonal, and dodecagonal tiles. Draw a tessellation they could use in their designs.

8-7 Translations

Objective

Create Escher-like drawings by using translations.

Words to Learn

translation

Maurits Cornelis Escher (1898–1972), a Dutch artist, was impressed by the Moorish mosaics he saw while traveling through southern Spain. He was inspired to create recognizable figures to fill space like the pieces of stone that filled the surface of the mosaic. His figures were often in the shapes of birds, fish, or reptiles.

© M.C. Escher/Cordon Art—Baarn—Holland
Collection Haags Gemeentemuseum—The Hague

Many of Escher's sketches began as tessellations of polygons. You can make Escher-like drawings by making changes in the polygons of the tessellation. One way to do this is by using a **translation.**

A translation is a slide. The square below has the left side changed. To make sure the pieces, or pattern units, will still tessellate, we are going to slide or translate that change to the opposite side and copy it.

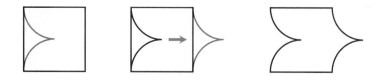

Now change all the squares in a tessellation the same way. The tessellation takes on Escher-like qualities when you use different colors.

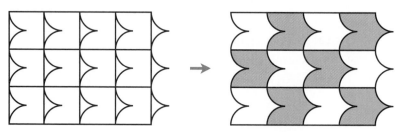

Example 1

Draw a tessellation using the change shown at the right.

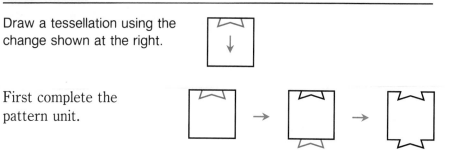

First complete the pattern unit.

A cardboard pattern unit can help in creating the tessellation.

Translate the change to all squares in the tessellation. Use color to complete the effect.

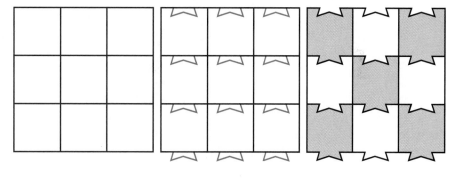

You can make more complex tessellations by doing two translations.

Example 2

Draw a tessellation using both changes shown at the right.

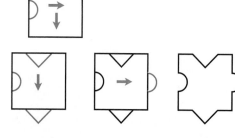

First complete the pattern unit.

Then complete the tessellation.

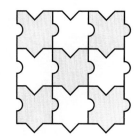

Checking for Understanding

Communicating Mathematics

Read and study the lesson to answer each question.

1. **Tell** how a translation can be used to form an Escher-like drawing.

2. **Write** a definition of translation.

Guided Practice

Complete the pattern unit for each translation. Then draw the tessellation.

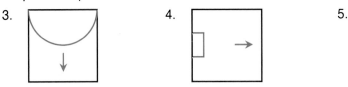

3.   4.   5.

Exercises

Independent Practice

Complete the pattern unit for each translation. Then draw the tessellation.

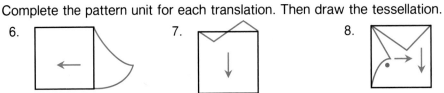

6.   7.   8.

9. Can the puzzle piece at the right tessellate? Explain.

Mixed Review

10. Divide $4\frac{3}{8}$ by $2\frac{1}{3}$. *(Lesson 5-10)*

11. **Art** Alex wishes to construct a tessellation for a wall-hanging made only from decagons. Is this possible? *(Lesson 8-6)*

Problem Solving and Applications

12. **Critical Thinking** The Escher work below is sometimes called *Pegasus*. It was created from a tessellation of squares. Study the print and locate the changes and the position of the squares.

13. **Critical Thinking** Is it possible to make a tessellation with translations by using equilateral triangles? Explain your answer and make a drawing.

14. **Design** The Art Club is making designs for wrapping paper. They want to use a tessellation of parallelograms as their basis. Create an Escher-like drawing using tessellated parallelograms.

15. **Data Search** Refer to pages 294 and 295. How much longer is the Lincoln Tunnel than the Liberty Tubes?

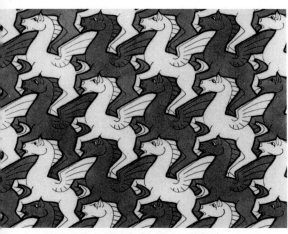

© M.C. Escher/Cordon Art—Baarn—Holland
Collection Haags Gemeentemuseum—The Hague

8-8 Reflections

Objective

Create Escher-like drawings by using reflections.

Words to Learn

line symmetry
line of symmetry
reflection

Have you ever made a valentine heart by folding a piece of paper and cutting half a heart? When you unfolded the paper, there was the heart with both sides evenly matched.

Figures that match exactly when folded in half have **line symmetry.** The figures below have line symmetry. Some figures can be folded in more than one way to show symmetry. Each fold line is called a **line of symmetry.**

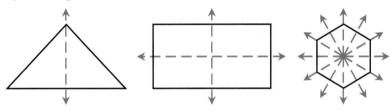

Examples

Determine which figures have line symmetry. Draw all lines of symmetry.

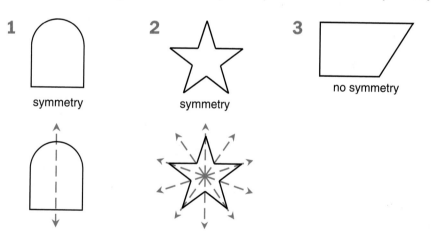

1   symmetry

2   symmetry

3   no symmetry

You can create figures that have line symmetry by using a **reflection.** A reflection is a mirror image of a figure across a line of symmetry.

Escher also used reflections in some of his works. You can create different types of drawings using reflections. However, in these tessellations, two pattern units are used.

Example 4

Complete an Escher-like drawing using the change shown at the right.

Complete the first pattern unit by drawing the reflection of the design on another side of the triangle.

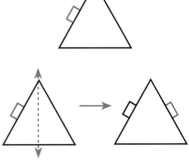

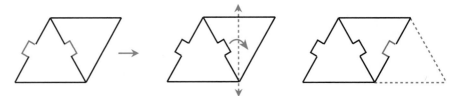

Now look at what happens when there are two triangles. The pattern on the left side of the second triangle is different from the pattern on the first triangle. Reflect the new pattern in the second triangle.

Problem-Solving Hint

● ● ● ● ● ● ● ● ● ● ● ● ●

Look for a pattern to determine the two different units of the tessellation.

Notice that the pattern on the right side of the second triangle now matches the pattern of the first triangle. Continue this process to complete the tessellation.

Checking for Understanding

Communicating Mathematics

Read and study the lesson to answer each question.

1. **Tell** how a reflection is different from a translation.

2. **Write** a brief description of how to do a tessellation by using reflections.

3. Copy the figure at the right. Draw all lines of symmetry.

4. Complete both pattern units for the reflection shown at the right. Then draw the tessellation.

Exercises

Copy each figure. Draw all lines of symmetry.

5. 6. 7.

Complete both pattern units for each reflection. Then draw the tessellation.

8. 9. 10.

11. Complete the tessellation described by the pattern shown at the right.

12. **Jobs** As part of his summer job, Andrew is responsible for mowing a soccer field that is rectangular in shape. It is 100 yards long and 45 yards wide. Find the total area of the grass Andrew mows. *(Lesson 6-7)*

13. **Art** Complete the pattern unit for the translation at the right. Then draw the tessellation. *(Lesson 8-7)*

14. **Critical Thinking** The double change at the right is reflected over the diagonal of the square instead of to the opposite side. Draw the two unit patterns.

15. **Biology** Some insects under the influence of certain substances alter their behavior in unusual ways. Some bees alter the way they build honeycombs. Create a honeycomb that involves a reflection.

16. **Journal Entry** Write your impressions of Escher's artwork.

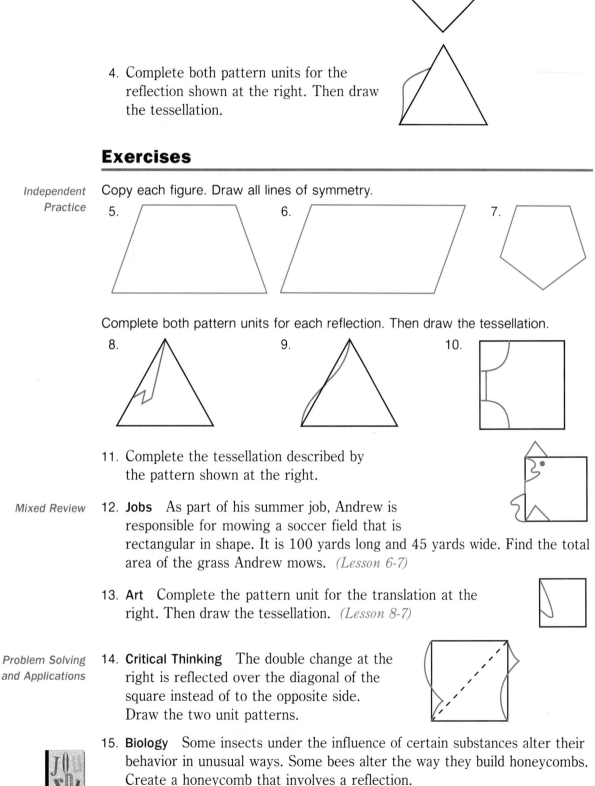

8 Study Guide and Review

Communicating Mathematics

State whether each sentence is true or false. If false, replace the underlined word to make the sentence true.

1. Angles that have a measure <u>greater</u> than 90° are called acute angles.
2. A decagon is a polygon having <u>10</u> sides.
3. A <u>scalene</u> triangle has 3 congruent sides.
4. A parallelogram which has 4 congruent sides is called a <u>trapezoid</u>.
5. A polygon that is both <u>equiangular</u> and equilateral is a regular polygon.
6. The sum of the angle measures at the vertex of any tessellation is <u>180°</u>.
7. A <u>reflection</u> is a mirror image of a figure across a given line.
8. In your own words, explain how a translation can be used in constructing a tessellation.

Skills and Concepts

Objectives and Examples	Review Exercises
Upon completing this chapter, you should be able to:	*Use these exercises to review and prepare for the chapter test.*

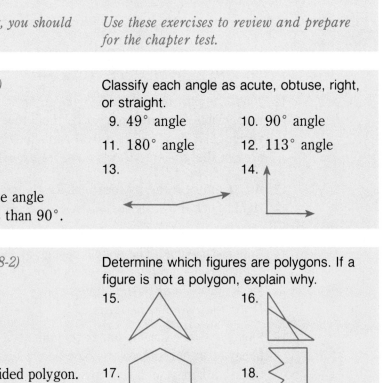

• classify angles *(Lesson 8-1)*

The angle above is an acute angle because its measure is less than 90°.

Classify each angle as acute, obtuse, right, or straight.

9. 49° angle   10. 90° angle

11. 180° angle   12. 113° angle

13.   14.

• identify polygons *(Lesson 8-2)*

The figure above is a six-sided polygon. It is called a hexagon.

Determine which figures are polygons. If a figure is not a polygon, explain why.

15.   16.

17.   18.

Objectives and Examples

- classify triangles and quadrilaterals *(Lesson 8-3)*

The figure above has 3 congruent sides and all of its angles are acute. It is an acute equilateral triangle.

- identify regular polygons *(Lesson 8-4)*

The figure above has 6 congruent sides and all of its angles are of equal measure. It is a regular hexagon.

- determine which regular figures can be used to form a tessellation *(Lesson 8-6)*

A tessellation cannot be made out of regular pentagons alone because each angle of a pentagon measures 108° and there is no whole number n such that $108n = 360$.

- create Escher-like drawings by using translations *(Lesson 8-7)*

Complete the pattern unit and then the tessellation.

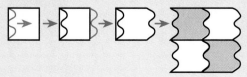

Review Exercises

Classify each triangle by its sides and by its angles.

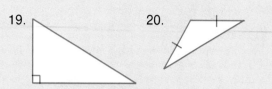

Name every quadrilateral that describes each figure. Then underline the name that best describes the figure.

Tell whether each polygon is a regular polygon. If not, tell why.

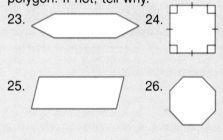

If each polygon is regular, determine if it can be used by itself to make a tessellation.

27. hexagon   28. decagon

Complete the pattern unit for each translation. Then draw the tessellation.

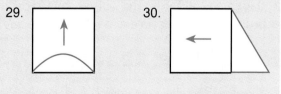

Objectives and Examples	*Review Exercises*

• create Escher-like drawings by using reflections *(Lesson 8-8)*

Complete the tessellation described by the pattern.

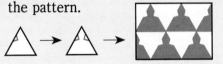

Complete both pattern units for each reflection. Then draw the tessellation.

31. 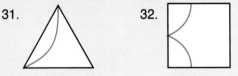 32.

Applications and Problem Solving

33. **Logic** Use the pattern at the right to draw the next two figures in the sequence. *(Lesson 8-5)*

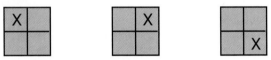

34. **Arts and Crafts** Edwyna is piecing together a quilt from fabric pieces in the shapes of hexagons and equilateral triangles. How many of each of the shapes will she need at each vertex in the tessellation created by the fabric pieces? Sketch the tessellation. *(Lesson 8-6)*

35. **Pizza Parlor** Angelo's Pizza Parlor shapes its pizzas as squares. After cooking, the pizzas are cut along the diagonal into two triangles. Describe completely the triangles that result. *(Lesson 8-3)*

Curriculum Connection Projects

• **Science**   Attach a weighted string to the midpoint of the straight edge of a protractor. With the straight edge of the protractor up, look down the edge at several objects in the room. Have another student record the angle measure where the string crosses the protractor. Subtract your measure from 90 degrees.

• **Geography**   Find acute, obtuse, and right angles on a street map. Use a protractor to measure and record each angle and location.

Read More About It

Cumming, Robert. *Just Look. . . . A Book About Paintings.*
Taylor, Barbara. *Bouncing and Bending Light.*
Spinelli, Jerry. *Space Station Seventh Grade.*

Chapter

8 Test

Classify each angle as acute, obtuse, right, or straight.

1.

2. 135° angle

3.

4. **Architecture**   Classify the angle made by a wall and the ceiling of your classroom.

Determine which figures are polygons. If a figure is not a polygon, explain why.

5.

6.

7.

Classify each triangle by its sides and by its angles.

8.

9.

Name every quadrilateral that describes each figure. Then underline the name that best describes the figure.

10.

11.

12.

Tell whether each polygon is a regular polygon. If not, tell why.

13.

14.

15. **Logic**   Nancy, Marti, Jessica, and Paul are studying music, journalism, physical education, and home economics. None of them is studying a subject that begins with the same letter of his or her first name. Jessica is practicing scales. Nancy has never taken physical education. Paul is reading about headlines. What subject is each student most likely studying?

16. If an octagon is regular, determine if it can be used by itself to make a tessellation.

Complete the pattern unit for each translation. Then draw the tessellation.

17.

18.

Complete both pattern units for each reflection. Then draw the tessellation.

19.

20.

Bonus   A square is separated into four triangles by drawing the diagonals. Are the resulting triangles equilateral? Explain your reasoning.

Chapter

9

Area

Spotlight on Oceans and Islands

Have You Ever Wondered. . .

- How large some of the world's islands are?
- How much area the world's oceans cover?

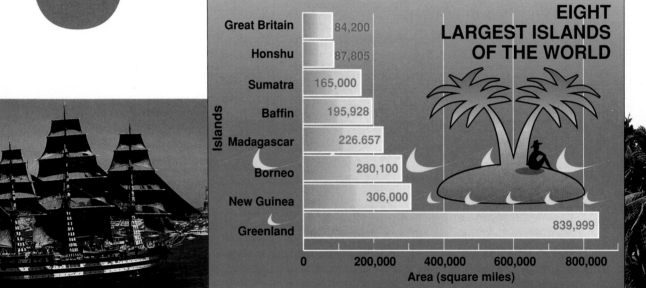

EIGHT LARGEST ISLANDS OF THE WORLD

Island	Area (square miles)
Great Britain	84,200
Honshu	87,805
Sumatra	165,000
Baffin	195,928
Madagascar	226.657
Borneo	280,100
New Guinea	306,000
Greenland	839,999

Oceans and Seas

Name	Area (Sq mi)	Average Depth (Feet)
Pacific Ocean	64,186,300	12,925
Atlantic Ocean	33,420,000	11,730
Indian Ocean	28,350,500	12,598
Arctic Ocean	5,105,700	3,407
South China Sea	1,148,500	4,802
Caribbean Sea	971,400	8,448
Mediterranean Sea	969,100	4,926
Bering Sea	873,000	4,893
Gulf of Mexico	582,100	5,297
Sea of Okhotsk	537,500	3,192
Sea of Japan	391,100	5,468

Chapter Project

Oceans and Islands

Work in a group.

1. Copy a simple world map.

2. Color in the oceans and indicate the approximate size of each.

3. Color in any islands that you see and indicate the approximate size of each.

4. Indicate where you live on the map. Find out how large your state is (in square miles). Add this information to your map.

YOU ARE HERE

Looking Ahead

In this chapter, you will see how mathematics can be used to answer questions about the area of the world's oceans and islands.

The major objectives of the chapter are to:

- find square roots
- use the Pythagorean Theorem
- estimate the area of irregular figures
- find the area of triangles, trapezoids, and circles

335

9-1 Guess and Check

Objective

Solve problems by using guess and check.

Mr. Andrews tells his math class that when two consecutive even numbers are multiplied, the product is 2,808. What are the two numbers?

Explore   What do you know?
The product of two consecutive even numbers is 2,808. Consecutive even numbers are pairs of numbers like 10 and 12 or 28 and 30.

What do you need to find?
You need to find the two numbers.

Plan   Multiply each multiple of 10 by the next consecutive even number to find a reasonable range for the number. Then guess numbers within that range.

Solve $40 \times 42 = 1,680$
$50 \times 52 = 2,600$
$60 \times 62 = 3,720$ $\leftarrow$ *2,808*

The two consecutive even numbers are between 50 and 60. Since 2,808 is closer to 2,500 than to 3,600, try two consecutive even numbers close to 50.

Try 52 and 54.   52 ⊠ 54 ⊟ 2808

The consecutive even numbers are 52 and 54.

Examine   Since $52 \times 54 = 2,808$, the two numbers are 52 and 54.

Checking for Understanding

Communicating Mathematics

1. **Tell** why, in the problem above, it is best to first find a range in which to guess numbers rather than just randomly choosing numbers.

Guided Practice Solve. Use the guess-and-check strategy.

2. Julie arranged square tables, each seating 4 people, into one long rectangular table so that her 16 dinner guests could eat together. How many tables did she use?

3. Masao is thinking of two whole numbers. When he adds them together, the sum is 107. When he subtracts the lesser number from the greater number, their difference is 17. What are the numbers?

4. Seth is the oldest of four children. Each of his sisters is 3 years older than the next oldest sibling. The combined age of Seth and his three sisters is 46. None of the children are over the age of 20. How old is Seth?

Problem Solving

Practice Solve. Use any strategy.

5. The Pike's Peak souvenir shop sells standard size postcards in packages of 5 and large size postcards in packages of 3. Bonnie bought 16 postcards. How many packages of each did she buy?

Strategies
●●●●●●●●●●
Look for a pattern.
Solve a simpler problem.
Act it out.
Guess and check.
Draw a diagram.
Make a chart.
Work backwards.

6. Find the length of a side of a square with an area of 625 square centimeters.

7. How many different ways can you arrange the numbers 1, 2, 3, and 4 as the last four digits in a telephone number?

8. The length of a rectangle is 8 inches longer than its width. What are the length and width if the area of the rectangle is 84 square inches?

9. Anne used her calculator to divide 4,567,325 by 326.4. She should expect the result to be about:

    a. 150   b. 15,000   c. 150,000

10. The product of a number and 47 is 1,081. Find the number.

9-2 Squares and Square Roots

Objective

Find square roots of perfect squares.

Words to Learn

square
perfect square
square root
radical sign

You can review exponents on page 32.

DID YOU KNOW

The first world championship in skydiving was held in Yugoslavia in 1951. Later world championships followed at two-year intervals with as many as 40 national teams competing.

Skydivers leap from an airplane at heights of up to 15,000 feet and fall freely at speeds of more than 100 miles an hour. In accuracy skydiving competitions, participants try to land on a square target that may measure only 5 centimeters across. The area of the target is 5×5 or 25 square centimeters.

Remember that an *exponent* tells how many times a number, called the *base,* is used as a *factor.* In the expression 5×5, 5 is used as a factor twice. When you compute 5×5 or 5^2, you are finding the **square** of 5.

$$5^2 = 5 \times 5$$
$$= 25 \qquad \text{The square of 5 is 25.}$$

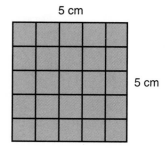

5 cm

5 cm

Examples

1 Evaluate 8^2.

$8 \times 8 = 64$

2 Evaluate 27^2.

27 $\boxed{y^x}$ 2 = $\mathsf{729}$

Numbers such as 25, 64, and 729 are called **perfect squares** because they are squares of whole numbers.

Mini-Lab

Work in pairs.
Materials: base-ten blocks

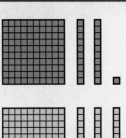

- Try to form a square with 121 unit blocks.

- Now try to form a square with 136 unit blocks.

Talk About It

a. Which number of unit blocks is a perfect square? What is the length of the side of its square?

b. Form three other perfect squares for numbers greater than 100. What is the length of each side?

Let's return to the skydiving problem at the beginning of this lesson. The target region has an area of 25 square centimeters. You can find the length of a side by arranging 25 base-ten blocks into a square as in the Mini-Lab, *or* you can find the **square root** of 25.

Square Root	If $a^2 = b$, then a is the square root of b.

Since $5^2 = 25$, one square root of 25 is 5. It is also true that $(-5)^2 = 25$, so another square root of 25 is -5. Since the length of a side of the target region must be a whole number, the answer is 5 centimeters.

The symbol used to represent a nonnegative square root is $\sqrt{}$. It is called a **radical sign.**

$\sqrt{25} = 5$ *The square root of 25 is 5.*

Calculator Hint
• • • • • • • • • • • • •
Some calculators may not have a $\sqrt{x}$ key. $\sqrt{x}$ may be written above the x^2 key. To find the square root of a number, press [INV] [x^2].

Examples

3 Find $\sqrt{49}$.
Since $7^2 = 49$, $\sqrt{49} = 7$.

4 Find $\sqrt{961}$.
961 [$\sqrt{x}$] $\exists\mathbf{1}$
So, $\sqrt{961} = 31$.

Example 5 *Connection*

Geometry   The area of a square is 16 square feet. Find the length of a side using the definition of square root.

Since $4^2 = 16$, $\sqrt{16} = 4$. The side of the square is 4 feet long.

Communicating Mathematics

Checking for Understanding

Read and study the lesson to answer each question.

1. **Tell** what it means to square a number.
2. **Draw** a picture that shows $\sqrt{196} = 14$.
3. **Tell** how finding the square root of a number is like finding the length of the side of a square given its area.

Guided Practice Find the square of each number.

4. 2   5. 3   6. 7

7. 10   8. 11   9. 12

Find each square root.

10. $\sqrt{4}$ 11. $\sqrt{36}$ 12. $\sqrt{81}$ 13. $\sqrt{100}$

Exercises

Find the square of each number.

14. 1   15. 5   16. 13

17. 14   18. 16   19. 20

20. 25   21. 30   22. 32

Find each square root.

23. $\sqrt{64}$ 24. $\sqrt{121}$ 25. $\sqrt{144}$

26. $\sqrt{225}$ 27. $\sqrt{400}$ 28. $\sqrt{625}$

29. $\sqrt{1,600}$ 30. $\sqrt{256}$ 31. $\sqrt{441}$

32. **Geometry** Find the length of a side of a square whose area is 784 square feet.

33. **Geometry** Find the area of a square whose side is 15 meters.

Determine whether each number is a perfect square. Write *yes* or *no*.

34. 36   35. 49   36. 136

37. 289   38. 645   39. 961

40. **Finances** Alexander had $2,345 in his checking account at the beginning of the month. During the month, he wrote checks in the amounts of $595, $75, and $123. Estimate his balance at the end of the month. *(Lesson 1-2)*

41. **Statistics** During a typical work week, Samantha records the number of minutes it takes her to drive to work each day. Find the mean and median for the following times: 12, 23, 10, 14, and 11. *(Lesson 3-5)*

42. Find the least common multiple of 35 and 49. *(Lesson 4-9)*

43. Solve $t - 3.6 = 4$. *(Lesson 6-2)*

44. **Geometry** Complete the pattern unit for the reflection. Then draw the tesselation. *(Lesson 8-8)*

45. **Board Games** A checkerboard has 8 squares on each side. How many small squares are there on the board?

46. **Sports** The backboard on a basketball hoop is a square with an area of 16 square feet. What is the length of a side?

47. **Critical Thinking** Numbers like 1, 8, and 27 are called perfect cubes because $1^3 = 1$, $2^3 = 8$, and $3^3 = 27$. Find two numbers that are both perfect squares and perfect cubes.

48. **Journal Entry** Make up a real-life problem where you need to find the square root of a number.

9-3 Estimating Square Roots

Objective

Estimate square roots.

Most professional baseball diamonds are covered with a square tarp to protect them when it rains. Suppose the tarp has an area of 1,000 square meters. What is the length of the side?

To find the answer, you need to find $\sqrt{1,000}$. Use your calculator.

1000 $\boxed{\sqrt{x}}$ **31.622777**

The length of a side is about 31.6 meters.

You know that the square root of a perfect square is a whole number. What happens when you try to find the square root of a number that is *not* a perfect square? Estimating the square root is often helpful.

Mini-Lab

Work in pairs.
Materials: base-ten blocks

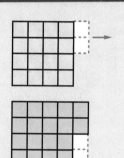

- Form the largest possible square using 18 unit blocks. This is a square of 16 unit blocks.

- Form the smallest possible square that has more than 18 unit blocks. This is a square of 25 unit blocks.

Talk About It
a. What is the area of each square?
b. What is the length of the side of each square?
c. Estimate $\sqrt{18}$.
d. Use base-ten blocks to estimate $\sqrt{41}$ and $\sqrt{150}$.

You can review
> and < on
page 48.

How does this method of estimating square roots compare to the one in the Mini-Lab?

Since 18 is not a perfect square, estimate $\sqrt{18}$ by finding the two perfect squares closest to 18.

$$16 < \ 18 < 25$$
$$\sqrt{16} < \sqrt{18} < \sqrt{25} \qquad \textit{Take the square root of each number.}$$
$$4 < \sqrt{18} < 5 \qquad \textit{Simplify, if possible.}$$

So, $\sqrt{18}$ is between 4 and 5. Since 18 is closer to 16 than 25, $\sqrt{18}$ is closer to 4 than to 5. The best whole number estimate for $\sqrt{18}$ is 4.

Example

Estimate $\sqrt{52}$.

Find the two perfect squares closest to 52.

$$49 < 52 < 64$$
$$\sqrt{49} < \sqrt{52} < \sqrt{64}$$
$$7 < \sqrt{52} < 8$$

Since 52 is closer to 49 than to 64, the best whole number estimate is 7.

Checking for Understanding

Communicating Mathematics

Read and study the lesson to answer each question.

1. **Tell** how the drawing at the right can help you estimate $\sqrt{10}$.

2. **Tell**, in your own words, why 4 is the best whole number estimate for $\sqrt{18}$.

Guided Practice

Estimate.

3. $\sqrt{11}$ 4. $\sqrt{23}$ 5. $\sqrt{27}$ 6. $\sqrt{34}$

7. $\sqrt{59}$ 8. $\sqrt{70}$ 9. $\sqrt{99}$ 10. $\sqrt{105}$

Exercises

Independent Practice

Estimate.

11. $\sqrt{12}$ 12. $\sqrt{21}$ 13. $\sqrt{56}$ 14. $\sqrt{85}$

15. $\sqrt{91}$ 16. $\sqrt{116}$ 17. $\sqrt{145}$ 18. $\sqrt{215}$

19. $\sqrt{350}$ 20. $\sqrt{500}$ 21. $\sqrt{721}$ 22. $\sqrt{1,050}$

23. Which is closer to 4, $\sqrt{14}$ or $\sqrt{24}$?

Mixed Review

24. Find the greatest common factor of 125, 240, and 375. *(Lesson 4-5)*

25. Find the difference of $1\frac{3}{4}$ and $\frac{4}{5}$. *(Lesson 5-4)*

26. Find the square root of 196. *(Lesson 9-2)*

Problem-Solving and Applications

27. **Critical Thinking** Tell what happens when you use a calculator to find $\sqrt{-25}$. Is it possible to take the square root of a negative number? Explain.

28. **Sightseeing** The distance you can see to the horizon in clear weather is given by the formula $d = 1.22\sqrt{h}$. In this formula, d represents the distance in miles and h represents the height in feet your eyes are from the ground. Estimate how far you can see if you are at the top of the Sears Tower in Chicago, 1,454 feet above the ground.

29. **Construction** City code requires that a party house must allow 4 square feet for each person on a dance floor. To have a square dance floor that is large enough for 50 people, how long should each side be?

9-4A **Pythagorean Theorem**

A Preview of Lesson 9-4

Objective

Find the relationship among the sides of a right triangle.

Materials

grid paper
straightedge
scissors

In this Lab, you will investigate the relationship that exists among the sides of a right triangle.

Try this!

Work in groups of three.

- Each member of your group should draw a segment that is 3 units long on a piece of grid paper.

- At one end of this segment, draw a perpendicular segment that is 4 units long.

- Draw a third segment to form a triangle. Cut out the triangle.

- Measure the length of the longest side in terms of units on the grid paper.

- Cut out three squares: one with 3 units on a side, one with 4 units on a side, and one with 5 units on a side.

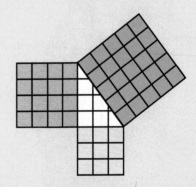

- Place the edges of the squares against the corresponding sides of the right triangle.

- Find the area of each square.

What do you think?

1. What relationship exists among the areas of the three squares?
2. Do you think the relationship you described in Exercise 1 is true for *any* right triangle? Repeat the activity for three other right triangles whose perpendicular sides are 8 units, 15 units; 6 units, 8 units; and 5 units, 12 units.
3. Write one or more sentences that summarize your findings.

Extension

4. Do you think your findings are true in other kinds of triangles? To test your theory, draw five different, non-right triangles on grid paper. Repeat the activity. Write one or more sentences to summarize your findings.

9-4 The Pythagorean Theorem

Objective

Find the length of a side of a right triangle using the Pythagorean Theorem.

Words to Learn

hypotenuse
leg
Pythagorean Theorem

The ancient Egyptians used mathematics to lay out their fields with square corners. About 2000 B.C. they discovered a 3-4-5 right triangle. They took a piece of rope and knotted it into 12 equal spaces. Taking three stakes, they stretched the rope around the stakes to form a right triangle. The sides of the triangle had lengths of 3, 4, and 5 units.

The longest side of a right triangle is called the **hypotenuse,** and is opposite the right angle. The other two sides, called **legs,** form the right angle.

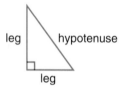

Several hundred years later, a Greek mathematician, Pythagoras, formalized a relationship between the sides of any right triangle. It became known as the **Pythagorean Theorem.**

In this theorem, a and b are the lengths of the legs and c is the length of the hypotenuse.

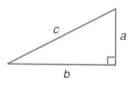

Pythagorean Theorem	**In words:** In a right triangle, the square of the measure of the hypotenuse is equal to the sum of the squares of the measures of the legs.
	Arithmetic $3^2 + 4^2 = 5^2$ **Algebra** $a^2 + b^2 = c^2$

Given the lengths of the two legs of a right triangle, you can use the Pythagorean Theorem to find the length of the hypotenuse.

Examples

1 Find the length of the hypotenuse of the triangle at the right.

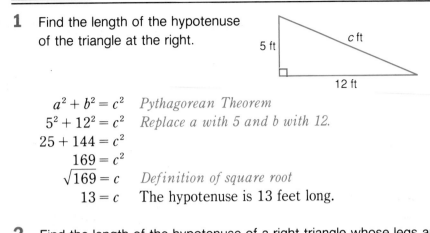

$$a^2 + b^2 = c^2 \quad \textit{Pythagorean Theorem}$$
$$5^2 + 12^2 = c^2 \quad \textit{Replace a with 5 and b with 12.}$$
$$25 + 144 = c^2$$
$$169 = c^2$$
$$\sqrt{169} = c \quad \textit{Definition of square root}$$
$$13 = c \quad \text{The hypotenuse is 13 feet long.}$$

2 Find the length of the hypotenuse of a right triangle whose legs are 7 meters and 9 meters.

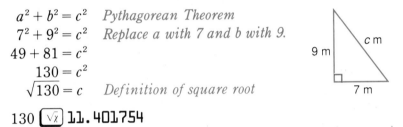

$$a^2 + b^2 = c^2 \quad \textit{Pythagorean Theorem}$$
$$7^2 + 9^2 = c^2 \quad \textit{Replace a with 7 and b with 9.}$$
$$49 + 81 = c^2$$
$$130 = c^2$$
$$\sqrt{130} = c \quad \textit{Definition of square root}$$

130 $\boxed{\sqrt{x}}$ 11.401754

The hypotenuse is about 11.4 meters long.

Estimation Hint

• • • • • • • • • • • •

$121 < 130 < 144$

$11 < \sqrt{130} < 12$

You can use the Pythagorean Theorem to find the length of a leg of a right triangle if you are given the lengths of the other leg and the hypotenuse.

Example 3

Find the length of the leg in the triangle at the right.

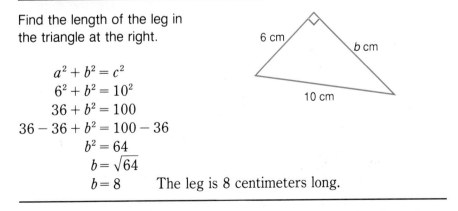

$$a^2 + b^2 = c^2$$
$$6^2 + b^2 = 10^2$$
$$36 + b^2 = 100$$
$$36 - 36 + b^2 = 100 - 36$$
$$b^2 = 64$$
$$b = \sqrt{64}$$
$$b = 8 \qquad \text{The leg is 8 centimeters long.}$$

You can also use the Pythagorean Theorem to determine if a triangle is a right triangle. The examples on the following page illustrate this.

Given the following lengths, determine whether each triangle is a right triangle.

4 7 inches, 24 inches, 25 inches

$$a^2 + b^2 = c^2$$
$$7^2 + 24^2 \overset{?}{=} 25^2$$
$$49 + 576 \overset{?}{=} 625$$
$$625 = 625$$

It is a right triangle.

5 8 inches, 13 inches, 16 inches

$$a^2 + b^2 = c^2$$
$$8^2 + 13^2 \overset{?}{=} 16^2$$
$$64 + 169 \overset{?}{=} 256$$
$$233 \neq 256$$

It is not a right triangle.

Remember the hypotenuse is always the longest side.

Checking for Understanding

Communicating Mathematics

Read and study the lesson to answer each question.

1. **Write** an equation that describes the relationship among the three large squares in the figure at the right.

2. **Draw** and label a right triangle with a hypotenuse of 17 units and legs of 8 units and 15 units.

3. **Write** the Pythagorean Theorem in your own words.

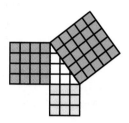

Guided Practice

State the lengths of the legs and hypotenuse of each triangle.

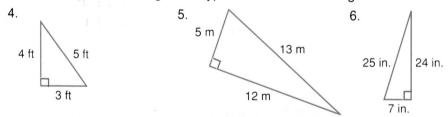

4. 4 ft, 5 ft, 3 ft

5. 5 m, 13 m, 12 m

6. 25 in., 24 in., 7 in.

State the equation you would use to find the length of the hypotenuse of each right triangle. Then solve. Round answers to the nearest tenth.

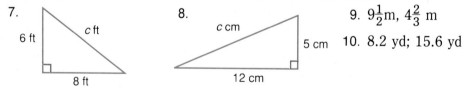

7. 6 ft, c ft, 8 ft

8. c cm, 5 cm, 12 cm

9. $9\frac{1}{2}$ m, $4\frac{2}{3}$ m

10. 8.2 yd; 15.6 yd

State the equation you would use to find the length of the leg of each right triangle. Then solve.

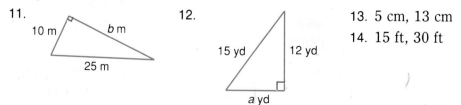

11. 10 m, b m, 25 m

12. 15 yd, 12 yd, a yd

13. 5 cm, 13 cm

14. 15 ft, 30 ft

Given the following lengths, determine whether each triangle is a right triangle. Write *yes* or *no*.

15. 2 m, 3 m, 4 m

16. 3 in., 4 in., 5 in.

17. 5 ft, 7 ft, 9 ft

Exercises

Independent
Practice

Use the Pythagorean Theorem to find the length of each hypotenuse given the lengths of the legs. Round answers to the nearest tenth.

18. 2 ft, 5 ft

19. 3 in., 7 in.

20. 9 cm, 40 cm

21. 14 ft, 8 ft

22. 13 mm, 9 mm

23. 11 yd, 17 yd

24. Draw a right triangle with legs of 4 centimeters and 7 centimeters. Find the length of the hypotenuse to the nearest tenth.

Find the missing lengths. Round decimal answers to the nearest tenth.

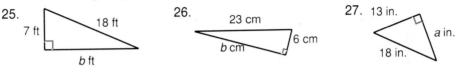

25. 7 ft, 18 ft, b ft

26. 23 cm, 6 cm, b cm

27. 13 in., 18 in., a in.

28. *b:* 9 yd; *c:* 15 yd

29. *a:* 13 cm; *c:* 27 cm

30. *b:* 24 m; *c:* 25 m

Given the following lengths, determine whether each triangle is a right triangle. Write *yes* or *no*.

31. 5 m, 12 m, 19 m

32. 7 ft, 24 ft, 25 ft

33. 8 cm, 11 cm, 19 cm

34. 9 in., 12 in., 15 in.

35. 20 ft, 25 ft, 30 ft

36. 30 yd, 40 yd, 50 yd

37. 19 m, 20 m, 21 m

38. 9 cm, 40 cm, 41 cm

Mixed Review

39. **Birthdays** Lisa's birthday is December 12. Julie's birthday is 19 days later. Solve mentally to find the date of Julie's birthday. *(Lesson 1-10)*

40. Use a factor tree to find the prime factorization of 720. *(Lesson 4-2)*

41. Use inverse operations to solve $\frac{n}{15} = 5$. *(Lesson 6-1)*

42. Solve the equation $m = 6 - (-12)$. *(Lesson 7-5)*

43. **Real Estate** Ms. Snyder owns a square plot of land that has an area of 8 square miles. Estimate to the nearest whole mile the length of a side of her plot of land. *(Lesson 9-3)*

Problem Solving
and
Applications

44. **Building Maintenance** A 15-foot ladder is propped against a wall. The base of the ladder is 3 feet from the base of the wall. How far up the wall does the ladder reach? Use a calculator and round to the nearest tenth.

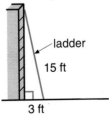

ladder

15 ft

3 ft

45. **Critical Thinking** On grid paper, draw a right triangle whose hypotenuse is 10 units long.

Lesson 9-4 The Pythagorean Theorem **347**

9-5 Using the Pythagorean Theorem

Objective

Solve problems using the Pythagorean Theorem.

In order to get that "in concert" feeling while watching a singing group on TV, Matt positions the television in a corner, puts the speakers along the wall 10 feet from the corner, and sits midway between the speakers. About how far is he from each speaker?

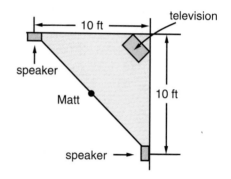

The two speakers and the television form a right triangle. To find the distance between the two speakers, use the Pythagorean Theorem.

$$a^2 + b^2 = c^2$$
$$10^2 + 10^2 = c^2 \quad \textit{Replace a and b with 10.}$$
$$100 + 100 = c^2$$
$$200 = c^2$$
$$\sqrt{200} = c$$

200 $\boxed{\sqrt{x}}$ **14.142136**

The speakers are about 14 feet apart. If Matt is sitting halfway between the speakers, he is about 7 feet from each speaker.

Example 1 *Problem Solving*

Sales When a TV is advertised as a 40-inch TV, the 40-inch label refers to the length of the diagonal of the screen. Find the length of a side of a 40-inch big-screen TV if the screen is square.

Since the screen is square, the sides will have the same length.

$$a^2 + a^2 = 40^2$$
$$2a^2 = 1{,}600 \quad a^2 + a^2 = 2a^2$$
$$\frac{2a^2}{2} = \frac{1{,}600}{2} \quad \textit{Divide each side by 2.}$$
$$a^2 = 800$$
$$a = \sqrt{800}$$

The first video aired on MTV at midnight on August 1, 1981.

800 $\boxed{\sqrt{x}}$ **28.284271**

The sides of the television screen are about 28 inches.

Example 2 *Connection*

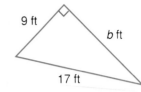

Geometry Find the perimeter of the triangle at the right.

First, find *b*.

$$a^2 + b^2 = c^2$$
$$9^2 + b^2 = 17^2 \quad \text{Replace a with 9 and c with 17.}$$
$$81 + b^2 = 289$$
$$81 - 81 + b^2 = 289 - 81 \quad \text{Subtract 81 from each side.}$$
$$b^2 = 208$$
$$b = \sqrt{208}$$

208 ⌷√x⌷ 14.422205   So, $b \approx 14.4$

To find the perimeter, add the lengths of the three sides.
$$P \approx 9 + 17 + 14.4$$
$$\approx 40.4$$
The perimeter of the triangle is about 40.4 feet.

Checking for Understanding

Communicating Mathematics

Read and study the lesson to answer each question.

1. **Draw** and label a right triangle that shows Bob and Mary sitting 5 feet apart. If Bob sits along the wall 3 feet from the corner of the room, how far from the corner of the room is Mary if she sits along the wall?

2. **Tell** if the hypotenuse of a right triangle is used to find the height. Explain.

Guided Practice

Write the equation you would use to find the value of x in each triangle. Then solve. Round decimal answers to the nearest tenth.

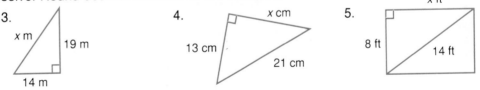

3. *x* m, 19 m, 14 m

4. *x* cm, 13 cm, 21 cm

5. *x* ft, 8 ft, 14 ft

6. Lidia walks 8 miles north and 5 miles east. How far is she from her starting point? Find the straight-line distance.

Exercises

Independent Practice

Solve. Round decimal answers to the nearest tenth.

7. Rob is flying a kite. Ellen is standing directly underneath the kite. If Rob and Ellen are 40 feet apart and Rob has let out 110 feet of string, how high is the kite?

8. Bo wants to put a diagonal brace across a gate that is 3 feet wide and 6 feet high. He has a board that is 8 feet long. About how much will he have left over after he cuts the board to make the brace?

9. **Smart Shopping** At the deli, 2.5 pounds of roast beef cost $7.50. What is the price per pound? *(Lesson 2-7)*

10. Is a triangle having sides of lengths 5 inches, 8 inches, and 10 inches a right triangle? *(Lesson 9-4)*

Problem Solving and Applications

11. **Computer Connection** The numbers 3, 4, and 5 are called *Pythagorean triples* because they are whole numbers that satisfy the Pythagorean Theorem. The computer program at the right will print all Pythagorean triples less than 21. Run the program and list the Pythagorean triples.

```
10 FOR A = 1 TO 21
20 FOR B = A TO 21
30 FOR C = B TO 21
40 IF A*A + B*B = C*
   C THEN PRINT A, B, C:
   GOTO 70
50 NEXT C
60 NEXT B
70 NEXT A
```

COMPUTER

CONNECTION

12. **Construction** A construction company is installing a roof on a new house. The width of the house is 26 feet. If the sides of the roof are to form a right angle and to be the same length, how long should each supporting board be, to the nearest tenth?

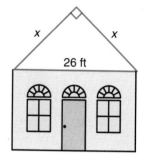

13. **Critical Thinking** Explain how to draw a right triangle with a hypotenuse of $\sqrt{10}$ units.

Mid-Chapter Review

1. The product of a number and itself is 1,296. What is the number? *(Lesson 9-1)*

Find the square of each number. *(Lesson 9-2)*

2. 9
3. 22
4. 31

Find each square root. If it is not a perfect square, estimate. *(Lessons 9-2, 9-3)*

5. $\sqrt{13}$
6. $\sqrt{64}$
7. $\sqrt{138}$

If the measures of the legs of a right triangle are a and b and the measures of the hypotenuse is c, find the missing length. Round decimal answers to the nearest tenth. *(Lesson 9-4)*

8. a: 9 feet; c: 41 feet
9. a: 11 inches; b: 15 inches

10. A helicopter rises vertically 800 feet and then travels west 1,200 feet. How far is it from its starting point? Round your answer to the nearest whole number. *(Lesson 9-5)*

9-6 Area of Irregular Figures

Objective

Estimate the area of irregular figures.

Words to Learn

irregular figure

Paul Melendez entered the "Design a Skateboard" contest sponsored by the Broadsports Supply Company. The only limitation was that the area of the board could not exceed 250 square inches. Will Paul's entry be accepted?

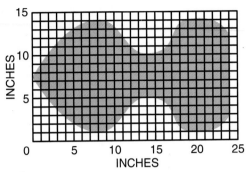

Paul's skateboard drawing is an **irregular figure.** Irregular figures do not necessarily have straight sides and square corners. One way to estimate the area of an irregular figure is to find the mean of the inner measure and outer measure. The inner measure is the number of whole squares within the figure. The outer measure is the number of squares inside of and touching the figure anywhere.

Example

Estimate the area of the skateboard.

inner measure: 187 in²
outer measure: 230 in²
mean: $\frac{(187 + 230)}{2} = 208.5$

An estimate of the area of Paul's skateboard is 208.5 square inches. Since the area is

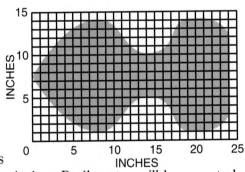

less than the limit of 250 square inches, Paul's entry will be accepted.

Mini-Lab

Work with a partner.
Materials: centimeter grid paper

- Draw an outline of your foot on a piece of grid paper.

Talk About It

a. Find the inner measure and outer measure of the outline of your foot.

b. Estimate the area by finding the mean.

c. Find another way you can estimate the area of the outline of your foot.

Checking for Understanding

Communicating Mathematics

Read and study the lesson to answer each question.

1. **Draw** an outline of your hand on a piece of centimeter grid paper. Estimate the area.

2. **Describe** an irregular shape that you have seen and estimate its area.

Guided Practice

Estimate the area of each figure.

3.

4.

5.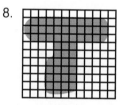

6.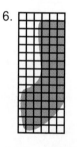

7.

8.

Exercises

Independent Practice

Estimate the area of each figure.

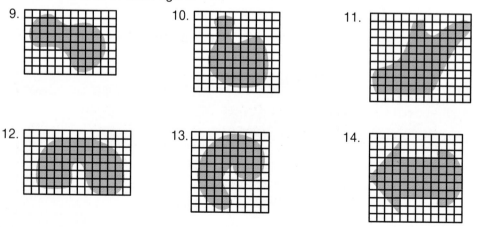

9.

10.

11.

12.

13.

14.

15. Draw a spoon on a piece of centimeter grid paper. Estimate its area.

16. Draw a hammer on a piece of centimeter grid paper. Estimate its area.

17. Classify the triangle at the right by sides and by angles. *(Lesson 8-3)*

18. **Sailing**   The main sail of a sailboat is in the shape of a right triangle. The hypotenuse of the sail is 32 feet long and the leg parallel to the water is 12 feet long. How tall is the sail? Round your answer to the nearest tenth. *(Lesson 9-5)*

19. **Geography**   Refer to the maps of the states shown below.
    a. Which states most closely resemble a rectangle?
    b. Which states most closely resemble a triangle?
    c. Use estimation to order the areas of the states below from largest to smallest.

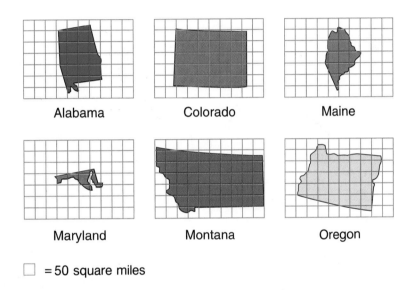

Alabama   Colorado   Maine

Maryland   Montana   Oregon

☐ = 50 square miles

20. **Critical Thinking**   Estimate the area of your state in square miles. Explain the process you used.

21. **Cooking**   Mrs. Rosales bakes cakes for special events. She has an order for a large sheet cake in the shape of Cookie Monster® for a child's birthday party. To figure out how much icing she needs to make, she estimates the area of the top of the cake. What is a good estimate of this area?

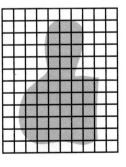

22. **Journal Entry**   Make a drawing of a skateboard that you would like to enter in a contest. Estimate its area. Is it less than 250 square inches?

9-7A Finding the Area of a Trapezoid

A Preview of Lesson 9-7

Objective

Find the area of a trapezoid.

In Lesson 8-3, you learned that a trapezoid is a quadrilateral with exactly one pair of parallel sides. In this lab, you will find the area of a trapezoid.

Materials

graph paper
scissors
tape

Try this!

Work in groups of three.

• Each student should draw a trapezoid on a piece of graph paper. Your trapezoid can be of any size or shape.

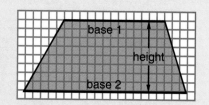

• Cut out your trapezoid. Label the bases and height as shown.

• Measure the length of base 1 and base 2. Measure the height. Record the measurements in a chart.

• Fold base 1 onto base 2. Unfold.

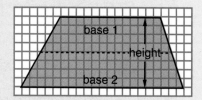

• Cut each trapezoid on the fold line. Then form a parallelogram.

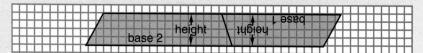

What do you think?

1. Find the length of the base of your parallelogram. How does it compare with the bases of your trapezoid?

2. Measure the height of your parallelogram. How does this height compare with the height of your trapezoid?

3. Find the area of your parallelogram.

4. What is the area of your trapezoid?

5. What conclusions can you make about the areas of all trapezoids?

6. Discuss in your group how to find the area of a trapezoid, given the lengths of base 1, base 2, and the height. Write a formula.

LOOK BACK

You can review area of parallelograms on page 244.

9-7 Area of Triangles and Trapezoids

Objective

Find the area of triangles and trapezoids.

Words to Learn

triangle
trapezoid

Delaware is nicknamed the Diamond State, but its shape looks more like a triangle. A triangle is a polygon that has three sides.

In this Mini-Lab, you will find the area of a triangle.

96 mi

39 mi

Mini-Lab

Work with a partner.
Materials: graph paper, scissors

- Draw a parallelogram of any shape or size on a piece of graph paper.

- Draw a diagonal in the parallelogram.

- Cut along the diagonal.

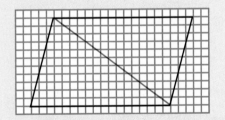

Talk About It

a. What two shapes are formed?

b. How do the two shapes compare?

c. What is the area of the original parallelogram?

d. What is the area of each triangle?

Area of a Triangle	**In words:** The area of a triangle is equal to half the product of the length of its base and height.
	In symbols: If a triangle has a base of b units and a height of h units, then the area, A square units, is $$A = \tfrac{1}{2}bh.$$

If the base measure of a triangle is an odd number, it is easier to multiply the base measure and height first rather than multiplying the base by $\frac{1}{2}$ and getting a mixed number.

Example 1

Find the area of the triangle at the right.

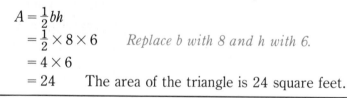

First identify the base and height.
base: 7 centimeters
height: 10 centimeters

$$A = \tfrac{1}{2}bh$$

$$= \tfrac{1}{2} \times 7 \times 10 \qquad \textit{Replace b with 7 and h with 10.}$$

$$= \tfrac{1}{2} \times 70 \qquad\qquad \textit{Associative property}$$

$$= 35$$

The area of the triangle is 35 square centimeters.

Example 2 *Connection*

Geometry Find the area of the triangle at the right.

In a right triangle, one leg is the base and the other leg is the height.

Find the height by finding the length of the missing side.

$$a^2 + b^2 = c^2 \qquad\qquad \textit{Pythagorean Theorem}$$
$$8^2 + b^2 = 10^2 \qquad\qquad \textit{Replace a with 8 and c with 10.}$$
$$64 + b^2 = 100$$
$$64 - 64 + b^2 = 100 - 64 \qquad \textit{Subtract 64 from each side.}$$
$$b^2 = 36$$
$$b = \sqrt{36} \qquad\qquad \textit{Definition of square root}$$
$$b = 6$$

Now find the area. The base is 8 feet and the height is 6 feet.

$$A = \tfrac{1}{2}bh$$
$$= \tfrac{1}{2} \times 8 \times 6 \qquad \textit{Replace b with 8 and h with 6.}$$
$$= 4 \times 6$$
$$= 24 \qquad \text{The area of the triangle is 24 square feet.}$$

The state of Nevada has a shape that looks like a **trapezoid.** A trapezoid is a quadrilateral with exactly one pair of parallel sides.

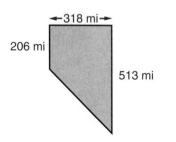

In Mathematics Lab 9-7A you found the area of a trapezoid by folding base *a* onto base *b*. Cut the trapezoid along this fold line and form a parallelogram.

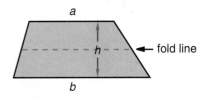

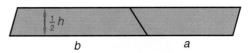

You can review the commutative property on page 204.

$A = \text{base} \times \text{height}$ *Area of a parallelogram*

$A = (a + b) \times \frac{1}{2}h$ *Substitute $a + b$ for the base and $\frac{1}{2}h$ for the height.*

$A = \frac{1}{2}h(a + b)$ *Commutative property*

Area of a Trapezoid	**In words:** The area of a trapezoid is equal to half the product of the height and the sum of the bases. **In symbols:** If a trapezoid has bases of *a* and *b* units and a height of *h* units, the area, *A* square units, is $$A = \frac{1}{2}h(a + b).$$

Example 3

Find the area of the trapezoid.

bases: 20 meters, 28 meters
height: 10 meters

$A = \frac{1}{2}h(a + b)$

$= \frac{1}{2}(10)(20 + 28)$ *Replace h with 10, a with 20, and b with 28.*

1 ÷ 2 × 10 × (20 + 28) = **240**

The area of the trapezoid is 240 square meters.

Checking for Understanding

Communicating Mathematics

Read and study the lesson to answer each question.

1. **Describe** the relationship between the area of a parallelogram and the area of a triangle with the same height and base. Explain.

2. **Draw** a trapezoid with at least one right angle.

3. **Draw** a trapezoid with bases of 6 centimeters and 12 centimeters and a height of 7 centimeters.
   a. Cut the trapezoid like you did on page 354 and form a parallelogram. Find the area of the parallelogram.
   b. Find the area of the trapezoid using the formula. Check by comparing it to the area of the parallelogram.

Lesson 9-7 Area of Triangles and Trapezoids   **357**

Find the area of each triangle.

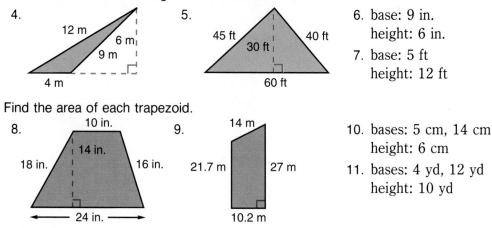

4.
12 m
6 m
9 m
4 m

5.
45 ft   30 ft   40 ft
60 ft

6. base: 9 in.
   height: 6 in.

7. base: 5 ft
   height: 12 ft

Find the area of each trapezoid.

8.
10 in.
14 in.
18 in.   16 in.
24 in.

9.
14 m
21.7 m   27 m
10.2 m

10. bases: 5 cm, 14 cm
    height: 6 cm

11. bases: 4 yd, 12 yd
    height: 10 yd

Exercises

Independent Practice Find the area of each triangle.

12. base: 16 m
    height: 12 m

13. base: 4 km
    height: 28 km

14. base: 7 ft
    height: 12 ft

15. base: 1.4 in.
    height: 1.1 in.

16. base: 8 cm
    height: 19 cm

17. base: $2\frac{1}{3}$ yd
    height: $1\frac{5}{6}$ yd

Find the area of each trapezoid.

18. bases: 4 in., 8 in.
    height: 6 in.

19. bases: 12 yd, 18 yd
    height: 10 yd

20. bases: 7.3 cm, 9.5 cm
    height: 8.8 cm

21. bases: $6\frac{1}{2}$ ft, $11\frac{2}{3}$ ft
    height: 14 ft

22. Find the area of a trapezoid that has bases of 11 yards and 19 yards and a height of $7\frac{1}{2}$ yards.

Mixed Review 23. **Statistics** Find the range for the following data. Then find an appropriate scale and interval. 25, 34, 16, 9, 22, 19, 31 *(Lesson 3-5)*

24. Find $4\frac{5}{8} \times 1\frac{2}{3}$. *(Lesson 5-5)*

25. Estimate the area of the irregular figure at the right. *(Lesson 9-6)*

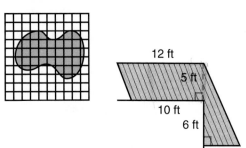

12 ft
5 ft
10 ft
6 ft
5 ft

Problem Solving and Applications 26. **Construction** The Deck and Porch Company has several designs for decks. One of them is shown at the right. Find the area of the deck.

27. **Critical Thinking** Show how the area of a triangle could be found using the trapezoid formula.

28. **Journal Entry** Give at least three examples where you would need to be able to find the area of a triangle or trapezoid.

9-8 Area of Circles

Objective

Find the area of circles.

Could the Houston Astrodome fit inside the Louisiana Superdome? The highest point of the Superdome is 273 feet and the Astrodome is only 208 feet tall. The distance across the Astrodome is about 214 yards, and the floor area of the Superdome is 41,000 square yards. Since both structures are in the shape of circles, you will need to find the area of the circular floor of the Astrodome to answer the question in Example 1. To find the area of a circle, we will need to use the formula for the area of a parallelogram.

Mini-Lab

Work in pairs.
Materials: paper, compass, straightedge, scissors, pencil

- Draw a circle and several radii that separate the circle into equal-sized sections.

  Let r units represent the length of the radius of the circle and let C units represent its circumference.

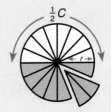

- Cut out each section of the circle.

- Reassemble the sections in the form of a parallelogram.

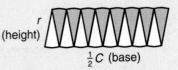

Talk About It

a. What is the height of this "parallelogram"? the length of the base?

b. What is the formula for the area of a parallelogram?

c. How could you use this formula to find the area of a circle?

You can review circumference on pages 197 and 198.

The base of the parallelogram shown on the previous page is equal to one half of the circumference of the circle ($\frac{1}{2}C$). The height of the parallelogram is the measure of the radius of the circle (r). Substitute this information into the formula for the area of a parallelogram.

$$A = bh \qquad \textit{Formula for area of a parallelogram}$$

$$= (\tfrac{1}{2}C)r \qquad \textit{Substitute } \tfrac{1}{2}C \textit{ for b and substitute r for h.}$$

$$= (\tfrac{1}{2} \times 2\pi r)r \qquad \textit{Substitute } 2\pi r \textit{ for C.}$$

$$= \pi r^2 \qquad \textit{Simplify: } \tfrac{1}{2} \times 2 = 1, \ r \times r = r^2.$$

Area of a Circle	**In words:** The area of a circle is equal to pi times the square of the radius.
	In symbols: If a circle has a radius of r units, then $A = \pi r^2$.

Example 1 *Problem Solving*

Sports   Find the floor area of the Astrodome if its diameter is 214 yards.

$$r = \tfrac{1}{2}d \qquad \textit{The radius is one half of the diameter.}$$

$$= \tfrac{1}{2} \times 214 \qquad \textit{Replace d with 214.}$$

$$= 107$$

You can review π, pi, on page 198.

Use the measure to find the area.

$$A = \pi r^2$$

$$= \pi \times 107^2 \qquad \textit{Replace r with 107.}$$

$$= 3.14 \times 11{,}449 \qquad \textit{Replace } \pi \textit{ with 3.14.}$$

$$\boxed{\pi}\ \boxed{\times}\ 11{,}449\ \boxed{=}\ \text{35968.094}$$

To the nearest whole number, the floor area of the Astrodome is 35,950 square yards. Could the Astrodome fit inside the Superdome?

You can also use the formula for the area of a circle to find the length of the radius when the area is known.

Example 2

Find the length of the radius of a circle if its area is 79 square inches.

$$A = \pi r^2$$

$$79 \approx 3.14 \times r^2 \qquad \textit{Replace A with 79 and } \pi \textit{ with 3.14.}$$

$$\frac{79}{3.14} \approx \frac{3.14 \times r^2}{3.14} \qquad \textit{Divide each side by 3.14.}$$

$$25.16 \approx r^2$$

$$\sqrt{25.16} \approx r \qquad \textit{Definition of square root}$$

$$5.02 \approx r \qquad \text{The radius of the circle is about 5 inches long.}$$

Checking for Understanding

Read and study the lesson to answer each question.

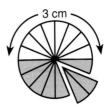

1. **Tell** how you can find the area of a circle given the length of the diameter of the circle.

2. **Draw** a circle with radius of 1 inch. Find its area.

3. **Tell** the base measures and the height of a parallelogram formed from the circle at the right. Find the area.

Guided Practice Find the area of each circle given the following information. Round answers to the nearest tenth.

4.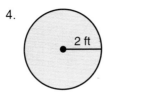

2 ft

5.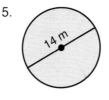

14 m

6. radius, 5 in.

7. diameter, 1.4 cm

Find the radius of each circle given the following areas. Round answers to the nearest tenth.

8. 12 cm^2 9. 27 ft^2 10. 75 m^2 11. 112 in^2

Exercises

Independent
Practice Find the area of each circle shown or described below. Round answers to the nearest tenth.

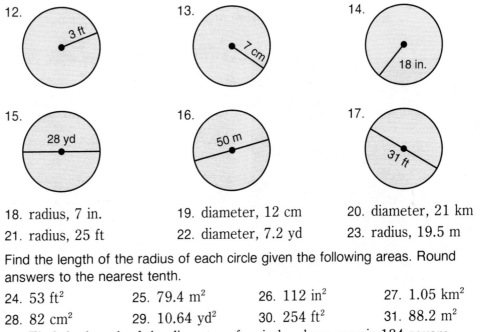

12. 3 ft

13. 7 cm

14. 18 in.

15. 28 yd

16. 50 m

17. 31 ft

18. radius, 7 in.   19. diameter, 12 cm   20. diameter, 21 km

21. radius, 25 ft   22. diameter, 7.2 yd   23. radius, 19.5 m

Find the length of the radius of each circle given the following areas. Round answers to the nearest tenth.

24. 53 ft^2 25. 79.4 m^2 26. 112 in^2 27. 1.05 km^2

28. 82 cm^2 29. 10.64 yd^2 30. 254 ft^2 31. 88.2 m^2

32. Find the length of the diameter of a circle whose area is 134 square meters.

33. Recycling Kimi collects 5.6 pounds of aluminum cans. The recycling center will pay her $0.42 per pound. How much will Kimi receive when she turns in the cans? *(Lesson 2-4)*

34. Order 6, −3, 0, 4, −8, 1, −4 from least to greatest. *(Lesson 7-2)*

35. Which quadrilaterals have four congruent sides? *(Lesson 8-3)*

36. Find the area of a trapezoid whose bases are 8 inches and 12 inches long and whose height is 6 inches. *(Lesson 9-7)*

37. History Stonehenge, an ancient monument in England, may have been used as a calendar. The stones are arranged in a circle 30 meters in diameter. Find the area of the circle.

38. Food The table at the right gives the diameter of three pizza sizes.
  a. Find the area of each size.
  b. Which has the greatest area: 1 large pizza or 2 medium pizzas?

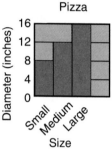

39. Critical Thinking The floor of the Superdome can be set up as a track for track and field events. Find the area inside the track.

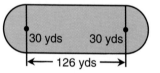

Hint: The track is formed by a rectangle and two semicircles.

40. Mathematics and Fine Art Read the following paragraphs.

A fresco is a special kind of wall painting that is made when the wall is plastered. Paints are mixed with the wet plaster as it is applied to the wall. The colors show up brightly when the plaster dries.

Diego Rivera was a Mexican painter whose bold murals stimulated a revival of fresco painting in Latin America and the United States. During the Great Depression, he was commissioned to paint a massive fresco honoring American auto workers. It fills an entire room in the Detroit Institute of Arts. His largest and most ambitious mural was an epic of the history of Mexico for the National Palace in Mexico City. It was unfinished when he died in 1957.

An artist wants to paint the sun in his fresco. He wants it to have an area of 40 square feet. What would be the diameter of this sun?

9-9A Probability and Area Models

A Preview of Lesson 9-9

Objective

Estimate the area of a figure using probability.

In this activity, you will investigate the relationship between area and probability.

Materials

inch grid paper
ruler
small counters

Try this!

Work in groups of three.

- Draw a square that has sides 8 inches long on your grid paper. Inside the square, draw a triangle with one side 6 inches long and the other side 5 inches long.

- Hold 20 counters about 5 inches above the paper and drop them onto the paper.

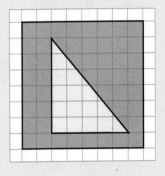

- Count the number of counters that landed completely within the square. (This includes those that landed within the triangle.) Count the number that landed completely inside the triangle. Do not count those that landed on a side of the triangle. These two numbers make up the first sample.

- Repeat the activity nine more times.

- Add the results of your ten samples to find the total number of counters that fell within the square and the total number that fell within the triangle.

- The probability that a counter will land inside the triangle is expressed by the fraction:

$$\frac{\text{total counters within the triangle}}{\text{total counters within the square}} = \text{probability}$$

Calculate this experimental probability based on your findings.

LOOKBACK

You can review probability on page 157.

- You can compute the following probability to estimate the area of the triangle.

$$\text{probability} = \frac{\text{area of triangle}}{\text{area of square}}$$

Substitute the probability into the left side of the equation. Calculate the area of the square and substitute it in the denominator of the right side. Then solve to find the area of the triangle.

What do you think?

1. Count the number of grid squares inside your triangle to get an estimate of the area. Since not all the squares are complete squares, you will sometimes need to combine two or three partial squares to get an estimate of the number of complete squares. How does this estimate of the area compare with the experimental probability estimate you found above?

2. Measure the base and height of the triangle. Find the area using the formula $A = \frac{1}{2}bh$. How does it compare with your experimental probability estimate?

Extension

3. Repeat this activity with a circle inside your square.
4. Repeat this activity with an irregular figure inside your square.

9-9 Area Models and Probability

Objective

Find the probability using area models.

Alberto Davilla designed the archery target shown at the right as part of a cooperative project for his art and physical education classes. An arrow is shot and hits the target. What is the probability that the arrow landed in the yellow region if it is equally likely to hit any square?

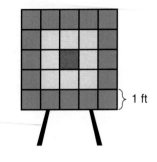

1 ft

In order to solve this problem, we need to know the area of the entire target and the area of the yellow region. Since the target is in the shape of a square, we can find the area of the target by either counting the small squares or by using the area formula.

$$A = s^2$$
$$= 5^2 \quad \textit{Substitute 5 for s.}$$
$$= 25$$

The target has an area of 25 square feet.

Next we need to find the area of the yellow region. We can count the yellow squares and find the area to be 8 square feet.

Now we can find the probability of an arrow landing in the yellow region. Remember the definition of probability.

$$\text{probability} = \frac{\text{number of ways an event can occur}}{\text{number of possible outcomes}}$$

In our target problem, the *event* is hitting the yellow region. This would be the area of the yellow region, 8 square feet. *All possible outcomes* would be hitting anywhere on the target. This would be the total area of the target, 25 square feet.

$$\text{probability} = \frac{\text{number of ways an event can occur}}{\text{number of possible outcomes}}$$

$$= \frac{8}{25}$$

The probability of hitting the yellow region is $\frac{8}{25}$ or 0.32.

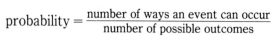

Determine the probability that a randomly-dropped counter will fall in the shaded area. Each unit square is 1 square foot.

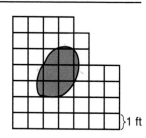

The area of the entire region is 42 square feet. The area of the shaded region is about 7 square feet.

$$\text{probability} = \frac{\text{number of ways an event can occur}}{\text{number of possible outcomes}}$$

$$= \frac{7}{42} \text{ or } \frac{1}{6}$$

Example 2 *Problem Solving*

Sports A golfer tees off and the ball lands in the rectangular region at the right. What is the probability that the ball lands on the green?

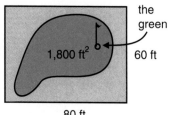

The area of the rectangular region is 80×60 or 4,800 square feet. The area of the green is 1,800 square feet.

$$\text{probability} = \frac{\text{number of ways an event can occur}}{\text{number of possible outcomes}} = \frac{1,800}{4,800} \text{ or } \frac{3}{8} \text{ or } 0.375$$

Checking for Understanding

Communicating Mathematics

Read and study the lesson to answer each question.

1. **Tell** how you would find the probability of an arrow landing in the red region of the target on page 365.

2. **Write** the equation you would use to estimate the area of the trapezoid at the right if 8 out of 20 counters landed in the trapezoid.

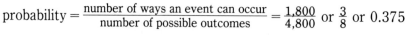

Guided Practice

Find the probability that a randomly-dropped counter will fall in the shaded region.

3.

4.

5.

Exercises

Independent Practice

Find the probability that a randomly-dropped counter will fall in the shaded region.

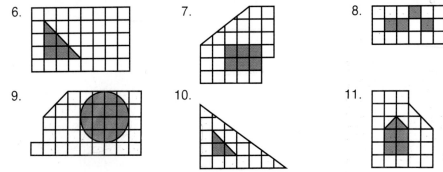

6.

7.

8.

9.

10.

11.

12. Draw a square on a piece of grid paper that is 6 units on a side. Shade in 4 squares. What is the probability that a randomly-dropped counter will fall in the shaded area?

Mixed Review

13. Classify the angle at the right as acute, obtuse, right, or straight. *(Lesson 8-1)*

14. **Coin Collection** The diameters of some United States coins are listed at the right. Find the area of each. *(Lesson 9-8)*

Coin	Diameter (cm)
penny	1.9
nickel	2.1
dime	1.8
quarter	2.4

Problem Solving and Applications

15. **Geology** Sam Tex owns an 18,000-acre ranch. A geologist told him that he can expect to find oil under 400 acres of his property. If Sam Tex randomly starts drilling, what is the probability of striking oil?

16. **Sports** A skydiver parachutes onto a square field that contains a pond. The field is 150 feet on a side and the pond has an area of 750 square feet. If there is an equal chance of landing at any point in the field, what is the probability that the diver has a dry landing?

17. **Data Search** The island of Greenland is larger than some seas. Use the information on page 335 to figure out which seas it is larger than.

DATA SEARCH

18. **Critical Thinking** What is the probability that a randomly-dropped counter will land on the green, yellow, or blue region?

9 Study Guide and Review

Communicating Mathematics

Choose the correct term to complete each sentence.
1. The number (49, 1,000) is a perfect square.
2. In a right triangle, the square of the measure of the hypotenuse is (equal to, greater than) the sum of the squares of the measures of the legs.
3. A trapezoid is a quadrilateral with exactly (one, two) pair(s) of parallel sides.
4. $A = \frac{1}{2}bh$ is the formula for the area of a (trapezoid, triangle).
5. The formula for the area of a circle includes the (height, radius).
6. Probability can be expressed as a (mixed number, fraction).
7. In your own words, explain how you would estimate the square root of 55.

Skills and Concepts

Objectives and Examples

Upon completing this chapter, you should be able to:

Review Exercises

Use these exercises to review and prepare for the chapter test.

- find the square roots of perfect squares *(Lesson 9-2)*

Evaluate $\sqrt{225}$.

Since $15^2 = 225$, $\sqrt{225} = 15$.

Find each square root.

8. $\sqrt{9}$ 9. $\sqrt{1}$
10. $\sqrt{100}$ 11. $\sqrt{169}$
12. $\sqrt{900}$ 13. $\sqrt{10,000}$

- estimate square roots *(Lesson 9-3)*

Estimate $\sqrt{75}$.

$64 < 75 < 81$ Since 75 is closer
$\sqrt{64} < \sqrt{75} < \sqrt{81}$ to 81 than to 64,
$8 < \sqrt{75} < 9$ $\sqrt{75}$ is closest to 9.

Estimate.

14. $\sqrt{5}$ 15. $\sqrt{35}$
16. $\sqrt{116}$ 17. $\sqrt{40}$
18. $\sqrt{435}$ 19. $\sqrt{399}$

- find the length of a side of a right triangle using the Pythagorean Theorem *(Lesson 9-4)*

Find the missing length.

$a^2 + b^2 = c^2$
$5^2 + 12^2 = c^2$
$169 = c^2$
$13 = c$

The hypotenuse is 13 inches long.

If the measures of the legs of a right triangle are *a* and *b* and the hypotenuse is *c*, find the missing length. Round decimal answers to the nearest tenth.

20. *a*: 5 ft; *b*: 7 ft
21. *b*: 10 yd; *c*: 15 yd
22. *a*: 12 in.; *b*: 3 in.
23. *c*: 18 m; *a*: 6 m

Objectives and Examples

- solve problems using the Pythagorean Theorem *(Lesson 9-5)*
  Judy leans a 6-foot mirror against a wall by placing the bottom 2 feet from the base of the wall. How far up the wall is the top of the mirror?

  $$2^2 + b^2 = 6^2$$
  $$4 + b^2 = 36$$
  $$b^2 = 32$$
  $$b = \sqrt{32}$$
  $$b \approx 5.7 \text{ ft}$$

  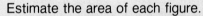

  The mirror is about 5.7 feet up the wall.

- estimate the area of irregular figures *(Lesson 9-6)*

  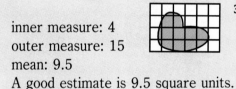
  369b

  inner measure: 4
  outer measure: 15
  mean: 9.5
  A good estimate is 9.5 square units.

- find the area of triangles and trapezoids *(Lesson 9-7)*
  Area of a triangle: $A = \frac{1}{2}bh$
  Area of a trapezoid: $A = \frac{1}{2}h(a+b)$
  $A = \frac{1}{2}h(a+b)$

  $\quad = \frac{1}{2}(4)(3+9)$

  $\quad = \frac{1}{2}(4)(12)$

  $\quad = 2(12)$

  $\quad = 24 \text{ cm}^2$ The area is 24 cm².

- find the area of circles *(Lesson 9-8)*
  Area of a circle: $A = \pi r^2$

  The area of a circle with a radius of 6 inches is: $A = \pi r^2$
  $$\approx (3.14)(6^2)$$
  $$\approx (3.14)(36)$$
  $$\approx 113.04 \text{ in}^2$$
  The area is about 113 square inches.

Review Exercises

Solve. Round answers to the nearest tenth.

24. Pete builds a swimming pool in the shape of a right triangle with legs that are 50 feet and 75 feet long. How far would a person walk if they walked the perimeter of Pete's pool?

25. While hiking in the woods, Angela walks 3 miles south and 5 miles west. How far is she from her starting point? Find the straight-line distance.

Estimate the area of each figure.

26. 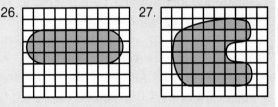 27.

Find the area of each triangle or trapezoid.

28. bases: 6 m, 9 m; height: 5 m
29. base: 15 yd; height: 8 yd
30. bases: 4 in., 8 in.; height: 5 in.
31. base: 5 ft; height: 12 ft

Find the area of each circle. Round answers to the nearest hundredth. Use 3.14 for π.

32. r, 7 ft
33. d, 18 mm
34. r, 25 in.
35. d, 11 yd

Objectives and Examples

Review Exercises

- find the probability using area models *(Lesson 9-9)*

probability of hitting a green square

$= \dfrac{\text{number of ways an event can occur}}{\text{number of possible outcomes}}$

$= \dfrac{16}{25}$

Find the probability that a randomly-dropped counter will fall in the shaded region.

36.

37.

38. Admission to the zoo is $5 for adults, $3.50 for children under 12, and $3 for seniors. Ten people paid a total of $37.50. If 3 adults attended, how many children and seniors were in the group? *(Lesson 9-1)*

39. **Treasure Island** An island has buried treasure under 50 of its 1,000 square feet area. If Amber randomly starts digging for treasure, what is the probability that she will find something? *(Lesson 9-9)*

Curriculum Connection Projects

- **Home Economics** Call your favorite pizza shop and ask for the diameters of their small, medium, and large pizzas. Find the area of each and the price per square inch.
- **History** Find the dimensions of an Aztec pyramid. Then find the area of each of its trapezoidal faces.

Read More About It

Wilcox, Charlotte. *A Skyscraper Story.*
Froman, Robert. *Rubber Bands, Baseballs, and Doughnuts: A Book About Topology.*
Paulson, Gary. *The Hatchet.*
Michener, James. *Journey.*

Test

1. What is the fewest number of square tables, each seating 4 people, that can be used to seat 20 people if the tables are arranged in one long rectangle?

Find each square root.

2. $\sqrt{49}$

3. $\sqrt{625}$

4. $\sqrt{22,500}$

5. **Physical Fitness**   Every morning, Ichiko jogs around a group of blocks that make up a square having an area of 4 square miles. How far does Ichiko jog each morning?

Estimate.

6. $\sqrt{18}$

7. $\sqrt{90}$

8. $\sqrt{490}$

If the measures of the legs of a right triangle are a and b and the measure of the hypotenuse is c, find the missing length. Round decimal answers to the nearest tenth.

9. a: 5 m,; b: 3 m

10. b: 12 in.; c: 25 in.

11. a: 4 yd; c: 8 yd

12. **Road Trip**   Sam gets lost while driving to a new vacation spot. After looking at a map, he sees that he is 18 miles too far east and 10 miles too far north. How far is Sam's drive to his intended target if he is able to drive a straight path?

Estimate the area of each figure.

13.

14.

Find the area of each triangle or trapezoid.

15. triangle: base: 12 ft; height: 6 ft

16. trapezoid: bases: 5 km, 11 km; height: 4 km

Find the area of each circle.

17. r, 11 cm

18. d, 23 in.

19. **Amusement**   The merry-go-round at an amusement park covers an area of 2,000 square feet. Find the length of its diameter. Round to the tenths place.

20. **Carnival Game**   A carnival game requires that a blindfolded contestant throw a dart at a wall partially covered with balloons. If a balloon is popped, the contestant wins. The wall has an area of 16 square feet. Six square feet are covered with balloons. What is the probability that a contestant wins?

Bonus   What is the formula for the area of a semi-circle?

9 Academic Skills Test

Directions: Choose the best answer. Write A, B, C, or D.

1. What is the value of $3b + 4a$ if $a = 2$ and $b = 6$?

   A 30   B 26

   C 17   D 12

2. Joyce is writing a book report. One page contains 175 words. If her report is 5 pages long and each page is about the same length, about how many words are in her report?

   A 40 words   B 175 words

   C 500 words   D 850 words

3. Kenny is saving to buy a computer that will cost about $1,200. He already has saved $500. If he can save $75 a month, it is reasonable to expect that he can buy the computer in

   A 3 months?   B 7 months?

   C 10 months?   D 16 months?

4.

   Judy's Budget

   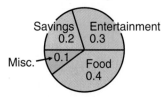

   If Judy has $50 for the week, about how much can she spend on food?

   A $20   B $30

   C $40   D $50

5. What is the prime factorization of 60?

   A $1 \cdot 60$ B $2 \cdot 3 \cdot 3 \cdot 5$

   C $2 \cdot 3 \cdot 10$ D $2 \cdot 2 \cdot 3 \cdot 5$

6. Rolito wants to find $\frac{3}{5}$ of $65 using a calculator. What decimal can he enter for $\frac{3}{5}$?

   A 0.3   B 0.5

   C 0.6   D 0.65

7. Which sentence is true?

   A $\frac{5}{8} < 0.6$ B $\frac{5}{8} = 0.6$

   C $\frac{5}{8} > 0.6$ D none of these

8. A person's weight on the moon is about $\frac{1}{6}$ of their weight on Earth. About how much would a person weigh on the moon if their weight on Earth is 125 pounds?

   A 20 lb   B 60 lb

   C 200 lb   D 600 lb

9. What is the reciprocal of $3\frac{3}{4}$?

   A $\frac{4}{3}$ B $\frac{4}{33}$

   C $\frac{15}{4}$ D $\frac{4}{15}$

10. Claire worked 20 hours last week. She earned $5.00 per hour. Which equation can be used to find her total earnings?

    A $x = 20 \div 5$ B $5x = 20$

    C $20 \times 5 = x$ D $20x = 5.00$

11. How much carpeting is needed to cover the floor of a 10 foot by 12 foot room?

    A 22 ft^2 B 44 ft^2

    C 120 ft^2 D 480 ft^2

12. $|24| =$

    A -24 B 0

    C 1   D 24

13. $-12 + 30 =$

    A 18   B 8

    C -18 D -42

14. $0.001 =$

    A 10^1 B 10^{-1}

    C 10^{-2} D 10^{-3}

15. Which polygon is *not* regular?

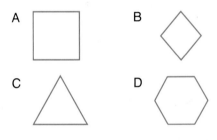

16. Which type of quadrilateral is an equilateral quadrilateral and an equiangular quadrilateral?

    A rectangle   B parallelogram

    C rhombus   D square

17.

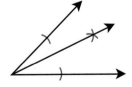

This drawing shows how to

A construct an angle bisector.
B construct an angle congruent to a given angle.
C construct perpendicular lines.
D construct a segment bisector.

18. Which is a perfect square?

    A 88   B 125

    C 181   D 625

19. What is the length of the hypotenuse of this triangle?

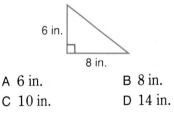

    A 6 in.   B 8 in.

    C 10 in.   D 14 in.

20. Jay walks 5 kilometers east and 5 kilometers south. To the nearest kilometer, how far is he from his starting point?

    A 25 km   B 10 km

    C 7 km   D 5 km

Surface Area and Volume

Spotlight on Volcanoes

Have You Ever Wondered . . .

- What the elevations are of some of the world's most famous volcanoes?

- How pressure and temperature change with depth below Earth's surface?

Some Famous Volcanoes Name	Location	Height (feet)
Cotopaxl	Ecuador	19,347
Hibokhibok	Philippines	4,363
Krakatoa	Indonesia	2,667
Lassen Peak	California	10,457
Mauna Loa	Hawaii	13,677
Mont Pelée	Martinique	4,583
Mount Etna	Sicily	11,122
Mount St. Helens	Washington	8,364
Nevado del Ruiz	Columbia	17,717
Vesuvius	Italy	4,190

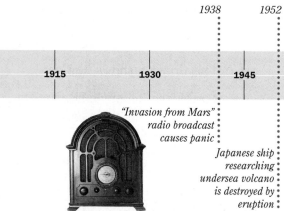

1906

1938   1952

1900   1915   1930   1945

"Invasion from Mars" radio broadcast causes panic

Mt. Vesuvious near Naples erupts

Japanese ship researching undersea volcano is destroyed by eruption

Chapter Project

Volcanoes
Work in a group.

1. Research the shape and formation of a cinder cone volcano.

2. Make a papiermâché three-dimensional model of the volcano.

3. Use a realistic scale for your model. Indicate measurements for the circular base, the sides, and the height.

Diamond Head

Looking Ahead

In this chapter, you will see how mathematics can be used to answer questions about volcanoes and other three-dimensional figures.

The major objectives of the chapter are to:

- sketch three-dimensional figures

- solve problems by making a model

- find the surface area and volume of rectangular prisms and cylinders

- solve problems by using a formula

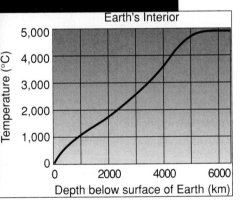

Earth's Interior

Temperature (°C) vs Depth below surface of Earth (km)

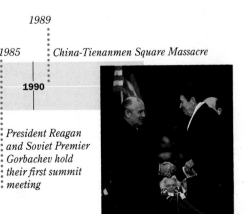

1969 — Woodstock music festival takes place

1980 — Mt. St. Helens erupts in Washington State

1985 — President Reagan and Soviet Premier Gorbachev hold their first summit meeting

1989 — China-Tienanmen Square Massacre

1960   1975   1990

375

10-1A Building Three-Dimensional Figures

A Preview of Lesson 10-1

Objective

Build and draw three-dimensional figures given the front, side, and top view.

The cubes you will be using in this lab are examples of three-dimensional figures. A three-dimensional figure has length, width, and depth. A two-dimensional figure has no depth, so it appears as a flat object.

Materials

cubes
pencil
paper

Activity One

Work with a partner.

- Use the cubes to build the three-dimensional model of the shape that is described by each drawing below. The front view, a side view, and the top view of each shape are given.

a.   front   side   top

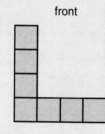

b.   front   side   top

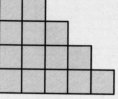

c.   front   side   top

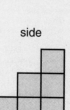   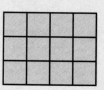

What do you think?

1. Share with other groups how you began building the figures.
2. Do you think you could have built the figures without one of the views? Explain.
3. Is there only one way to build the figures from the drawings given? If no, build another model. If yes, explain.

Activity Two

- After you have built each figure and both partners have agreed that it is correct, draw a three-dimensional figure that shows depth as well as length and width.

What do you think?

4. Compare your models and drawings with other groups. Are they the same? Tell how they are different.
5. Describe a real-life situation where it might be necessary for you to draw a three-dimensional figure.

Extension

By using one-point perspective, you can create the illusion of depth. Study the example of one-point perspective below. The following characteristics give the illusion of depth.

- The front of the building is drawn like a rectangle.
- The sides of the building are drawn along lines to the vanishing point.

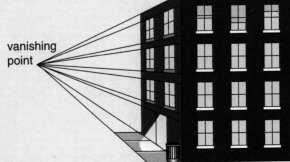

vanishing point

6. **Art** Make a drawing of your own using one-point perspective.

Mathematics Lab 10-1A Building Three-Dimensional Figures   **377**

10-1 Drawing Three-Dimensional Figures

Objective

Draw three-dimensional figures.

The photograph at the right has two dimensions, width and height. Yet when you look at it, you can visualize what the real totem pole must look like in three dimensions.

As in Mathematics Lab 10-1A, we often need to make a two-dimensional drawing of a three-dimensional object. How you go about that may depend on which view is the most important. For example, photographing the back or top of the totem pole would have eliminated many of the important features of the pole.

Think about drawing a two-dimensional cylinder.

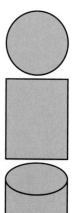

- If you simply look down from directly above, you would see a *circle*. Drawing a circle would not indicate that the figure has three dimensions.
- Looking at the cylinder directly from the side, you can see a *rectangle*. Again this does not give any indication that this is a view of a cylinder.
- However, if you make a drawing that is somewhere between a top and side view, you are able to see that the figure has three dimensions.

Example 1

Make a two-dimensional drawing of a figure by using the top, front, and side views of the figure below.

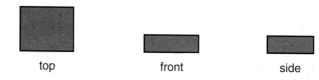

| top | front | side |

By drawing the figure as shown, you are able to see how all three views are correct.

In geometry, we study three-dimensional figures called *solids*. Some common solids are shown below.

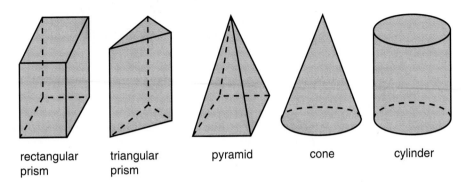

rectangular   triangular   pyramid   cone   cylinder
prism   prism

Checking for Understanding

Communicating Mathematics

Read and study the lesson to answer each question.

1. **Tell** why there can be more than one two-dimensional drawing of a three-dimensional object.

2. **Make a model** of a solid figure that would have the same view from the top and the bottom.

3. **Write** the names of two solids that would have a different view from the top and bottom.

Guided Practice

Draw a top, front, and side view of each figure.

4. 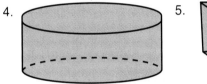   5. 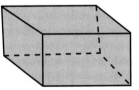   6.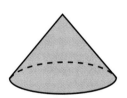

Draw a three-dimensional figure given the top, front, and side views shown below.

7.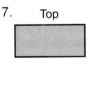

Top   Front   Side

Exercises

Independent Practice

Draw top, front, and side views of each figure.

8. 9. 10.

Draw top, front, and side views of each figure.

11. 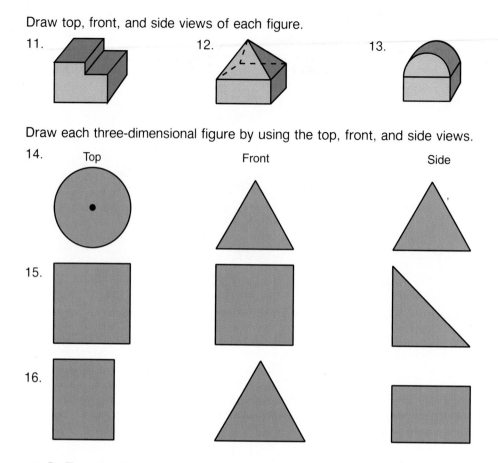   12.   13.

Draw each three-dimensional figure by using the top, front, and side views.

14. Top   Front   Side

15.

16.

17. In Exercise 7, would you expect your drawings to be the same as others in your class? Why or why not?

Mixed Review 18. The number of sick days Aaron accumulated at his job during the first three years were 23, 17, and 9. Compute mentally the total number of sick days Aaron had accumulated at the end of his third year. *(Lesson 1-4)*

19. **Geometry** Find the area of a parallelogram having a base of 2.3 centimeters and a height of 1.6 centimeters. *(Lesson 6-7)*

20. **Probability** Find the probability of a counter that is randomly dropped falling in the shaded region. *(Lesson 9-9)*

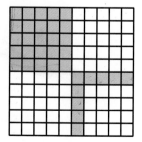

Problem Solving and Applications 21. **Construction** Mr. Sims installs windows in homes. Which views of a house would be most helpful to Mr. Sims? Why?

22. **Architecture** Why would it be necessary for an architect to draw several different views of a building before starting construction?

 23. **Critical Thinking** Explain what two-dimensional and three-dimensional means. What do you think the fourth dimension might be?

24. **Journal Entry** Draw three different views of a cube, a pyramid, and a cone.

10-2 Make a Model

Objective

Solve problems by making a model.

Materials

sugar cubes

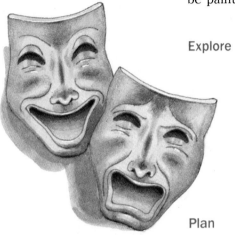

Concheta Lopez is building props for an upcoming school play. She is making a large display stand made from 20 boxes that are cubes that will be placed up against a wall during Scene I of the play. To save time and money, she wants to paint only the sides of each box that the audience will see. How many sides will be painted?

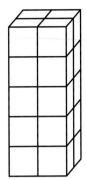

Explore What do you know?
You know that the display is made from 20 boxes that are cubes. Concheta will paint only the sides of the boxes that the audience will see.

What are you trying to find?
You are trying to find the number of sides on the boxes that Concheta will paint.

Plan Use sugar cubes to make a model of the display stand shown above. Count the sides of each box that could be visible to the audience.

Solve

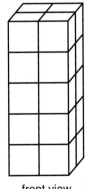

front view

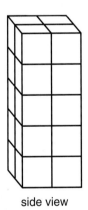

side view

The display stand will have 34 sides painted.

Examine The stack has 5 layers with 2 cubes in each layer. There are 3 sides visible to the audience plus the 4 sides on the top of the display stand.
$3 \times (5 \times 2) + 4 = 34$

Checking for Understanding

Communicating Mathematics

Read and study the lesson to answer each question.

1. **Tell** what the advantages are of using a model instead of a drawing in the example on page 381.
2. **Write** a sentence that explains how to use the make-a-model strategy.

Guided Practice

Solve by making a model.

3. A publisher packages six small books for a children's collection in a decorated 4-inch cube. They are shipped to bookstores in cartons. Twenty cubes fit in a carton. What are the dimensions of the carton?
4. During a special on repair work, eight customers lined up outside Brian's Bicycle Shop with either a bicycle or a tricycle that needed repair. When Brian looked out the window, he counted 21 wheels outside the shop. How many tricycles and bicycles are outside the shop?

Problem Solving

Practice

Solve. Use any strategy.

Strategies
●●●●●●●●●●
Look for a pattern.
Solve a simpler problem.
Act it out.
Guess and check.
Draw a diagram.
Make a chart.
Work backwards.

5. José is building a triangular-shaped display of facial tissues in the supermarket. Each box of facial tissue is in the shape of a cube. There are 10 boxes of tissue in the 10th and bottom row of the display. If there is one less box in each of the rows above, how many boxes does José use to make the display?
6. Fred, Sarah, and Greg take French, Spanish, and German. No person's language class begins with the same letter of their first name. Sarah's best friend takes French. Which language does each person take?
7. How many different-shaped rectangular prisms can be formed using exactly 16 cubes?
8. Jenna spent 3 hours addressing 50 graduation announcements. At this rate, how long will it take her to address 125 announcements?
9. A designer wants to arrange 36 glass bricks in a rectangular shape and in a single layer with the smallest perimeter possible. How many bricks are in each row?
10. Al bought $5\frac{1}{4}$-inch floppy diskettes for $1.19 each and labels for $0.59 per package. He spent $15.44. How many of each did he buy?

10-3 Surface Area of Prisms

Objective

Find the surface area of rectangular prisms.

Words to Learn

surface area
face
rectangular prism
base

Many department stores offer gift-wrapping services. If you don't have time to wrap the gift yourself, or if you're all thumbs when you try, you can pay an employee to do it for you.

Suppose your gift is in a box that is 2 feet by 1 foot by 3 feet. If you know the surface area of the box, you can find out how much paper it would take to cover the box.

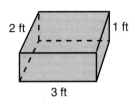

The **surface area** is the sum of the areas of all of the surfaces or **faces** of the box. Boxes such as the one we want to wrap are called **rectangular prisms** since all of the faces are in the shape of a rectangle. The faces on the top and the bottom are called the **bases.**

You can find the surface area of a prism by finding the area of each face. The sum of the six areas will be the surface area of the box.

Mini-Lab

Work with a partner.
Materials: gift box, graph paper, pencil.

Unfold or cut apart the gift box. It should resemble the frame at the right.

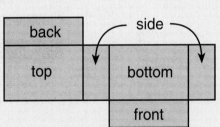

- Trace each side of the box onto your graph paper to make a figure like the one at the right.

- Label the dimensions of each rectangle on the graph paper.

Talk About It

a. What is the area of each base and the other four faces? To help you, copy and complete the following chart.

b. What is the total surface area of the gift box?

	Dimensions	Area
front		
back		
top		
bottom		
left side		
right side		
		Total:

You can see that the length and height were multiplied to find the areas of the front and back. You multiplied length and width to find the areas of the top and bottom, and you multiplied width and height to find the areas of the sides.

You can write this as a formula.

$$A = \ell h + \ell h + \ell w + \ell w + wh + wh$$

Since there are two of each face that are exactly the same, you can write the formula a shorter way.

$$A = 2\ell h + 2\ell w + 2wh$$
$$A = 2(\ell h + \ell w + wh) \qquad \textit{Distributive property}$$

Example 1 *Connection*

Algebra   Use the formula, $A = 2(\ell h + \ell w + wh)$ to find the surface area of a rectangular prism that is 3 inches by 4 inches by 5 inches.

$$A = 2(\ell h + \ell w + wh)$$
$$= 2\,(3 \times 4 + 3 \times 5 + 5 \times 4) \quad \textit{Replace } \ell \textit{ with 3, h with 4,}$$
$$= 2\,(12 + 15 + 20) \qquad \textit{and w with 5.}$$
$$= 2(47) \qquad \textit{Add first. Then multiply.}$$
$$= 94 \text{ in}^2 \qquad \textit{The area of a 2-dimensional figure uses the}$$
$$\textit{exponent 2 with the unit of measure.}$$

The surface area is 94 square inches.

You can review order of operations on page 24.

You can use a calculator to compute surface area. Remember to enter a multiplication sign in front of the parentheses.

Example 2

Find the surface area of the rectangular prism at the right by using a calculator.

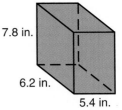

7.8 in.

6.2 in.

5.4 in.

$$A = 2(\ell h + \ell w + wh)$$
$$= 2(5.4 \times 6.2 + 5.4 \times 7.8 + 6.2 \times 7.8)$$

2 $\boxed{\times}$ $\boxed{(}$ 5.4 $\boxed{\times}$ 6.2 $\boxed{+}$ 5.4 $\boxed{\times}$ 7.8 $\boxed{+}$ 6.2 $\boxed{\times}$ 7.8 $\boxed{)}$ $\boxed{=}$ 247.92

The surface area of the rectangular prism is about 248 square inches.

Checking for Understanding

Communicating Mathematics   Read and study the lesson to answer each question.

1. **Draw** a rectangular prism that is unfolded with the dimensions of 3 inches, 4 inches, and 5 inches.

2. **Tell** which property allows you to state $\ell h + \ell h + \ell w + \ell w + wh + wh$ as $2(\ell h + \ell w + wh)$.

3. **Write** a sentence explaining why it is important to know how to find surface area when you paint the walls in a room.

Guided Practice Explain how you would find the surface area of each rectangular prism. Then find the surface area.

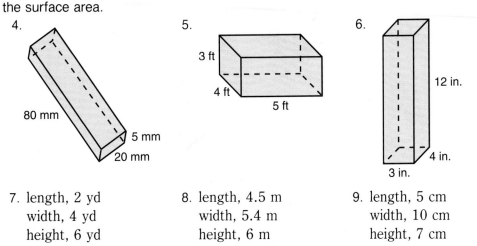

4. 80 mm 5 mm 20 mm

5. 3 ft 4 ft 5 ft

6. 12 in. 4 in. 3 in.

7. length, 2 yd
   width, 4 yd
   height, 6 yd

8. length, 4.5 m
   width, 5.4 m
   height, 6 m

9. length, 5 cm
   width, 10 cm
   height, 7 cm

Exercises

Independent Practice Find the surface area of each rectangular prism. Round answers to the nearest tenth.

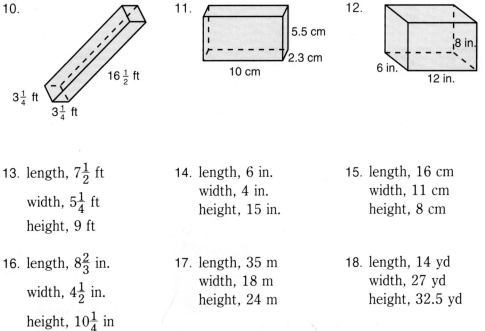

10. $16\frac{1}{2}$ ft $3\frac{1}{4}$ ft $3\frac{1}{4}$ ft

11. 5.5 cm 2.3 cm 10 cm

12. 8 in. 6 in. 12 in.

13. length, $7\frac{1}{2}$ ft

    width, $5\frac{1}{4}$ ft

    height, 9 ft

14. length, 6 in.
    width, 4 in.
    height, 15 in.

15. length, 16 cm
    width, 11 cm
    height, 8 cm

16. length, $8\frac{2}{3}$ in.

    width, $4\frac{1}{2}$ in.

    height, $10\frac{1}{4}$ in

17. length, 35 m
    width, 18 m
    height, 24 m

18. length, 14 yd
    width, 27 yd
    height, 32.5 yd

19. Each face of a cube has an area of 8 square inches. What is the surface area of the cube?

20. A cube has a surface area of 42 square feet. What is the area of one face?

Mixed Review

21. **Smart Shopping** Belinda pays $5.89 for 1.89 pounds of salami at the local delicatessen. Estimate the price per pound of the salami. *(Lesson 2-3)*

22. Solve $y = 18 + (-17)$ *(Lesson 7-4)*

23. Draw top, front, and side views of each three-dimensional figure. *(Lesson 10-1)*

a.   b.   c.

Problem Solving and Applications

24. **Manufacturing** Tissues come in boxes that are rectangular prisms.
    a. Use two sizes of tissue boxes to estimate the length, width, and height. Then estimate the surface area of each type of box.
    b. Measure the dimensions on the tissue boxes and calculate the actual surface area. Compared to the estimate, are your answers reasonable?

25. **Gifts** Matías wants to wrap a package that is 8 inches by 11 inches by 2 inches.
    a. What is the surface area of his package?
    b. Matías purchased a roll of wrapping paper that is 1 foot wide and 2 feet long. Will it be large enough to wrap his package?

26. **Geometry** A rectangular prism has a square base of $\frac{1}{2}$ yard on a side. If the prism is 2 yards high, what is the surface area?

27. **Critical Thinking** The surface area of a cube is 54 square feet. What is the length of the edge of the cube?

CULTURAL KALEIDOSCOPE

Maya Lin

In 1981, at the age of 22, Chinese-American Maya Lin won the national design competition for the Vietnam Veterans Memorial in Washington, D.C. The memorial is a symbol of United States honor and recognition of the men and women who served in the armed forces in the Vietnam War. It is inscribed with the names of the more than 58,000 persons who gave their lives for their country and those that are still missing.

Ms. Lin is the daughter of Taiwanese immigrants who make their home in Ohio. She viewed herself as just an American student until criticism of her design and heated opposition to the memorial changed her life and her attitudes. She believed that some of the criticism stemmed from prejudice toward her as a person of Chinese descent. The competition was anonymous but Maya often wonders what would have happened if names were allowed on the entries. She said, "Until that time, I hope I am an example for people who are young and who have a chance to say something when they should."

10-4A Introduction to Surface Area of a Cylinder

A Preview of Lesson 10-4

Objective

Develop the formula for surface area of a cylinder by constructing a cylinder.

Materials

centimeter
 grid paper
compass
scissors
tape

In this lab, you will be constructing a cylinder that will help you to develop the formula for surface area of a cylinder.

Try this!

- On centimeter grid paper, draw a circle whose radius is 5 centimeters.

- Cut out the circle.

- Repeat the process so that you have two circles of exactly the same size.

- Cut out a rectangle that is 12 centimeters wide and 33 centimeters long.

- Roll the rectangle into a cylinder so that each circle fits into the end. You should have enough to be able to overlap the rectangle. Tape it so that the grid side of the paper is showing.

- Tape the circles onto each end, forming a cylinder. Make sure that the grid side of the paper is showing.

Let's investigate how we might find the surface area of the cylinder we have constructed.

What do you think?

You can review area of a circle on page 359.

1. What is the shape of each end of the cylinder?
2. Find the area of each end.
3. Now find the area of the curved surface. Before you rolled it into a cylinder, what was its shape?
4. What is the total surface area? How did you find it?
5. Write a formula for the surface area of a cylinder.
6. Will a cylinder *always* be made up of the same three shapes? Explain.

10-4 Surface Area of Cylinders

Objective

Find the surface area of cylinders.

Words to Learn

cylinder

A 12-ounce serving of a soft drink contains 9 teaspoons of sugar, or about 150 calories.

If you live in New Jersey, you probably say "soda." If you live in Ohio, you probably say "pop." What are we talking about? Carbonated beverages, or soft drinks.

Carbonated beverages have become very popular in the past three decades, mostly because of modern bottling methods. Today, most carbonated beverages are "bottled" in aluminum cans.

Aluminum cans are in the shape of **cylinders.** You can find the surface area of a cylinder by finding the area of all of the surfaces.

You can see that the top and the bottom of a cylinder are in the shape of a circle. If you take the remainder of the can and unfold it, you see the curved surface is in the shape of a rectangle. By finding the area of these three surfaces, you can calculate the surface area.

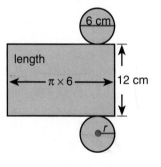

The length equals the circumference of the circle.

Example 1   *Problem Solving*

Find the surface area of a soft drink can.

The height of a soft drink can is about 12 centimeters, and the diameter is about 6 centimeters.

- First, find the area of the top and bottom. Use the formula for the area of a circle.

  $A = \pi r^2$

  Since the diameter is 6 centimeters, the radius is 3 centimeters.

  ⊡ ⊠ 3 ⊡ ⊟ **28.274334**

The area of one circle is about 28.3 square centimeters.

The area of the two circles is about 28.3×2 or 56.6 square centimeters or 56.6 cm².

Calculator Hint

● ● ● ● ● ● ● ● ● ● ● ● ●

You can use the π key when you need to find area or circumference of a circle.

- Now calculate the area of the curved surface. When unfolded, the curved surface has the shape of a rectangle. The width of the rectangle is the height of the can, and the length is the circumference of the base.

$A = \ell w$

$\quad = 2\,\pi r \times w$ *Replace ℓ with $2\pi r$, the circumference.*

$\quad = 2\pi r \times h$ *Replace w with h.*

$\quad = 2 \times \pi \times 3 \times 12$ *Replace r with 3 and h with 12.*

2 ⊠ π ⊠ 3 ⊠ 12 ▣ 226.19467

To the nearest tenth, the area of the curved surface is 226.2 square centimeters.

- Then add the area of the curved surface to the area of the two circles.

226.2 ⊕ 56.6 ▣ 282.8

The surface area of a soft drink can is about 282.8 square centimeters.

You can combine the steps to find the surface area of a cylinder.

Example 2

Find the surface area of the cylinder at the right.

$A = \begin{matrix} areas\ of \\ bases \end{matrix} + \begin{matrix} area\ of \\ curved\ surface \end{matrix}$

$A = (2\pi r^2) + (2\pi rh)$

$A = (2 \cdot \pi \cdot 4^2) + (2 \cdot \pi \cdot 4 \cdot 7)$ *Replace r with 4 and h with 7.*

(2 ⊠ π ⊠ 4 x^2) ⊕ (2 ⊠ π ⊠ 4 ⊠ 7)

▣ 276.46015

To the nearest tenth, the surface area is 276.5 square feet.

Checking for Understanding

Communicating Mathematics

Read and study the lesson to answer each question.

1. **Tell** which two measurements must be known to find the surface area of a cylinder.

2. **Draw a model** of an unfolded cylinder.

3. **Write** one sentence explaining why we use the formula for the area of a rectangle in finding the surface area of a cylinder.

4. a. Find the area of each base of the cylinder.

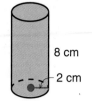

8 cm

2 cm

   b. Find the area of the curved surface in the cylinder. Record your answer.

   c. What is the surface area of the cylinder above?

Find the surface area of each cylinder. Use 3.14 for π.

5. height, 2.5 m
   radius, 3.4 m

6. height, 4 ft
   radius, 5 ft

7. height, 5 cm
   diameter, 20 cm

Exercises

Find the surface area of each cylinder. Use 3.14 for π. Round answers to the nearest tenth.

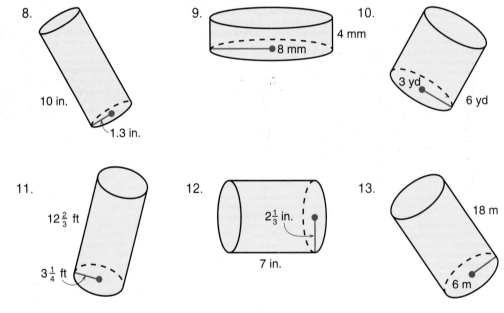

8. 10 in. 1.3 in.

9. 4 mm 8 mm

10. 3 yd 6 yd

11. $12\frac{2}{3}$ ft $3\frac{1}{4}$ ft

12. $2\frac{1}{3}$ in. 7 in.

13. 18 m 6 m

Find the surface area of each cylinder. Use $\frac{22}{7}$ for π.

14. height, 5 in.
    radius, 11 in.

15. height, 4.2 cm
    diameter, 12.6 cm

16. height, 8 mm
    radius, 9 mm

17. height, 7 m
    radius, 16 m

18. height, 10 ft
    diameter, 24 ft

19. height, $12\frac{1}{2}$ yd
    radius, $8\frac{3}{4}$ yd

20. Find the surface area of a cylinder whose height is 14 inches and whose base has a diameter of 16 inches.

21. Find the surface area of a cylinder whose height is 12 inches and whose base has a circumference of 37.68 inches. Use 3.14 for π.

Mixed Review

22. **Statistics** Construct a line plot for the following set of data: 65, 72, 83, 81, 65, 71, 72, 70, 81. *(Lesson 3-4)*

23. **Geometry** Describe the quadrilateral at the right. *(Lesson 8-3)*

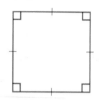

24. Nicole has a cushion of an old chair that she would like to cover with a new fabric. The cushion is in the shape of a rectangular prism with a length of 1.5 feet, a width of 1.5 feet, and a height of 0.25 feet. How much fabric will she need to cover the cushion? *(Lesson 10-3)*

Problem Solving and Applications

25. **Critical Thinking** If you double the height of a cylinder, will its surface area double? Explain.

26. **Marketing** A can of vegetables is 5 inches high, and its base has a radius of 2 inches. How much paper is needed to make the label on the can?

27. **Agriculture** A cylindrical gasoline storage tank on Mr. Baker's farm needs to be painted. The tank is 8 feet long and has a diameter of 4 feet. If one gallon of paint covers 350 square feet, how many cans of paint will Mr. Baker need?

Mid-Chapter Review

1. **Draw** three different views of a cylinder and a rectangular prism. *(Lesson 10-1)*

2. During the Pee-Wee race, Jonah saw 4 cyclists cross the finish line. Jessie counted 11 wheels go by. How many bicycles and tricycles did they see cross the finish line? *(Lesson 10-2)*

Find the surface area of each rectangular prism. *(Lesson 10-3)*

3. length, 6 cm
   width, 4 cm
   height, 2 cm

4.

8 m, 4 m, 3 m

Find the surface area of each cylinder. *(Lesson 10-4)*

5. height, 3 ft
   radius, 2.2 ft

6.

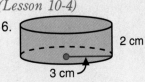

2 cm, 3 cm

DECISION MAKING

Choosing a Scholarship Prize

Situation

A bowling association is holding a two-day single elimination scholarship tournament for middle school students. The winners receive scholarships to a college of their choice. As secretary of the bowling association, you must decide which of the scholarship funds shown below to select for the tournament.

Hidden Data

How much, if anything, will each participant have to pay for an entry fee?
How much of the entry fee will go towards the scholarship fund?
Will the scholarship be awarded at the end of the tournament or upon registering for college?
How many students will participate?

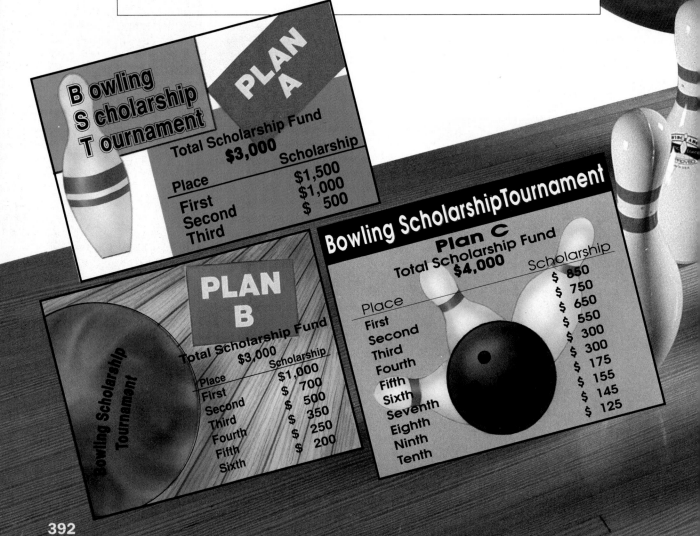

Bowling Scholarship Tournament

PLAN A

Total Scholarship Fund **$3,000**

Place	Scholarship
First	$1,500
Second	$1,000
Third	$ 500

PLAN B

Total Scholarship Fund $3,000

Place	Scholarship
First	$1,000
Second	$ 700
Third	$ 500
Fourth	$ 350
Fifth	$ 250
Sixth	$ 200

Bowling Scholarship Tournament

Plan C

Total Scholarship Fund **$4,000**

Place	Scholarship
First	$ 850
Second	$ 750
Third	$ 650
Fourth	$ 550
Fifth	$ 300
Sixth	$ 300
Seventh	$ 175
Eighth	$ 155
Ninth	$ 145
Tenth	$ 125

Analyzing the Data

1. How much more could a student in first place win under Plan A than under Plan C?
2. If a student is in fourth place under plan C, how much more would the student win than in Plan B?
3. What is the difference in scholarship funds among Plans A, B, and C?
4. How many times more scholarship money could you win if you won first place rather than third place under Plan A?

Making a Decision

5. **Under which plan** do you think you would have more success at attracting a larger amount of students to participate? Explain.
6. **What would be** the advantage of choosing a plan after all the participants have entered the tournament?
7. **Do you think** the competition will be more enjoyable if there were more scholarships of lesser values or fewer scholarships with greater values?
8. **If each** participant had to pay a $50 entry fee, which plan would you choose? Explain.
9. **Which plan** did you choose? Why?

Making Decisions in the Real World

10. **Research** Contact a local bowling alley and inquire about the tournaments available to students.

10-5 Volume of Prisms

Words to Learn

volume

When water turns into ice, it expands. When it expands inside a closed container, it can cause so much pressure to build up that the container cracks or explodes. This is what causes water pipes and car radiators to crack in the winter. In this lesson, you will learn how to find the amount of space inside a prism.

Volume is the measure of the space occupied by a solid figure. It is measured in cubic units. You can use cubes to make models of solid figures.

The container at the right has a length of 6 inches, a width of 2 inches, and a height of 4 inches. The model is made of 4 layers. Each layer has 12 cubes. The area of the base is 12 square inches, the product of the length and width.

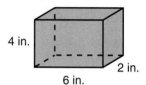

Since the container is 4 layers high and has a base of 12 one-inch cubes, it will take 4 · 12 or 48 one-inch cubes to fill the container. The volume of the container is 48 cubic inches.

Volume of a Rectangular Prism	**In words:** The volume (v) of a rectangular prism is found by multiplying the length (ℓ), the width (w), and the height (h).
	In symbols: $V = \ell wh$

Example 1

Draw and label a rectangular prism whose length is 5 centimeters, width is 2 centimeters, and height is 8 centimeters. Find its volume.

$V = \ell wh$

$\quad = 5 \cdot 2 \cdot 8$ *Replace ℓ with 5, w*

$\quad = 80$ *with 2, and h with 8.*

The prism has a volume of 80 cm³.
The volume of a 3-dimensional figure uses the exponent 3 with the unit of measure.

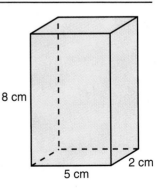

Example 2

Find the volume of the rectangular prism at the right.

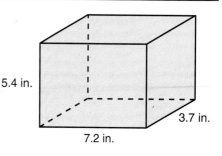

5.4 in.

3.7 in.

7.2 in.

$V = \ell w h$

$= 7.2 \cdot 3.7 \cdot 5.4$

7.2 ⊠ 3.7 ⊠ 5.4 ⊟ 𝟣𝟦𝟥.𝟪𝟧𝟨

The prism has a volume of 146.52 cubic inches.

Mini-Lab

Work with a partner.
Materials: 20 × 20 grid paper, scissors, tape

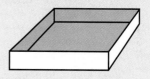

- Cut off square corner sections from each corner of the 20 × 20 grid paper to make an open box that is 14 × 14 × 3.

- Fold the paper to make the box. Then tape the corners together.

- Find the volume of the box.

- Continue making boxes by cutting off square corners from 20 × 20 grids until you have found a box that has the greatest volume. How do you know when you have found the box with the greatest volume?

- Once you have found a box with the greatest volume, convince another group that yours has the greatest volume.

Checking for Understanding

Communicating Mathematics

Read and study the lesson to answer each question.

1. **Tell** the difference between volume and surface area.

2. **Write** an example to show that you can find the volume of a rectangular prism by multiplying the height, width, and length.

Find the volume of each rectangular prism.

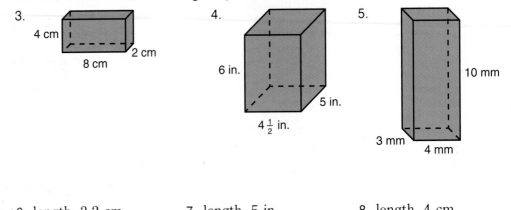

3. 4 cm, 8 cm, 2 cm

4. 6 in., 5 in., $4\frac{1}{2}$ in.

5. 10 mm, 3 mm, 4 mm

6. length, 2.2 cm
   width, 4.4 cm
   height, 5.5 cm

7. length, 5 in.
   width, 3 in.
   height, 10 in.

8. length, 4 cm
   width, 6 cm
   height, 1 cm

Exercises

Find the volume of each rectangular prism.

9. $7\frac{1}{2}$ in., 6 in., $5\frac{1}{4}$ in.

10. 7 cm, 14 cm, 12 cm

11. 3.5 ft, 7.2 ft, 9 ft

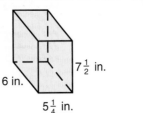

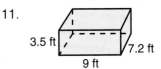

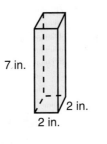

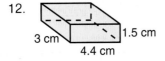

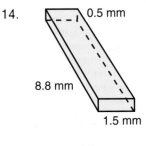

12. 3 cm, 1.5 cm, 4.4 cm

13. 7 in., 2 in., 2 in.

14. 0.5 mm, 8.8 mm, 1.5 mm

15. length, 5 mm
    width, 7 mm
    height, 10 mm

16. length, 12 in.
    width, 9 in.
    height, 7 in.

17. length, 12.1 cm
    width, 8.2 cm
    height, 10.6 cm

18. Find a volume of a rectangular prism whose length is 14 centimeters, width is 7 centimeters, and height is 12 centimeters.

19. A cube has sides that are 7 inches long.
    a. What is the volume of the cube?
    b. Write a formula for finding the volume of a cube.

20. Use divisibility rules to determine whether 4,500 is divisible by 2, 3, 4, 5, 6, 9, or 10. *(Lesson 4-1)*

21. Find the best whole number estimate for $\sqrt{167}$. *(Lesson 9-3)*

22. **Construction** The Blue Mountain Oil Company stores the heating oil it produces in tanks that are in the shape of cylinders with heights of 10 meters, and radii of 2.5 meters. How much steel is needed to construct each tank? *(Lesson 10-4)*

23. **Algebra** Write a formula for the volume of a cube that has sides x units long.

24. **Exercise** A swimming pool is 75 feet long, 45 feet wide, and 8 feet deep. It is filled to a depth of 5 feet.
    a. How much water is in the pool?
    b. Water weighs about 62 pounds per cubic foot. What is the weight of the water in the swimming pool?
    c. In the winter, the water in the pool freezes and the volume expands to 20,250 cubic feet without the pool breaking. How much does the volume increase?

25. **Critical Thinking** Refer to Exercise 24c.
    a. What would be the dimensions of the frozen water?
    b. Which dimension changed? Explain.

26. **Data Search** Refer to pages 374 and 375.
    Find the height of the tallest volcano in the chart and the height of the shortest volcano in the chart. What is the difference in their heights?

27. **Mathematics and History** Read the following paragraph.

Ever since he was a boy, Frederick McKinley Jones (1892–1961) enjoyed taking machines apart and putting them together again. In 1935, he began working on an invention that made him famous. After listening to farmers complain about losing truckloads of crops because they spoiled during shipping, Mr. Jones began putting odds and ends of machinery together. When he finished building his machine, he attached it to a truck and created the first mechanically-refrigerated truck. Food was shipped longer distances across the country without spoiling. Soon his mechanical refrigerating system was placed in ships and railway cars.

Find the volume of a refrigerated compartment 12 meters by 5 meters by 8 meters.

10-6 Volume of Cylinders

Objective

Find the volume of cylinders.

Ms. Eng makes candles and sells them at arts and crafts festivals. She is planning to make one that is 8 inches high and 4 inches in diameter. The candle wax comes in 72 cubic-inch blocks. Will one block of wax be enough to make the candle?

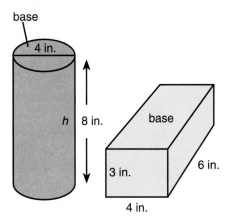

In Lesson 10-5, you found the volume of a rectangular prism by using the formula $V = \ell w h$. Remember that ℓw square units is equal to the area of the base.

To find the volume of a cylinder, such as the candle, multiply the area of the base by the height. The area of the base of a cylinder is the area of a circle, πr^2. So, the volume of a cylinder is $(\pi r^2)h$ or $\pi r^2 h$.

$V = \pi r^2 h$
$\quad = \pi \cdot 2^2 \cdot 8$ *Since the diameter is 4 inches,*
$\quad = \pi \cdot 4 \cdot 8$ *the radius is 2 inches.*

$\boxed{\pi}\ \boxed{\times}\ 4\ \boxed{\times}\ 8\ \boxed{=}\ 100.53096$ *Round to the nearest whole number.*

$V \approx 101$

The candle has a volume of *about* 101 cubic inches. Since the block of wax is only 72 cubic inches, one block will not be enough.

Volume of a Cylinder	**In words:** The volume of a cylinder is found by multiplying the area of the base (πr^2) times the height (h).
	In symbols: $V = \pi r^2 h$

Example 1

Find the volume of a cylinder with a radius of 3 inches and a height of 5 inches.

$V = \pi r^2 h$
$\quad \approx 3.14 \cdot 3^2 \cdot 5$ *Replace r with 3 and h with 5. Use 3.14 for π.*
$\quad \approx 3.14 \cdot 9 \cdot 5$
$\quad \approx 141.3$ The cylinder has a volume of *about* 141.3 cubic inches.

Estimation Hint

• • • • • • • • • • • • •

You can estimate the volume of a cylinder by squaring the radius and multiplying by 3 ($\approx \pi$) and the height. In Example 1, the volume of the cylinder is about $3 \times 3 \times 3 \times 5$ or 135 cubic inches.

Example 2 *Problem Solving*

Manufacturing   Suppose you are designing a glass that has a radius of 3.5 centimeters and a height of 15 centimeters. If one cubic centimeter can contain one milliliter of liquid, how many milliliters of liquid will it hold?

Find the volume of the glass.

$V = \pi r^2 h$

$= \pi \cdot 3.5^2 \cdot 15$ *Replace r with 3.5 and h with 15.*

$\boxed{\pi}\ \boxed{\times}\ 3.5\ \boxed{x^2}\ \boxed{\times}\ 15\ \boxed{=}\ \textbf{577.26765}$

The glass has a volume of *about* 577.3 cubic centimeters. So it will hold about 577.3 milliliters of liquid.

Checking for Understanding

Communicating Mathematics

Read and study the lesson to answer each question.

1. **Tell** how the formula for the volume of a cylinder is similar to the formula for the volume of a prism.

2. **Write** a sentence explaining why you express the volume of a cylinder in cubic units.

3. **Tell** why the volume of a cylinder is an approximation.

4. **Tell** how many blocks of wax are needed to make 4 of the candles described on the previous page.

Guided Practice

Find the volume of each cylinder. Round answers to the nearest tenth.

5. 4 in. 2 in.

6. 2 in. $10\frac{1}{3}$ in.

7. 3 m 6 m

8. radius, 3 cm
   height, 6.5 cm

9. radius, 5 in.
   height, 3 in.

10. radius, 2 mm
    height, 3 mm

Exercises

Independent Practice

Find the volume of each cylinder. Round answers to the nearest tenth.

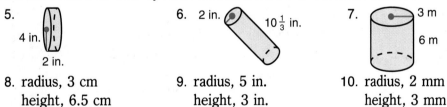

11. 2.2 cm 5 cm

12. 6 in. 14 in.

13. 8 cm 2 cm

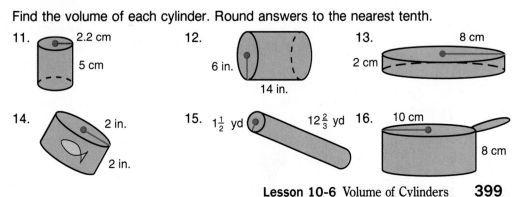

14. 2 in. 2 in.

15. $1\frac{1}{2}$ yd $12\frac{2}{3}$ yd

16. 10 cm 8 cm

Find the volume of each cylinder. Round answers to the nearest tenth.

17. radius, 3 in.
    height, 10 in.

18. radius, 6 mm
    height, 6.8 mm

19. radius, $2\frac{1}{2}$ ft
    height, $8\frac{1}{4}$ ft

20. A can of potato chips is 8 inches high and has a radius of 1.5 inches. Find the volume of the can.

Mixed Review

21. Typing paper measures $8\frac{1}{2}$ inches wide and 11 inches long. Find the perimeter of typing paper. *(Lesson 5-6)*

22. Solve $z = 360 \div (-6)$. *(Lesson 7-8)*

23. Find the volume of a rectangular prism having a length of 6 centimeters, a width of 4.9 centimeters, and a height of 5.2 centimeters. *(Lesson 10-5)*

Problem Solving and Applications

24. **Measurement** If a 12-ounce soft drink can is *about* $4\frac{3}{4}$ inches tall and has a diameter of *about* $2\frac{1}{2}$ inches, to the nearest whole number, how many cubic inches of a soft drink fills a can?

25. **Measurement** If a quart is equal to 57.75 cubic inches, how many 12-ounce cans of soft drinks would you have to drink to drink a quart? Refer to Exercise 24.

Use the graph to answer Exercises 26–28.

26. In which country does each person consume the most soft drinks? the least?

27. How many more 8-ounce soft drinks does a person in Belgium consume than a person in France?

28. If the numbers on the graph represent 8-ounce soft drinks, how many gallons of soft drinks does each American drink in a year? *(1 gallon = 128 fluid ounces)*

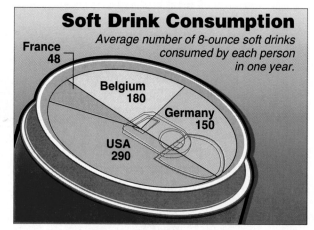

Soft Drink Consumption
Average number of 8-ounce soft drinks consumed by each person in one year.

France 48
Belgium 180
Germany 150
USA 290

29. **Critical Thinking** If you double the height of a cylinder, how does that affect its volume? Explain.

30. **Journal Entry** Draw and label a cylinder to help you write an explanation of how to find the volume of a cylinder.

10-6B Volume of Cylinders

A Follow-Up of Lesson 10-6

Objective

Estimate and compare volumes of cylinders of various sizes.

Materials

cylinder-shaped objects
rulers
paper
pencil
calculator

Cylinder-shaped objects come in many sizes and shapes. Companies often do research to determine which shape is best for their product.

Try this!

Work in small groups.

- Place all of the cylinders so that everyone can see them. Number each item so that it can be identified easily.
- Rank the cylinders according to their volume from smallest to largest using estimation.
- Display your rankings and compare with other groups in the class.
- Measure the height and the radius of the base of each cylinder to the nearest tenth of a centimeter.
- Calculate the volume of each cylinder.
- Rank the cylinders according to their volume from smallest to largest based on your calculations.
- Display your rankings. Compare your estimated rankings with your calculated rankings.

What do you think?

1. How close were your estimates to your actual calculations?
2. Was it more difficult to estimate when the cylinders had different relative shapes (such as short and fat compared to tall and thin)?
3. Which shape of the cylinders was the most misleading as far as what you expected the volume to be?

Extension

4. Without measuring, determine which is greater, the circumference of a tennis ball or the height of the tennis ball can. Explain how you arrived at your answer.

10-7 Use a Formula

Objective

Solve problems by using a formula.

The Eppersteins decided to landscape a rectangular part of their front lawn with rocks. First they covered the 40-foot by 12-foot section with heavy plastic to prevent weeds from growing through the cracks. Then they spread rocks on top of the plastic. If they ordered 120 cubic feet of rocks, how deep will the rocks be?

Explore   What do you know?
You know the space to be covered with rocks is in the shape of a rectangle. It is 40 feet long and 12 feet wide. The volume is 120 ft³.

What do you need to find?
You need to find the depth of the rectangular space.

Plan   Use a formula for the volume of a rectangular prism, $V = \ell wh$. Substitute the given measure into the formula. Then divide to find the value of h.

Solve   To find the depth, use the formula $V = \ell wh$.

$$V = \ell wh$$
$$120 = 40 \times 12 \times h$$
$$120 = 480h$$
$$\frac{120}{480} = \frac{480h}{480}$$
$$\frac{1}{4} = h$$

Examine   The rocks will be $\frac{1}{4}$ foot or 3 inches deep. Check your answer by replacing the variables in the formula, $V = \ell wh$.

$$V = \ell wh$$
$$120 = 40 \times 12 \times \frac{1}{4}$$
$$120 = 120 \checkmark$$

Checking for Understanding

Communicating Mathematics   Read and study the lesson to answer each question.

1. **Draw** a labeled diagram of the rectangular prism in the example.

2. Suppose the Eppersteins wanted the rocks to be 4 inches in depth. How many cubic feet of rocks would they need to order?

Solve by using a formula.

3. Chris is covering all sides of a cylinder-shaped can with contact paper. The radius of the can is 4 centimeters, and its height is 10 centimeters. How much contact paper does he need?

4. Donna Mendez's new office is 20 feet long, 15 feet wide, and 12 feet high. If it costs 9 cents per year to air condition one cubic foot of space, how much does it cost to air condition her office for one month?

Problem Solving

Practice

Solve. Use any strategy.

5. A number is doubled and then 10 is added. The result is −8. What is the number?

6. What are the next two numbers in this sequence: 5, 9, 7, 11, 9, . . . ?

7. A gardener is digging a rectangular space for a new flower bed that is 6 feet by 6 feet by $1\frac{1}{2}$ feet. How many cubic yards of topsoil does he need to fill this space?

8. Kayla spent 3 hours calling 25 potential magazine subscribers. At this rate, how many hours will it take her to call 100 people?

9. Ralph has 5,000 cm² of wrapping paper. Does he have enough to wrap a present in a box that measures 30 cm long by 40 cm wide by 20 cm high?

Strategies
●●●●●●●●●●
Look for a pattern.
Solve a simpler problem.
Act it out.
Guess and check.
Draw a diagram.
Make a chart.
Work backwards.

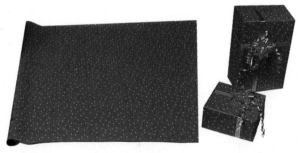

COMPUTER

CONNECTION

10. **Computer Connection**   Suppose a gardener has 80 feet of fencing with which to enclose a vegetable garden. What is the greatest area she can enclose?

The BASIC computer program at the right will test all possible integer values for the length and width, and compute each area.

```
10 INPUT P
20 FOR W = 1 TO P/2
30 L = (P - 2 * W)/2
40 A = W * L
50 PRINT W, L, A
60 NEXT W
```

a. Run the program.
b. What is the greatest area?
c. What is the shape of the garden?
d. Run the program again for a perimeter of 120 feet.
e. What is the greatest area?
f. What is the shape of the garden?

10 Study Guide and Review

Communicating Mathematics

Choose the correct term or expression to complete each sentence.

1. Another name for a box is a(n) ___?___.

2. The formula for the surface area of a rectangular prism is ___?___.

3. The surface area of a(n) ___?___ can be found by finding the area of both a circle and a rectangle.

4. ___?___ is the measure of the space occupied by a solid figure.

5. The formula for the volume of a cylinder is ___?___.

6. In your own words, explain the relationship between the area of a square and the surface area of a cube.

> area
> cube
> $\pi r^2 h$
> rectangular prism
> $\ell w h$
> $2(\ell h + \ell w + wh)$
> cylinder
> volume
> $2\pi r$
> square
> πr^2

Skills and Concepts

Objectives and Examples

Upon completing this chapter, you should be able to:

- draw three-dimensional figures
  (Lesson 10-1)

  Draw a three-dimensional figure by using the top, front, and side views.

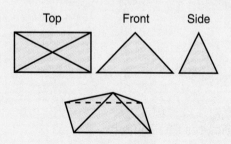

Review Exercises

Use these exercises to review and prepare for the chapter test.

Draw each three-dimensional figure by using the top, front, and side views.

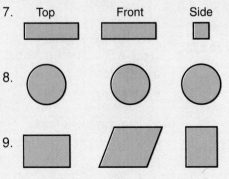

- find the surface area of rectangular prisms *(Lesson 10-3)*

Find the surface area of a prism having a length of 3 centimeters, a width of 8 centimeters, and a height of 2 centimeters.

$A = 2(\ell h + \ell w + wh)$
$\;\;\; = 2(3 \times 2 + 3 \times 8 + 8 \times 2)$
$\;\;\; = 2(46)$
$\;\;\; = 92$

The surface area is 92 cm².

Find the surface area of each prism. Round answers to the nearest tenth.

10. length, $4\frac{1}{3}$ in.
    width, $2\frac{1}{4}$ in.
    height, 6 in.

11. length, 2.6 yd
    width, 2.6 yd
    height, 1 yd

12.

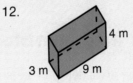

3 m   9 m   4 m

- find the surface area of cylinders *(Lesson 10-4)*

Find the surface area of a cylinder with a height of 6 millimeters and a radius of 2 millimeters.

$A = 2\pi r^2 + 2\pi rh$
$\;\;\; \approx 2 \times 3.14 \times 4 + 2 \times 3.14 \times 2 \times 6$
$\;\;\; \approx 100.48$ square millimeters

The surface area is about 100.5 mm².

Find the surface area of each cylinder. Use 3.14 for π. Round answers to the nearest tenth.

13. height, $5\frac{3}{8}$ in.
    radius, $3\frac{1}{2}$ in.

14. height, 6 ft
    radius, 0.5 ft

15.

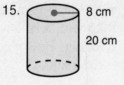

8 cm
20 cm

- find the volume of rectangular prisms *(Lesson 10-5)*

Find the volume of a rectangular prism having a length of 5 inches, a width of 3 inches, and a height of 2 inches.

$V = \ell wh$
$\;\;\; = 5 \times 3 \times 2$
$\;\;\; = 30$

The volume is 30 in³.

Find the volume of each rectangular prism.

16. length, 6.3 mm
    width, 2.5 mm
    height, 1.2 mm

17. length, $4\frac{1}{2}$ ft
    width, $6\frac{1}{4}$ ft
    height, 6 ft

18.

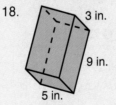

3 in.
9 in.
5 in.

Objectives and Examples

Review Exercises

- find the volume of cylinders
  (Lesson 10-6)

Find the volume of a cylinder with a radius of 4 centimeters and a height of 8 centimeters.

$$V = \pi r^2 h$$
$$\approx 3.14 \times 4^2 \times 8$$
$$\approx 401.92$$

The volume is about 402 cm³.

Find the volume of each cylinder. Round answers to the nearest tenth.

19. radius, 1 ft
    height, 3 ft
20. radius, $3\frac{2}{3}$ yd
    height, $6\frac{4}{5}$ yd
21.

1.7 in.

5.2 in.

Applications and Problem Solving

22. **Design**   A large cube is made up of 27 small cubes. The outside of the cube is painted blue. If the large cube is taken apart, how many small cubes would have none of their sides painted? *(Lesson 10-2)*

23. **Manufacturing**   A cereal box has a length of 11 inches, a height of 14 inches, and a depth of 1.5 inches. *(Lesson 10-7)*
    a. What is the volume of the box?
    b. If each cubic inch holds about 0.08 ounce of cereal, about how many ounces of cereal can one box hold?

24. **Pet Supplies**   The Pets Are Us Company wants to make a fish tank that is open on the top and has a length of 2.5 feet, a height of 1 foot, and a width of 1.25 feet. How much glass is needed to make this tank? *(Lesson 10-3)*

25. **Pottery**   In her art class, Andrea made a vase in the shape of a cylinder. The diameter is 5 inches, and the height is 10 inches. Find the maximum volume of water the vase can hold. *(Lesson 10-6)*

Curriculum Connection Projects

- **Automotive**   Find the approximate surface area and volume of your family's car or a friend's car.
- **Art**   Construct a three-dimensional drawing of your home from at least two different views.

Read More About It

Rinaldi, Ann. *The Last Silk Dress.*
Johnson, Neil. *Fire and Silk: Flying in a Hot Air Balloon.*
McCauley, David. *Pyramid.*

Test

Draw each three-dimensional figure by using the top, front, and side views.

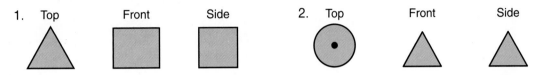

1. Top   Front   Side   2. Top   Front   Side

Find the surface area of each rectangular prism. Round answers to the nearest tenth.

3. length, $2\frac{2}{3}$ ft

    width, $1\frac{3}{4}$ ft

    height, $4\frac{1}{2}$ ft

4. length, 3.6 cm
    width, 2.1 cm
    height, 8 cm

5.
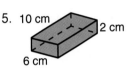
10 cm, 2 cm, 6 cm

Find the surface area of each cylinder. Round answers to the nearest tenth.

6. height, 3.7 yd
    radius, 0.4 yd

7. height, $\frac{1}{3}$ yd
    diameter, 14 yd

8.   15 in.

    $2\frac{1}{2}$ in.

9. The Andersons' backyard pool is in need of a paint job. The pool is 30 feet long, 18 feet wide, and 6 feet deep. One gallon of pool paint will cover 12 square feet. How many gallons of paint are needed? *Remember, a pool does not have a top.*

Find the volume of each rectangular prism.

10. length, 3 mm
    width, 2 mm
    height, 1 mm

11. length, 10.4 ft
    width, 2.5 ft
    height, 3 ft

Find the volume of each cylinder. Round answers to the nearest tenth.

12. radius, 6.3 cm
    height, 3.1 cm

13. radius, 2 yd
    height, $1\frac{1}{2}$ yd

14. A rectangular prism is formed using exactly 20 cubes. How many different prisms can be formed?

15. The standard-size drinking straw has a radius of $\frac{1}{8}$ inch and a height of $7\frac{3}{4}$ inches. What is the maximum volume of liquid that can be contained in the straw at any given time?

Bonus   If one view of a figure is △ and another view of the same figure is ▢, what is the third view of the same figure?

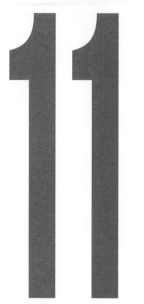

Ratio, Proportion, and Percent

Spotlight on Health and Safety

Have You Ever Wondered. . .

- If fewer people are drinking and driving now than in the past?

- What percent of high school students smoke?

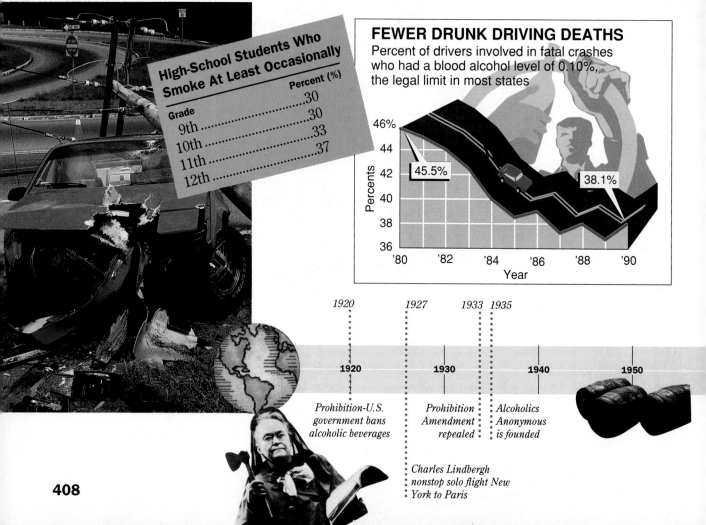

High-School Students Who Smoke At Least Occasionally

Grade	Percent (%)
9th	30
10th	30
11th	33
12th	37

FEWER DRUNK DRIVING DEATHS
Percent of drivers involved in fatal crashes who had a blood alcohol level of 0.10%, the legal limit in most states

45.5%

38.1%

Percents / Year

1920 — Prohibition-U.S. government bans alcoholic beverages

1927 — Charles Lindbergh nonstop solo flight New York to Paris

1933 — Prohibition Amendment repealed

1935 — Alcoholics Anonymous is founded

Chapter Project

Health and Safety

Work in a group.

1. Conduct a poll about smoking. Interview at least 10 people who used to smoke cigarettes but have quit.

2. Find out why these people started smoking, at what age, and how long they smoked. Also ask them why and how they quit.

3. Make an oral presentation of your findings.

Looking Ahead

In this chapter, you will see how mathematics can be used to answer the questions about health and safety.

The major objectives of the chapter are to:

- express ratios as fractions

- solve proportions

- solve problems involving scale drawings and diagrams

- express percents as fractions and decimals

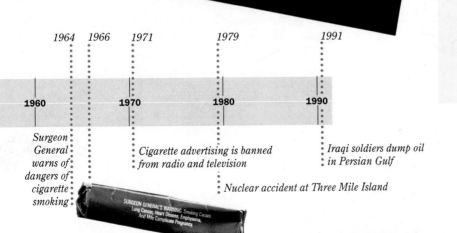

1964	1966	1971	1979	1991

1960   **1970**   **1980**   **1990**

Surgeon General warns of dangers of cigarette smoking

Cigarette advertising is banned from radio and television

Iraqi soldiers dump oil in Persian Gulf

Nuclear accident at Three Mile Island

SURGEON GENERAL'S WARNING: Smoking Causes Lung Cancer, Heart Disease, Emphysema, And May Complicate Pregnancy.

409

11-1A Equal Ratios

A Preview of Lesson 11-1

Objective

Explore the meaning of ratio and proportion.

Words to Learn

ratio

Materials

dried beans or squares of paper

A **ratio** is the comparison of two numbers. Often ratios are used to show how large one quantity is compared to another. For example, for every two squares in pile X there is one square in pile Y. We say that piles X and Y have the ratio 2 to 1.

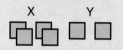

Try this!

Work in groups of two.

● Use dried beans or small squares of paper to make the arrangements shown below.

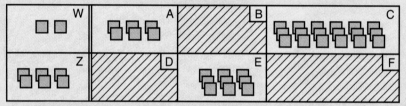

● Work together to place piles of squares in spaces B, D, and F. The ratio of the piles in each column should be equal to the ratio of piles W and Z.

What do you think?

1. How did you decide how many squares to put in space D? B? F?

2. If space D has 99 squares, how many squares will be in space A?

3. Copy the table shown below. Place squares in spaces B, C, D, F, G, and H so that the ratios in each column are equal to the ratio of P to Q to R.

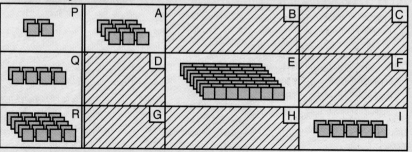

4. Did you use multiplication or division to decide on your answers?

11-1 Ratios

Objective

Express ratios as fractions and determine whether two ratios are equivalent.

Words to Learn

ratio

Where did people move in 1991? The ratio of people moving in to people moving out is greatest in Spokane, Washington.

During the first six months of 1991, for every 100 families who moved out of Nashville, 137 families moved in. The **ratio** of the number of families who moved out of Nashville to the number of families who moved in was 100 to 137.

Where People Moved in '91
1. Spokane, WA
2. Springfield, MO
3. Greenville-Washington, NC
4. Nashville, TN
5. Las Vegas, NV
6. San Antonio, TX
7. Fresno-Visalia, CA
8. Richmond, VA
9. Albuquerque, NM
10. Raleigh-Durham, NC

Ratio	**In words:**	A ratio is a comparison of two numbers by division.		
		Arithmetic		**Algebra**
	100 to 137	100:137	$\frac{100}{137}$	a to b $\quad a{:}b$ $\quad \frac{a}{b}$

Since a ratio can be written as a fraction, ratios are often written in simplest form.

Example 1 *Problem Solving*

Real Estate Mr. McLevy, a real estate agent, knew that last year 480 families had moved out of Broome County and 560 families had moved in. Write the ratio of the number of families moving out to the number of families moving in.

Write the ratio in simplest form.

$$\begin{array}{l} \text{families moving out} \rightarrow \\ \text{families moving in} \rightarrow \end{array} \frac{480}{560} = \frac{480 \div 80}{560 \div 80} \quad \begin{array}{l}\textit{The GCF of 480}\\ \textit{and 560 is 80.}\end{array}$$

$$= \frac{6}{7}$$

The ratio in simplest form is $\frac{6}{7}$, or 6 to 7, or 6:7.

Ratios can also be expressed as decimals. The ratio in Example 1 can be expressed as a decimal in the following way.

$$480 \;\boxed{\div}\; 560 \;\boxed{=}\; \mathsf{0.857143}$$

To simplify a ratio that compares measurements, you must be sure that the measurements have the same unit of measure.

Example 2

Write the ratio 4 inches:2 feet in simplest form.

$$\frac{4 \text{ inches}}{2 \text{ feet}} = \frac{\overset{1}{\cancel{4}} \text{ inches}}{\underset{6}{\cancel{24}} \text{ inches}} \qquad 2 \text{ feet} = 24 \text{ inches}$$

$$= \frac{1}{6} \qquad\qquad \text{The GCF of 4 and 24 is 4.}$$

The ratio in simplest form is $\frac{1}{6}$, or 1:6.

Two ratios are equivalent if they have the same value.

Example 3

Are 12:16 and 21:28 equivalent ratios?

Express each ratio as a fraction in simplest form.

$$\frac{12}{16} = \frac{12 \div 4}{16 \div 4} \qquad\qquad \frac{21}{28} = \frac{21 \div 7}{28 \div 7}$$

$$= \frac{3}{4} \qquad\qquad\qquad = \frac{3}{4}$$

Since the ratios in simplest form are equal, 12:16 and 21:28 are equivalent ratios.

Calculator Hint

●●●●●●●●●●●●●

Another way to find out whether 12:16 and 21:28 are equivalent is to express the ratios as decimals and then compare.

12 ÷ 16 = 0.75

21 ÷ 28 = 0.75

Since the decimals are the same, the ratios are equivalent.

Checking for Understanding

Communicating Mathematics

Read and study the lesson to answer each question.

1. **Draw** the chart at the right and draw squares in space D such that A : B and C : D are equivalent.

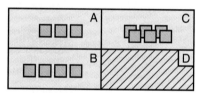

2. **Write** the ratio *19 students out of 24 students* in three different ways.
3. **Tell** how to simplify a ratio written as a fraction.

Guided Practice

Express each ratio as a fraction in simplest form.

4. $\frac{9}{12}$ 5. 6 to 12   6. 24:4

7. 45 minutes out of 1 hour   8. 2 pounds:24 ounces

Tell whether the ratios in each pair are equivalent. Show your answer by simplifying.

9. $\frac{65}{100}$ and $\frac{5}{8}$ 10. $\frac{4}{9}$ and $\frac{36}{81}$ 11. $\frac{12}{9}$ and $\frac{15}{12}$

12. **Family** Alma is 12 years old. Her brother Javier is 4 years old and their mother is 36 years old. Is the ratio of Javier's age to Alma's age equivalent to the ratio of Alma's age to their mother's age? Show your answer by simplifying ratios.

Exercises

Express each ratio as a fraction in simplest form.

13. 27 to 15

14. 21:45

15. 49:14

16. 125 to 25

17. 11 weeks out of 33 weeks

18. 64 inches to 18 inches

19. 2 feet to 6 yards

20. 5 pounds to 10 ounces

21. 36 to 27

22. 21 minutes:66 minutes

23. 48 hours:21 hours

24. 625 to 25

Tell whether the ratios in each pair are equivalent. Show your answer by simplifying.

25. 2 pounds:24 ounces and 6 pounds:72 ounces

26. 13 to 39 and 26 to 78

27. 4 hours to 3 days and 12 hours to 9 days

28. 150 to 15 and 3 to 1

29. 6 to 39 and 3 to 13

30. $\frac{65}{5}$ and $\frac{1}{13}$

Mixed Review
31. **Algebra** Evaluate $6m - 2(m - n) + mn$ if $m = 10$ and $n = 5$. *(Lesson 1-8)*

32. Describe the pattern in the sequence 4, 12, 36, 108, ... Then find the next three terms. *(Lesson 4-3)*

33. Order the integers 5, −1, 3, and −5 from least to greatest. *(Lesson 7-2)*

34. **Packaging** Production specialists suggest that a good way to package flour is in containers shaped as cylinders. The package they suggest has a radius of 2.5 inches and a height of 6 inches. Find the volume of this package. *(Lesson 10-6)*

*Problem Solving
and
Applications*
35. **Geometry** Complete the following for the rectangles shown at the right.

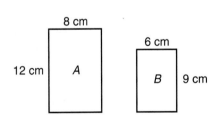

a. Write ratios comparing the widths, the lengths, and the perimeters of rectangle A to rectangle B.

b. Show whether or not these ratios are equivalent.

c. Write a ratio comparing the areas of rectangle A to rectangle B. Describe this ratio in terms of the other three ratios.

36. **Probability** The probability of rolling a sum of 7 with two number cubes is 6 out of 36. The probability of rolling a 5 with one number cube is 1 out of 6. Are these probabilities equivalent? Show your answer by simplifying.

37. **Critical Thinking** Susanne found that 6 students out of 18 students she surveyed liked rock music. Sam found that 9 students out of 24 students he surveyed liked rock music. Which result shows a greater preference for rock music?

11-2 Rates

Objective

Determine unit rates.

Words to Learn

rate
unit rate
unit price
population
  density

George is in charge of buying paper plates, napkins, and cups for the middle school dance. Fourth Street Market sells 100 plates for $1.39. Ben's Supermarket sells the same plates at 50 for $0.74. Which store has the better buy? *This problem will be solved in Example 2.*

Often you must compare quantities with different units. For example, the ratio $1.39 for 100 plates compares a number of dollars to a number of plates. This type of ratio is called a **rate**.

Rate	A rate is a ratio of two measurements with different units.

Example 1 *Problem Solving*

Nutrition   There are 264 calories in an 8-ounce serving of soup. How many calories are there in one ounce of soup?

$$\frac{calories}{ounces} \begin{array}{c} \rightarrow \\ \rightarrow \end{array} \frac{264}{8} = \frac{264 \div 8}{8 \div 8} = \frac{33}{1}$$

There are 33 calories in one ounce of soup, or 33 calories *per ounce*.

A rate such as *33 calories per ounce* is an example of a **unit rate**.

Unit Rate	A unit rate is a rate in which the denominator is 1 unit.

Some other common unit rates are given at the right.

miles per gallon	mi/gal (or mpg)
miles per hour	mi/h (or mph)
price per pound	dollars/lb
meters per second	m/s

In some grocery stores, the labels on the shelves give the cost of the item and its unit rate, which is called the **unit price**. This information helps consumers to make good buying decisions.

Example 2 *Problem Solving*

Smart Shopping   Refer to the problem in the lesson introduction. Which store, Fourth Street Market or Ben's Supermarket, has the better buy on paper plates?

Explore   To determine which store has the better buy, George needs to find the unit price for the plates at each store.

Plan   Find the unit price for plates at each store and compare.

Solve   **Fourth Street:** $\frac{\$1.39}{100 \text{ plates}}$

1.39 ÷ 100 = 0.0139

Ben's Supermarket: $\frac{\$0.74}{50 \text{ plates}}$

0.74 ÷ 50 = 0.0148

The price per plate, or unit price, at Fourth Street Market is $0.0139. The unit price at Ben's Supermarket is $0.0148. Since 0.0139 < 0.0148, the Fourth Street Market has the better buy on plates.

Examine To check your answer, multiply. One hundred plates at $0.0139 per plate cost $1.39. One hundred plates at $0.0148 per plate cost $1.48. So the Fourth Street Market has the better buy.

Community planners use a unit rate called **population density,** which is the population per square mile.

Example 3 *Problem Solving*

Social Studies   In 1991, New York State had a population of 17,950,000 people and an area of 49,108 square miles. What was the population density of New York State in 1991?

$\frac{17,950,000 \text{ people}}{49,108 \text{ square miles}}$ → 17,950,000 ÷ 49,108 = 365.520893

In 1991, New York State had a population density of about 366 people per square mile.

Checking for Understanding

Communicating Mathematics

Read and study the lesson to answer each question.
1. **Tell** whether the rate is a unit rate.
   a. 240 km in 5 hours   b. $3.20 per pound   c. 8 miles per hour
2. **Write,** in your own words, the difference between a ratio and rate.
3. **Tell** why unit rates are more helpful than rates that are not unit rates.

Guided Practice

Express each rate as a unit rate.
4. $2.37 for 3 pounds
5. 200 miles in 5 hours
6. 6 cups for 3 pounds
7. $11.90 for 10 disks
8. $350 for 5 days
9. 12 people in 3 cars
10. 12 pounds in 3 weeks
11. 150 tickets in 5 days

Exercises

Independent Practice

Express each rate as a unit rate.

12. 20 people in 4 rows
13. 480 miles in 6 days
14. $6.20 for 5 pounds
15. $29.00 for 20 disks
16. 1,200 tickets in 8 days
17. 12 cups for 24 pounds
18. $960 for 16 days
19. 24 people in 8 cars

20. The Fourth Street Market sells napkins at $2.29 for 300 and cups at $1.75 per 50. At Ben's Supermarket, the same napkins cost $1.49 for 200, and cups are $0.89 for 25.

    a. Find the unit price for each item at each store.

    b. Which store has the better buy for napkins and for cups?

Mixed Review

21. **Chemistry** Bob is performing a chemistry experiment that requires that 3 milligrams of potassium be added to the solution. How many grams of potassium chloride will Bob add? *(Lesson 2-9)*

22. Express $\frac{35}{4}$ as a mixed number. *(Lesson 5-1)*

23. **Geometry** What is the name of a polygon having 6 sides? *(Lesson 8-2)*

24. Write the ratio 45:81 in simplest form. *(Lesson 11-1)*

Problem Solving and Applications

Population Find the population density, to the nearest person, for each country.

	Country	Population	Square Miles
25.	Brazil	153,771,000	3,286,470
26.	Canada	26,527,000	3,558,096
27.	China	1,130,065,000	3,705,390
28.	Egypt	54,139,000	386,650
29.	Hong Kong	5,693,000	409

30. **Entertainment** The chorus has 5 days to sell 195 tickets to ensure a sell-out at their spring concert. At what rate must they sell the tickets?

31. **Critical Thinking** The heart beat of an average adult human is 72 beats per minute. The average number of heart beats for an elephant is 35 beats per minute.

    a. Whose heart will beat more in an hour?

    b. About how many days will it take for a human's heart to beat 1,000,000 times?

    c. About how many days will it take an elephant's heart to beat 1,000,000 times?

32. **Journal Entry** Suppose the Fourth Street Market and Ben's Supermarket are on opposite sides of a large town. Should George go to each store and purchase the items that gave him the better buy or should he go to one store to purchase everything? Explain.

11-3 Proportions

Objective

Solve proportions.

Words to Learn

proportion
cross products

The first jumbo jet was the Boeing 747. It began airline service in 1970. Because of its long, wide body, it can seat nearly 500 passengers. The length of the Boeing 747 is 70.5 meters, or 7,050 centimeters. The wingspan is about 60 meters, or 6,000 centimeters.

A model of this plane has a wingspan of 80 centimeters and a length of 94 centimeters.

You can find the ratios of the wingspan to the length for the plane and then for the model. How do they compare?

plane: $\dfrac{wingspan}{length}$ $\longrightarrow$ $\dfrac{6,000}{7,050} = \dfrac{6,000 \div 150}{7,050 \div 150} = \dfrac{40}{47}$

model: $\dfrac{wingspan}{length}$ $\longrightarrow$ $\dfrac{80}{94} = \dfrac{80 \div 2}{94 \div 2} = \dfrac{40}{47}$

The ratios $\dfrac{6,000}{7,050}$ and $\dfrac{80}{94}$ are equivalent. So, you can write $\dfrac{6,000}{7,050} = \dfrac{80}{94}$. This equation is an example of a **proportion.**

DID YOU KNOW

The fastest jet is the U.S. Lockheed SR-71. It can reach a speed of 2,200 miles per hour.

Proportion	**In words:** A proportion is an equation that shows that two ratios are equivalent.
	Arithmetic $\dfrac{3}{4} = \dfrac{9}{12}$ **Algebra** $\dfrac{a}{b} = \dfrac{c}{d}, \; b \neq 0, \; d \neq 0$

In a proportion, the two **cross products** are equal. The cross products in the proportion below are 4×9 and 3×12.

$$\dfrac{3}{4} = \dfrac{9}{12}$$

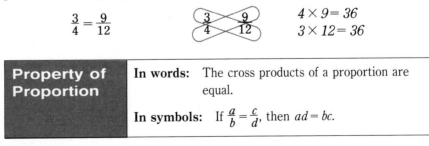

$4 \times 9 = 36$
$3 \times 12 = 36$

Property of Proportion	**In words:** The cross products of a proportion are equal.
	In symbols: If $\dfrac{a}{b} = \dfrac{c}{d}$, then $ad = bc$.

In a proportion like $\dfrac{2}{5} = \dfrac{3}{n}$, if one of the terms is not known, you can use cross products to find the unknown, n. This is known as *solving the proportion.*

Example 1 *Connection*

Algebra Solve $\frac{2}{5} = \frac{3}{n}$.

$$\frac{2}{5} = \frac{3}{n}$$

$2 \times n = 5 \times 3$ *Find the cross products.*

$2n = 15$

$\frac{2n}{2} = \frac{15}{2}$ *Divide each side by 2.*

$n = 7\frac{1}{2}$ The solution is $7\frac{1}{2}$.

Example 2 *Problem Solving*

Sports During the first week of baseball season, the Minnesota Twins won 6 games out of 8 games. How many of the 162 regular season games must they win to maintain this ratio?

Explore   In the first week, they won 6 games out of 8 games. The regular season has 162 games.

Plan   Let n represent the number of games they must win. Write a proportion.

$$
\begin{array}{ccccc}
 & & \text{first week} & & \text{season} \\
wins \;\rightarrow & & \dfrac{6}{8} & = & \dfrac{n}{162} & \rightarrow \; wins \\
games \;\rightarrow & & & & & \rightarrow \; games
\end{array}
$$

Find the cross products. Then solve the proportion.

Solve $\frac{6}{8} = \frac{n}{162}$

$6 \times 162 = 8n$ *Find the cross products.*

$6 \boxed{\times} 162 \boxed{\div} 8 \boxed{=} \; \mathbf{121.5}$

$n = 121.5$

Since the number of games must be expressed as a whole number, the Minnesota Twins must win 122 games to maintain the ratio of 6 to 8.

Examine   Express each ratio as a decimal and compare.

$\frac{6}{8}$: $6 \boxed{\div} 8 \boxed{=} \mathbf{0.75}$

$\frac{121.5}{162}$: $121.5 \boxed{\div} 162 \boxed{=} \mathbf{0.75}$

Since the decimals are equal, the ratios are equivalent. So, 121.5 is correct.

Mental Math Hint
• • • • • • • • • • • • •

Sometimes you can solve a proportion mentally by using equivalent fractions.

$$\frac{5}{9} = \frac{20}{m}$$

$\times 4$

$$\frac{5}{9} = \frac{20}{36}$$

$\times 4$

So, $m = 36$.

Problem Solving Hint
• • • • • • • • • • • • •

Often there is more than one way to solve a problem that involves proportions. Just be sure that each side of the proportion compares the quantities in the same order. For instance, here is another way to write a proportion for the problem in Example 2.

wins   games

1st week → $\dfrac{6}{n} = \dfrac{8}{164}$ ← *1st week*
season → ← *season*

Checking for Understanding

Communicating Mathematics

Read and study the lesson to answer each question.

1. **Tell** how you can determine whether two ratios are equivalent.

2. **Write** the cross products for $\frac{1}{3} = \frac{5}{n}$

3. **Tell** how to solve $\frac{n}{4} = \frac{9}{17}$ with a calculator.

Guided Practice

Solve each proportion.

4. $\frac{3}{4} = \frac{n}{8}$

5. $\frac{6}{9} = \frac{4}{m}$

6. $\frac{5}{t} = \frac{2}{6}$

7. $\frac{x}{3} = \frac{18}{27}$

8. $\frac{x}{36} = \frac{15}{24}$

9. $\frac{r}{3} = \frac{5}{9}$

10. $\frac{6}{5} = \frac{k}{4}$

11. $\frac{2.5}{4} = \frac{10}{y}$

12. **Finance** Paul saves 10 cents of every dollar of his allowance. Paul's allowance is $2.50 a week. How much does he save each week?

Exercises

Independent Practice

Solve each proportion.

13. $\frac{3}{4} = \frac{9}{n}$

14. $\frac{8}{12} = \frac{a}{3}$

15. $\frac{10}{t} = \frac{15}{9}$

16. $\frac{5}{n} = \frac{6}{3}$

17. $\frac{n}{7} = \frac{18}{42}$

18. $\frac{2.6}{13} = \frac{8}{h}$

19. $\frac{8}{20} = \frac{30}{n}$

20. $\frac{12}{4} = \frac{y}{12}$

21. $\frac{5}{9} = \frac{n}{5.4}$

22. $\frac{21}{m} = \frac{10}{20}$

23. $\frac{1,200}{s} = \frac{6}{1}$

24. $\frac{0.1}{x} = \frac{3}{10}$

25. Do $\frac{7}{8}$ and $\frac{13}{15}$ form a proportion? Explain why or why not.

Mixed Review

26. Express 2,000,000 in scientific notation. *(Lesson 2-6)*

27. **Statistics** The number of days that it rains each month in Cincinnati, Ohio, is recorded for a complete year with the following results: 12, 17, 9, 21, 15, 7, 14, 7, 15, 22, 14, 19. Construct a stem-and-leaf plot for this data. *(Lesson 3-6)*

28. Solve $5t = 125$. *(Lesson 6-3)*

29. **Consumer Math** Find the unit price for cheese that is sold in 12-ounce packages priced at $3.72. *(Lesson 11-2)*

Problem Solving and Applications

30. **School Planning** Smallwood Middle School has 1,000 students, 40 teachers, and 5 administrators. If the school grows to 1,200 students and the ratios are maintained, find the number of teachers and administrators that will be needed.

31. **Measurement**   Elmer uses an old spring scale to weigh his fish. The markings have rusted off the scale, but Elmer knows that a 3-pound fish pulls the spring down $1\frac{7}{8}$ inches. Today his fish pulled the spring down $3\frac{1}{2}$ inches. How much does his fish weigh?

32. **Nutrition**   For her health class Noelle must record what she eats for breakfast and compute the total number of calories consumed. Write and solve the proportions that you can use to find the number of calories in Noelle's breakfast.

Noelle's Breakfast
½ banana
6 oz orange juice
¾ cup corn flakes
½ cup milk

Calories:

1 medium banana	100
8 oz orange juice	100
1 cup corn flakes	112
$\frac{3}{4}$ cup milk	126

33. **Critical Thinking**   Use the digits 1 through 9 to write as many proportions as possible. Each digit may be used only once in a proportion. The numbers that make up the proportion can consist of only one digit. For example, one proportion could be $\frac{1}{2} = \frac{3}{6}$.

34. **Mathematics and Design**   Read the following paragraph.

> Throughout the ages, designers and architects have attempted to establish ideal proportions. The most famous of all principles about proportion was the **golden section** established by the ancient Greeks. According to this principle, a line segment can be separated into two parts so that the ratio of the shorter section to the longer section is equal to the ratio of the longer section to the whole line. This proportion is believed to be pleasing to the eye and has been used in the design of many buildings, both old and new, as well as sculptures and paintings.

Show the golden section by separating a line segment into two sections so that the ratio of the shorter section to the longer section is equal to the ratio of the longer section to the whole line. Label the line segment and write the proportion.

11-3B Capture and Recapture

A Follow-Up of Lesson 11-3

Objective

Use the capture-recapture technique to estimate.

Materials

bowl
lima beans
marker

One method of estimating a population is the **capture-recapture** technique. Naturalists often use this method to monitor the population of animals and fish. In this lab, you will model this technique. Lima beans will represent deer and the bowl will represent the forest.

Try this!

Work in small groups.

- Fill a small bowl with dried lima beans.
- Grab a small handful of the beans. Count the number of beans selected. These represent the captured deer. Mark each bean selected with an X on both sides.
- Return the beans to the bowl and mix them in well with the rest.
- Grab another handful of beans from the bowl. Count the number of beans selected. This represents the number of deer recaptured. Count the number of beans marked with an X. This represents the number of tagged deer recaptured.
- Use the proportion shown below to estimate the total number of beans in the bowl.

$$\frac{\text{original number captured}}{\text{total population }(P)} = \frac{\text{tagged in sample}}{\text{recaptured}}$$

  Record the value of *P*.
- Return the beans to the bowl.
- Repeat this process nine more times.

What do you think?

1. Do you think that a good estimate for the population *P* is the average of your ten estimates for *P*?
2. Do you think the population could be greater than any of your estimates?
3. Count the number of beans in the bowl. How does the actual count compare with your estimates?
4. Why is it important to return the beans to the bowl and mix each time you repeat the experiment?
5. Why is it a good idea to base an estimate on several samples rather than just one sample?
6. What would happen to your estimate if some of the Xs wore off? How could something like this happen with deer?

11-4 Similar Polygons

Objectives

Identify corresponding parts of similar polygons. Find missing measures by using lengths of corresponding sides.

Words to Learn

similar polygons

Jamie sizes and positions photos for the Menden Junior High Yearbook. She enlarges a photo that is 3 inches wide and 5 inches long. The enlargement is 6 inches wide and 10 inches long.

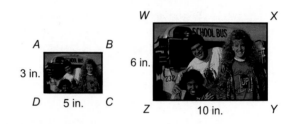

When you compare the dimensions of the photos, you get a proportion.

$$\frac{photo\ width}{enlargement\ width} \rightarrow \frac{3}{6} = \frac{5}{10} \leftarrow \frac{photo\ length}{enlargement\ length}$$

The photo and its enlargement are said to be **similar polygons.** Similar polygons have the same shape but may not have the same size.

Similar Polygons	**In words:**	Two polygons are similar if their corresponding angles are congruent and their corresponding sides are in proportion.
	In symbols:	$ABCD \sim WXYZ$ The symbol $\sim$ means *is similar to.*

Proportions are useful in finding the missing length of a side in any pair of similar polygons.

Example 1

If $\triangle ABC \sim \triangle DEF$, find the length of $\overline{DE}$.

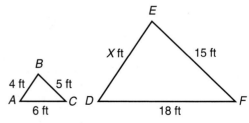

$\overline{AB}$ and $\overline{DE}$ are corresponding sides. $\overline{AC}$ and $\overline{DF}$ are corresponding sides. You can write a proportion using the measures of corresponding sides.

$$\overline{AB} \rightarrow \frac{4}{x} = \frac{6}{18} \leftarrow \overline{AC}$$
$$\overline{DE} \rightarrow \frac{4}{x} = \frac{6}{18} \leftarrow \overline{DF}$$

$4 \times 18 = 6x$ *Find the cross products.*

$4 \boxed{\times} 18 \boxed{\div} 6 \boxed{=} 12$

$x = 12$ The length of $\overline{DE}$ is 12 feet.

Example 2

Rectangles *A* and *B* are similar. The ratio of rectangle *B*'s width to rectangle *A*'s width is 3:2. Rectangle *A* has a length of 24 inches and a width of 16 inches. What is the perimeter of rectangle *B*?

First use proportions to find the length ℓ and width w of rectangle *B*.

$\frac{length\ of\ B}{length\ of\ A} \rightarrow \frac{\ell}{24} = \frac{3}{2}$ $\frac{width\ of\ B}{width\ of\ A} \rightarrow \frac{w}{16} = \frac{3}{2}$

$\qquad\qquad 2\ell = 3 \times 24$ $2w = 3 \times 16$

$\qquad\qquad 2\ell = 72$ $2w = 48$

$\qquad\qquad \frac{2\ell}{2} = \frac{72}{2}$ $\frac{2w}{2} = \frac{48}{2}$

$\qquad\qquad \ell = 36$ $w = 24$

Now use the formula for the perimeter of a rectangle.

$P = 2\ell + 2w$

$\quad = 2(36) + 2(24)$ *Substitute 36 for ℓ and 24 for w.*

$\quad = 72 + 48$

$\quad = 120$

The perimeter of rectangle *B* is 120 inches.

Checking for Understanding

Communicating Mathematics

Read and study the lesson to answer each question.

1. **Draw** two rectangles whose corresponding sides have measures in the ratio 3 to 1.

2. **Write** the pairs of sides that correspond in similar triangles *KLM* and *TSR* shown at the right.

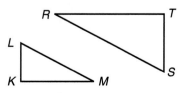

3. **Tell** the difference between similar polygons and congruent polygons.

Tell whether each pair of polygons is similar. Justify your answer.

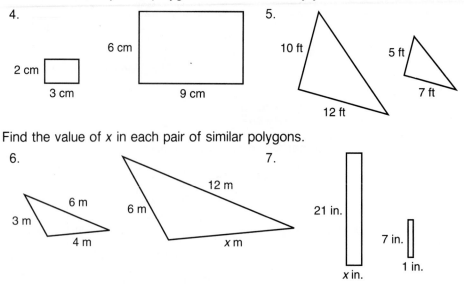

4.

6 cm
2 cm
3 cm
9 cm

5.

10 ft
12 ft
5 ft
7 ft

Find the value of x in each pair of similar polygons.

6.

6 m
3 m
4 m
12 m
6 m
x m

7.

21 in.
x in.
7 in.
1 in.

8. Triangles A and B are similar. The ratio of a side of triangle B to a corresponding side of triangle A is 5:3. The sides of triangle A measure 18 feet, 27 feet, and 30 feet. Find the perimeter of triangle B.

Exercises

Tell whether each pair of polygons is similar. Justify your answer.

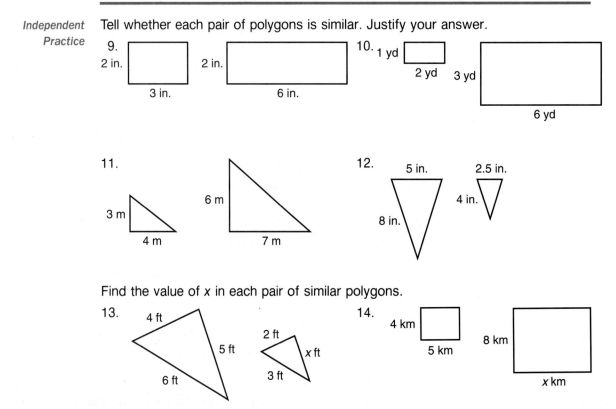

9.
2 in.
3 in.
2 in.
6 in.

10.
1 yd
2 yd
3 yd
6 yd

11.
3 m
4 m
6 m
7 m

12.
5 in.
8 in.
2.5 in.
4 in.

Find the value of x in each pair of similar polygons.

13.
4 ft
5 ft
6 ft
2 ft
3 ft
x ft

14.
4 km
5 km
8 km
x km

15.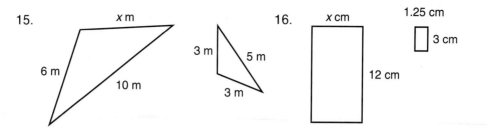

x m

6 m

10 m

3 m 5 m

3 m

16. *x* cm

12 cm

1.25 cm

3 cm

17. Rectangles *F* and *G* are similar. The ratio of rectangle *F*'s width to rectangle *G*'s width is 2:3. The length of rectangle *F* is 15 inches and its width is 10 inches. Find the perimeter of rectangle *G*.

Mixed Review

18. **Physical Fitness** Heather does 47 sit-ups in 98 seconds. Estimate the number of seconds it takes her to do one sit-up. *(Lesson 1-3)*

19. **Statistics** Takeo surveyed 50 of his classmates to find out their favorite winter sport. Which sport was the most popular? *(Lesson 3-1)*

20. Solve $\frac{15}{32} = \frac{5}{p}$. *(Lesson 11-3)*

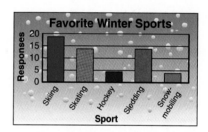

Favorite Winter Sports

Responses

Skiing Skating Hockey Sledding Snow-mobiling

Sport

Problem Solving and Applications

21. **Photography** Juana wants to have an enlargement made of a photo she took at the Grand Canyon. The negative is 1.5 centimeters by 2.2 centimeters. What will be the perimeter of the enlargement if its width is 22 centimeters? Solve mentally.

22. **Measurement** Sara's shadow is 60 inches long. A nearby bush casts a shadow 40 inches long. If Sara is 48 inches tall, what is the height of the bush?

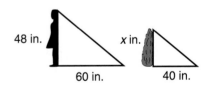

48 in.

x in.

60 in.

40 in.

23. **Critical Thinking**

a. Copy and complete this table.

	Length	Width	Perimeter	Area
Rectangle I	5	3	?	?
Rectangle II	20	12	?	?

b. Compute $\frac{\text{length I}}{\text{length II}}$, $\frac{\text{perimeter I}}{\text{perimeter II}}$, and $\frac{\text{area I}}{\text{area II}}$.

c. What pattern do you see?

d. Does your pattern apply to the rectangles in Example 2? Explain.

Lesson 11-4 Geometry Connection: Similar Polygons **425**

11-5 Scale Drawings

Objective

Solve problems involving scale drawings.

Words to Learn

scale drawing

Jack and Jessie Anderson are planning a trip from Knoxville to Memphis to attend a family reunion. A map of Tennessee is shown below. The scale of the map is 1 inch:152 miles. The distance between Knoxville and Memphis on the map is $2\frac{1}{4}$ inches. What is the actual distance between Knoxville and Memphis?

1 inch = 152 miles

A map is an example of a scale drawing. A **scale drawing** is used to present something that is too large or too small to be conveniently drawn to actual size.

The scale on a map is the ratio of the distance on the map of the actual distance. When you know the scale of a map, you can find actual distances by writing and solving proportions.

Example 1 *Problem Solving*

Geography Refer to the problem in the lesson introduction. Find the actual distance between Knoxville and Memphis.

When am I ever going to use this?

If your family is planning to travel this summer, you will probably use a map to help you get to your destination. You can find out how many miles you will be traveling by using the scale on the map and writing a proportion.

Let n represent the actual distance between the cities. Write and solve a proportion.

$$\begin{array}{c} map\ distance \\ actual\ distance \end{array} \begin{array}{c} \rightarrow \\ \rightarrow \end{array} \frac{1\ inch}{152\ miles} = \frac{2\frac{1}{4}\ inches}{n} \begin{array}{c} \leftarrow \\ \leftarrow \end{array} \begin{array}{c} map\ distance \\ actual\ distance \end{array}$$

$$1 \times n = 2\frac{1}{4} \times 152$$

$$n = 342$$

The actual distance between Knoxville and Memphis is 342 miles.

Example 2 *Problem Solving*

Calculator Hint

• • • • • • • • • • • • •

When a scale drawing involves several measurements, use a calculator and enter the scale factor into its memory. For instance, if the scale is 1 inch : 12 inches, enter 12 [STO].

Then whenever you need to use 12 in your calculations, you only need to press [RCL].

Decorating   A building is 275 feet long. On a scale drawing, 1 inch represents 25 feet. What is the length of the building in the scale drawing?

Let ℓ represent the length of the building in the scale drawing.

$$\frac{scale}{actual} \; \substack{\rightarrow \\ \rightarrow} \; \frac{1 \text{ in.}}{25 \text{ ft}} = \frac{\ell \text{ in.}}{275 \text{ ft}} \; \substack{\leftarrow \\ \leftarrow} \; \frac{scale}{actual}$$

$$1 \times 275 = 25 \, \ell \quad \textit{Find the cross products.}$$

$$275 \; \boxed{\div} \; 25 \; \boxed{=} \; 11$$

$$11 = \ell$$

On the scale drawing, the building will be 11 inches long.

Mini-Lab

Work with a partner.

Materials: measuring tape, $\frac{1}{4}$-inch graph paper, ruler

- Use the measuring tape to find the length of each wall, door, window, and chalkboard in your classroom.

- Round each length to the nearest inch. Record the lengths.

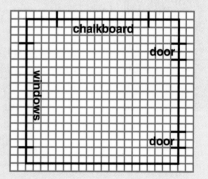

- On a sheet of grid paper, make a scale drawing of your classroom like the one above. Use $\frac{1}{4}$ inch:12 inches as the scale.

Talk About It

a. What proportion did you use to find the length of the chalkboard on your scale drawing?

b. Describe how a scale drawing with $\frac{1}{4}$ inch:24 inch would differ from your scale drawing with $\frac{1}{4}$ inch:12 inch.

Checking for Understanding

*Communicating
Mathematics*

Read and study the lesson to answer each question.

1. **Tell**, in your own words, what a scale drawing is.

2. **Draw** a scale drawing of your bedroom on $\frac{1}{4}$-inch graph paper. Use $\frac{1}{4}$ inch:12 inches as the scale. Include the closet(s), the window(s), and the door(s) in your drawing.

3. **Tell** what important information must be given on a scale drawing in order to use it.

Guided Practice

Find the actual distance between each pair of cities, given the map distance. Use the scale 1 inch:152 miles.

4. Chattanooga and Memphis, $1\frac{3}{4}$ inches

5. Knoxville and Chattanooga, $\frac{3}{4}$ inch

6. **Geography** Choose two cities in Tennessee, using the map on page 426. Find the actual distance between them.

Find the length of each object on a drawing with the given scale.

7. a room whose length is 50 feet; 1 inch:2 feet

8. a window whose height is 30 inches; $\frac{1}{2}$ inch:6 inches

9. a car that is 135 inches long; $\frac{1}{4}$ inch:2 feet

10. **Structural Engineering** The longest suspension bridge in the United States is the Verrazano-Narrows Bridge connecting Staten Island and Brooklyn, New York. Its length is 4,260 feet. On a scale drawing, 1 inch represents 60 feet. Find the length of the bridge in the scale drawing.

Exercises

*Independent
Practice*

On a map, the scale is 1 inch:120 miles. For each map distance, find the actual distance.

11. 4 inches

12. $2\frac{1}{2}$ inches

13. $\frac{7}{8}$ inch

14. $4\frac{3}{8}$ inches

15. $3\frac{1}{4}$ inches

16. $\frac{1}{2}$ inch

17. $\frac{1}{4}$ inch

18. $5\frac{3}{4}$ inches

19. On a scale drawing, 1 centimeter represents 2 meters. What length on the drawing would be used to represent 3.2 meters?

On a scale drawing, the scale is $\frac{1}{2}$ inch:1 foot. Find the dimensions of each room in the scale drawing.

20. 30 feet by 20 feet

21. 18 feet by 12 feet

22. 15 feet by 7 feet

23. 10 feet by 11 feet

24. Express $\frac{144}{180}$ in simplest form. *(Lesson 4-6)*

25. **Stock Market** A particular stock on the New York Stock Exchange lost a total of 6 points over a period of 2 days. What was the change in points per day? *(Lesson 7-8)*

26. **Geometry** Tell whether the polygons at the right are similar. *(Lesson 11-4)*

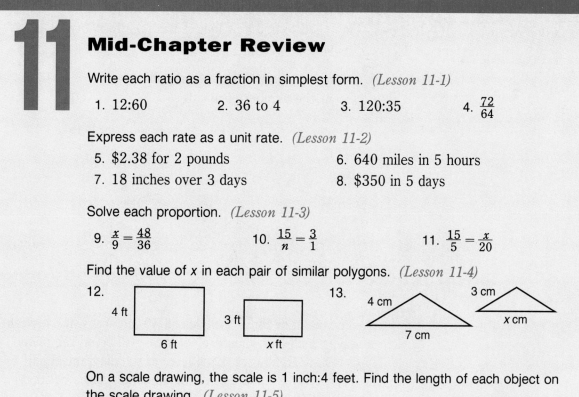

27. **Horticulture** Ms. Lee wants to plant some flowers and bushes in her front yard. The width of her yard is 40 feet, and the length is 35 feet.

    a. Draw a scale drawing of her yard using a scale of $\frac{1}{2}$ inch:5 feet.

    b. Assuming she has nothing in her front yard right now, use your scale drawing and your creativity to come up with a design for her front yard.

28. **Critical Thinking** A computer chip measures $\frac{1}{2}$ inch by $\frac{3}{8}$ inch. A scale drawing of the chip measures 6 inches by $4\frac{1}{2}$ inches. What is the scale of the drawing?

11 Mid-Chapter Review

Write each ratio as a fraction in simplest form. *(Lesson 11-1)*

1. 12:60
2. 36 to 4
3. 120:35
4. $\frac{72}{64}$

Express each rate as a unit rate. *(Lesson 11-2)*

5. $2.38 for 2 pounds
6. 640 miles in 5 hours
7. 18 inches over 3 days
8. $350 in 5 days

Solve each proportion. *(Lesson 11-3)*

9. $\frac{x}{9} = \frac{48}{36}$
10. $\frac{15}{n} = \frac{3}{1}$
11. $\frac{15}{5} = \frac{x}{20}$

Find the value of *x* in each pair of similar polygons. *(Lesson 11-4)*

12.

4 ft · 6 ft   3 ft · *x* ft

13.

4 cm · 7 cm   3 cm · *x* cm

On a scale drawing, the scale is 1 inch:4 feet. Find the length of each object on the scale drawing. *(Lesson 11-5)*

14. a desk 60 inches long
15. a car 10 feet long

11-6 Draw a Diagram

Objective

Solve problems by drawing a diagram.

The Science Club at Crestview Middle School is planning an end-of-year picnic at Whetstone Park. If it looks like it will rain, the picnic will be rescheduled for the following weekend. The club president decides to set up a "telephone tree" so that everyone in the club will be notified quickly in the event that the picnic has to be rescheduled. The club president plans to call three members and have each of them call three other members and so on. If each phone call takes 1 minute, how long will it take to notify all 40 members of the club?

Explore   What do you know?
Each person will call 3 people and each phone call will take 1 minute. There are 40 members in the club. A hidden assumption is that there will be no busy signals.

What do you need to find?
You need to find how long it will take to notify all 40 members.

Plan   Make a diagram of the phone calls.
Count the number of minutes.

Solve   Use an "o" to stand for each person who is called.

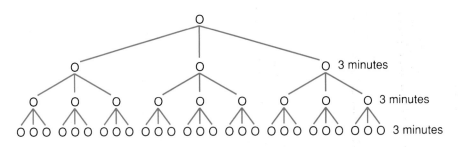

The diagram above includes all 40 members. If each call takes 1 minute, then each student will be on the phone 3×1, or 3 minutes. Each level of the diagram, then, represents 3 minutes. All the students can be notified in $3 + 3 + 3$, or 9 minutes.

Examine $3 \times 3 = 9$ ✓   The answer checks.

Juan and Diego are brothers sharing the same bedroom. Juan's alarm is set for 6:30 A.M., and it has a snooze alarm that goes off every 9 minutes. Diego's alarm is set for 6:50 A.M., and its snooze alarm goes off every 5 minutes. If both Juan and Diego hit the snooze alarm several times in one morning, at what time would both alarms go off at the same time?

Explore   What do you know?
Juan's alarm is set for 6:30 A.M. and has a snooze alarm that goes off every 9 minutes. Diego's alarm is set for 6:50 A.M. and has a snooze alarm that goes off every 5 minutes.

What do you need to find?
You need to find when the two alarms will go off at the same time.

Plan   Draw a diagram that includes the times that the alarms go off. The solution will be the time when both alarms go off at the same time.

Solve

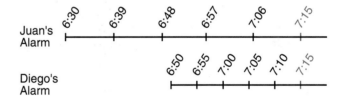

Both alarms will go off at 7:15 A.M.

Examine   The number of minutes from 6:30 to 7:15 is 45. 45 is a multiple of 9.

The number of minutes from 6:50 to 7:15 is 25. 25 is a multiple of 5.

Checking for Understanding

Communicating Mathematics

1. **Tell** at what time Juan's and Diego's alarm clocks will go off together if Diego's snooze alarm went off every 4 minutes.

2. **Write** one or two sentences explaining why a diagram can be a useful strategy in solving problems.

Solve by drawing a diagram.

3. A shuttle bus at Cedar Point Amusement Park holds 30 passengers. It starts out empty and picks up 1 passenger at the first stop, 2 passengers at the second stop, 3 at the third stop, and so on. After how many stops will the bus be full?

4. After a student council meeting, each of the five members shook hands with each other. How many handshakes were there in all?

Exercises

Practice   Solve. Use any strategy.

5. Marie buys T-shirts for $12.95 and gym shorts for $6.99. She spends a total of $72.77. How many of each did she buy?

6. An airplane flew 6,000 miles in 12 hours. What was its rate?

   a. 72,000 mph   b. 500 mph   c. 6,262 mph

7. Estrella mails a recipe to four of her friends. Each of the four friends mails the recipe to four of their friends and so on. How many recipes are in the fifth mailing?

8. Sixteen softball teams are participating in a single-elimination contest; that is, only the winners of each game go on to play the next game. How many games will the winning team have played?

9. Chairs are to be set up in a meeting room so that each row has 1 more chair than the previous row. This way, none of the chairs will be directly behind another.

   If there are 5 chairs in the first row, how many chairs will be in the sixth row?

10. The product of two consecutive whole numbers is 3,906. What are the numbers?

11. Marla has a total of 44 compact discs and cassette tapes. If she has three times as many compact discs as tapes, how many of each does she have?

12. A ball is dropped from 10 feet above ground. It hits the ground and bounces up half as high as it fell.

    a. What is the height of the ball after the fourth bounce?

    b. What is the total up and down distance the ball has traveled when it hits the ground the fifth time?

11-7 Percent

Objective

Illustrate the meaning of percent using models or symbols.

Words to Learn

percent

Does it seem like you're always buying pens or pencils? In the United States, people buy about $1.9 billion worth of pens and pencils each year. Of all these pens and pencils, 33 of every 100 are ballpoint pens. The shaded area in the grid at the right shows the ratio 33 out of 100. Another name for this ratio is 33 **percent**.

Percent	In words:	A percent is a ratio that compares a number to 100.
	In symbols:	$\frac{n}{100} = n\%$ The symbol % means *percent*.

Examples

Express each ratio as a percent.

1 $\frac{33}{100} = 33\%$ **2** 62.5 out of $100 = 62.5\%$ **3** $8\frac{1}{2}$ per $100 = 8\frac{1}{2}\%$

Example 4

Write a percent to represent the number of shaded squares.

The grid has one hundred squares in all. Count the number that are shaded.

$(3 \times 8) + (3 \times 4) + (5 \times 1) = 41$

There are 41 squares shaded. So, 41% represents the shaded area.

Mini-Lab

Work with a partner.
Materials: grid paper, markers

- Draw nine 10×10 squares on your grid paper.
- For each percent below, shade three different 10×10 grids, each in a different way.

   a. 80%   b. 35%   c. $41\frac{1}{2}\%$

Talk About It

d. Compare your shaded areas with others in your class. Do the shaded areas need to be the same shape in order to represent the same percent? Explain why or why not.

You can review area of triangles on page 355.

e. How many different ways can you shade a 10×10 grid in order to represent 100%?

f. How can you find the percent represented by the shaded area at the right if you don't count squares?

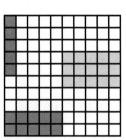

Checking for Understanding

Communicating Mathematics

Read and study the lesson to answer each question.

1. **Tell**, in your own words, what percent means.

2. **Draw** a diagram to show 30%.

3. **Write** a percent that means 3 out of 100.

4. **Write** a percent to show the ratio of the number of squares shaded to the total number of squares in the figure at the right.

Guided Practice

Express each ratio as a percent.

5. $\frac{45}{100}$ 6. 37 out of 100   7. 13 hundredths

8. 18.5 out of 100   9. $12\frac{1}{2}$:100   10. 98.5 to 100

Write a percent to represent the shaded area. If necessary, round answers to the nearest percent.

11.   12.   13.

14.   15.   16.

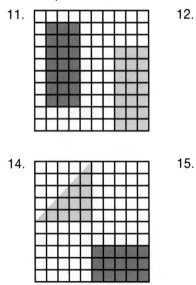
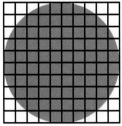

Exercises

Independent Practice

Express each ratio as a percent.

17. 22 people out of 100

18. 1 clown out of 100

19. 98:100

20. 60 of 100 flowers

21. $11 per $100

22. 9 rows out of 100 rows

Write a percent to represent the shaded area. If necessary, round answers to the nearest percent.

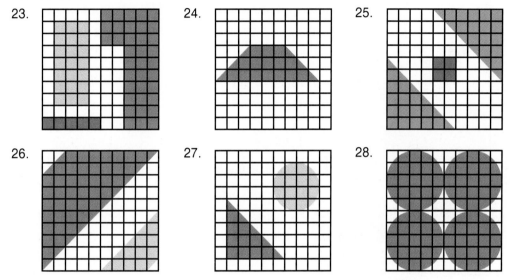

23. 24. 25.

26. 27. 28.

29. **Geometry** Square *A* measures 10 units on a side. Square *B* measures 6 units on a side. Write a percent for the ratio of the area of square *B* to the area of square *A*.

Mixed Review

30. Round 125.0765 to the underlined place-value position. *(Lesson 2-2)*

31. Multiply 4 and $4\frac{3}{8}$. *(Lesson 5-5)*

32. **Geometry** Complete the pattern unit for the translation at the right. Then draw the tessellation. *(Lesson 8-7)*

Problem Solving and Applications

33. **Critical Thinking** Copy and complete each diagram so that it shows 60%.

a. ←——+——+——→
     0   1

b. ☐ ☐ ☐ ☐ ☐

c.

d.

e.

34. **Journal Entry** Find an advertisement that uses percents. Draw a diagram to illustrate the percent. What does the percent mean?

11-8 Percents and Fractions

Objective

Express fractions as percents, and vice versa.

Did you know that bananas do not grow on trees? They grow on plants that can be as tall as 30 feet. A banana is $\frac{3}{4}$ water. What percent of a banana is water?

You can express any fraction as a percent. One way to change the fraction $\frac{3}{4}$ to a percent is to find an equivalent fraction with a denominator of 100.

$$\frac{3}{4} = \frac{3 \times 25}{4 \times 25} = \frac{75}{100} = 75\% \qquad \text{So, 75\% of the banana is water.}$$

Another way to express a fraction as a percent is to use a proportion.

Examples

Express each fraction as a percent.

1 $\frac{3}{8}$

$$\frac{3}{8} = \frac{n}{100} \qquad \textit{Find the cross products.}$$
$$300 = 8n$$
$$\frac{300}{8} = \frac{8n}{8} \qquad \textit{Divide each side by 8.}$$
$$37.5 = n$$

So, $\frac{3}{8} = 37.5\%$.

2 $\frac{5}{24}$

$$\frac{5}{24} = \frac{n}{100} \qquad \textit{Find the cross products.}$$
$$5 \times 100 = 24n$$

$5 \boxed{\times} 100 \boxed{\div} 24 \boxed{=}$
20.833333

So, $\frac{5}{24}$ is about 20.8%.

To write a percent as a fraction, write a fraction with a denominator of 100. Then write the fraction in simplest form.

Examples

Express each percent as a fraction.

3 65%

$$65\% = \frac{65}{100}$$
$$= \frac{65 \div 5}{100 \div 5} \qquad \textit{The GCF is 5.}$$
$$= \frac{13}{20}$$

So, $65\% = \frac{13}{20}$.

4 $16\frac{2}{3}\%$

$$16\frac{2}{3}\% = \frac{16\frac{2}{3}}{100}$$
$$= 16\frac{2}{3} \div 100$$
$$= \frac{50}{3} \div 100$$
$$= \frac{50}{3} \times \frac{1}{100} \qquad \textit{To divide by 100, multiply by } \frac{1}{100}.$$
$$= \frac{50}{300} \text{ or } \frac{1}{6}$$

So, $16\frac{2}{3}\% = \frac{1}{6}$.

Estimation Hint

● ● ● ● ● ● ● ● ● ● ● ● ●

Sometimes you only
need to find a
fraction that is
reasonably close to
a given percent.
When this happens,
try to work with one
of the common
percents from the
table.

38.9% is close to 40%.
So, 38.9% is about $\frac{2}{5}$.

In everyday situations, some percents are used much more frequently
than others. For this reason, you probably will find it helpful to
memorize the equivalent percents and fractions listed in the table
below.

$20\% = \frac{1}{5}$	$25\% = \frac{1}{4}$	$12\frac{1}{2}\% = \frac{1}{8}$	$16\frac{2}{3}\% = \frac{1}{6}$
$40\% = \frac{2}{5}$	$50\% = \frac{1}{2}$	$37\frac{1}{2}\% = \frac{3}{8}$	$33\frac{1}{3}\% = \frac{1}{3}$
$60\% = \frac{3}{5}$	$75\% = \frac{3}{4}$	$62\frac{1}{2}\% = \frac{5}{8}$	$66\frac{2}{3}\% = \frac{2}{3}$
$80\% = \frac{4}{5}$		$87\frac{1}{2}\% = \frac{7}{8}$	$83\frac{1}{3}\% = \frac{5}{6}$
$100\% = 1$			

Mini-Lab

Work with a partner.

Materials: paper and pencil

What percent of the students in your class do you think are in
each category? Estimate by using one of the choices listed at the
right.

a. left-handed

b. male

c. wear glasses or contact lenses

d. less than 2 years old

e. at school today

0%
less than 10%
about 25%
about 50%
at least 75%
100%

Talk About It

f. How does your group's estimate for each category compare
with the estimates of the other groups in your class?

g. How can you tell if your estimates are reasonable?

Checking for Understanding

Communicating Mathematics

Read and study the lesson to answer each question.

1. **Show** one way to express $\frac{7}{20}$ as a percent.

2. **Tell** what percent of the students in your classroom are wearing something
that has blue in it: less than 50%, about 50%, more than 50%.

Guided Practice

Express each fraction as a percent.

3. $\frac{9}{10}$ 4. $\frac{9}{20}$ 5. $\frac{1}{3}$ 6. $\frac{7}{8}$ 7. $\frac{5}{16}$

Express each percent as a fraction in simplest form.

8. 30%   9. 1%   10. 45%

11. 23%   12. $22\frac{1}{2}\%$

13. **Entertainment** Three-fifths of the students at the dance were seventh graders.

   a. What percent were seventh graders?

   b. What percent were *not* seventh graders?

Exercises

Independent Practice

Express each fraction as a percent.

14. $\frac{24}{25}$ 15. $\frac{43}{50}$ 16. $\frac{18}{25}$ 17. $\frac{2}{5}$ 18. $\frac{11}{20}$

19. $\frac{3}{10}$ 20. $\frac{19}{20}$ 21. $\frac{1}{4}$ 22. $\frac{2}{3}$ 23. $\frac{1}{8}$

24. $\frac{5}{6}$ 25. $\frac{7}{16}$ 26. $\frac{10}{12}$ 27. $\frac{40}{125}$ 28. $\frac{3}{3}$

Express each percent as a fraction in simplest form.

29. 25%   30. 15%   31. 72%   32. 10%   33. 70%

34. 50%   35. 80%   36. 34%   37. $12\frac{1}{2}\%$ 38. $11\frac{1}{2}\%$

39. $66\frac{2}{3}\%$ 40. $62\frac{1}{2}\%$ 41. $17\frac{1}{2}\%$ 42. $3\frac{1}{3}\%$ 43. $6\frac{1}{4}\%$

44. Express $\frac{9}{24}$ as a percent.

45. Express *fifty-four percent* as a fraction in simplest form.

Mixed Review

46. Compare the fractions $\frac{15}{24}$ and $\frac{17}{32}$ using the least common denominator. *(Lesson 4-10)*

47. Write 0.00005 in scientific notation. *(Lesson 7-10)*

48. **Geometry** Classify the triangle at the right by its sides and by its angles. *(Lesson 8-3)*

49. Alfonso places a 10-foot ladder 6 feet from his house and leans the ladder against the house. How high off the ground does the ladder reach on the side of the house? *(Lesson 9-4)*

10 ft

b ft

6 ft

50. **Geometry**   Find the volume of a rectangular prism having a length of 5 centimeters, a width of 3 centimeters, and a height of 8 centimeters. *(Lesson 10-5)*

51. Express *34 hits out of 100* as a percent. *(Lesson 11-7)*

Problem Solving and Applications

Animals   Use the circle graph at the right for Exercises 52–55.

52. What fraction of people surveyed get their pets from animal shelters?

53. What fraction of people surveyed get their pets from either breeders or friends?

54. What fraction of people surveyed get their pets from places other than pet stores?

55. Where do you think is the best place to get a pet? Explain your reasons.

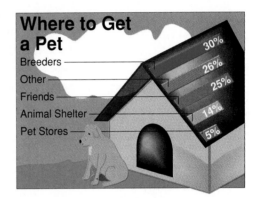

Where to Get a Pet

Breeders — 30%
Other — 26%
Friends — 25%
Animal Shelter — 14%
Pet Stores — 5%

56. **Statistics**   Gail, Jil, and Michael are practicing free throws on the basketball court. Ned kept a tally of their shooting. What percent of shots attempted did each person make?

	Attempts	Made
Gail	ⅢⅢ Ⅲ II	Ⅲ II
Michael	Ⅲ Ⅲ IIII	Ⅲ II
Jil	Ⅲ Ⅲ Ⅲ	Ⅲ III

57. **Critical Thinking**   The best player on the girls' basketball team made 12 out of 20 free throws. The best player on the boys' basketball team has made 14 out of 25 free throws. Who is the better free throw shooter? Explain your answer.

DATA SEARCH

58. **Data Search**   Refer to pages 408 and 409. Express the number of students that smoke at each grade level as a fraction.

59. **Journal Entry**   Describe an everyday situation where it may be more convenient to use a fraction and another situation where it may be more convenient to use a percent. Explain your reasoning for each choice.

11-9   Percents and Decimals

Objective

Express decimals as percents, and vice versa.

An iceberg is a huge mass of ice that comes from a glacier. Icebergs are made of fresh water even though they float in salt water. If a ship's crew runs out of fresh water, they can get it from an iceberg. The captain must be careful in approaching an iceberg because only about 0.125 of it is above water. What percent of the iceberg is above water?

You can review changing decimals to fractions on page 155.

In Chapter 4, you learned that any decimal can be expressed as a fraction. You can use that fact to express any decimal as a percent.

Example 1

Express 0.07 as a percent.

$0.07 = \frac{7}{100}$ *First, express the decimal as a fraction.*

$\quad = 7\%$ *Then, express the fraction as a percent.*

So, $0.07 = 7\%$.

Example 2   *Problem Solving*

Navigation   Refer to the problem in the lesson introduction. What percent of the iceberg is above water?

$0.125 = \frac{125}{1,000}$ *First, express 0.125 as a fraction.*

$\quad = \frac{125 \div 10}{1,000 \div 10}$ *Divide the numerator and the denominator by 10 to get a denominator of 100.*

$\quad = \frac{12.5}{100}$

$\quad = 12.5\%$

So, 12.5% of the iceberg is above water.

To express a percent as a decimal, express the percent as a fraction with a denominator of 100. Then express the fraction as a decimal.

Examples

Express each percent as a decimal.

3 72%

$$72\% = \frac{72}{100}$$
$$= 0.72$$

So, $72\% = 0.72$.

4 83.5%

$$83.5\% = \frac{83.5}{100}$$
$$= \frac{83.5 \times 10}{100 \times 10}$$
$$= \frac{835}{1,000}$$
$$= 0.835$$

Multiply the numerator and denominator by 10 so that the numerator is a whole number.

So, $83.5\% = 0.835$.

5 $37\frac{1}{2}\%$

$$37\frac{1}{2}\% = \frac{37\frac{1}{2}}{100}$$
$$= \frac{37.5}{100}$$

Rewrite $37\frac{1}{2}$ as 37.5.

37.5 ÷ 100 = **0.375**

$$= 0.375$$ So, $37\frac{1}{2}\% = 0.375$.

Mini-Lab

Work with a partner.

Materials: tape measure, yardstick, or meterstick

- Measure your partner's height (h) in inches.
- With your partner's arms outstretched, measure the distance from fingertip to fingertip (r) in inches.
- Compute the ratio r to h for your partner.
- Write the ratio as a percent.
- Repeat these steps as your partner measures you.

Talk About It

a. What does the percent represent?

b. If you know the distance from fingertip to fingertip of a person, do you think you could predict that person's height? Explain.

Checking for Understanding

Communicating Mathematics

Read and study the lesson to answer each question.

1. **Write** the steps you would use to express a percent as a decimal.

2. **Tell** how the diagram illustrates that any number can be expressed in three ways.

Guided Practice

Express each decimal as a percent.

3. 0.46   4. 0.05   5. 0.6   6. 0.565

Express each percent as a decimal.

7. 39%   8. 4%   9. 70%   10. $23\frac{1}{4}\%$

11. Which is greater: 17% or 1.7?

Exercises

Independent Practice

Express each decimal as a percent.

12. 0.39   13. 0.75   14. 0.875   15. 0.325

16. 0.4   17. 0.03   18. 0.07   19. 0.01

20. 0.075   21. 0.999   22. 0.099   23. 1

Express each percent as a decimal.

24. 43%   25. 89%   26. 7%   27. 2%

28. 17%   29. 90%   30. 34.5%   31. 13.4%

32. 6.2%   33. $62\frac{1}{2}\%$ 34. $33\frac{1}{4}\%$ 35. 100%

Replace each ⬤ with $<$, $>$, or $=$.

36. 35% ⬤ 3.5   37. 7.8 ⬤ 78%   38. 0.05 ⬤ 50%

39. 100% ⬤ 1.1   40. 57.8% ⬤ 0.0578   41. 0.3 ⬤ 30%

42. 0.09 ⬤ 1%   43. 2.4% ⬤ 0.0204   44. $1\frac{3}{4}\%$ ⬤ 0.175

45. Which is greater: 0.63 or 6.3%?

46. *True* or *false:* 0.425 = 42.5%

Mixed Review

47. Evaluate 5^3. *(Lesson 1-9)*

48. Change 56 ounces to pounds. *(Lesson 6-6)*

49. Find the absolute value of -21. *(Lesson 7-1)*

50. Find $\sqrt{144}$. *(Lesson 9-2)*

51. Andrea must wrap a package shaped as a rectangular prism having length 12 inches, width 8 inches, and height 4 inches. How much paper will Andrea need? *(Lesson 10-3)*

52. Express 35% as a fraction in simplest form. *(Lesson 11-8)*

Problem Solving and Applications

53. **Sports**   Tadashi has a batting average of 0.344.

   a. Write this number as a percent.

   b. About how many hits could he expect to have out of his next 100 times at bat?

54. **Consumer Math**   The standard rate for tipping in a restaurant is 15% of your total bill.

   a. Write this percent as a decimal.

   b. A family of six has a bill of $100 at a restaurant. What should their tip be?

   c. What is the total amount they should expect to pay at the restaurant?

55. **Critical Thinking**   Write 5.4×10^{-2} as a percent.

Save Planet Earth

Up in Smoke   Over 467,000 tons of tobacco are burned each year. Smoking is the largest cause of indoor air pollution. It affects not only the smoker, but also those that live and work around smokers. Those that inhale the smoke of other smokers are called "passive smokers."

Because smoking is a health hazard for everyone involved, people and businesses are beginning to take steps to reduce the amount of smoke in the workplace and home.

How You Can Help

- Don't smoke and discourage smoking in your home.
- Encourage smokers to go outdoors to smoke.
- If someone is smoking indoors, make sure the room is ventilated.

Percents Greater Than 100% and Less Than 1%

Objectives

Express percents greater than 100% and percents less than 1% as fractions and as decimals, and vice versa.

In 1990, the world's population was 5,292,177,000. Experts estimate that the population will increase to 8,466,516,000 by the year 2025, which is about 160% of the 1990 population. How many times greater than the 1990 population will the 2025 population be? *This question will be answered in Example 1.*

People who analyze and interpret data often use percents greater than 100%. These percents represent numbers that are greater than 1. You can also use percents less than 1%. These precents represent numbers that are less than 0.01 or $\frac{1}{100}$.

Mini-Lab

Work with a partner.

Materials: grid paper, markers

- Draw two 10×10 squares on your grid paper. Each large square represents 100%, and each small square represents 1%. Shade 160 squares.

- Draw another 10×10 square. Shade four tenths of one square.

Talk About It

a. Which drawing represents a percent greater than 100%? What is the percent?

b. Which drawing represents a percent less than 1%. What is the percent?

Example 1 *Connection*

Statistics Refer to the problem in the lesson introduction. Express 160% as a decimal to find out how many times greater the population will be in 2025.

$$160\% = \frac{160}{100}$$
$$= 1.6 \quad \text{So, in the year 2025, the world population will be about 1.6 times the 1990 population.}$$

Examples

Express each decimal as a percent.

2 1.25

$$1.25 = 1\frac{25}{100}$$
$$= \frac{125}{100}$$
$$= 125\%$$

So, $1.25 = 125\%$.

3 6.3

$$6.3 = 6\frac{3}{10}$$
$$= \frac{63}{10} \qquad \textit{Multiply}$$
$$= \frac{630}{100} \qquad \begin{array}{l}\textit{numerator and}\\\textit{denominator}\\\textit{by 10.}\end{array}$$
$$= 630\%$$

So, $6.3 = 630\%$.

Example 4 *Connection*

Statistics Recently, it was estimated Nissan and Mitsubishi together had 0.9% of the mini-van market. What is this percent as a fraction and as a decimal?

$$0.9\% = \frac{0.9}{100}$$
$$= \frac{0.9 \times 10}{100 \times 10} \qquad \textit{Multiply numerator and denominator by 10.}$$
$$= \frac{9}{1,000}$$
$$= 0.009 \quad \text{Nissan and Mitsubishi together had about } \frac{9}{1,000}, \text{ or } 0.009, \text{ of the market.}$$

Saab and Volvo are Swedish-made cars. They made up 40% of all Swedish goods imported to the United States in 1987.

Examples

Express each fraction as a percent.

In Example 5, explain why the numerator and denominator of $\frac{3}{400}$ are divided by 4.

5 $\frac{3}{400}$

$$\frac{3}{400} = \frac{3 \div 4}{400 \div 4}$$
$$= \frac{0.75}{100} \qquad 3 \div 4 = 0.75$$
$$= 0.75\%$$

So, $\frac{3}{400} = 0.75\%$.

6 $\frac{12}{2,000}$

$$\frac{12}{2,000} = \frac{12 \div 20}{2,000 \div 20}$$

$12 \boxed{\div} 2000 \boxed{=} 0.006$

$0.006 = 0.6\%$

So, $\frac{12}{2,000} = 0.6\%$.

Checking for Understanding

Communicating Mathematics

Read and study the lesson to answer each question.

1. **Tell** why 175% of 80 must be greater than 80.

2. **Tell** how you know if a decimal or fraction will be a percent greater than 100%.

3. **Tell** the percent represented by each diagram below.

   a.  b.

4. a. **Draw** a diagram to represent 130%.

   b. **Draw** a diagram to represent $\frac{1}{2}$%.

Guided Practice

Express each percent as a decimal.

5. 400%	6. 180%	7. 130%	8. 145%
9. 0.75%	10. 0.24%	11. $\frac{1}{5}$%	12. $\frac{1}{8}$%

Express each number as a percent.

13. 1.8	14. 1.1	15. 0.005	16. 0.0035
17. $9\frac{1}{4}$	18. $7\frac{1}{2}$	19. 0.0092	20. 0.00116

Tell whether each of the following is reasonable. Explain your answer.

21. The 1990 population of California is 115% of its 1980 population.

22. Wayne Gretzky makes 125% of his shots on goal.

23. **Statistics** The population of Mexico in 1990 was 88,597,000. Experts estimate that by the year 2025, the population will be 150,061,000. Use a calculator to find the ratio of the 2025 population to the 1990 population as a percent. Round your answer to the nearest whole percent.

Exercises

Independent Practice

Express each percent as a decimal.

24. 100%	25. 0.068%	26. 325%	27. 200%
28. 0.0025%	29. 0.012%	30. 240%	31. 0.032%

Express each number as a percent.

32. $3\frac{1}{2}$	33. 5	34. $4\frac{3}{4}$	35. $5\frac{1}{4}$
36. $3\frac{2}{5}$	37. $1\frac{9}{10}$	38. 80	39. 285
40. 1.7	41. 0.001	42. 2.25	43. 0.009
44. 18	45. 3.1	46. 0.0025	47. 4

Replace each ● with <, >, or =.

48. 1.5 ● 150%

49. 1.25 ● 125%

50. 14,000% ● 14

51. 560 ● 5,600%

52. $12 \times \frac{1}{4}$ ● $12 \times 25\%$

53. $15 \times 1\frac{1}{3}$ ● $133\frac{1}{3}\% \times 15$

Tell whether each of the following is reasonable. Explain your answer.

54. 140% of the M&M® candies in a one-pound bag are brown.

55. An antique toy car is now worth 2,300% of its original price.

56. Nina makes 105% of her free throws.

57. John gave away 130% of his coin collection.

58. The school's enrollment is 118% of last year's enrollment.

59. A pine tree grows to 150% of its present height within two years.

Mixed Review
60. **Statistics** Construct a line plot for the following set of data: 14, 18, 15, 14, 17, 16, 17, 14, 15. *(Lesson 3-4)*

61. Find the prime factorization of 140. *(Lesson 4-2)*

62. **Geometry** Find the area of a rectangle having a length of 8.5 inches and a width of 4 inches. *(Lesson 6-7)*

63. **Geometry** Complete the pattern unit for the reflection at the right. Then draw the tessellation. *(Lesson 8-8)*

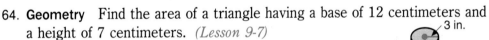

64. **Geometry** Find the area of a triangle having a base of 12 centimeters and a height of 7 centimeters. *(Lesson 9-7)*

65. **Geometry** Find the surface area of the cylinder at the right. *(Lesson 10-4)*

3 in.

8 in.

66. **Personal Finance** Wendy spends 45% of her earnings on clothing. Write this percent as a decimal. *(Lesson 11-9)*

Problem Solving and Applications
67. **Statistics** The population of Kenya in 1990 was 25,129,000. Experts estimate that by the year 2025 the population will be 77,615,000. Use a calculator to express the ratio of the 2025 population to the 1990 population as a percent. Round your answer to the nearest whole percent.

68. **Critical Thinking** In 1990, the population of Hong Kong was 5,840,000. Experts estimate that by the year 2025 the population will be 119% of the 1990 population. What is the estimated population of Hong Kong in the year 2025?

Study Guide and Review

Communicating Mathematics

Choose the letter that best matches each phrase.

1. a comparison of two numbers by division

2. a ratio of two measurements with different units

3. an equation that shows that two ratios are equivalent

4. a ratio that compares a number to 100

5. $\frac{3}{5}$ written as a percent

6. In your words, explain how the scale on a map and a ruler can be used to find the actual distance between two cities.

a. percent
b. 60%
c. decimal
d. ratio
e. equation
f. 30%
g. unit rate
h. fraction
i. proportion
j. rate

Skills and Concepts

Objectives and Examples	Review Exercises
Upon completing this chapter, you should be able to:	*Use these exercises to review and prepare for the chapter test.*

• express ratios as fractions and determine whether two ratios are equivalent *(Lesson 11-1)*

Express the ratio 6:18 as a fraction in simplest form.

$$\frac{6}{18} = \frac{6 \div 6}{18 \div 6} = \frac{1}{3}$$

The simplest form is $\frac{1}{3}$.

Express each ratio as a fraction in simplest form.

7. 25 to 10   8. 14:70
9. 11:66   10. 12 to 64
11. 90 to 33   12. 50:100
13. 63:9   14. 5 to 10

• determine unit rates *(Lesson 11-2)*

Find the unit price for a 16-ounce box of pasta on sale for 96 cents.

$$\frac{cents}{ounces} \rightarrow \frac{96}{16} = \frac{96 \div 16}{16 \div 16} = \frac{6}{1}$$

The unit price is 6 cents per ounce.

Express each rate as a unit rate.

15. 16 cups for 4 people
16. 150 people for 5 classes
17. $23.75 for 5 pounds
18. 810 miles in 9 days
19. $38 in 4 hours
20. 24 gerbils in 3 cages

Objectives and Examples

- solve proportions *(Lesson 11-3)*

  Solve $\frac{6}{9} = \frac{x}{12}$.

  $$\frac{6}{9} = \frac{x}{12}$$

  $$6 \times 12 = 9x$$

  $6\ \boxed{\times}\ 12\ \boxed{\div}\ 9\ \boxed{=}\ 8$

  $8 = x$ The solution is 8.

- identify corresponding parts of similar polygons *(Lesson 11-4)*

  1 cm $\boxed{}$ 2 cm $\boxed{}$
  3 cm   6 cm

  The polygons above are similar because corresponding angles are congruent and corresponding lengths are in proportion: $\frac{1}{2} = \frac{3}{6}$.

- solve problems involving scale drawings *(Lesson 11-5)*

  On a map, the scale is 1 inch:80 miles. Find the actual distance for a map distance of $3\frac{1}{4}$ inches.

  $$\frac{map}{actual} \xrightarrow{} \frac{1}{80} = \frac{3\frac{1}{4}}{n} \xleftarrow{} \frac{map}{actual}$$

  $n = 80 \times 3\frac{1}{4}$ or 260 miles

- illustrate the meaning of percent *(Lesson 11-7)*

  Express 47:100 as a percent.

  $$\frac{47}{100} = 47\%$$

- express fractions as percents, and vice versa *(Lesson 11-8)*

  Express $\frac{18}{20}$ as a percent.

  $$\frac{18}{20} = \frac{x}{100} \rightarrow 1{,}800 = 20x$$
  $$90 = x$$

  So, $\frac{18}{20} = 90\%$.

Review Exercises

Solve each proportion.

21. $\frac{13}{25} = \frac{39}{m}$ 22. $\frac{w}{6} = \frac{12}{8}$

23. $\frac{350}{p} = \frac{2}{10}$ 24. $\frac{45}{5} = \frac{x}{7}$

Tell whether each pair of polygons is similar. Justify your answer.

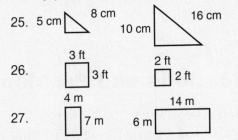

25. 5 cm 8 cm   10 cm   16 cm

26. 3 ft / 3 ft   2 ft / 2 ft

27. 4 m / 7 m   6 m / 14 m

On a map, the scale is 1 cm:36 km. For each map distance, find the actual distance.

28. 6 cm   29. 4 cm
30. 45 cm   31. 10 cm
32. 2 cm   33. 12 cm
34. 3 cm   35. 100 cm

Express each ratio as a percent.

36. 12 out of 100
37. 63 out of 100 days
38. 99 out of 100 students

Express each fraction as a percent, or vice versa. Express fractions in simplest form.

39. $\frac{3}{5}$ 40. 65%

41. $13\frac{1}{2}\%$ 42. $\frac{150}{200}$

43. $\frac{5}{8}$ 44. $33\frac{1}{3}\%$

Objectives and Examples	Review Exercises
• express decimals as percents, and vice versa *(Lesson 11-9)* Express 27% as a decimal. $27\% = \frac{27}{100} = 0.27$	Express each decimal as a percent, or vice versa. 45. 0.47 46. 43.5% 47. 75% 48. 0.375 49. 0.995 50. $22\frac{1}{2}\%$
• express percents greater than 100% and percents less than 1% as fractions and as decimals, and vice versa *(Lesson 11-10)* Write 2.35 as a percent. $2.35 = 2\frac{35}{100} = \frac{235}{100} = 235\%$	Express each percent as a decimal, or vice versa. 51. 125% 52. 0.25% 53. 0.002 54. 0.05% 55. 4.75 56. 0.0095

Applications and Problem Solving

57. **Population Density** In 1988, Memphis, Tennessee had a population of 645,190 people and an area of 264 square miles. To the nearest whole number, how many people per square mile are there in Memphis? *(Lesson 11-2)*

58. The advisor for the Spanish Club tells three students about a club meeting. It takes 1 minute for her to tell three students and 1 minute for each of those three students to tell three other students, and so on. How many students will know about the meeting in three minutes? *(Lesson 11-6)*

Curriculum Connection Projects

• **Sports** Find the fastest times for at least three Olympic running events. Write a ratio for each, comparing the length of the race (in meters) to the running time (in seconds). Find equivalent ratios, comparing kilometers to seconds and centimeters to seconds.

• **Consumer Awareness** Find the unit price of the different-sized boxes of several kinds of cereal. Make comparisons and write out your conclusions about how to be a smart shopper.

Read More About It

Du Bois, William Penn. *Giant.*
Phillips, Louis. *Brain Busters: Just How Smart Are You Anyway?*
Renner, A.G. *How to Build a Better Mousetrap, Car, and Other Experiments.*

Test

Express each ratio as a fraction in simplest form.

1. 35:15

2. 42 out of 60 days

Express each ratio as a unit rate.

3. 24 cards for $4.80

4. 330 miles on 15 gallons of gas

5. In 1988, Richmond, Virginia had a population of 213,300 people and an area of 60 square miles. How many people per square mile were there in Richmond in 1988?

Solve each proportion.

6. $\frac{2}{3} = \frac{x}{42}$

7. $\frac{9}{m} = \frac{12}{36}$

8. **Physical Fitness** Alissa swims 3 laps in 12 minutes. At this same rate, how many laps will she swim in $10\frac{1}{2}$ minutes?

Tell whether each pair of polygons is similar. Justify your answer.

9.

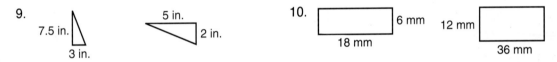

10.

On a map, the scale is 1 inch:150 miles. For each map distance, find the actual distance.

11. 5 inches

12. $3\frac{5}{6}$ inches

13. The express bus arrives at the Maple Street bus stop every 25 minutes, beginning at 5:40 A.M. The local bus arrives at this same bus stop every 15 minutes, beginning at 6:00 A.M. What is the first time during the day that the buses will arrive at the same time?

Express each ratio as a percent.

14. 15:100

15. 95 points out of 100

Express each fraction as a percent, or vice versa.

16. $\frac{3}{8}$

17. $24\frac{1}{2}\%$

Express each decimal as a percent, or vice versa.

18. 0.65

19. $\frac{1}{4}\%$

20. **Economics** Analysts predict that the minimum wage in the year 2000 will be 130% of the 1991 minimum wage, which was $4.25. What is the predicted minimum wage for the year 2000? Round to the nearest cent.

Bonus Express $\frac{1}{25}\%$ in scientific notation.

Applications with Percent

Spotlight on Mail

Have You Ever Wondered. . .

- What percent of the mail is delivered early or late?

- How the cost of a postage stamp has changed over the years?

Price of a First-Class Postage Stamp

Year	Price ($)
1974	0.10
1975	0.13
1978	0.15
1981	0.20
1985	0.22
1988	0.25
1991	0.29

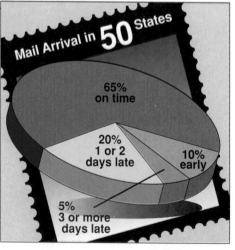

Mail Arrival in **50** States

65% on time

20% 1 or 2 days late

10% early

5% 3 or more days late

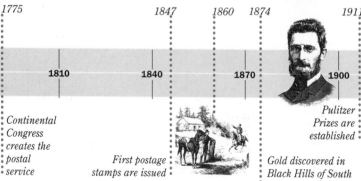

1775   1847   1860   1874   1911

1810   1840   1870   1900

Continental Congress creates the postal service

First postage stamps are issued

Gold discovered in Black Hills of South Dakota

Pulitzer Prizes are established

Looking Ahead

In this chapter, you will see how mathematics can be used to answer questions about the mail.

The major objectives of the chapter are to:

- find the percent of a number
- estimate using fractions, decimals, and percents
- solve problems using the percent proportion or percent equation
- solve problems involving discounts, sales tax, and simple interest

Chapter Project

Mail
Work in a group.

1. For three weeks, keep track of some of the mail that comes into your home. Record the date stamped on the envelope as well as the date it is received.

2. Group your data according to the number of days it took to arrive. Present all of your data in a chart or a graph.

3. Make a circle graph to show what percent of the mail fell into each group.

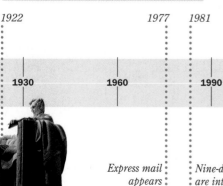

1922		1977	1981
1930	1960		1990

Express mail appears

Nine-digit zip codes are introduced

453

12-1 Percent of a Number

Objective

Find the percent of a number.

Words to Learn

percent
percentage
base
rate
percent
 proportion

"Hello, we can't come to the phone right now. . . ." Have you ever called a friend and gotten a message like this on their answering machine? It can be frustrating to get the answering machine rather than the person. And yet these devices can also be helpful. Rather than calling someone over and over, you can just leave a message for them to call you back.

Americans are realizing the convenience of answering machines. *About* 40% of American households now have answering machines. If there are 92,800,000 households in the United States, *about* how many have answering machines?

Results of surveys are often reported as percents. A **percent** is a ratio that compares a number to 100. So 40% stated in the problem above means that 40 out of every 100 households have answering machines.

Let x represent the number of households with answering machines. Write a proportion.

You can review proportions on page 414.

$$\frac{x}{92,800,000} = \frac{40}{100}$$

$$x \cdot 100 = 92,800,000 \cdot 40$$

$$x = \frac{3,712,000,000}{100} \qquad \textit{Divide each side by 100.}$$

$$x = 37,120,000$$

In the United States, 37,120,000 households have answering machines.

In the proportion above, x is called the **percentage (P).** The number 92,800,000 is called the **base (B).** The ratio $\frac{40}{100}$ is called the **rate.**

$$\frac{40}{100} \quad \rightarrow \quad \frac{\textit{Percentage}}{\textit{Base}} = \textit{Rate}$$

If r represents the number per hundred, the proportion can be written as $\frac{P}{B} = \frac{r}{100}$. This proportion is called the **percent proportion.**

Example *Problem Solving*

Music   Compact disc players are on sale for 20% off. If the regular price is $200, what is the discount?

You can draw a diagram to show 20% of 200.

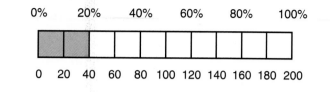

You can also use the percent proportion to find 20% of 200.

$$\frac{P}{B} = \frac{r}{100}$$

$$\frac{P}{200} = \frac{20}{100}$$ *Replace B with 200 and r with 20.*

$$P \cdot 100 = 200 \cdot 20$$ *Write the cross products.*

$$\frac{P \cdot 100}{100} = \frac{4,000}{100}$$ *Divide each side by 100.*

$$P = 40$$

The discount is $40.

Checking for Understanding

Communicating Mathematics

Read and study the lesson to answer each question.

1. **Tell** what is meant by the rate.
2. **Write** a proportion that can be used to find 42% of 386.
3. **Write** a percent proportion that represents the diagram at the right.

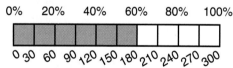

4. **Show** how you would express a fraction as a percent.

Guided Practice

Complete each proportion or write a proportion for each problem. Then solve. Round answers to the nearest tenth.

5. Find 93% of 215.

$$\frac{P}{B} = \frac{r}{100}$$

$$\frac{P}{215} = \frac{?}{?}$$

6. Find 64% of 88.

$$\frac{P}{B} = \frac{r}{100}$$

$$\frac{?}{?} = \frac{64}{100}$$

7. What number is 40% of 220?

8. 12% of 16.5 is what number?

Exercises

Independent Practice

Use a proportion to solve each problem. Round answers to the nearest tenth.

9. What number is 25% of 560?
10. Find $37\frac{1}{2}$% of 64.
11. 50% of 128 is what number?
12. What number is 25% of 36?
13. Find 80% of $90\frac{1}{2}$.
14. What number is 75% of 92?

15. 30% of 96 is what number?

16. 28% of 14 is what number?

17. Twenty-five percent of 32 is what number?

18. What number is 20% of twenty?

19. The purchase price of an answering machine is $140. The state tax rate is 6.5% of the purchase price.
    a. How much state tax is charged?
    b. What is the total cost?

Mixed Review

20. Find 2.6×0.3. *(Lesson 2-4)*

21. Find $12 - (-4)$. *(Lesson 7-5)*

22. **Measurement** A can of mixed vegetables is in the shape of a cylinder with a radius of 1.5 inches and a height of 6 inches. Find the amount of paper necessary to make a label for the can. Remember, the label on a can does not cover the ends. *(Lesson 10-4)*

23. Express 0.00065 as a percent. *(Lesson 11-9)*

Problem Solving and Applications

24. **Education** In the United States, approximately 30% of the students entering ninth grade do not graduate from high school. Of the 345 people entering the ninth grade class at Jefferson High, how many of these students are likely to graduate from high school?

25. **Ecology** On an average day, approximately 93.3 million aluminum cans are produced. Of these, *about* 50% are recycled.
    a. To the nearest million, how many aluminum cans are recycled each day?
    b. If an aluminum can weighs *about* 0.5 ounce, how many pounds of aluminum are recycled on an average day?

26. **Critical Thinking** Ngan paid $224 for a remote-control, color television. Tim paid 110% of this amount for the same television at another store. How much more did Tim pay for the same television?

27. **Immigration** On an average day 1,648 persons immigrate to the United States. Nearly 41% of these will become naturalized American citizens. On an average day, how many of the immigrants are likely to become citizens?

28. **School** If 43% of the 219 students in the seventh grade bring lunch, *about* how many students buy their lunch in the school cafeteria?

29. **Consumer Math** Sam Davis bought a couch that cost $899. The store required a 25% down payment to hold the couch. How much was the down payment?

30. **Travel** The travel club is going to the Grand Canyon. There are 44 members. If 75% of the members sign up for the trip, how many members are going to the Grand Canyon?

12-2 **Solve a Simpler Problem**

Objective

Solve problems by solving a simpler problem.

A 1989 survey of 58,000 households found that 11 states west of the Mississippi River had the highest percent of high school graduates, while the Washington, D.C. area had the highest percent with college degrees.

Suppose the number of people over the age of 25 in a certain county is 14,757,899. If 78.6% of the people over 25 have high school diplomas, *about* how many people have diplomas?

Explore What do you know?
You know that the county's population is 14,757,899 and that the percent of people over 25 who graduated from high school is 78.6%.

What are you trying to find?
You are trying to find *about* how many people over the age of 25 in this county have high school diplomas.

Plan Round the numbers to make a simpler problem. Then use patterns to find the product.

Solve Round each number to its greatest place value.

14,757,889 → 10,000,000
78.6% → 80%

Think:
$$80\% \text{ of } 10 = 8$$
$$80\% \text{ of } 100 = 80$$
$$80\% \text{ of } 1,000 = 800$$
$$80\% \text{ of } 10,000 = 8,000$$
$$80\% \text{ of } 100,000 = 80,000$$
$$80\% \text{ of } 1,000,000 = 800,000$$
$$80\% \text{ of } 10,000,000 = 8,000,000$$

About 8,000,000 people have diplomas.

Examine You could also solve the problem by changing 80% to a fraction and multiply mentally to find $\frac{4}{5}$ of 10,000,000.

$$\frac{4}{5} \times 10,000,000 = 8,000,000 \checkmark$$

Checking for Understanding

Communicating
Mathematics
Read and study the lesson to answer each question.

1. **Tell** how you would use a calculator to find the exact number of people in California who have high school diplomas.

Guided Practice
Solve by solving a simpler problem.

2. Rochelle sent out 24 invitations to her party and asked each guest to RSVP by Saturday. Twenty-five percent of the guests could not attend. How many people were at her party?

3. Michael made a 30% down payment on a $12,000 car. How much was his down payment?

Problem Solving

Practice
Solve. Use any strategy.

Strategies
● ● ● ● ● ● ● ● ● ●
Look for a pattern.
Solve a simpler problem.
Act it out.
Guess and check.
Draw a diagram.
Make a chart.
Work backwards.

4. Two thirds of the student body voted in the Student Council election. If there are 600 students, how many people voted?

5. There are 16 dancers trying out for a musical. Twenty-five percent of them will receive a part. How many parts are there for dancers?

6. There are 24,624 high school seniors applying to a certain college for enrollment. If the college only accepts 2 out of every 9 applicants, how many of the seniors will be accepted?

7. A number is doubled and -9 is added to it. The result is -1. What is the number?

CULTURAL KALEIDOSCOPE

Henry Cisneros

Henry Cisneros was aware of his Mexican-American heritage and keenly interested in helping poorer citizens. His doctorate in public administration took him to San Antonio State University in Texas, where he began his teaching and political careers.

At 27 he became the youngest councilman in San Antonio history (1975–1981). During his term, he became involved with the Mexican-American community action group called Communities Organized for Public Service (COPS). He gained national attention as a leading Hispanic politician and was appointed by President Reagan to serve on the National Bi-Partisan Commission on Central America.

In 1981, he was elected Mayor of San Antonio and retired from office in 1989. During his administration, San Antonio benefitted from his vision and desire for improvement in the lives of all citizens. He brought revenue, jobs, and revitalization to San Antonio.

12-3 Percent and Estimation

Objective

Estimate by using fractions, decimals, and percents interchangeably.

Suppose that ElectroWorld is having a 25% off sale on all CD players in February. You decide to buy one that regularly costs $239. *About how much money could you save during this sale?*

You can estimate the savings by rounding $239 to $240 and using one of the methods below.

Fraction Method	1% Method	Meaning of Percent Method
25% is the same as $\frac{1}{4}$. $\frac{1}{4}$ of $240 is $60.	1% means $\frac{1}{100}$. 1% of 240 is $\frac{1}{100}(240)$ or 2.40. $2.40 rounds to $2. Now find 25% or 25(1%). $25 \times \$2 = \$50$	25% means $25 for every $100 and $2.50 for every $10. $240 = 2(\$100) + 2(\$20)$ $25(2) + \$2.50(4) = \$60$
Estimate: $60	**Estimate: $50**	**Estimate: $60**

Calculator Hint

● ● ● ● ● ● ● ● ● ● ● ●

You can use the %
key on your calculator to compute with percents. The percent key replaces the percent with its decimal equivalent.

You can use a calculator to find the exact savings.   25 [%] [×] 239 [=] **59.75**

The exact savings would be $59.75. Compare this to the estimates to see if the answer is reasonable.

The following Mini-Lab will help you relate area and percent.

Mini-Lab

Work with a partner.
Materials: grid paper, marker

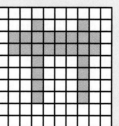

- Estimate the percent of the shaded portion of each figure.
- Then count grid squares to find the percent shaded.
- Draw a design on a 10×10 grid and shade it. Estimate the percent shaded. Then count to find the exact percent.

Talk About It

How do the estimates compare with the actual percents?

Example 1

Estimate 38% of 500.

38% is about 40% or $\frac{2}{5}$.

$\frac{1}{5}$ of 500 is 100.

$\frac{2}{5}$ of 500 is 200.

So, 38% of 500 is *about* 200.

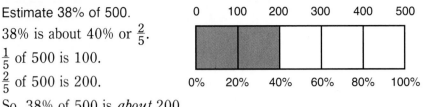

Example 2 *Problem Solving*

Smart Shopping Estimate the savings on a $349 item with 30% off.

10% of $349 is $34.90

$34.90 rounds to $35.00.

3 times $35.00 is $105.00   The savings is *about* $105.00.

You can also estimate percents of numbers when the percent is less than 1 or the percent is greater than 100.

Examples

3 Estimate 112% of $36.

112% is more than 100%, so 112% of 36 is greater than 36.

112% is *about* 110%.

110% = 100% + 10%

36(100% + 10%) = 36 + 3.6

                   = 39.6

112% of 36 is *about* 39.6.

4 Estimate 0.5% of 521.

0.5% is half of 1%.

521 is about 500.

1% means $\frac{1}{100}$.

$\frac{1}{100} \cdot 500 = 5$

$\frac{1}{2}$ of 5 is 2.5.

0.5% of 521 is *about* 2.5.

Checking for Understanding

Communicating Mathematics

Read and study the lesson to answer each question.

1. **Draw** a figure or design on a 10×10 grid. Shade $\frac{2}{10}$ of the figure or design. What percent is shaded?

2. **Show** how you would find a percent of a number using a calculator.

Estimate the percent shaded. Then count to find the exact percent.

3.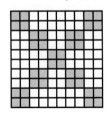

4.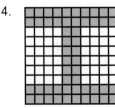

5.

Estimate.

6. 25% of 408   7. 0.3% of 425   8. 121% of $56

Exercises

Independent Practice

Write the fraction, decimal, mixed number, or whole number equivalent of each percent that could be used to estimate.

9. 37%   10. 25%   11. 200%   12. 87%
13. 13%   14. $4\frac{1}{2}$%   15. 0.8%   16. 43.5%
17. $\frac{7}{8}$%   18. 16.97%   19. $12\frac{2}{5}$%   20. 350%

Estimate.

21. 40% of 62   22. 25% of 18   23. 16% of 32.6
24. 1% of 89   25. $6\frac{1}{2}$% of 236   26. 30.5% of 50
27. 0.6% of 220   28. 150% of 52   29. 75% of 125
30. 8% of $12\frac{3}{4}$ 31. 0.3% of 35   32. 200% of 540
33. What number is 20% of $16.21? Estimate.
34. Estimate 50% of 89.

Mixed Review

35. **Statistics** Find an appropriate scale and interval for the following data. Then construct a number line using the scale and interval. 25, 39, 15, 48, 33, 27, 14. *(Lesson 3-3)*

36. Solve $p - 14 = 27$. *(Lesson 6-2)*

37. **Geometry** Classify the angle at right as acute, obtuse, right, or straight. *(Lesson 8-1)*

38. **Travel** On his summer vacation, Matthew drove 350 miles in 5 hours on the first day. He continued driving at the same rate the second day and drove for 8 hours. How many miles did Matthew drive the second day of his vacation? *(Lesson 11-2)*

39. Find 45% of 1,600. *(Lesson 12-1)*

Problem Solving and Applications

40. **Consumer Math** When the McGraw family went out for pizza, their bill was $21.97. They wanted to leave a tip of approximately 15%. What is a reasonable estimate of the tip?

41. **Horoscope** On an average day, more than 105 million Americans read a newspaper. Of these, 26% of the people read their horoscope. Estimate the number of newspaper readers who read their horoscope.

42. **Animals** A flying squirrel ranges from 20 to 37 inches in length. Its tail is 40% of the squirrel's total length. Estimate the length of the tail of a flying squirrel.

43. **Journal Entry** Draw a figure or design on a 10×10 grid that represents 35%. Then express 35% as a fraction with a denominator of 100 and simplify. Write its decimal equivalent.

The Percent Proportion

Objective

Solve problems using the percent proportion.

What do Betsy King, Steffi Graf, and Ingrid Kristiansen have in common? They are all women athletes.

Women are competing in more and more sporting events these days. And yet women's sporting events only get 5% of the total television air time. If there are a total of 140 hours of air time in an average week, how many of these hours are in women's sporting events? Use the percent proportion to solve this problem.

$$\frac{P}{B} = \frac{r}{100}$$

$$\frac{P}{140} = \frac{5}{100} \qquad B = 140,\ r = 5$$

$$P \cdot 100 = 140 \cdot 5 \qquad \textit{Find the cross products.}$$

$$\frac{P \cdot 100}{100} = \frac{700}{100} \qquad \textit{Divide each side by 100.}$$

$$P = 7 \qquad \text{Women's sporting events are aired 7 hours per week.}$$

There are many competitors for each position in professional and amateur sports. Competition for the Olympics is very fierce. The graph shows an estimated number of athletes who competed for Olympic teams and the number of athletes who were selected to attend the games.

Olympic Hopefuls
Summer Olympics
Barcelona, Spain

1,050 Wrestling
20
900 Track and Field
125
800 Swimming
50
684 Boxing
12
400 Cycling
20
120 Baseball
20
120 Gymnastics
15

■ Athletes trying out
■ Athletes attending games

Example 1 *Connection*

Statistics   What percent of the athletes trying out for boxing in the 1992 Olympics actually got to attend the games?

$$\frac{P}{B} = \frac{r}{100}$$

$$\frac{12}{684} = \frac{r}{100} \qquad \textit{Replace B with 684 and P with 12.}$$

$$12 \cdot 100 = 684 \cdot r \qquad \textit{Find the cross products.}$$

$$1,200 = 684\,r$$

$$1200 \; \boxed{\div} \; 684 \; \boxed{=} \; \text{1.754386} \qquad \textit{Divide each side by 684.}$$

$$1.754386 \approx r$$

To the nearest tenth of a percent, *about* 1.8% of the athletes who tried out for the 1992 Olympic boxing team got to attend the games.

You can also use the percent proportion to find the base when the percentage and rate are known.

Example 2 *Problem Solving*

Smart Shopping   A combination speakerphone/answering machine is on sale for $140. This is 70% of the regular price. What is the regular price?

$$\frac{P}{B} = \frac{r}{100}$$

$$\frac{140}{B} = \frac{70}{100} \qquad P = 140,\ r = 70$$

$$140 \cdot 100 = B \cdot 70 \qquad \textit{Find the cross products.}$$

$$\frac{14,000}{70} = \frac{B \cdot 70}{70}$$

$$200 = B$$

The regular selling price is $200.   *You can check by finding 70% of $200.*

Checking for Understanding

Communicating Mathematics

Read and study the lesson to answer each question.

1. **Tell** what P, B, and r represent in the percent proportion.
2. **Tell** why the percent proportion is helpful.
3. **Show** how you would use the percent proportion to find the rate if the percentage is 18 and the base is 54.

Guided Practice

Match each rate with its corresponding proportion.

4. 48 is 75% of what number?   a. $\frac{P}{48} = \frac{75}{100}$

5. 48 is what percent of 75?   b. $\frac{48}{B} = \frac{75}{100}$

6. What number is 75% of 48?   c. $\frac{48}{75} = \frac{r}{100}$

Write a proportion for each problem. Then solve. Round answers to the nearest tenth.

7. What number is 45% of 60?   8. 3 is what percent of 40?

9. 80 is 75% of what number?   10. What percent of 24 is 12?

11. Find 42.5% of 48.   12. 20% of what number is 25?

13. The class picture included 95% of the students. Seven students were missing from the picture. How many students were in the class?

Exercises

Independent Practice

Name the percentage, base, or rate.

14. 8 is what percent of 16?

15. 20% of what number is 18?

16. 58.2% of 50 is what number?

17. What number is 105% of 36?

18. 5% of what number is $6\frac{1}{2}$?

19. What percent of 30 is 15?

Write a proportion for each problem. Then solve. Round answers to the nearest tenth.

20. What number is 38% of 70?

21. 14 is what percent of 49?

22. 61 is 35% of what number?

23. 7.5% of 48 is what number?

24. What percent of 180 is 30?

25. $12\frac{1}{2}$% of what number is 24?

26. 63 is what percent of 42?

27. 50% of what number is 15.8?

28. $6\frac{1}{4}$% of 235 is what number?

29. What percent of 250 is 25?

30. What number is 40% of 86?

31. 20% of what number is 12?

32. Use the graph on page 462 to find the percent of the 1992 Olympic hopefuls in track and field that actually got to attend the games.

Mixed Review

33. **Finances**   The Andrews' savings account has a balance of $8,750. Mrs. Andrews deposits a check in the amount of $2,175. Estimate the new balance. *(Lesson 1-2)*

34. Use divisibility rules to determine whether 2,350 is divisible by 2, 3, 4, 5, 6, 9, or 10. *(Lesson 4-1)*

35. Find $\frac{9}{12} - \frac{3}{8}$. *(Lesson 5-3)*

36. **Geometry**   Find the value of X for the right triangle. *(Lesson 9-5)*

5 in.

X in.

3 in.

37. Estimate 28% of 160. *(Lesson 12-3)*

Problem Solving and Applications

38. **Critical Thinking**   Use the graph on page 462. The teams of wrestling, baseball, and cycling have the same number of Olympic competitors. Order the sports from highest to lowest percent of hopefuls who made the cut. Explain how you can determine the order without determining the actual percents.

39. **Defense**   On an average day, 715 men and 99 women enlist in the armed forces. What percent of the enlistees are women?

40. **Olympics**   Use the graph on page 462 to determine what percent of the Olympic hopefuls attended the Olympics in each sport.
    a. swimming   b. cycling   c. wrestling

41. **Library**   About 32% of the 1,290 library books have been checked out during the past month. How many library books were checked out?

42. **School**   Twenty-six of the 168 students in Mrs. Johnson's math classes received As on the last test. *About* what percent of the class earned As?

12-5 **The Percent Equation**

Objective

Solve problems using the percent equation.

When am I ever going to use this?

Retail Sales

An interest in working with people and an outgoing personality are an asset to any salesperson in retail sales.

Mathematical knowledge and skills are essential in retail. Calculating sales tax, price reductions, commissions, and making change are all in a day's work.

For additional information, contact:
Professional Salespersons of America
3801 Monaco NE
Albuquerque, NM 87111

Salespeople often work on commission. Mr. O'Donnell sells sports equipment on commission. He sold $14,207 of equipment last month. If he earns 12% commission, how much did he earn that month?

To find out how much commission Mr. O'Donnell earns, find 12% of $14,207.

You can write an equation to solve this problem.

Commission is 12% of $14,207.

$$c = 0.12 \cdot \$14,207 \qquad 12\% = \frac{12}{100} \ or \ 0.12$$
$$c = 0.12 \cdot 14{,}207$$
$$.12 \ \boxed{\times} \ 14207 \ \boxed{=} \ \mathbf{1704.84}$$
$$c = 1{,}704.84$$

Mr. O'Donnell earned $1,704.84 in commission.

Compared to the estimate, is the answer reasonable?

The percent proportion could also have been used to solve this problem.

Steps	Arithmetic	Algebra
Use the percent proportion.	$\frac{P}{14{,}000} = \frac{12}{100}$	$\frac{P}{B} = \frac{r}{100}$
Multiply each side by the base.	$\frac{P}{14{,}000} \cdot 14{,}000 = \frac{12}{100} \cdot 14{,}000$	$\frac{P}{B} \cdot B = \frac{r}{100} \cdot B$
Simplify.	$P = \frac{12}{100} \cdot 14{,}000$	$P = \frac{r}{100} \cdot B$

Lesson 12-5 Algebra Connection: The Percent Equation   **465**

Remember that $\frac{r}{100}$ is called the *rate*. Let R represent $\frac{r}{100}$.

$$P = R \cdot B, \text{ where } R = \frac{r}{100}$$

Percentage = rate · base

Mental Math Hint

● ● ● ● ● ● ● ● ● ● ● ● ●

When solving a percent problem where the base and rate are given, using the equation $P = R \cdot B$ is very convenient.

Examples

1 What number is 32% of 870? *Estimate: $\frac{1}{3} \cdot 900 = 300$*

$P = R \cdot B$

$P = 0.32 \cdot 870$ *Replace R with 0.32 and B with 870.*

$P = 278.4$

32% of 870 is 278.4. Compare to the estimate.

2 28 is what percent of 86? *Estimate: $\frac{28}{86} = \frac{30}{90}$*

$$= \frac{1}{3} \text{ or } 33\frac{1}{3}\%$$

$P = R \cdot B$

$28 = R \cdot 86$ *Replace P with 28 and B with 86.*

$28 \; \boxed{÷} \; 86 \; \boxed{=} \; \text{0.3255813}$ *Divide each side by 86.*

$0.33 \approx R$ *Round to the nearest hundredth.*

28 is about 33% of 86. Compare to the estimate.

3 44 is 55% of what number? *Estimate: 44 is 50% or $\frac{1}{2}$ of 88.*

$P = R \cdot B$

$44 = 0.55 \cdot B$ *Replace P with 44 and R with 0.55.*

$44 \; \boxed{÷} \; 0.55 \; \boxed{=} \; \text{80}$ *Divide each side by 0.55.*

$80 = B$

44 is 55% of 80.

Compared to the estimate, 55% is close to 50% or $\frac{1}{2}$.

Checking for Understanding

Communicating Mathematics

Read and study the lesson to answer each question.

1. **Tell** if the percentage is greater than or less than the base if the rate is less than 100%.

2. **Write** a sentence explaining why the percent proportion is sometimes easier to use than $P = R \cdot B$.

3. **Tell** why percent is equal to the ratio of a number compared to 100.

Write each equation in $P = R \cdot B$ form. Then solve. Round answers to the nearest tenth.

4. $24 = 60\%$ of ▨
5. $22 = $ ▨ $\%$ of 50
6. 16% of $32 = $ ▨
7. $17 = $ ▨ $\%$ of 68
8. 30% of ▨ $= 27$
9. ▨ $= 28\%$ of 32

Write an equation for each problem. Then solve. Round answers to the nearest tenth.

10. Find 26% of 119.
11. 29 is what percent of 61?
12. 17 is 40% of what number?
13. What percent of 87 is 57?
14. 26% of 48 is what number?
15. 75 is 78% of what number?

Exercises

Write an equation for each problem. Then solve. Round answers to the nearest tenth.

16. 15% of what number is 21?
17. 45 is what percent of 36?
18. Find 8% of 38.
19. 55% of what number is 1.265?
20. 70% of what number is 42?
21. Find 20% of 68.
22. 24 is what percent of 25?
23. 18.5% of what number is 11.84?
24. 33% of 72 is what number?
25. 75% of what number is 93?
26. 25 is what percent of 75?
27. 6% of what number is 30?
28. Find 36% of 228.
29. 42.5% of what number is 36?
30. 44 is what percent of 62?
31. $12\frac{3}{4}\%$ of 54 is what number?
32. What is 15% of $\$9.00$?
33. 30 is what percent of 45?

34. A class picnic was attended by 85% of the students. Nine students did not attend. How many students were in the class?

35. If you copy a picture on the photocopying machine at 85%, are you enlarging or reducing the picture? Explain.

36. Find the greatest common factor of 24 and 42. *(Lesson 4-5)*

37. Multiply -13 and -4. *(Lesson 7-6)*

38. **Geometry** Explain why the figure at the right is not a polygon. *(Lesson 8-2)*

39. **Geometry** Find the volume of a cylinder having a radius of 4 centimeters and a height of 6.5 centimeters. *(Lesson 10-6)*

40. Alex answered 23 of the 30 questions on his French test correctly. Find the percent of the questions that Alex answered correctly. *(Lesson 12-4)*

41. **Critical Thinking** Is a 20% discount on a $\$35$ item the same as a 35% discount on a $\$20$ item? Explain.

42. **Photocopying** Suppose you enlarge a drawing to 120% of its original size on the photocopy machine. If the drawing is 2 inches long and 3 inches wide, what are the dimensions of the copy?

43. **Sports** If a pro basketball player makes free throws 92% of the time, how many free throws would you expect him to make if he attempted 18?

44. **Photocopying** Suppose you copy a newspaper ad at 86% on the photocopy machine. If the ad measures 8 inches long and 4 inches wide, what are the dimensions of the copy?

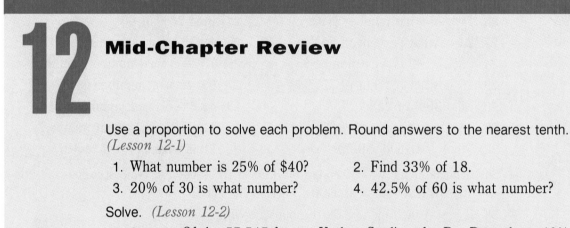

Mid-Chapter Review

12

Use a proportion to solve each problem. Round answers to the nearest tenth. *(Lesson 12-1)*

1. What number is 25% of $40?
2. Find 33% of 18.
3. 20% of 30 is what number?
4. 42.5% of 60 is what number?

Solve. *(Lesson 12-2)*

5. **Sports** Of the 57,545 fans at Yankee Stadium for Bat Day, about 40% of them were children 14 and under and received a free bat. About how many children received free bats?

Estimate. *(Lesson 12-3)*

6. 22% of 19
7. $8\frac{1}{2}$% of 32
8. 175% of 370

Write a proportion for each problem. Then solve.
Round answers to the nearest tenth. *(Lesson 12-4)*

9. What number is 27% of 29?
10. 22 is what percent of 88?
11. 40 is 50% of what number?
12. What percent of 220 is 100?

Write an equation for each problem. Then solve.
Round answers to the nearest tenth.
(Lesson 12-5)

13. 10% of what number is $12\frac{1}{2}$?
14. Find 61% of 83.
15. 35 is what percent of 105?
16. 200% of what number is 720?

12-6A Jelly Bean Statistics

A Preview of Lesson 12-6

Objective

Make a circle graph.

Materials

jelly beans
needles
thread
compass
straightedge

LOOKBACK

You can review
frequency tables on
page 93.

What is your favorite flavor of jelly bean? Almost everyone has a favorite.

Try this!

Work with a partner.

- Take a survey of the people in your class. Tally responses by flavor in a frequency table.

- Sort the jelly beans to reflect the results of the survey. For example, if there are 3 people whose favorite flavor is licorice (black), you would select 3 black jelly beans, and so on.

- String the jelly beans with like flavors together.

- Arrange the jelly beans in a circle. Use a compass to draw a circle the same size.

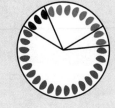

- On the circle, mark sections to indicate the separation by flavor.

- Draw a radius from each mark on the circle to the center.

- Identify each section by flavor.

What do you think?

1. Write a short paragraph describing the circle graph. Include a description of the sizes of the sections in relation to each other.

2. Is there a relationship between the number of tally marks and the size of a section by flavor? If so, write a sentence to describe that relationship.

3. Explain how you can use the percent proportion to find the percent represented by each flavor.

4. Find the percent represented by each flavor. Label each section by flavor and by the percentage it represents.

5. The circle graph represents the same information as the frequency table. Discuss the advantages and disadvantages of each.

12-6 Circle Graphs

Objective

Construct circle graphs.

Words to Learn

circle graph

In *The Wizard of Oz,* Dorothy discovers that "there's no place like home." Most people today would agree with that. Whether you live in an apartment, a mobile home, or a house, there is something special about being home.

American households, or homes, can be broken down into the categories shown in the chart at the right.

American Households 1990	
House	71%
Apartment	15%
Mobile Home	6%
Condominium	3%
Other	5%

You can draw a circle graph to show this information. A **circle graph** is used to compare parts of a whole.

Make a circle graph as follows.

a. Find the number of degrees for each section of the graph. There are 360° in a circle.

House	$71\% \times 360° = 0.71 \times 360° = 255.6°$
Apartment	$15\% \times 360° = 0.15 \times 360° = 54°$
Mobile Home	$6\% \times 360° = 0.06 \times 360° = 21.6°$
Condominium	$3\% \times 360° = 0.03 \times 360° = 10.8°$
Other	$5\% \times 360° = 0.05 \times 360° = 18°$

b. Use a compass to draw a circle. Then draw a radius as shown.

c. You can start with the least number of degrees, in this case, 10.8°. Use your protractor to draw an angle of 10.8°.

d. Repeat for the remaining sections. Label each section of the graph with the category and percent. Give the graph a title.

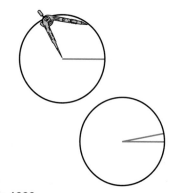

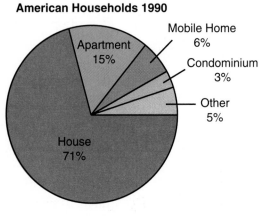

American Households 1990

Mobile Home 6%

Apartment 15%

Condominium 3%

Other 5%

House 71%

Sales Bicycles sold in your local bike shop may be made in the United States or imported from another country. Make a circle graph to represent the types of bicycles sold in 1989 in the U.S that were made in the U.S.

U.S. Bicycle Sales (in millions)				
	U.S. Made		Foreign Made	
Type of bicycle	1980	1989	1980	1989
Twenty-inch wheels	3.7	2.3	0	1.5
Lightweight	3.2	1.1	1.7	2.3
Other	0.1	1.9	0.3	1.5

- Find the total number of U.S. made bicycles sold in 1989.

  Twenty-inch wheels: 2.3 million
  Lightweight: 1.1 million
  Other: <u>1.9</u> million
  Total sold: 5.3 million

- Find the ratio that compares the number sold of each type with the total number sold. Round to the nearest hundredth.

  Twenty-inch wheels: $\frac{2.3}{5.3} = 0.43$

  Lightweight: $\frac{1.1}{5.3} = 0.21$

  Other: $\frac{1.9}{5.3} = 0.36$

- Find the number of degrees for each section of the graph.

  Twenty-inch wheels: $0.43 \times 360° = 155°$ *Note that the sum of the*
  Lightweight: $0.21 \times 360° = 76°$ *degrees is not 360° due*
  Other: $0.36 \times 360° = 130°$ *to rounding.*

- Make the circle graph.

  U.S. Sales of Bicycles Made in U.S., 1989

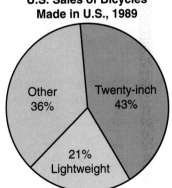

Checking for Understanding

Read and study the lesson to answer each question.

1. **Tell** how to make a circle graph when you know the percent represented by different parts of the whole.

2. **Write** a sentence explaining why it would not be appropriate to make a circle graph called "U.S. Bicycle Sales 1980–1989" using the information in the chart on page 471.

Guided Practice

Write a ratio that compares the number of students who chose each type of pizza to the total number of students surveyed. Use the graph.

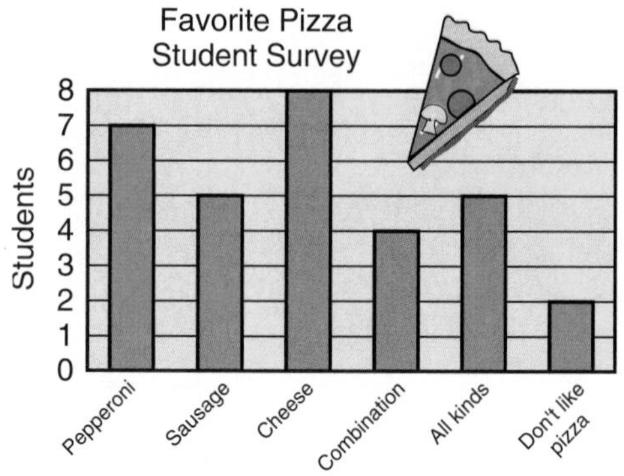

3. pepperoni   4. sausage

5. cheese   6. combination

7. all kinds   8. don't like pizza

Favorite Pizza Student Survey

Students (y-axis: 0–8)

Pepperoni, Sausage, Cheese, Combination, All kinds, Don't like pizza

Use the ratios from Exercises 3-8 to find the number of degrees each choice of pizza would be on a circle graph.

9. pepperoni   10. sausage   11. cheese

12. combination   13. all kinds   14. don't like pizza

15. Make a circle graph that shows the results of the pizza survey.

Exercises

independent Practice

16. Use the chart on page 471 to make a circle graph that shows the types of bicycles sold in 1989 that were foreign made.

17. Use the information from the chart at the right to make a circle graph of the colors of bicycles sold in 1989.

18. Suppose you are a domestic manufacturer of bicycles. Explain how the information in Exercise 17 would affect the color of bicycles you produce.

Bicycle Sales by Color for 1989	
Color	Percent
Blue	24
Black	23
Red	22
White	8
Silver	5
Yellow	2
All others	16

Mixed Review

19. **Smart Shopping**   Philip purchases 0.70 pounds of cheese priced at $3.15 per pound. What is the cost of his purchase? Round to the nearest cent. *(Lesson 2-4)*

20. **Geometry**   Determine whether the polygon at the right is a regular polygon. *(Lesson 8-4)*

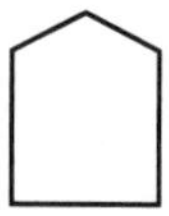

21. Find the number that is 35% of 20 using the percent equation. *(Lesson 12-5)*

Problem Solving and Applications

22. **Critical Thinking** Using the chart for bicycles sales in 1980 and 1989 on page 471, explain how this information would be useful to you as the owner of a bicycle shop.

23. **Collect Data** Make a circle graph that represents the colors of your classmates' eyes.

24. **Business** Use the chart on page 471.

a. Make a circle graph that compares the U.S. made bicycle sales and foreign-made bicycle sales for 1989.

b. Make a circle graph that compares the U.S. made bicycle sales and foreign-made bicycle sales for 1980.

c. What similarities or differences do you notice between the graphs for 1980 and 1989?

25. **Journal Entry** Write a sentence explaining how to construct a circle graph for a given set of data.

26. **Mathematics and Energy** Read the following paragraphs.

Energy is usually measured in millions (or even larger quantities) of British thermal units (Btus). One Btu is approximately the energy released in burning a wooden match. An automobile engine burning eight gallons of gasoline releases one million Btus.

Historically, three fossil fuels (coal, crude oil, natural gas) have accounted for most of the U.S. energy production, which totaled 66 quadrillion Btus in 1989. Transportation, residential, and commercial use accounted for the increase in energy usage during 1949–1989. Following the decline in energy prices in 1986 and 1988, residential and commercial consumption grew to a record 29.6 quadrillion Btus.

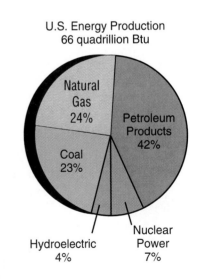

U.S. Energy Production
66 quadrillion Btu

Natural Gas 24%

Petroleum Products 42%

Coal 23%

Hydroelectric 4%

Nuclear Power 7%

Using the circle graph above, what percent of the energy production comes from sources other than the three fossil fuels?

12-7A Dot Paper and Percent

A Preview of Lesson 12-7

Objective

Use dot paper to show percent increase or percent decrease.

Materials

dot paper or graph paper
pencil

You can use dot paper or graph paper to help you understand the meaning of percent increase or percent decrease.

Try this!

Work in groups of three.

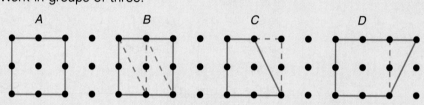

- Make a 2×2 square like the one shown in Figure A.

- Suppose you want to decrease the area of square A by 25%. Think: $25\% = \frac{1}{4}$. Separate the square into 4 equal parts as in Figure B.

- Remove 25% or $\frac{1}{4}$ from the original figure to show a decrease of 25%. See Figure C.

- Figure D shows an increase of 25% from the original figure.

What do you think?

1. Once you showed a 25% decrease in Figure A, what percent remains?

2. Describe other ways you could show a 25% decrease in the area of Figure A.

3. Explain how an increase of 25% was shown in Figure D.

4. Use Figures E, F, and G below. Explain how you can determine by what percent the original figure was increased or decreased.

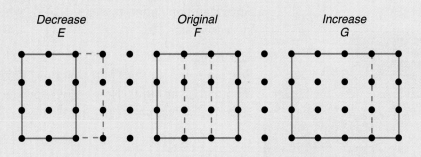

Decrease
E

Original
F

Increase
G

5. Use Figure H to draw figures that show an increase of $16\frac{2}{3}$% and a decrease of $16\frac{2}{3}$%. *(Hint: What fraction is equal to $16\frac{2}{3}$%?)*

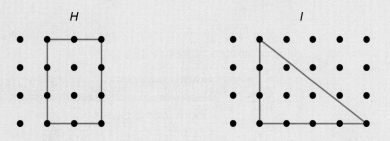

H

I

6. For Figure I, draw figures to indicate an increase of 50% and a decrease of 50%.

7. Copy and complete the chart for Figures A, E, H, and I.

Decreased area (units²)	% of original area	Original area (units²)	Increased area (units²)	% of original area

8. When finding the increase or decrease, explain what number is used as your base.

Extension

9. Construct or draw a 3×3 square. Remove $33\frac{1}{3}$% from the figure to show a decrease of $33\frac{1}{3}$%.

10. Construct or draw a 5×5 square. Add or draw an area to the original figure to show an increase of 20%.

12-7 **Percent of Change**

Objective

Find the percent of increase or decrease.

In the past few decades, the United States has become increasingly multicultural. For example, between 1980 and 1990, the Hispanic-American population increased 44%, while the entire population increased just 10.2%.

According to the 1990 census, there are 21,032,471 people of Spanish origin in the United States. This is equal to *about* 144% of the number in 1980. How many Hispanic-Americans were there in 1980? To solve, write an equation.

$\underbrace{144\% \text{ of }}\ \underbrace{\text{the Hispanics in 1980}}\ \underbrace{\text{is equal to}}\ \underbrace{21{,}032{,}471.}$

$$1.44 \cdot h = 21{,}032{,}471$$

$$\frac{1.44 \cdot h}{1.44} = \frac{21{,}032{,}471}{1.44} \quad \textit{Divide each side by 1.44.}$$

21,032,471 $\boxed{\div}$ 1.44 $\boxed{=}$ 𝟣𝟦𝟨𝟢𝟧𝟪𝟪𝟥

$h = 14{,}605{,}883$

There were *about* 14,605,883 Hispanic-Americans living in the United States in 1980.

Mini-Lab

Work with a partner.
Materials: ruler, paper, pencil

- Draw a segment that you estimate to be 25% longer than $\overline{MN}$.

  ●————————————————————————●
  M   N

- Measure the length of $\overline{MN}$. Use this number as the base, *B*.
- Measure the length of your segment. Use this as the percentage, *P*.

Talk About It

a. Will 50% of the length of $\overline{MN}$ be greater or less than its length?
b. Will 100% of the length of $\overline{MN}$ be greater or less than its length?
c. Do you think the length of your segment is greater or less than 100% of the length of $\overline{MN}$?
d. Write a proportion or equation to find the percent the length of $\overline{MN}$ is of the length of your segment. Solve.
e. The segment you drew is actually what percent longer than $\overline{MN}$?

When using the percent proportion to find the percent of increase or decrease, compare the amount of the increase to the original amount.

Example 1 *Problem Solving*

Health   When Lisa started using the exercise machine, she could only work out for 8 minutes. Now she can work out on it for 15 minutes. Find the percent of increase.

$15 - 8 = 7$ *Find the amount of increase.*

$\frac{7}{8} = \frac{r}{100}$ *Write the percent proportion.*

$7 \cdot 100 = 8r$ *The original time was 8 minutes.*

$\frac{700}{8} = \frac{8r}{8}$ *Find the cross products.*

$87.5 = r$ *Divide each side by 8.*

The percent of increase is 87.5%.

You can find the percent of decrease in a similar way.

Example 2 *Problem Solving*

Retail Sales   In 1970, a desktop calculator sold for *about* $100. Today the same type of calculator sells for as little as $25. What was the percent of decrease in the cost of desktop calculators?

$100 - 25 = 75$ *Find the amount of decrease.*

$\frac{75}{100} = \frac{r}{100}$ *The original cost was $100.*

$75 \cdot 100 = 100r$ *Find the cross products.*

$\frac{7,500}{100} = \frac{100r}{100}$ *Divide each side by 100.*

$75 = r$

The percent of decrease was 75%.

Checking for Understanding

Communicating Mathematics

Read and study the lesson to answer each question.

1. **Draw** a picture on dot paper to show an increase in area of 75%.
2. **Tell** what amount is used as a base in the percent proportion when finding the percent of change.
3. **Write** a sentence explaining the first step in finding the percent of increase or decrease.

Guided Practice

Estimate the percent of increase or decrease.

4. old: $4
   new: $6

5. old: $30
   new: $24

6. old: $0.36
   new: $0.18

Find the percent of change. Round to the nearest whole percent.

7. old: $60
   new: $38

8. old: $456
   new: $500

9. old: 0.76
   new: 0.9

10. Use the figure at the right.
    a. Draw a figure 25% larger.
    b. Draw a figure decreased
       in size by $66\frac{2}{3}$%.

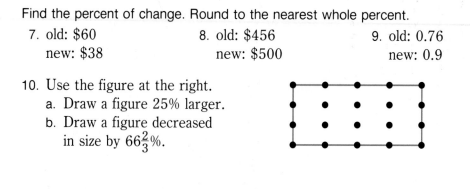

Exercises

*Independent
Practice*

Find the percent of change. Round to the nearest whole percent.

11. old: $85
    new: $68

12. old: $126
    new: $150

13. old: 1.6
    new: 0.95

14. old: 20.5
    new: 35.5

15. old: $62
    new: $50

16. old: 275
    new: 150

17. old: 40
    new: 80

18. old: 35
    new: 45

19. old: 87.5
    new: 36

20. If the figure at the right represents
    75% of something, draw a diagram to
    represent 100%.

Mixed Review

21. **Probability**   A fair die is rolled. Find the probability that the result is an
    even number. *(Lesson 4-8)*

22. **Algebra**   Write an algebraic expression for the phrase *6 less than w.*
    (Lesson 6-4)

23. **Geometry**   Find the surface area of a prism that has a length of
    8 centimeters, a width of 6.5 centimeters, and a height of
    10.4 centimeters. *(Lesson 10-3)*

*Problem Solving
and
Applications*

24. **History**   In 1991, the country of Georgia declared independence from the
    Soviet Union. The population was given as 5.5 million. The area of the
    country of Georgia was 26,910 square miles.
    The area of the state of Georgia in the United States is 58,910 square
    miles. The population is 6.8 million.
    a. Which Georgia has the greater population? By what percent is it higher?
    b. Which Georgia has the greater area? By what percent is it higher?

25. **Critical Thinking**   A book is on sale for $24. This is 25% off the original
    price. Can you find the original price by adding 25% of $24 to the sale
    price? Explain.

26. **Data Search**   Refer to pages 452 and 453. What was the percent of
    increase in the price of postage stamps between the years 1974 and 1975?
    Between 1981 and 1991?

12-8 Discount and Sales Tax

Objective

Solve problems involving sales tax and discount.

Words to Learn

sales tax

Do you know the sales tax rate in your community? **Sales tax** is the main way we pay for many state and city services.

Shalonda goes to the check-out counter with a pair of in-line skates that cost $85. If the sales tax in her city is 6%, find the total cost of her purchase.

You can use two methods to find the total cost. You get the same result using either method.

Method One

First find the amount of the tax.
6% of $85 = t

0.06 $\boxed{\times}$ 85 $\boxed{=}$ **5.1**

The sales tax is $5.10.

Then add to find the total cost.
$85.00 + $5.10 = $90.10

Method Two

First add the percent of tax to 100%.

Since 100% + 6% = 106%, Shalonda will pay 106% of the market price of the skates.

Then multiply to find the total cost including tax.

1.06 $\boxed{\times}$ 85 $\boxed{=}$ **90.1**

The total cost of the in-line skates will be $90.10.

Percents are often used to show discounts during store sales.

Estimation Hint

• • • • • • • • • • • • •

0.06 of 90 is 5.4. The sales tax is *about* $5.40. The total cost is $85.00 + $5.40 or $90.40.

Example 1 *Problem Solving*

Smart Shopping   Saki plans to buy a pair of jeans that are on sale for 25% off. If the regular price is $27, how much will she have to pay?

What is 25% of $27?

0.25 $\boxed{\times}$ 27 $\boxed{=}$ **6.75**

The discount is $6.75.

Then subtract to find the discount price.

27 $\boxed{-}$ 6.75 $\boxed{=}$ **20.25**

Saki will have to pay $20.25 for the jeans.

Example 2 *Problem Solving*

Smart Shopping   A shirt that normally sells for $18.95 is on sale for $16.50. What is the rate of discount?

Subtract to find the amount of discount.
$18.95 - $16.50 = $2.45

Then find what percent $2.45 is of $18.95.

$$\frac{2.45}{18.95} = \frac{r}{100}$$ *Use the percent proportion.*
                   P = 2.45; B = 18.95

$2.45 \cdot 100 = 18.95r$ *Find the cross products.*
$\quad\quad 245 = 18.95r$ *Divide each side by 18.95.*
$12.928759 = r$ *Round 12.928759 to 13.*

The rate of discount is *about* 13%.

Checking for Understanding

Communicating Mathematics

Read and study the lesson to answer each question.

1. **Tell** how you would use a different method in Example 1 to find the sale price.

2. **Show** two different ways that you could find the total purchase price including tax on a sweater selling for $45 if the tax rate is 5%.

Guided Practice

Find the sales tax or discount to the nearest cent.

3. $28 shoes; 5% tax
4. $17.42 book; $5\frac{1}{2}$% tax
5. $38.50 sweater; 15% off
6. $25 watch; 30% discount

Find the total cost or sale price to the nearest cent.

7. $135.59 speakers; 33% off
8. $175.95 suit; 6% tax
9. $9.95 cassette; $6\frac{1}{2}$% tax
10. $2.50 socks; 25% off

Find the rate of discount to the nearest percent.

11. regular price, $24
    sale price, $20
12. regular price, $224
    sale price, $180

Exercises

Independent Practice

Find the sales tax or discount to the nearest cent.

13. $37 radio; 6% tax
14. $16.58 gloves; $6\frac{1}{2}$% tax
15. $49.50 drill; 35% off
16. $145 chair; 22% discount

Find the total cost or sale price to the nearest cent.

17. $15.99 T-shirt; 20% off
18. $32 coat; $5\frac{1}{2}$% tax
19. $3.99 toy; 7% tax
20. $40 video; 20% off

Find the rate of discount to the nearest percent.

21. regular price, $35
    sale price, $30

22. regular price, $44
    sale price, $34

23. regular price, $18.99
    sale price, $13.29

24. regular price, $70
    sale price, $52.50

25. Find the total purchase price to the nearest cent if a $65 dress is on sale for 20% off and the sales tax is 6%.

Mixed Review   26. Evaluate $12(5) - 16 \div 8 + 5$. *(Lesson 1-7)*

27. **Geometry**   Find the circumference of a circle having a diameter of 5 inches. Use 3.14 for π. Round to the nearest tenth of an inch. *(Lesson 5-7)*

28. Solve $\frac{t}{16} = 8$. *(Lesson 6-1)*

29. **Health**   Before beginning his new diet, Peter weighed 195 pounds. After dieting for 12 weeks, Peter's new weight was 170 pounds. Find the percent of decrease in Peter's weight. *(Lesson 12-7)*

Problem Solving and Applications

30. **Critical Thinking**   A shirt regularly sells for $22.50. It is on sale at a 15% discount. The sales tax is $5\frac{1}{2}$%.
    a. Does it matter in which order the discount and the sales tax are applied? Explain.
    b. Would the result change if the sales tax were added before the discount was subtracted? Explain.

31. **Consumer Math**   The Wilsons went out for hamburgers and salad. The bill was $17.70 before tax and tip were added.
    a. What is the total including 5% tax?
    b. If a 15% tip is left on the bill including tax, how much is the tip?

32. **Smart Shopping**   Derrick is planning to buy a new watch that normally sells for $38.50. If it is on sale for 20% off, what is the sale price?

33. **Consumer Math**   James Weaver bought a car for $7,800. He had to pay sales tax on the car.
    a. If the sales tax is $6\frac{1}{2}$%, how much sales tax did he pay?
    b. What was the cost of the car including sales tax?

34. **Smart Shopping**   Ms. Collins bought a new suit that originally cost $175. She bought it on sale for 45% off.
    a. What was her discount?
    b. How much did she pay for the suit on sale?

35. **Journal Entry**   Make up a problem involving the discount rate where the regular price and the sale price are given. Show how to use the percent equation to solve the problem.

12-9 Simple Interest

Objective

Solve problems involving simple interest.

Words to Learn

interest
principal
rate
time

Credit cards are a convenient way to pay for purchases. But you need to be careful not to buy more than you can afford. Also you will be charged interest if you do not pay the bill off when it is due.

Suppose your parents have a credit card and their monthly balance is $800. If they are charged 18% interest, how much interest will they pay in a year?

Simple **interest** (I) is calculated by finding the product of the **principal** (p), which is the amount borrowed, the **rate** (r), which is a rate of interest as a percent, and the **time** (t), which is given in years. This is expressed by the formula $I = prt$.

$I = prt$
$I = 800 \cdot 0.18 \cdot 1$ *p = $800, r = 18%, t = 1 year*
$I = 144$ The interest is equal to $144 in a year.

The formula, $I = prt$, can also be used to find the simple interest when you deposit money in a savings account. In this case, the principal is the amount in your savings account.

Example *Problem Solving*

Saving Money   Ms. Sung deposited $600 in her savings account. Her account earns $6\frac{3}{4}\%$ interest annually. If she does not deposit or withdraw any money, how much will be in her account after 6 months?

$I = prt$
$I = 600 \cdot 0.0675 \cdot 0.5$ *p = $600, r = $6\frac{3}{4}$%, t = 6 months or 0.5 year*

600 ⊠ 6.75 ⟦%⟧ ⊠ 0.5 ⟦=⟧ **20.25**

$I = 20.25$

The interest earned on $600 in 6 months was $20.25. So, Ms. Sung will have $600 + $20.25 or $620.25 in her account.

Checking for Understanding

Communicating Mathematics

Read and study the lesson to answer each question.

1. **Explain** how to find the interest on $550 at 8% for one year.

2. **Tell** how to write time in terms of years when it is given in months.

Find the interest to the nearest cent for each principal, interest rate, and time.

3. $200, 6%, 2 years

4. $340.10, 12%, 1.5 years

5. $121, 16%, 2 months

6. $4,200, $9\frac{1}{4}$%, 3 years

Find the interest to the nearest cent on credit cards for each credit card balance, interest rate, and time.

7. $325, 18.5%, 1 year

8. $1,200, 19%, 9 months

Exercises

Find the interest to the nearest cent for each principal, interest rate, and time.

9. $2,250, 7%, 3 years

10. $175.80, 12%, 1.25 years

11. $875, 15%, 4 months

12. $98.50, $6\frac{1}{2}$%, 16 months

13. $3,186, 10%, 2 years

14. $514, 8.75%, 6 months

Find the interest to the nearest cent on credit cards for each credit card balance, interest rate, and time.

15. $1,000, $20\frac{1}{2}$%, 1 year

16. $5,096, 17%, 2 years

17. $400, 19%, 6 months

18. $839, 21%, 1 year

19. Find the interest on $443 in a savings account at 6.5% interest for 1 year.

20. **Statistics**   Find the mean, median, and mode for 5, 2, 3, 6, 4, 3, 3, 4, 2, and 8. *(Lesson 3-5)*

21. **Geometry**   Find the area of a circle having a radius of 6 inches. Use 3.14 for π. *(Lesson 9-8)*

22. **Smart Shopping**   At a sale, Barb finds a $125 sweater marked down to $87.50. What percent of decrease is this? *(Lesson 12-8)*

23. **Critical Thinking**   Find the amount of simple interest earned on $1,000 at the end of 4 years at 8% per year, if the interest is added to the principal at the end of each year.

24. **Consumer Math**   Mitch Lowe bought a watch for $95. He used his credit card, which charges 21% annual interest from the moment of purchase. If he does not make any payments or any additional charges, how much would he owe at the end of the first month?

COMPUTER CONNECTION

25. **Computer Connection**   A spreadsheet can be used to generate a simple interest table for various account balances. Change cell B2 to 6 in the table below. What is the new balance for a principal of $1,000?

	A	B	C	D	E
1	PRINCIPAL	RATE	TIME	INTEREST	NEW BALANCE
2		5	2		
3	500	0.05	2	50	550
4	1000	0.05	2	100	1100

12 Study Guide and Review

Communicating Mathematics

State whether each sentence is *true* or *false*. If false, replace the underlined word or number to make a true sentence.

1. A percent is a ratio that compares a number to <u>100</u>.
2. In the proportion $\frac{1}{4} = \frac{25}{100}$, $\frac{25}{100}$ is called the <u>percentage</u>.
3. A circle graph is used to <u>compare</u> parts of a whole.
4. There are <u>300°</u> in a circle.
5. When finding a percent of increase, compare the amount of the increase to the <u>new</u> amount.
6. The formula for simple interest is <u>$I = prt$</u>.
7. In your own words, explain each of the three methods for estimating percents.

Skills and Concepts

Objectives and Examples

Upon completing this chapter, you should be able to:

Review Exercises

Use these exercises to review and prepare for the chapter test.

- find the percent of a number
  (Lesson 12-1)

  Find 60% of 300.

  $$\frac{P}{B} = \frac{r}{100}$$
  $$\frac{P}{300} = \frac{60}{100}$$
  $$100P = 18,000 \quad \textit{Find the cross-products.}$$
  $$P = 180$$

Use a proportion to solve each problem. Round answers to the nearest tenth.

8. What number is 30% of 250?
9. Find 42% of 850.
10. $12\frac{1}{2}$% of 145 is what number?
11. Seventy percent of 504 is what number?

- estimate by using fractions, decimals, and percents interchangeably
  (Lesson 12-3)

  Estimate 57% of 453.
  57% is about 60% or $\frac{3}{5}$.
  $\frac{3}{5}$ of 450 is 270.
  So, 57% of 453 is *about* 270.

Estimate.

12. 12% of 75   13. 89% of 500
14. $8\frac{3}{4}$% of 15   15. 148% of 20
16. 0.95% of 800   17. 65% of 1,000
18. 99% of 1   19. 14.6% of 78

Objectives and Examples

- solve problems using the percent proportion *(Lesson 12-4)*

What percent of 80 is 12?

$$\frac{P}{B} = \frac{r}{100}$$

$$\frac{12}{80} = \frac{r}{100}$$

$1,200 = 80r$ *Find the*
$15 = r$ *cross products.*
15% of 80 is 12.

- solve problems using the percent equation *(Lesson 12-5)*

What number is 18% of 120?

$P = R \cdot B$
$P = 0.18 \cdot 120$
$P = 21.6$ 18% of 120 is 21.6.

- construct circle graphs *(Lesson 12-6)*

Favorite Season:
Spring, 35%, 126°; Summer, 29%, 104.4%; Fall, 24%, 86.4°; Winter, 12%, 43.2°

Favorite Seasons

- find the percent of increase or decrease *(Lesson 12-7)*

old: $2.95   new: $3.45
$3.45 − $2.95 = $0.50

$$\frac{0.50}{2.95} = \frac{r}{100}$$

$50 = 2.95r$ *Find the*
            cross products.

$$\frac{50}{2.95} = \frac{2.95r}{2.95}$$ *Divide each side by 2.95*

$16.95 = r$ The percent of increase to the nearest whole percent is 17%.

Review Exercises

Write a proportion for each problem. Then solve. Round answers to the nearest tenth.
20. Find 36.5% of 150.
21. What percent of 95 is 5?
22. 105 is $33\frac{1}{3}$% of what number?

Write an equation for each problem. Then solve.
23. 42 is what percent of 80?
24. 65% of what number is 91?
25. Find 62% of 350.
26. 58% of 450 is what number?
27. 12.5% of what number is 93.75?

Use the information from the chart below to construct a circle graph.

28.

Favorite Soft Drink	
Soft Drink	**Percent**
cola	38%
diet cola	26%
lemon-lime	18%
root beer	8%
other	10%

Find the percent of change. Round to the nearest whole percent.
29. old: 42   30. old: $23,500
    new: 65   new: $26,400
31. old: $212   32. old: 299
    new: $180   new: 216

Objectives and Examples	Review Exercises

- solve problems involving sales tax and discount *(Lesson 12-8)*

  Find the sales tax on a $35 pair of shoes at 5% tax. Let t represent the sales tax.

  $5\% \times \$35 = t$

  $0.05 \times 35 = t$

  $\$1.75 = t$

Find the sales tax or discount to the nearest cent.

33. $175 bicycle; 25% off
34. $7,500 car; 7% tax
35. $50 sweater; $\frac{1}{3}$ off

- solve problems involving simple interest *(Lesson 12-9)*

  The interest on $500 at 8% for 5 years is:

  $I = prt$

  $I = 500 \times 0.08 \times 5$

  $I = \$200$

Find the interest to the nearest cent for each principal, interest rate, and time given.

36. $6,000, 9%, 2 years
37. $75, $7\frac{1}{2}$%, 8 months
38. $2,450, 12%, $4\frac{1}{2}$ years
39. $675, 18%, 32 months

Applications and Problem Solving

40. **Consumer Math**   Debbie bought a ski outfit for $325. The sales tax rate was 5%. How much sales tax did she pay? *(Lesson 12-8)*

41. **School**   During the 1990–91 school year, Juanita attended school 90% of the days school was in session. She was in class a total of 171 days. Find the total number of days in the school calendar. *(Lesson 12-5)*

Curriculum Connection Projects

- **Transportation**   Find the percent of your classmates that ride the school bus. Use that percent to find the number of students in the entire school that ride the bus.
- **Economics**   Use newspapers from the past week to find the number of points the Dow Jones Averages went up or down each day. Find the percent of change for each day.

Read More About It

Byers, Patricia, and Julia Preston, and Patricia Johnson. *The Kid's Money Book: Great Money Making Ideas.*

Adler, David A. *Banks.*

Bethancourt, T. Ernesto. *The Me Inside of Me.*

486   **Chapter 12** Study Guide and Review

Study Guide and Review

Chapter

12 Test

Use a proportion to solve. Round answers to the nearest tenth.

1. Find 25% of 145.
2. 96 is 40% of what number?
3. **College Tuition** State University presently charges $6,500 per year for tuition. Tuition will be going up 6% for the coming academic year. Find the amount of increase.
4. **Health** Experts recommend that at most 30% of your total calorie intake should come from fat. If you eat 1,600 calories per day, what is the maximum number of calories that should be from fat?

Estimate.

5. 19% of 248
6. 149% of 79
7. **Smart Shopping** Grace finds a pair of jeans originally priced $39.95 on a rack labeled 33% off. Estimate the sale price.

Write a proportion for each problem. Then solve. Round answers to the nearest tenth.

8. 50% of what number is 334.8?
9. What percent of 48 is 7?

Write an equation for each problem. Then solve.

10. Find 12% of 75.
11. 150 is what percent of 120?
12. Use the information in the bar graph at the right to make a circle graph.

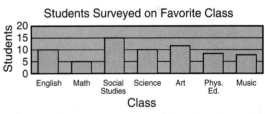

Find the percent of change. Round to the nearest whole percent.

13. old: $59, new: $42
14. old: 145, new: 183

Find the sales tax or discount to the nearest cent.

15. $29.99 book, 15% off
16. $16.99 CD, $6\frac{1}{2}$% tax
17. Find the total purchase price to the nearest cent if an $85 pair of boots is on sale for 20% off and the sales tax is 7%.

Find the interest to the nearest cent for each principal, interest rate, and time given.

18. $1,350, 12%, 2 years
19. $2,400, $11\frac{3}{4}$%, 9 months
20. **Finances** Jorge borrows $3,500 to buy a new motorcycle. The loan is for 3 years at an annual interest rate of 9.5%. Find the total amount Jorge will pay over the 3 year life of the loan.

Bonus Find 1,000% of 100.

12 Academic Skills Test

Directions: Choose the best answer. Write A, B, C, or D.

Academic Skills Test

1. What is the value of $8 + x^2$ if $x = 12$?

   A 32   B 152
   C 400   D none of these

2. How could you calculate the perimeter of an $8\frac{1}{2}$ by 11 inch piece of paper?

   A Add $8\frac{1}{2}$ and 11.

   B Multiply 2 times $8\frac{1}{2}$ and add 11.

   C Add $8\frac{1}{2}$ and 11 and multiply by 2.

   D Multiply $8\frac{1}{2}$ and 11.

3. Which is an equivalent equation, using the inverse operation, for $x - 3.2 = 1.7$?

   A $x = 1.7 + 3.2$ B $x + 3.2 = 1.7$
   C $x - 1.7 = 3.2$ D $x = 3.2 - 1.7$

4. What is the probability that a randomly-dropped counter will fall in the shaded region?

   A $\frac{1}{8}$ B $\frac{1}{4}$
   C $\frac{1}{3}$ D $\frac{1}{2}$

5. The top, front, and side views of a figure are given below. What is the figure?

   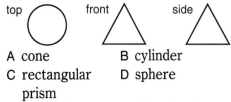

   A cone   B cylinder
   C rectangular   D sphere
     prism

6. Jessie has a paper cup that is shaped like a cylinder with a radius of 1.5 in. and a height of 5 in. To the nearest cubic inch, what is the volume of the cup?

   A 7.5 in³   B 30 in³
   C 35 in³   D 45 in³

7. In the proportion $\frac{2}{3} = \frac{x}{8}$, what is the value of x?

   A 12   B 7
   C $6\frac{1}{2}$ D $5\frac{1}{3}$

8. $\triangle LMN$ is similar to $\triangle PQR$. What is the length of side PQ?

   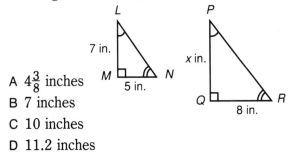

   A $4\frac{3}{8}$ inches
   B 7 inches
   C 10 inches
   D 11.2 inches

9. 30% of a number is 24. What is the number?

   A 80   B 72
   C 8   D 7.2

10. Colleen wants to buy a pair of shoes that cost $54.98. She must also pay 6% sales tax. To the nearest cent, what is the total cost of the shoes?

    A $3.30   B $51.68
    C $55.31   D $58.28

The questions on this page involve comparing two quantities, one in Column A and one in Column B. In some questions, information related to one or both quantities is centered above them.

Directions: Write A if the quantity in Column A is greater. Write B if the quantity in Column B is greater. Write C if the quantities are equal. Write D if there is not enough information to decide.

Column A	Column B
11. $0.3 + 0.8 + 1.2$	0.23
12. $1\frac{1}{4} + 2\frac{2}{3}$	$1\frac{1}{8} + 2\frac{1}{2}$

13.

the number represented by point X	the number represented by point Y
14. the sum of any three negative integers	the product of any three negative integers

15.

measure of $\angle AXB$	measure of $\angle AXC$
16. $\sqrt{16} + \sqrt{16}$	$\sqrt{32}$

Column A	Column B

17.

8 ft

surface area of top and bottom	surface area of curved surface
18. ratio of 54 to 45	ratio of 84 to 70
19. 20%	$\frac{1}{5}$
20. 5% simple interest on $100 for 2 years	8% simple interest on $100 for 1 year

Discrete Math and Probability

Spotlight on Earthquakes

Have You Ever Wondered. . .

- What it means when an earthquake measures 6.5 on the Richter scale?

- How much energy is released during an earthquake?

Earthquake Scales

Richter Scale

2.5	Generally not felt, but recorded on seismometers.
3.5	Felt by many people
4.5	Some local damage may occur.
6.0	A destructive earthquake.
7.0	A major earthquake. About 10 occur each year.
8.0 and above	Great earthquakes. These occur once every five to 10 years.

Mercalli Earthquake Intensity Scale

I. Felt by few.
II. Felt only by persons at rest.
III. Felt quite noticeably indoors.
IV. Sensation like heavy truck striking building.
V. Felt by nearly everyone. Disturbances of trees, poles, and other tall objects.
VI. Felt by all; Some heavy furniture moved.
VII. Damage slight to well built structures.
VIII. Damage considerable in ordinary substantial buildings with partial collapse.
IX. Damage considerable. Buildings shifted off foundations. Ground cracked conspicuously.
X. Most masonry and frame structures destroyed with foundations. Ground badly cracked.
XI. Few, if any (masonry) structures remain standing. Broad fissures in ground.
XII. Damage total. Waves seen on ground surfaces.

1810   1840   1846   1870   1879   1900   1899

Scientific study of earthquakes begins

United States Geological Survey is established

William McKinley first president to ride in an automobile

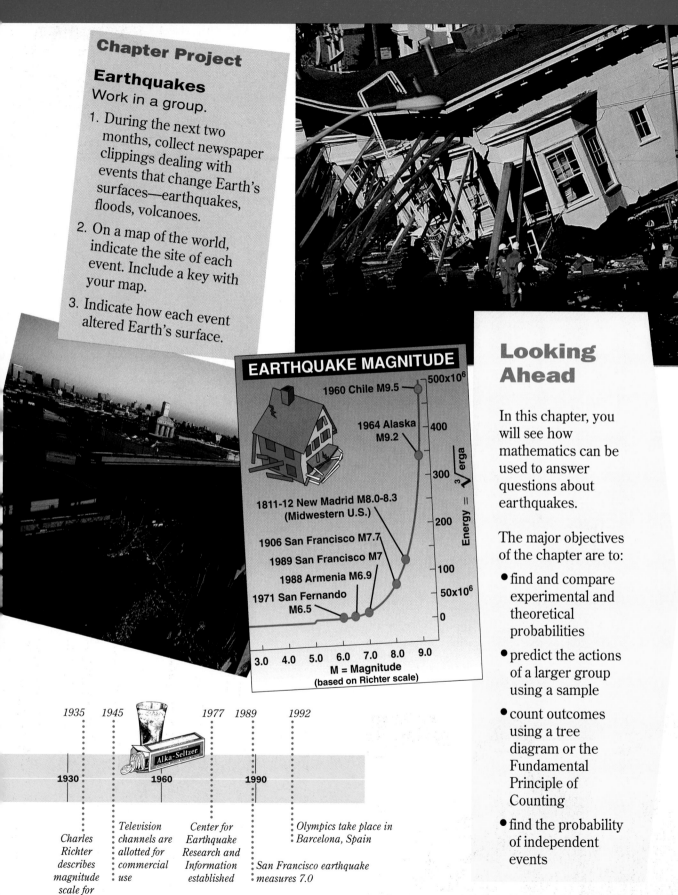

Chapter Project

Earthquakes

Work in a group.

1. During the next two months, collect newspaper clippings dealing with events that change Earth's surfaces—earthquakes, floods, volcanoes.

2. On a map of the world, indicate the site of each event. Include a key with your map.

3. Indicate how each event altered Earth's surface.

EARTHQUAKE MAGNITUDE

1960 Chile M9.5 — 500x10⁶

1964 Alaska M9.2 — 400

Energy = $\sqrt[3]{\text{erga}}$

300

1811-12 New Madrid M8.0-8.3 (Midwestern U.S.)

200

1906 San Francisco M7.7

1989 San Francisco M7

1988 Armenia M6.9

100

1971 San Fernando M6.5

50x10⁶

0

3.0 4.0 5.0 6.0 7.0 8.0 9.0

M = Magnitude
(based on Richter scale)

Looking Ahead

In this chapter, you will see how mathematics can be used to answer questions about earthquakes.

The major objectives of the chapter are to:

- find and compare experimental and theoretical probabilities

- predict the actions of a larger group using a sample

- count outcomes using a tree diagram or the Fundamental Principle of Counting

- find the probability of independent events

1935   1945   1977 1989   1992

1930   1960   1990

Charles Richter describes magnitude scale for earthquakes

Television channels are allotted for commercial use

Center for Earthquake Research and Information established

Olympics take place in Barcelona, Spain

San Francisco earthquake measures 7.0

491

13-1 Tree Diagrams

Objective
Use tree diagrams to count outcomes.

Words to Learn
outcomes
sample space
tree diagram

Do you dread surprise quizzes? Most students don't like them. But if you do your homework, you'll be more likely to answer the questions correctly and get a good grade.

Imagine that you walk into science class one day and your teacher hands you a pop quiz with three true-false questions on it. You're not prepared this time, so you guess on the answers and then hand in the quiz.

Is guessing a good strategy? To find out, figure how many possible sets of answers, or **outcomes,** there would be by guessing.

When you guess at the answers to a test, there are several possible outcomes. The set of all possible outcomes is called the **sample space.** Often a **tree diagram** is used to picture a sample space.

What is the sample space for the science quiz? You can guess T or F for the first question, T or F for the second question, and T or F for the third question. The tree diagram below shows the sample space.

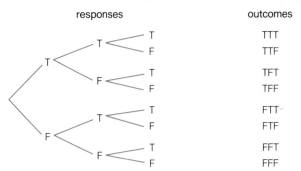

responses	outcomes
T	TTT
F	TTF
T	TFT
F	TFF
T	FTT
F	FTF
T	FFT
F	FFF

LOOK BACK

You can review probability on page 157.

There are eight different sets of guesses, but only one is the correct set of guesses. The probability that you will get all three answers correct by guessing is only $\frac{1}{8}$. Guessing is not a very good strategy.

Example | *Problem Solving*

Food Service A concession stand sells hot dogs, hamburgers, and ham barbecues. They also sell cola, diet cola, and lemon-lime drinks. How many different sandwich/drink choices are sold at the stand?

Make a tree diagram. List the kinds of sandwiches. For each kind of sandwich, list the three kinds of drinks. Altogether there are nine different choices of sandwiches and drinks.

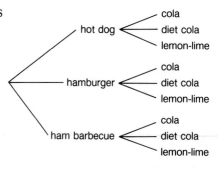

Mini-Lab

Work with a partner.

Materials: three plain chips and a marker

- Mark one side of a chip A. Mark the other side B. Mark one side of another chip A. Mark the other side C. Mark one side of the last chip B. Mark the other side C.
- Decide which student will be Player 1 and which student will be Player 2.
- Player 1 tosses the chips. If two chips show the same letter, Player 1 wins. Otherwise, Player 2 wins. Who won? Record the result.
- Repeat the chip toss experiment 20 more times. Record who won each time.

Talk About It

a. Is there a pattern to the results? What is it?

b. Compare the results with those of other pairs of students.

c. Would you rather be Player 1 or Player 2? Why?

d. Make a tree diagram to show all the possible outcomes for the toss of the chips. In how many branches does a letter occur twice? Does this explain your results?

Checking for Understanding

Communicating Mathematics

Read and study the lesson to answer each question.

1. **Tell** what an outcome and a sample space are.

2. **Tell** whether guessing TTF on a three-question true-false quiz is the same as guessing TFT. Explain.

3. **Write** a problem that can be solved by using the tree diagram at the right.

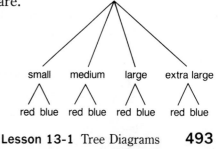

For each situation, make a tree diagram to show all the outcomes in the sample space. List the outcomes. Then give the total number of outcomes.

4. choosing one blouse from a red blouse and a white blouse and one skirt from a blue skirt and a black skirt

5. flipping a coin and rolling a number cube

6. spinning the spinner below and choosing a card

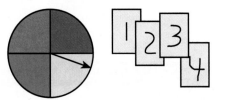

7. Diana is planning a dinner party. She plans to offer three kinds of salad: potato salad, tossed salad, and cole slaw. There will be three kinds of meat: ham, beef, and turkey. For dessert, she plans to offer ice cream, pie, and cake. How many salad/meat/dessert selections are there?

Exercises

Make a tree diagram to show all the outcomes in the sample space. Then give the total number of outcomes.

8. flipping a penny and flipping a dime

9. choosing a letter from among the letters A, B, and C and a number from among the numbers 1, 2, and 3

10. choosing cereal, French toast, or pancakes and choosing orange, tomato, or grapefruit juice

11. choosing a bicycle having one speed, three speeds, or ten speeds and either red, blue, green, or white in color

12. rolling a number cube, flipping a coin, and choosing a card from among cards marked W, X, Y, and Z

13. **Statistics** Find the mean, median, and mode for the following set of data: 42, 35, 52, 63, 41, 38, and 44. *(Lesson 3-5)*

14. **Finance** Margie borrowed $3,500 to help pay for her college tuition. The loan was made at 12% interest for 18 months. Find the amount of interest Margie will pay on the loan. *(Lesson 12-9)*

15. **Telephones** In a certain area, a telephone number begins with a three-digit number consisting of 2, 3, and 4, where a digit may be used more than once. How many three-digit numbers are possible?

16. **Critical Thinking** Without drawing a tree diagram, tell how many sets of guesses there are for a four-question true/false quiz. Use a tree diagram to check.

13-1B Fair and Unfair Games

A Follow-Up of Lesson 13-1

Objective

Count outcomes to determine whether a game is fair or unfair.

Materials

spinners
paper
pencil
ruler

Many people enjoy playing games like board games. One of the reasons they enjoy playing games is because players of equal skill have the same chance of winning. That is, each player has a fifty-fifty chance. Such games are called fair games. In an unfair game, players having equal skill do not have an equal chance of winning.

In this mathematics lab, you will play several games to determine whether or not they are fair games.

Activity One

Game 1

Play with a partner.

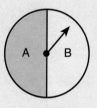

- Use the spinner shown at the right.
- Spin the spinner twice. Player 1 scores 1 point if the spinner lands on the same letter twice. Player 2 scores 1 point if the spinner lands on different letters.
- Play 50 rounds. The winner is the player with more points.
- Play the game three or four times.

Game 2

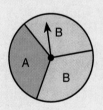

- Follow the same rules as Game 1, except use the spinner shown at the right.

What do you think?

1. Based on your data, is Game 1 a fair game?
2. Based on your data, is Game 2 a fair game?

One way to explain the results of your games is to make a drawing.

Activity Two

Game 1

For the first spin, there is an equal chance of landing in A or B. We can represent this by a square that is divided in half. For the second spin, you can further divide the square in half.

Mathematics Lab 13-1B Fair and Unfair Games **495**

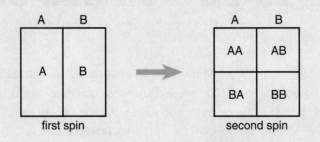

first spin → second spin

	A	B
A	AA	AB
B	BA	BB

Now you can see four outcomes, AA, AB, BA, and BB. For two of these four outcomes, the letter matches. Therefore, the probability of the spinner landing on the same letter twice is $\frac{2}{4}$ or $\frac{1}{2}$. The game is fair.

Game 2

For the first spin, there are three outcomes: A, B, and B. Divide a square into three sections, as shown at the right.

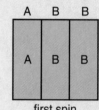

first spin

What do you think?

3. How would you show the results of the second spin on the square?
4. List all the outcomes for this game.
5. What is the probability that the spinner lands on the letter A twice? letter B twice?

Extension

Suppose your teacher has two tickets to a local concert. She devises a game to award the tickets.

- You have five marbles: two red, two green, and one yellow, and two small brown paper bags.
- You can place the marbles any way you want into the bags.
- Your partner then chooses two marbles from one bag. If your partner chooses two red marbles, you and your partner win the tickets.

6. First list all of the possible ways you can distribute the marbles in the bags.
7. Use drawings like the ones shown above to devise a "winning" strategy.
8. Write a paragraph explaining how you would distribute the marbles and why you believe your method is the best strategy.

13-2 Counting Using Multiplication

Objective

Use multiplication to count outcomes.

Words to learn

Fundamental Counting Principle

TEEN SCENE

Have you ever wondered why boys' shirts button on the right side and girls' shirts button on the left side? In the 1400s, only wealthy people wore buttons. Women usually had maids to dress them, and the maids buttoned from right to left, but they were facing the shirt, not looking down at it. Men usually didn't have maids, so they buttoned their own shirts.

You may remember when the United States aided Kuwait in the war against Iraq in January 1991. This war was referred to as "Desert Storm" because it was fought in the deserts of the Middle East. Special camouflage clothing was made to blend in with the desert environment. It has been calculated that there were a total of 483 different combinations of tops and bottoms!

Suppose there are 2 choices for tops: small and large. And there are 3 choices for bottoms: short, average, and tall. Draw a tree diagram to determine the number of possible outcomes.

tops	bottoms	outcomes
small	short	(small, short)
	average	(small, average)
	tall	(small, tall)
large	short	(large, short)
	average	(large, average)
	tall	(large, tall)

The total number of possible outcomes is 2×3 or 6. The **Fundamental Counting Principle** gives a way of counting all outcomes by using multiplication instead of a tree diagram.

Fundamental Counting Principle	If there are m ways of selecting an item from set A and n ways of selecting an item from set B, then there are $m \times n$ ways of selecting an item from set A and an item from set B.

Example 1   *Problem Solving*

Food Service   The Bowl 'N' Ladle Restaurant advertises that you can have a different lunch every day of the year. They offer 13 different kinds of soups and 24 different kinds of sandwiches. If the restaurant is open every day of the year, is their claim valid? Explain.

$$\binom{number\ of\ choices}{for\ sandwiches} \times \binom{number\ of\ choices}{for\ soup} = \binom{number\ of\ choices}{for\ lunch}$$

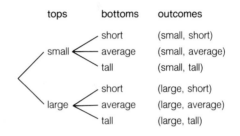

24   ⊠   13   ⊟   312

Their claim isn't valid, because the number of selections is 312, and $312 < 365$.

Lesson 13-2 Counting Using Multiplication   **497**

Example 2 *Problem Solving*

Highways   There are three highways connecting Tomville and Greeburg. There are two roads connecting Greeburg and Morristown. There is one superhighway from Morristown to Hodge. How many ways are there to drive from Tomville to Hodge?

Explore   3 highways from Tomville to Greeburg
2 roads from Greeburg to Morristown
1 superhighway from Morristown to Hodge

Plan   Draw a diagram showing the highway system.

Solve   Apply the Fundamental Counting Principle.

$$3 \times 2 \times 1 = 6$$

There are six ways to drive from Tomville to Hodge.

Examine   Count all the different routes from Tomville to Hodge on the drawing. Did you count six?

Checking for Understanding

Communicating Mathematics   Read and study the lesson to answer each question.

1. **Tell** how to use the Fundamental Counting Principle to find the number of ways of selecting 1 shirt from among 3 different shirts and 1 tie from among 6 different ties.

2. **Tell** when the Fundamental Counting Principle is more useful than a tree diagram to count outcomes.

3. **Draw** a diagram to show the three roads connecting Tumba City and Birdville and the four roads connecting Birdville and Meclaville.

Guided Practice   Use multiplication to find the total number of outcomes in each situation.

4. choosing an exterior color and an interior color for a new car if there are 5 choices for exterior color and 6 choices for interior color

5. making a sandwich with raisin bread, whole wheat bread, white bread, or an English muffin and choosing a filling from among peanut butter, jelly, or cream cheese

6. choosing the first two letters/digits for a license plate if the license plate begins with a letter of the alphabet and is followed by a digit

Exercises

Independent Practice

Find the total number of outcomes in each situation.

7. tossing a penny, a nickel, a dime, and a quarter

8. spinning the spinners shown at the right

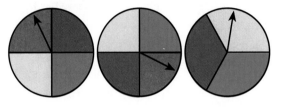

9. choosing a way to drive from Milton to Harper's Township if there are three roads that lead from Milton to Westwood, three highways that connect Westwood to Morgantown, and two streets that join Morgantown to Harper's Township

Mixed Review

10. Solve $6p = 72$. *(Lesson 6-3)*

11. **School** Jen must take a science class, a math class, and an English class next semester. She can choose from two science classes, three math classes, and two English classes. Make a tree diagram to show all of the possible schedules she can arrange. *(Lesson 13-1)*

Problem Solving and Applications

12. **Food Service** The Soup 'N' Salad Restaurant advertises that you can have a different soup-and-salad lunch every day of the year. They sell 20 different kinds of soup and 18 different kinds of salad. Is their claim valid? Explain.

13. **Algebra** In literature class, each student must read a short story, a poem, and a newspaper article. If there are g short stories, h poems, and n newspaper articles, how many different selections can be made?

14. **Critical Thinking** How many outcomes are there if you toss
    a. one coin?   b. two coins?   c. three coins?   d. n coins?

15. **Journal Entry** Write about the strong points and weak points of using tree diagrams and using the Fundamental Counting Principle.

13-3 Theoretical and Experimental Probability

Objective
Find and compare experimental and theoretical probabilities.

Words to Learn
theoretical probability
experimental probability

Backgammon is a game played with a board, counters, and dice. It is one of the most ancient games, dating back to 3000 B.C. The game is played by two players, who take turns rolling the dice and moving their counters. Rolling doubles is highly desirable in this game. If a player rolls doubles, that player gets to double the value of the roll. That is, if double 4s are rolled, the value of the roll is four 4s, rather than two 4s.

The sample space, or all the possible outcomes, for a roll of two dice is shown below.

1, 1	1, 2	1, 3	1, 4	1, 5	1, 6
2, 1	2, 2	2, 3	2, 4	2, 5	2, 6
3, 1	3, 2	3, 3	3, 4	3, 5	3, 6
4, 1	4, 2	4, 3	4, 4	4, 5	4, 6
5, 1	5, 2	5, 3	5, 4	5, 5	5, 6
6, 1	6, 2	6, 3	6, 4	6, 5	6, 6

The probability of rolling doubles at any time is $\frac{6}{36}$ or $\frac{1}{6}$. This is called the **theoretical probability.** You can also do an experiment by rolling two dice 100 times and recording the number of times you roll doubles. The ratio $\frac{\text{number of doubles}}{100}$ is called the **experimental probability.**

44 When am I ever going to use this? 77

Suppose you are a basketball player. You have made 88 out of your last 100 free throws. You are ready to shoot a free throw that will determine the outcome of a game. Based on your past record, the probability you will make the shot is 0.88. What other factors would you want to take into consideration?

Mini-Lab

Work with a partner.
Materials: two dice

- Roll two dice 36 times. Record each time doubles occur.
- Compute the ratio $\frac{\text{number of times doubles occur}}{36}$.

Talk About It

a. How does your ratio compare to the ratio of others?

b. How does your ratio compare to $\frac{1}{6}$? If it is different, why do you think it is different?

c. Combine the results of all your classmates. Find the ratio of the total number of doubles to the total number of rolls. How does it compare to $\frac{1}{6}$?

Find the theoretical probability of getting heads if you toss a coin.

$P(H)$ represents the probability of getting heads.

$$P(H) = \frac{1}{2} \quad \begin{matrix} \leftarrow & \textit{number of ways to toss heads} \\ \leftarrow & \textit{number of possible outcomes} \end{matrix}$$

Checking for Understanding

Communicating Mathematics

Read and study the lesson to answer each question.

1. **Tell** the difference between experimental probability and theoretical probability.

2. a. **Tell** whether the experimental probability of an event is always the same.
   b. **Tell** whether the theoretical probability of an event is always the same.
   c. **Show** examples to support your answers to parts a and b.

Guided Practice

3. Find the theoretical probability of choosing a boy's name from 20 boy's names and 10 girl's names.

4. Find the theoretical probability of rolling a sum greater than 7 on two dice.

5. Larry tosses a coin 30 times. It lands on heads 16 times.
   a. What is his experimental probability of getting heads?
   b. How does the experimental probability compare to the theoretical probability of getting heads?

Exercises

Independent Practice

6. A soft drink machine contains cola, ginger ale, root beer, orange, and diet cola. Without looking, choose a soft drink.
   a. Find P(cola or diet cola).
   b. Write C, G, R, O, and D on slips of paper to represent each type of soft drink. Without looking, choose a slip of paper. Record the letter. Replace the slip. Repeat this 9 times. Compute $\frac{\text{number of Cs or Ds.}}{10}$. How does the ratio compare to your answer in part a?

Two dice are rolled. Find each theoretical probability.

7. a sum of 2   8. a sum less than 4   9. a sum of 1

10. Estimate $\sqrt{236}$. *(Lesson 9-3)*

11. **Probability**   On a television game show, a contestant selects one of three doors. A great prize is behind only one of the doors. What percent chance does the contestant have of selecting the right door? *(Lesson 12-3)*

Problem Solving and Applications

12. **Games**   In a game of backgammon, Lorena rolls doubles 8 times.
    a. If she rolls the dice a total of 56 times during the game, what is her experimental probability of rolling doubles?
    b. How does her experimental probability compare to the theoretical probability of rolling doubles?

13. **Coin Toss**   Ten students in Ms. Imhoff's class believe that a coin is not fair. That is, they believe there is not a fifty-fifty chance that it will land on heads if it is tossed. To find out, each student tosses the coin 10 times and records the results as shown in the table below.

Student	1	2	3	4	5	6	7	8	9	10	Total
Number of heads	4	6	2	3	4	4	7	2	3	3	38
Number of tails	6	4	8	7	6	6	3	8	7	7	62

    a. Find the experimental probability of tossing heads for the total number of tosses.
    b. Based on your answer to part a, does the coin appear to be fair?

14. **Critical Thinking**   Is the experimental probability of an event ever equal to the theoretical probability? Explain.

15. **Make up a problem** similar to the one in Exercise 6.
    a. Determine the theoretical probability.
    b. Perform an experiment and determine the experimental probability.
    c. Write a report about the experiment and the results.

COMPUTER CONNECTION

16. **Computer Connection**   The BASIC program at the right simulates tossing a fair coin. A sample output for 10 trials is shown below.

```
HOW MANY TIMES? 10
THHTTTHTHT
YOU TOSSED 4 HEADS
AND 6 TAILS.
```

```
10 INPUT "HOW MANY TRIALS?";N
20 FOR I = 1 TO N
30 C = INT(RND(1)*2)
40 IF C = 0 THEN PRINT "H";:
   H = H + 1
50 IF C = 1 THEN PRINT "T";:
   T = T + 1
70 NEXT I
80 PRINT:PRINT "YOU TOSSED";
   H; " HEADS AND";T;"TAILS."
```

    a. Use the program to simulate tossing a coin for 50 trials and for 100 trials.
    b. For which number of trials is the probability of tossing tails closest to $\frac{1}{2}$?

13-4 Act It Out

Objective

Solve problems by acting them out.

A baseball card manufacturer is holding a contest. Each package of baseball cards in a limited edition series contains a puzzle piece. If you collect all 6 different pieces, you win two tickets to any home game of your choice. There is an equally likely chance of getting a different puzzle piece each time. How many packages of cards would you need to buy to win the contest?

Explore   What do you know?
If you collect all 6 different puzzle pieces, you will win the contest. There is an equally likely chance of getting a different piece each time.

What do you need to find?
You need to find how many packages of cards you would need to buy to win the contest.

Plan   Act out the problem. Work with a partner.

Use a spinner divided into 6 sections to simulate the problem. Each section represents one of the pieces. Spin the spinner until you have one of each number.

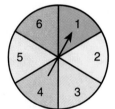

Record the data in a frequency table like the one below.

Outcome	Tally	Frequency
1	|	1
2	|||	3
3	⸝⸝⸝⸝	5
4	|	1
5	||	2
6	|	1

Solve   Find the sum of the frequencies of spinning each number.

$$1 + 3 + 5 + 1 + 2 + 1 = 13$$

According to this simulation, you would have to buy at least 13 packages of cards to collect all 6 puzzle pieces.

Examine Compare your results to other pairs of students. Were your results reasonable?

Example

During a fire drill, Ryan, Debbie, Miguel, and Carianne each left one book in the school library. When they returned to their classroom, their homeroom teacher randomly handed out the four books to the students. What is the probability that Carianne receives the same book she was reading in the library?

You can act out this problem using index cards and books. Work in groups of five.

Write each student's name on two separate index cards. Let one set of cards represent the books. Place the book cards face down on a flat surface and mix them up. Place the other set of index cards in a brown paper bag.

To act out the situation, draw a card out of the bag and choose a book card. Was it the correct book for the first person? the second person? Make a chart like the one below and record this information on the chart.

Trial	Ryan	Debbie	Miguel	Carianne
1	no	no	yes	no
2				
3				

Continue until each book card has been paired with a student. Repeat this activity 20 times.

To solve the problem, find the experimental probability that Carianne receives the correct book.

$$\text{Experimental Probability} = \frac{\text{number of times she received correct book}}{20 \text{ trials}}$$

If Carianne received the correct book 4 times out of 20, the experimental probability is $\frac{4}{20}$, or 0.2.

Checking for Understanding

Communicating Mathematics

Read and study the lesson to answer each question.

1. **Tell** how you can find the answer to the problem in the Example without acting it out.
2. **Tell** one advantage of finding the answer by acting it out.

Guided Practice

Solve by acting it out.

3. Charo has 3 different-colored winter hats. What is the probability that she wears the same hat more than once in a 5-day school week?
4. Suppose the puzzle in the baseball card contest on page 503 had 8 puzzle pieces. How many packages of cards would you need to buy to win the contest?
5. How many times do you need to roll a die to get all 6 numbers?

Problem Solving

Practice

Solve. Use any strategy.

Strategies

• • • • • • • • • •

Look for a pattern.

Solve a simpler problem.

Act it out.

Guess and check.

Draw a diagram.

Make a chart.

Work backward.

6. A number is multiplied by 4 and 12 is added. The result is 20. What is the number?
7. Yolanda has 8 different-colored pens in a pencil cup. If she chooses 6 times, each time replacing the pen before choosing the next one, what is the probability that she chooses a different color each time without looking?
8. Becky buys birthday wrapping paper and cards at the store. Wrapping paper costs $1.75 per package, and cards cost $1.25 each. She spends $12.75. How many of each did she buy?

Mid-Chapter Review

1. Jeremy has a choice of two juices (orange or apple) and three cereals (wheat, rice, or corn) for breakfast. Make a tree diagram to show all the possible outcomes. *(Lesson 13-1)*
2. Find the total number of outcomes if you toss a penny and spin the spinner at the right. *(Lesson 13-2)*
3. Out of 30 rolls of a die, Allison rolls a 4 three times. What is the experimental probability? *(Lesson 13-3)*
4. Five students forgot to write their names on their test papers. Before the end of the period, the teacher randomly hands back the test papers to these students. What is the probability that each of these five students receives his or her own test? Solve by acting it out. *(Lesson 13-4)*

13-5 Using Statistics to Predict

Objective

Predict actions of a larger group by using a sample.

Words to Learn

population
sample
random

DID YOU KNOW

In A.D. 1185 an astronomer, Johannes of Toledo (Toledo, a city in Spain both then and now), predicted that a terrible wind would bring famine and destruction to Europe in the following year. People were so scared that they built underground homes to protect themselves. However, the prediction never came true.

Have you ever received a phone call from someone taking a survey? More than likely your phone number was randomly selected out of a list of thousands, or possibly millions, of phone numbers. The reason the surveyors don't call everyone is because it would be too time-consuming and too expensive.

Actually, it is not necessary to call everyone. You can predict the responses of an entire **population** by surveying a representative **sample** of that population. A sample is called **random** if the members of the sample are selected purely on the basis of chance.

Example 1

Members of the student council wanted to know if there was enough time between classes. They asked fifty students leaving gym class what they thought. Forty students responded that there was not enough time. Is this sample random? Explain.

This sample is not random because it does not represent the total student population. Also, no teachers or administrators were surveyed.

If you survey a random sample of the population, you can use the results to make predictions about the actions of the entire population.

Example 2 *Problem Solving*

Marketing One hundred people chosen at random in a town of 7,500 people were asked what TV station they watched for the morning news. If 15 out of 100 people surveyed responded that they watch Channel 4, how many people in the town can be expected to watch Channel 4?

This ratio, 15 out of 100, is 15%. Find 15% of 7,500.

$$15\% \text{ of } 7{,}500 = 0.15 \times 7{,}500$$
$$= 1{,}125$$

You can predict that about 1,125 people in the town watch Channel 4 for the morning news.

Calculator Hint

••••••••••••

To find 15% of 7,500 on a calculator, enter:

15 % × 7500 = .

The result is 1,125.

Checking for Understanding

Read and study the lesson to answer each question.

Communicating Mathematics

1. **Tell** about a situation in which you might be part of a sample.

2. **Tell** about a situation in which you might be part of a population.

3. **Tell** whether a prediction can ever be exact.

4. **Show** an example of a sample that is not random. Explain why the sample is not random.

Guided Practice

Social Studies Of 42,000 registered voters, the voting preferences of a sample of 1,200 are listed in the table below.

Candidate	Number of Votes
Jenkins	640
Caldarolla	460
Undecided	100
Total	**1,200**

5. How many voters out of the 42,000 might you expect to vote for Jenkins?

6. How many voters out of the 42,000 might you expect to be undecided at election time?

7. If the undecided voters choose Caldarolla, how many votes might Caldarolla expect to receive?

Exercises

Independent Practice

School Student council members surveyed seniors to find out what they would be willing to pay for a copy of the school yearbook. The results are shown at the right.

Price Range	Number of Students
not more than $18	7
not more than $20	11
not more than $22	33
not more than $24	19

8. What was the sample size?

9. To the nearest percent, what percent of students responded that they would buy the yearbook if the cost was not more than $20?

10. If there are 250 seniors at the school, about how many can be expected to buy the yearbook at $19.95?

Social Studies Of 32,500 registered voters, 740 were surveyed. Their voting preferences are listed at the right.

Candidate	Number of Votes Received
Memphez	117
Jeniac	276
Steinmetz	235
Undecided	112
Total	**740**

11. To the nearest percent, what percent of the registered voters can be expected to vote for Steinmetz?

12. To the nearest percent, what percent of the registered voters can be expected to vote for Jeniac or Memphez?

13. If the undecided choose to vote for Memphez, *about* how many can be expected to vote for Memphez?

14. If the undecided give their votes to Memphez, who is expected to win the election?

15. On the evening before the election, a pollster called 500 people whose names were selected at random from the telephone book. Is the sample a random sample?

Mixed Review

16. Multiply 0.0004 and 10^6.
    (Lesson 2-5)

17. **Geometry** Classify the triangle at the right by its sides and by its angles.
    (Lesson 8-3)

18. Find the probability of observing one head and one tail when two fair coins are tossed. *(Lesson 13-3)*

Problem Solving and Applications

19. **Population** In 1989, the total number of unemployed in the United States was about 3,499,000 people. The circle graph below shows the breakdowns by percent.

Unemployment in the United States, 1989

16% married women

22% married men

27% single men

18% single women

10% women: widowed, divorced, separated

7% men: widowed, divorced, separated

Estimate each number of unemployed.

a. married women

b. single men

c. married women or married men

20. **Consumer Math** The chart shows the ownership of video equipment in U.S. households.

Year	1985	1988
Video Cassette Recorders (VCRs)	21%	67%
Video Camera/Camcorders	2%	7%

a. In 1985, there were 87 million households. About how many of them owned a VCR?

b. In 1988, there were 91 million households. About how many of them owned a video camera or camcorder?

c. Do the data suggest that more VCRs were sold in 1988 than in 1985? Explain.

d. Suppose that you visited a friend in 1985. What is the probability that there was a video camera or a camcorder in your friend's household?

21. **Critical Thinking** In a recent survey of radio listeners, a researcher concludes that about 18% of the audience enjoy talk shows. The researcher concludes that this is 1,825 people. Estimate the population of the city.

22. **Journal Entry** Write a paragraph about different ways that people can make sure that a sample is random.

23. **Mathematics and Weather Forecasting** Read the following paragraphs.

The United States Weather Service makes about 2 million weather forecasts each year. It claims that more than 75% of their one-day forecasts are accurate.

Forecasts are used every day to help people decide what to wear and where to go. In addition, they are vital to pilots, sailors, and farmers who need to know exactly what kind of weather to expect in order to carry out their jobs effectively.

A forecaster is like a detective gathering information and clues. Detailed information about the weather at a certain time of day is collected and plotted on a map called a synoptic chart. Using a computer, the forecasters can predict what the next day's weather will be like.

Mount Wai-'ale-'ale in Hawaii gets more rain than any other place in the world. It rains 92% of the year. To the nearest day, how many days of rain can it expect to get every year?

13-6 Probability of Two Events

Objective

Find the probability of independent and dependent events.

Words to Learn

independent events
dependent events

At the Creolia County Fair, there is a booth that has two spinning wheels like the ones shown below.

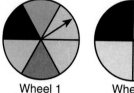

Wheel 1   Wheel 2

The winner gets a big stuffed animal if Wheel 1 shows black and Wheel 2 shows white. What is the probability of winning?

When the outcome of one event does *not* influence the outcome of a second event, the events are called **independent events.** The results of spinning Wheel 1 and Wheel 2 are independent of one another.

To find the total number of outcomes, you can use the Fundamental Counting Principle. Of the 6×4, or 24, outcomes, there is 1 winning pair (black, white). Compute the probability.

$$P(\text{black, white}) = \frac{\text{favorable}}{\text{total}} = \frac{1}{24}$$

Note that $P(\text{black})$ on Wheel 1 is $\frac{1}{6}$ and $P(\text{white})$ on Wheel 2 is $\frac{1}{4}$. Also notice the following.

$$P(\text{black, white}) = \frac{1}{24}$$
$$= \frac{1}{6} \cdot \frac{1}{4}$$
$$= P(\text{black on 1}) \cdot P(\text{white on 2})$$

This suggests that the probability of two independent events can be found by multiplying.

Probability of Two Independent Events	**In words:** The probability of two independent events can be found by multiplying the probability of one event by the probability of the second event.
	In symbols: If A and B are independent events, then $P(A \text{ and } B) = P(A) \cdot P(B)$.

Example 1

A green die and a red die are rolled. Find the probability that an odd number is rolled on the green die and a multiple of 3 is rolled on the red die.

There are three odd numbers: 1, 3, and 5.
So, $P(\text{odd number}) = \frac{3}{6}$ or $\frac{1}{2}$.

There are two multiples of 3: 3 and 6.
So, $P(\text{multiple of 3}) = \frac{2}{6}$ or $\frac{1}{3}$.

The two rolls are independent of each other.
So $P(\text{odd and multiple of 3}) = \frac{1}{2} \cdot \frac{1}{3}$ or $\frac{1}{6}$.

The probability that an odd number is rolled on the green die and a multiple of 3 is rolled on the red die is $\frac{1}{6}$.

If the result of one event affects the result of a second event, the events are called **dependent events.**

Example 2

A bag contains 5 white, 4 blue, and 3 red marbles. Two marbles are drawn, but the first marble drawn is not replaced. Find $P(\text{blue, then red})$.

$P(\text{blue}) = \frac{4}{12}$ ← *There are 4 blue marbles.*
                    ← *There are a total of 12 marbles.*

The result of the first draw affected the probability of the second draw.

$P(\text{red after blue}) = \frac{3}{11}$ ← *There are 3 red marbles left.*
                                ← *There are a total of 11 marbles left.*

$P(\text{blue, then red}) = P(\text{blue}) \cdot P(\text{red after blue})$

$$= \frac{4}{12} \cdot \frac{3}{11}$$

$$= \frac{\overset{1}{\cancel{4}}}{\underset{1}{\cancel{12}}_3} \cdot \frac{\overset{1}{\cancel{3}}}{11}$$

$$= \frac{1}{11}$$

The probability of drawing a blue marble and then a red marble is $\frac{1}{11}$.

Checking for Understanding

Communicating Mathematics

Read and study the lesson to answer each question.

1. **Tell** when two events are independent.
2. **Write** a paragraph describing two dependent events.
3. **Tell** how to find the probability of an outcome of two independent events.

Guided Practice

Tell whether the events are independent or dependent. Explain.

4. choosing a card from a hat and then choosing a second card without replacing the first one
5. selecting a name from the Chicago telephone book and a name from the Houston telephone book

Find each probability.

6. A blue die and a yellow die are rolled. Find the probability that an odd number is rolled on the blue die and a multiple of 6 is rolled on the yellow die.

7. A bag contains 10 white, 8 blue, and 6 red marbles. Two marbles are drawn, but the first marble drawn is not replaced. Find P(blue, then blue).

8. A coin is tossed and a die is rolled. Find the probability of getting heads and a multiple of 2.

Exercises

Independent Practice

Tell whether the events are independent or dependent.

9. tossing a coin twice
10. selecting a computer disk from a file box and choosing a second disk without replacing the first one

Find each probability.

11. A wallet contains five $5 bills, three $10 bills, and two $20 bills. Two bills are selected without the first selection being replaced. Find P($10, then $10).

12. A wallet contains five $5 bills, three $10 bills, and two $20 bills. Two bills are selected without the first selection being replaced. Find P($5, then $20).

13. A blue die and a green die are rolled. Find the probability that a multiple of 2 is rolled on the blue die and a multiple of 3 is rolled on the green die.

Boxes A and B contain pecans, cashews, and peanuts. The table shows how many of each are in Boxes A and B.

Box	Pecans	Cashews	Peanuts
A	12	20	24
B	30	18	50

Find the probability of each choice.

14. a peanut from Box A and a peanut from Box B

15. a peanut from Box A and a pecan from Box B

Mixed Review 16. Add $\frac{3}{15}$ and $\frac{5}{8}$. *(Lesson 5-3)*

17. **Geometry** Sketch a three-dimensional figure given the front, side, and top view. *(Lesson 10-1)*

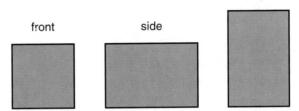

front   side   top

18. **Sports** It is predicted that the Wilson Junior High baseball team will win 65% of their games this season. If they are scheduled to play 60 games, how many games would you expect them to win? *(Lesson 13-5)*

Problem Solving and Applications

19. **Traffic Lights** One traffic light is red 60% of the time. The next traffic light is red 50% of the time. If the lights operate independently of one another, find P(red, then red).

20. **Clothing** In one drawer, Kwag has 2 pairs of brown socks, 3 pairs of black socks, and 4 pairs of blue socks. In another drawer, he has 3 red sweaters, 2 brown sweaters, and 2 blue sweaters. Suppose Kwag makes a selection from each drawer without looking. What is the probability that he will have brown socks and a brown sweater?

21. **Critical Thinking** Suppose that events A and B are independent. How would you find the probability that neither A nor B will happen?

22. **Journal Entry** List the things you do to get ready in the morning, such as take a shower, eat breakfast, and so on. Name which events are dependent and which are independent. The dependent events are the ones that can't be done until you have done an earlier event.

Choosing a Camcorder

Situation

The Drama Club earned $1,000 in a fundraiser. They plan to purchase a compact camcorder to videotape rehearsals and actual performances of their theater productions. You have been chosen to be on the committee that will make a recommendation to club members and faculty advisors on which kind of camcorder to buy.

Hidden Data

How many hours of use do you get with the battery that is included in the price of the camcorder? How much will it cost to buy a rechargeable battery?

What kind of warranty comes with the camera?

Does the price of the extended service protection plan vary depending on the selling price of an item? If so, which one can you afford?

Is a storage case included in the purchase price or must it be purchased separately?

CAMCORDERS

FROM YOUR ELECTRONICS *STORE*

Manu-facturer	Format	Power Zoom	Low Light Rating	Flying Erase Head	A/V Dub	Self Timer Time Lapse Recording	Titler	Ship Wt.	($) Price
A	8mm	8:1	4 lux	yes	no	no	no	6	897.79
B	8mm	6:1	2 lux	yes	yes	no	no	4	887.78
C	VHS-C	6:1	2 lux	yes	yes	no	no	5	793.97
D	VHS-HQ	12:1	1 lux	yes	yes	yes	no	12	998.83
E	VHS-HQ	16:1	3 lux	yes	yes	yes	no	12	1,299.94
F	VHS-HQ	8:1	2 lux	yes	yes	yes	yes	15	898.82
G	VHS-HQ	8:1	2 lux	yes	yes	yes	no	10	869.96
H	VHS-HQ	8:1	2 lux	no	no	yes	no	7	849.94
I	VHS-HQ	8:1	1 lux	yes	yes	yes	no	12	829.92

Extended Service Protection Plan
Cost varies depending on the selling price of the item.

Protect IT

Selling Price of Item	1 Yr Plan Price	2 Yr Plan Price
Up to $99.99	9.99	N/A
$100 - 199.99	19.99	36.99
$200 - 299.99	29.99	49.99
$300 - 999.99	39.99	69.99
$1,000 - 2,000	59.99	99.99

Analyzing the Data

1. If you have $1,000 to spend, which camcorders can you afford to buy?
2. What is the difference in price between the most expensive 8mm camcorder and the least expensive VHS-HQ?
3. What kind of camcorder is the least expensive?
4. If you buy manufacturer B's camcorder and buy the extended warranty for 1 year, what is the total cost?

Making a Decision

5. **Is** a VHS-HQ, a VHS-C, or an 8mm right for your clubs needs?
6. Each camcorder has a power zoom option. The higher the ratio, the closer the image. **How important** is this feature to your club?
7. **What factors** should you consider in selecting a camcorder's weight?
8. **What compromises,** if any, did you have to make because of your budget?
9. **Would your decision** be affected if your club could rent the camcorder to other departments or clubs in your school?
10. **Could a profit** be made if videotapes of the plays were sold to the cast members and their families?
11. **Ask retailers** for special prices for your drama club in exchange for free advertising in the play programs.

Choosing the Right Camcorder Format

VHS: Larger camcorder that rests on your shoulder for extra stability when recording. Uses standard VHS tape that you simply pop into any VHS VCR for viewing.

VHS-C: Lightweight and designed for hand-held use. Uses compact-sized VHS-C tape that can be viewed in VHS VCR by using a cassette adapter, included with the camcorder.

8mm: Lightweight and designed for hand-held use. Uses 8mm video tape, producing better quality video recordings than VHS or VHS-C. To view tape, connect camcorder to TV and use it like a VCR, or connect it to your VHS VCR and make a copy on standard VHS tape.

Making Decisions in the Real World

12. **Consult** consumer magazines to see how different camcorder brands compare. Which camcorder would you buy if you were making this purchase for yourself or your family?

HIGH STANDARD
FULL LIFETIME WARRANTY

13-7A Exploring Permutations

A Preview of Lesson 13-7

Objective

Explore permutations.

Words to Learn

permutation

Materials

four index cards
marker

An arrangement of letters, names, or objects in a particular order is called a **permutation.** For instance, if you use E, D, and N, three permutations result in English words.

DEN   END   NED

There are three more permutations of these letters, but they are not English words.

DNE   EDN   NDE

Try this!

Work with a partner.

- Place the digits 1, 4, 7, and 9 on four cards, one digit on each card. Shuffle the cards and place them face down. Choose three of the cards and turn them face up. For example, you may have chosen the cards shown below.

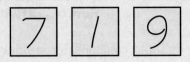

- Record the number shown; for example, 719.
- Now rearrange the three cards to make another three-digit number. Record that number.
- Continue rearranging the cards and recording the numbers shown until you have listed all the three-digit numbers you can with the cards chosen. Record the total number.
- Shuffle the four cards. Place them face down on the desk. Choose three cards at random. Count how many three-digit numbers you got from that selection of cards. Did you get the same number of arrangements for your second selection of cards as you did for the first selection?
- Now use all four cards. Arrange the cards to form all the four-digit numbers that you can. How many arrangements are there?

What do you think?

1. Could a tree diagram be helpful in listing all the permutations of three or four digits? Why or why not?
2. Suppose that you had five cards with five different digits. How many five-digit numbers do you think can be formed?

13-7 Permutations

Objective

Find the number of permutations of a set of objects.

Words to Learn

permutation
factorial

The flag of Mali, Africa, has three vertical stripes. The stripes are green, yellow, and red. The flag of Italy is very similar. It has three vertical stripes that are green, white, and red.

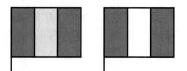

How many different flags can be made from the colors green, yellow, white, and red if each flag has three vertical stripes? *This question will be answered in Example 2.*

Problem Solving Hint

● ● ● ● ● ● ● ● ● ● ● ● ●

Make a list.

A **permutation** is an arrangement, or listing, of objects in which order is important. For example, suppose that the high school band has three seats for their three trumpeters. In how many different ways can they be seated?

There are three different choices for the first seat, then two choices for the next seat, and only one choice for the third seat. So, the number of permutations is $3 \cdot 2 \cdot 1 = 6$.

In this example, the symbol $P(3, 3)$ represents the number of permutations of 3 things taken 3 at a time.

$$P(3, 3) = 3 \cdot 2 \cdot 1$$

In general, $P(n, r)$ means the number of permutations of n things taken r at a time.

Example 1

In how many ways can 5 people be seated in a row of 5 chairs?

Find $P(5, 5)$, the number of permutations of 5 things taken 5 at a time. There are 5 choices for the first seat. After that seat is taken, there are 4 choices for the second seat, and so on.

$P(5, 5) = 5 \cdot 4 \cdot 3 \cdot 2 \cdot 1$
$\qquad = 120$

Five people can be seated in 120 ways.

The expression $5 \cdot 4 \cdot 3 \cdot 2 \cdot 1$ can be written as $5!$, which is read "five **factorial**." In general, $n!$ is the product of all the counting numbers beginning with n and counting backwards to 1. We define $0!$ to be 1. For example, this is how to compute $4!$.

$$4! = 4 \cdot 3 \cdot 2 \cdot 1$$
$$= 24$$

Some arrangements involve only part of a group.

Example 2 *Problem Solving*

Geography   Refer to the problem in the lesson introduction. How many different flags consisting of three vertical stripes can be made from green, yellow, white, and red?

Find $P(4, 3)$, the number of permutations of four colors taken three at a time.

For the first vertical stripe, there are 4 possible choices. After that stripe is chosen, there are 3 possible choices. Finally, there are 2 possible choices for the third stripe.

$$4 \cdot 3 \cdot 2 = 24$$

There are 24 different flags with three vertical stripes that can be made from green, yellow, white, and red. *To check your answer, use colored pencils and actually draw each possible flag.*

Example 3 *Problem Solving*

Vacations   The DiGrazzias want to visit Arizona, Florida, San Francisco, Chicago, and Boston on their next five vacations. In how many ways can they visit three of these places over the next three years?

Find $P(5, 3)$, the number of permutations of 5 things taken 3 at a time.

For the first vacation, there are 5 possible choices. After they take that vacation, there are 4 possible choices. Then for the third vacation, there are 3 possible choices.

$$5 \cdot 4 \cdot 3 = 60$$ There are 60 ways to visit three of these places.

Checking for Understanding

Read and study the lesson to answer each question.

1. **Tell** how to find the number of permutations of 5 colors taken 5 at a time.

2. **Tell** how to find the number of permutations of 5 colors taken 2 at a time.

3. **Tell** whether the following list is a list of permutations. Explain.

    123,   132,   213,   231,   312,   321

Guided Practice

Find the value of each expression.

4. 3!   5. 0!   6. $P(4, 1)$ 7. $P(5, 2)$

8. At a pet show, first, second, and third prizes will be awarded to Eau, Tabbie, and Luv. In how many ways can the prizes be awarded?

9. A license plate begins with three letters. If the possible letters are A, B, C, D, E, and F, how many different permutations of these letters can be made if no letter is used more than once?

10. A flag consists of three horizontal stripes. If the colors can be chosen from among white, black, green, red, and blue, how many flags can be made?

Exercises

Independent Practice

Find the value of each expression.

11. 1!   12. 5!   13. $P(6, 4)$ 14. $P(5, 1)$

15. How many different five-letter "words" can be formed from the letters A, B, C, D, and E if no letter may be used more than once?
    "Words" means any arrangement of letters, not just English words.

16. How many different five-letter "words" can be formed from all the letters of the alphabet if no letter may be used more than once?

17. In how many ways can seven different books be arranged on a shelf?

18. In how many ways can a president, a treasurer, and a secretary be chosen from among 8 candidates?

19. A zip code contains 5 digits. How many different zip codes can be made with the digits 0–9 if no digit is used more than once and the first digit is not 0?

Mixed Review

20. Find the least common multiple of 12 and 27. *(Lesson 4-9)*

21. Solve $6 = \frac{w}{-3}$. *(Lesson 7-9)*

22. **Medicine**   The effect of a drug used to control epilepsy is independent from patient to patient. The probability that the drug is successful is 0.88. What is the probability that the drug is a success with two randomly selected patients? *(Lesson 13-6)*

23. **Celebrations** Parade organizers want to place a fire engine, a police car, and an ambulance after the grand marshal's car. In how many ways can the fire engine, the police car, and the ambulance be arranged?

24. **Architecture** An architectural firm has drawn up six different plans for a new city park. In how many ways can the architects present the plans to the city council?

25. **Probability** A flag consists of three vertical stripes. The colors are chosen from among red, blue, green, black, white, and gray. No color can be used more than once. Find the probability that a flag chosen at random will have a red stripe in the middle.

26. **Critical Thinking** In a certain area, a license plate consists of three letters followed by three digits. No letter may be used twice, no digit may be used twice, and the letters I and O are not used. How many different license plates are possible?

Save Planet Earth

Lawn Care

One of the most popular activities of American homeowners is maintaining and beautifying their lawns. Many homeowners put a lot of effort into having the perfect green lawn. This "perfect green lawn," however, comes with a price tag—$6 billion annually.

In addition, lawn care can often lead to problems in the environment. Each American uses about 3 to 10 pounds of chemical products on their lawns each year. Many of these products are intended to kill insects that can damage a lawn. However, the chemicals in these products can also have a hazardous effect on birds, amphibians, and fish (when the chemicals get into the ground water). Humans have also been known to react to the chemicals. Some people have experienced dizziness, rashes, headaches, respiratory illness, and other problems from these chemicals.

How You Can Help

- Take a sample of your grass to a local garden center. Find out the type of grass and how to care for it. Is your family's current lawn care program correct? Are there organic pesticides available that you could use instead of chemicals?
- Eliminate harmful pesticides.
- Water your lawn either in the morning or evening to save water loss through evaporation.

13-8A Exploring Combinations

A Preview of Lesson 13-8

Objective

Explore combinations.

Materials

index cards
marker

The student council at Arlington Junior High School is planning to have a pizza sale. They plan to offer eight different kinds of pizza.

extra cheese	mushroom
pepperoni	sausage
meatball	onion
pepper	anchovy

The price of a pizza depends on the *combination* of toppings chosen.

Try this!

Work in groups of two or three.

- Write the names of the eight toppings on eight index cards.
- To make a pizza, select any pair of cards. Make a list of all the different combinations that are possible. Note that the order is not important.

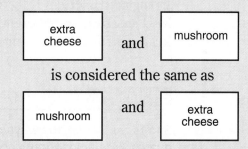

- Suppose that the student council decides to offer a deluxe pizza with three toppings on it. Make a list of all the different toppings that are possible if three of the eight toppings are used. How many different combinations did you list?
- Suppose that the student council decides to offer a small pizza with one topping on it. Make a list of all the different toppings that are possible if one of the eight toppings is used.

What do you think?

1. What is the difference between a combination (like a combination of toppings) and a permutation?
2. Would a tree diagram be a good method for listing all the combinations of two toppings? Why or why not?

13-8 Combinations

Objective

Find the number of combinations of a set of objects.

Words to Learn

combination

In October 1991, Clarence Thomas was confirmed as a member of the United States Supreme Court. He replaced Justice Thurgood Marshall who announced his retirement in July of the same year.

There are 9 Supreme Court justices in all. If at least 5 of the 9 justices agree on a decision, they can issue a majority opinion. How many different combinations of 5 justices can there be? *This question will be answered in Example 2.*

An arrangement of objects in which order is unimportant is called a **combination.** Since order is not important, the number of combinations of a set of objects is less than the number of permutations of those objects. For example, suppose you are arranging the letters R, E, and D. In a permutation, the arrangements RED and DER are different. But in a combination, the arrangements RED and DER are the same because order is not important.

Example 1 *Problem Solving*

Sports   In how many ways can a basketball coach choose two starting guards from among four capable players?

Call the players A, B, C, and D.
List *all* arrangements of A, B, C, and D.

AB	AC	AD	BC	BD	CD
BA	CA	DA	CB	DB	DC

Problem Solving Hint

●●●●●●●●●●●●●

Make a list.

Count all the *different* arrangements. Arrangements AB and BA are not different here. Altogether, there are six different arrangements.

AB	AC	AD	BC	BD	CD

The number of combinations of n things taken r at a time is written as $C(n, r)$. In Example 1, you saw how to find the number of combinations by listing. List all the permutations of a set of objects, then eliminate the arrangements that are the same except for order. Use a formula to find the number of combinations.

In general, $C(n, r)$ means the number of combinations of n things taken r at a time.

$$C(n, r) = \frac{P(n, r)}{r!}$$

Example 2 _Problem Solving_

Law Refer to the problem in the lesson introduction. In how many ways can five of the nine justices of the Supreme Court agree?

Use the formula to find the number of combinations of nine things taken five at a time.

$$
\begin{aligned}
C(9, 5) &= \frac{P(9, 5)}{5!} \\
&= \frac{9 \cdot 8 \cdot 7 \cdot 6 \cdot 5}{5 \cdot 4 \cdot 3 \cdot 2 \cdot 1} \\
&= \frac{\overset{3}{\cancel{9}} \cdot \overset{2}{\cancel{8}} \cdot 7 \cdot \overset{3}{\cancel{6}} \cdot \overset{1}{\cancel{5}}}{\underset{1}{\cancel{5}} \cdot \underset{1}{\cancel{4}} \cdot \underset{1}{\cancel{3}} \cdot \underset{1}{\cancel{2}} \cdot 1} \\
&= 126
\end{aligned}
$$

Five of the nine justices can agree in 126 ways.

Example 3 _Connection_

Probability At Sanchez Taco Restaurant, customers may choose three fillings for their tacos from a list of five fillings. The fillings are beef, chicken, cheese, lettuce, and tomato. If it is equally likely that a customer will choose any combination of three fillings, find the probability that a customer will choose beef, cheese, and lettuce.

Explore There are five fillings and customers may choose three fillings from among them.

Plan Use the formula to find the number of combinations of five fillings taken three at a time.

Solve
$$
\begin{aligned}
C(5, 3) &= \frac{P(5, 3)}{3!} \\
&= \frac{5 \cdot 4 \cdot 3}{3 \cdot 2 \cdot 1} \\
&= \frac{5 \cdot \overset{2}{\cancel{4}} \cdot \overset{1}{\cancel{3}}}{\underset{1}{\cancel{3}} \cdot \underset{1}{\cancel{2}} \cdot 1} \\
&= 10
\end{aligned}
$$

(continued on next page)

Of the 10 combinations, one of them is beef, cheese, and lettuce.

So, P(beef, cheese, lettuce) $= \frac{1}{10}$ or 0.1.

Actually list all the possible combinations of the fillings to check your answer.

Checking for Understanding

Communicating Mathematics

Read and study the lesson to answer each question.

1. **Write** how to find the number of combinations of 3 letters taken 2 at a time.
2. **Tell** whether the following is a list of permutations or a list of combinations. Explain.

    ABC, ACB, BAC, BCA, CAB, CBA

3. **Tell** whether there are more permutations of a set of objects taken 3 at a time or more combinations of the objects taken 3 at a time.
4. **Write** a paragraph about why a combination lock really should be called a permutation lock.

Guided Practice

Find the value of each expression.

5. $\frac{5!}{2!}$ 6. $\frac{8!}{6!}$ 7. $C(6, 2)$ 8. $C(4, 3)$

9. List all combinations of the letters R, S, T, U, and V taken 3 at a time.
10. List all combinations of Bob, Debbie, Julio, Terese, and Yoki taken 2 at a time.
11. At Mario's Pizza Parlor, customers may choose 2 pizza toppings from the following: cheese, pepper, onion, pepperoni, salami, meatball, mushroom, or sausage. If it is equally likely that a customer will choose any combinations of 2 toppings, find the probability that a customer will choose cheese/pepperoni or sausage/mushroom.

Exercises

Independent Practice

Find the value of each expression.

12. $\frac{7!}{1!}$ 13. $\frac{8!}{4!}$ 14. $C(7, 4)$ 15. $C(8, 6)$

16. An 11-member city council makes its decisions by simple majority vote. That is, if 6 out of the 11 members vote for an issue, the issue is passed. In how many ways can a simple majority of the council decide on an issue?
17. In a lottery, each ticket has 5 one-digit numbers 0–9 on it. You win if your ticket has the digits *in any order*. What are your chances of winning?

Tell whether each problem involves a permutation or a combination.
Then solve the problem.

18. In how many ways can three swimmers for a team be chosen from six swimmers?

19. In how many ways can four cars line up for a race?

20. In how many ways can seven people make a decision by a simple majority?

Mixed Review 21. The two graphs below show that the number of farms in the United States is decreasing, but the average size of each farm is increasing. What do you predict the number of farms will be in the year 2000, and what will be the average size of each farm in the year 2000? *(Lesson 3-7)*

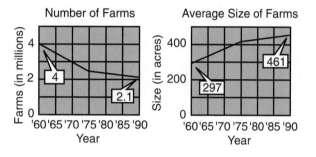

22. Solve $\frac{3}{x} = \frac{25}{15}$. *(Lesson 11-3)*

23. **Sailing** When Bob purchased his sailboat, it came with 6 different-colored flags to be used for sending signals. The specific signal depended on the order of the flags. How many different 3-flag signals can Bob send? *(Lesson 13-7)*

Problem Solving and Applications 24. **Sales** In how many ways can 30 different books be displayed 5 at a time if the order of the arrangement is not important?

25. **Food Service** At the Grill and Skillet Restaurant, customers may choose 3 toppings for their hamburgers from a list of 6 toppings. The toppings are lettuce, tomato, onion, cheese, pickle, and mushroom. Find the probability that a customer will choose onion, pickle, and cheese.

26. **Critical Thinking**

  a. How many combinations of A, B, C, and D are there taken 1 at a time? 2 at a time? 3 at a time? 4 at a time?

  b. What is the sum of the answers to part a?

  c. How is your answer related to 2^4?

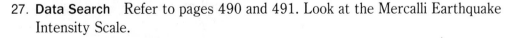

27. **Data Search** Refer to pages 490 and 491. Look at the Mercalli Earthquake Intensity Scale.

  a. When an earthquake registers 8.0 and above, what kind of damage occurs?

  b. How often does this kind of earthquake occur?

13 Study Guide and Review

Communicating Mathematics

Choose the correct term to complete the sentence.

1. The set of all possible outcomes for an experiment is called the (sample space, combination).

2. The Fundamental Counting Principle counts the number of possible outcomes using the operation of (addition, multiplication).

3. The ratio of the number of times an event occurs to the number of trials done is called the (theoretical, experimental) probability.

4. When the outcome of one event influences the outcome of a second event, the events are called (independent, dependent).

5. A (permutation, combination) is an arrangement of objects in which order is important.

6. In your own words, explain the relationship between a sample and a population.

Skills and Concepts

Objectives and Examples	Review Exercises
Upon completing this chapter, you should be able to:	*Use these exercises to review and prepare for the chapter test.*

• use tree diagrams to count outcomes *(Lesson 13-1)*

When a coin is tossed twice,

there are 4 possible outcomes.

Make a tree diagram and then give the total number of outcomes.

7. choosing a red, black, or white car with either black or gray interior

8. tossing a coin and rolling a die

• use multiplication to count outcomes *(Lesson 13-2)*

There are 2 possible outcomes each time a coin is tossed. If a coin is tossed 2 times, there are $2 \times 2 = 4$ outcomes.

Use multiplication to find the total number of outcomes in each situation.

9. rolling 3 dice

10. selecting a house from 3 styles, 2 locations, and 2 exterior colors

Objectives and Examples

- find and compare experimental and theoretical probabilities *(Lesson 13-3)*

  If 1 marble is drawn from a bag containing 6 red and 4 black marbles, the theoretical probability that it is red is $\frac{6}{10}$ or $\frac{3}{5}$.

- predict actions of a larger group by using a sample *(Lesson 13-5)*

  If 5% of a random sample of 100 students at Roosevelt Junior High have after-school jobs, then we could predict how many out of all 500 students have after-school jobs.

  5% of 500 = 0.05 × 500
  $\qquad$ = 25 students

- find the probability of independent and dependent events *(Lesson 13-6)*

  If A and B are independent events, then $P(A \text{ and } B) = P(A) \cdot P(B)$.

- find the number of permutations of objects *(Lesson 13-7)*

  $P(4, 2) = 4 \cdot 3$
  $\qquad = 12$

Review Exercises

A bowl contains the names of 30 middle school students. Nine are 6th graders, 14 are 7th graders and 7 are 8th graders. One name is randomly selected. Find the probability that the student selected is:

11. a 6th grader

12. not an 8th grader

13. at least a 7th grader

Of 25,000 registered voters, the preferences of 1,000 are listed in the table below.

Candidate	Number of Votes Received
Chung	240
Brown	380
Armas	300
Undecided	80

14. How many of the 25,000 voters might you expect to vote for Brown?

15. How many of the 25,000 voters might you expect to be undecided at election time?

A bag contains 8 blue, 6 white, and 4 red marbles. Two marbles are randomly drawn. Find P(white, white) if

16. the first marble drawn is replaced

17. the first marble drawn is not replaced

Find the value of each expression.

18. 5! $\qquad$ 19. 0!

20. $P(8, 3)$ $\qquad$ 21. $P(6, 6)$

Objectives and Examples

• find the number of combinations of a set of objects *(Lesson 13-8)*

$$C(5, 3) = \frac{P(5, 3)}{3!}$$

$$= \frac{5 \cdot \overset{2}{\cancel{4}} \cdot \overset{1}{\cancel{3}}}{\underset{1}{\cancel{3}} \cdot \underset{1}{\cancel{2}} \cdot 1}$$

$$= 10$$

Review Exercises

Find the value of each expression.

22. $\frac{4!}{3!}$

23. $\frac{7!}{7!}$

24. $C(3, 3)$

25. $C(8, 4)$

Applications and Problem Solving

26. How many times do you need to spin this spinner to spin all seven numbers? Solve by acting it out. *(Lesson 13-4)*

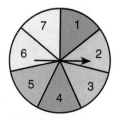

27. **Raffle** In how many ways can the grand-prize ticket, second-prize ticket, and third-prize ticket be selected from the 20 raffle tickets sold? *(Lesson 13-7)*

28. **Games** Hoshi has 12 different games that she keeps in her bedroom. She is allowed to bring 2 of these games into the family room for the evening. In how many ways can Hoshi select the 2 games? *(Lesson 13-8)*

Curriculum Connection Projects

• **Home Economics** Make a list of different flavors of potato chips and survey your classmates to find out how many people prefer each flavor. Predict how many students in the school would prefer each flavor.

• **Sports** List the catchers and pitchers on your favorite baseball team. Find how many different catcher-pitcher combinations are possible.

Read More About It

Rubinstein, Gillian. *Beyond the Labyrinth.*
Riedel, Manfred. *Odds and Chances for Kids: A Look at Probability.*
Phillips, Jo. *Exploring Triangles: Paper Folding Geometry.*

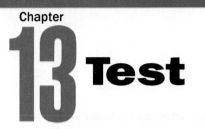
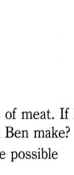

1. Ben is making a sandwich. He has 2 kinds of bread and 3 kinds of meat. If he only uses one kind of each, how many different sandwiches can Ben make?

2. A couple has 3 children. Make a tree diagram to show all of the possible orders of children if each child is listed as a boy or a girl.

3. How many different local telephone numbers (XXX-XXXX) are possible?

The spinner at the right has an equal chance of landing on each number. Find the probability that the spinner lands on:

4. an even number.   5. a 3 or a 5.   6. a number greater than 6.

7. Ryan works at The Eggroll House. He has 6 different kinds of shirts to wear with his uniform. What is the probability that he wears the same shirt more than once in a 3-day work week?

The opinions of 1,000 voters on an upcoming school levy are listed at the right. The entire city has 25,000 registered voters.

Opinion	Number of Votes
Oppose	280
Neutral	200
Favor	520

8. How many of the 25,000 voters might you expect to oppose the levy?

9. How many of the 25,000 voters might you expect to be neutral?

10. Would you expect the levy to pass?

Tell whether the events are independent or dependent.

11. rolling a pair of dice and getting a 6 on the first die and a sum of 9 for both dice

12. having brown hair and owning a brown car

13. The two fire engines in a small town operate independently. The probability that each engine is available when needed is 0.96. Find the probability that both engines are available at one time.

Find the value of each expression.

14. $6!$ 15. $P(7, 3)$ 16. $\frac{5!}{3!}$ 17. $0!$ 18. $C(9, 2)$

19. **Sports** Six runners are racing in a 100-meter sprint. In how many ways can the gold, silver, and bronze medals be awarded?

20. **School Government** From a student council consisting of 12 students, 3 are selected to represent the student body at a school board meeting. In how many ways can these students be selected?

Bonus Evaluate $C(n, n)$ for any positive integer n.

Functions and Graphs

Spotlight on Animals

Have You Ever Wondered. . .

- During what time of the year the most sea creatures are stranded on beaches?

- How many species of animals are threatened or endangered?

THREATENED AND ENDANGERED WILDLIFE

Item	Mammals	Birds	Reptiles	Fishes
Endangered species, total	290	221	74	58
U.S. only	31	61	8	45
U.S. and foreign	19	15	7	2
Foreign only	240	145	59	11
Threatened species, total	30	10	32	30
U.S. only	5	7	14	24
U.S. and foreign	2	3	4	6
Foreign only	23	0	14	0

BEACHED SEA CREATURES

155

102

46

Number

120
80
40
0

Jan   Month   Dec

number of whales, dolphins, and
porpoises beached in 1990

1859   1872   1889   1896

1864

1840   1865   1890   1915

*Philadelphia Zoological
Society marks beginning
of American zoos*

Central Park Zoo opens in New York City

Yellowstone National Park opens

*First American
motion pictures
appear in theaters*

*National Zoological Park
opens in Washington, DC*

Chapter Project

Animals

Work in a group.

1. Several times each month, for the next several months, observe the animals you see each day.

2. Make a list of the animals you see.

3. Make a bar graph showing how many of each animal you see. Indicate how the number of animals you see changes with the time of the year, if it does. Indicate if the number changes with the weather. Explain why.

Looking Ahead

In this chapter, you will see how mathematics can be used to answer the questions about animals.

The major objectives of the chapter are to:

- solve problems by working backward

- solve two-step equations

- graph ordered pairs and transformations on a coordinate plane

- graph linear equations and functions by plotting points

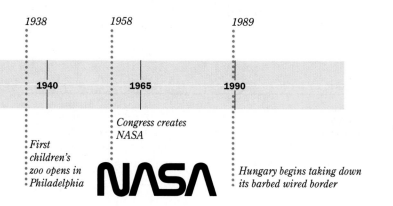

1938	1958	1989
1940	**1965**	**1990**

First children's zoo opens in Philadelphia

Congress creates
NASA

NASA

Hungary begins taking down its barbed wired border

14-1 Work Backward

Objective

Solve problems by working backward.

Sherry and her sister, Ann, each own an equal number of shares of Glaxon stock. Sherry sells one third of her shares for $2,700. What was the total value of Sherry's and Ann's stock just before the sale?

Explore What do you know?
Each sister owns an equal number of shares of stock. Sherry sells one third of her half for $2,700.

What do you need to find?
The total value of Sherry's and Ann's stock before the sale.

Plan Multiply to find the value of Sherry's stock. Then double the amount to find the total value of Sherry's and Ann's stock.

Solve One third of Sherry's half is $2,700. So, Sherry's shares are worth $3 \times 2,700$ or $8,100. Ann has the same number of shares as Sherry. The value of their stock is the same. The total value is $2 \times 8,100$ or $16,200.

The value of Sherry's and Ann's stock before the sale was $16,200.

Examine Assume Sherry's and Ann's stock was worth $16,200. The value of Sherry's stock was half of $16,200, or $8,100. She sold one third of her stock: $8,100 \div 3$ or $2,700. Since $2,700 matches the information given, the answer is correct.

Example

Smart Shopping Sam is planning a luncheon. He goes to the grocery store and buys a ham for $24.98 and a vegetable tray for $17.49. There is no tax on food. He gives the cashier one bill and receives less than $10 in change. What was the denomination of the bill Sam gave the cashier?

You can estimate the denomination of the bill by working backward. His total purchase was about $25 + $20 or $45. His change was less than $10. Estimate by adding $10 to $45. $55 is an overestimate. The denomination of the bill must be between $45 and $55. So, Sam gave the cashier a $50 bill.

Checking for Understanding

Communicating Mathematics

Read and study the lesson to answer the question.

1. **Tell** how at least two problem-solving strategies are used to solve the problem in the example.

Guided Practice

Solve by working backward.

2. A can of evaporated milk weighs 15 ounces. Mrs. Foster uses half of the milk to make pumpkin pudding. The can and the milk that is left weigh 9 ounces. How much does the can weigh?

3. Jim rented 3 times as many videotapes as Phyllis last month. Phyllis rented 4 fewer videotapes than Ed, but 4 more than Matsu. Ed rented 10 videotapes. How many videotapes did each person rent?

Problem Solving

Practice

Solve. Use any strategy.

4. Copy and complete the table.

	4		5	
	12		13	15
Subtract 5		3	8	
Multiply by 2			16	20

Strategies
• • • • • • • • • •
Look for a pattern.
Solve a simpler pattern.
Act it out.
Guess and check.
Draw a diagram.
Make a chart.
Work backward.

5. Ms. Lia orders an oil delivery since her oil tank reads $\frac{1}{5}$ full. After the tank is filled, it reads $\frac{9}{10}$ full. Ms. Lia receives a receipt for a delivery of 105 gallons of oil. How much oil was in the tank before the delivery?

6. Suppose your locker number has three digits. The digits appear in ascending order. If the product of the digits is 216 and the sum is 19, what is your locker number?

7. Carla is playing a game with her brother. She says that she has some quarters, dimes, and nickels in her pocket. She has two more dimes than quarters and three fewer nickels than dimes. How much money does she have if she has three quarters?

8. Sandy, Toi, and Jessie have careers as a teacher, a doctor, and an actor. If Toi doesn't want to act, and Sandy likes to grade papers, what career does each person have?

9. Draw as many different patterns to make a cube as you can. Cut out your patterns to make the cubes. How many different patterns did you make?

Lesson 14-1 Problem-Solving Strategy: Work Backward **533**

10. Look at the models of the triangular numbers below. How many dots would be in a triangle that has 10 dots on a side?

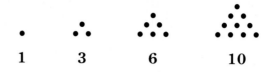

1   3   6   10

11. Mr. Rogers is delivering cartons of breakfast cereal to supermarkets. At the first supermarket, he drops off half of the cartons he has in the truck. At each of the other markets, he drops off half of the cartons he has left in the truck. At the eleventh market, he drops off 1 carton, which is the last one. How many cartons were originally in the truck?

12. Antonio sends letters and postcards to his friends while he is on vacation. He spends $2.48 on postage. If a stamp for a letter costs 29¢ and postcards require 19¢ postage, how many letters and how many postcards did Antonio send?

13. A car's gas tank contains 0.0454 m³. The capacity of the *Pierre Guillaumat,* the world's largest oil tanker, is 687,000 m³. How many times could the car fill up its tank from the fully-loaded tanker? Round your answer to the nearest tenth.

14. The rectangular solid shown is painted blue and then cut along the lines to make 1-inch cubes. How many cubes will have exactly two blue faces?

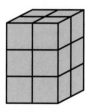

14-2A Two-Step Equations

A Preview of Lesson 14-2

Objective

Solve two-step equations with integers using models.

Materials

cups
counters
mats

LOOKBACK

You can review solving two-step equations on page 232 and solving one-step equations with integers on page 284.

Remember how you used cups and counters to solve two-step equations such as $3x + 1 = 7$? You can also use models to solve two-step equations with integers.

Let's start with the equation $2x + (-3) = 1$. You want to find the value of x that makes this equation true.

Try this!

Work in groups of three.

- Make a model of the equation using cups and counters. Remember, a cup represents x.

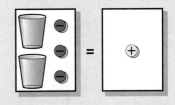

- Next add 3 positive counters to each side of the equation to create zero pairs on the left side. Remove the zero pairs, since their value is 0. The new equation is $2x = 4$.

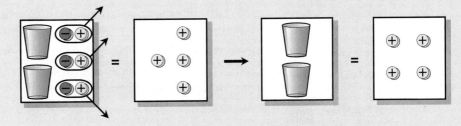

- Now pair up an equal number of counters with each cup. Since each cup can be paired with 2 counters, the solution is 2.

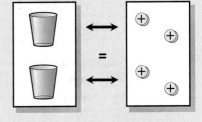

What do you think?

1. Why is it important to get the cups on one side by themselves?
2. Why do you need to pair up an equal number of counters with each cup?

14-2 Solving Two-Step Equations

Objective
Solve two-step equations.

Have you ever ordered anything from a mail-order catalog? It can save you time and energy as well as money. But you need to make sure you always include the shipping cost in the total price.

Suppose you want to purchase two tie-dyed T-shirts from a catalog. The total price including shipping is $17. If the total shipping cost is $1, how much does each T-shirt cost?

First write an equation using the information given. Let t represent the cost of each T-shirt.

Problem-Solving Hint
• • • • • • • • • • • •
Work backward.

Two times the cost of a T-shirt plus shipping cost is $17.

$$2 \quad \times \quad t \quad + \quad 1 \quad = \quad 17$$

TEEN SCENE

In the 1909 Sears Roebuck and Company catalog, a boy's cotton shirt cost 35¢. Today, it can cost as much as $35—100 times as much!

The equation is $2t + 1 = 17$. This is a two-step equation because it involves two different operations, multiplication and addition. To solve this equation, we need to "undo" the operations, or work backward. In the equation $2t + 1 = 17$, the order of operation is: multiplication, then addition. To undo these operations in reverse order, we need to subtract first and then divide.

LOOKBACK

You can review order of operations on page 24.

Here is how you solve the equation.

$$2t + 1 = 17$$
$$2t + 1 - 1 = 17 - 1 \quad \textit{Subtract 1 from each side.}$$

$$2t = 16$$
$$\frac{2t}{2} = \frac{16}{2} \quad \textit{Divide each side by 2.}$$
$$t = 8$$

The cost of each T-shirt is $8.

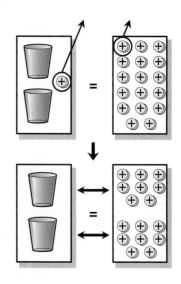

You can graph the solution of the T-shirt problem by drawing a
number line and marking a dot on 8.

Examples

1 Solve $-2r + 8 = -4$. Graph the solution.

$$-2r + 8 = -4$$
$$-2r + 8 - 8 = -4 - 8 \qquad \textit{Subtract 8 from each side.}$$
$$-2r = -12$$
$$\frac{-2r}{-2} = \frac{-12}{-2} \qquad \textit{Divide each side by } -2$$
$$r = 6$$

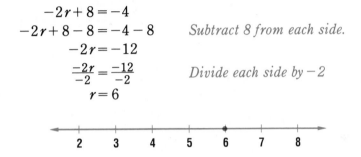

2 Solve $\frac{1}{5}(k - 8) = -3$. Graph the solution.

$$\frac{1}{5}(k - 8) = -3$$
$$5 \times \frac{1}{5}(k - 8) = 5 \times (-3) \quad \textit{Multiply each side by 5.}$$
$$k - 8 = -15$$
$$k - 8 + 8 = -15 + 8 \quad \textit{Add 8 to each side.}$$
$$k = -7$$

Example 3   *Problem Solving*

Weather   On a January day in Buffalo, New York, the temperature
dropped to $-5°F$. Find this temperature in degrees Celsius by using the
formula $F = \frac{9}{5}C + 32$.

$$F = \frac{9}{5}C + 32$$
$$-5 = \frac{9}{5}C + 32 \qquad \textit{Replace F with } -5.$$
$$-5 - 32 = \frac{9}{5}C + 32 - 32 \qquad \textit{Subtract 32 from each side.}$$
$$-37 = \frac{9}{5}C$$
$$\frac{5}{9} \times (-37) = \frac{5}{9} \times \frac{9}{5}C \qquad \textit{Multiply each side by } \frac{5}{9}.$$
$$\frac{5}{9} \times (-37) = C$$

5 ⌈÷⌉ 9 ⌈×⌉ 37 ⌈+/-⌉ ⌈=⌉ ⁻20.555556

The temperature is about $-20.6°C$.

Checking for Understanding

Read and study the lesson to answer each question.

1. **Show** how you would solve the equation represented by the model below.

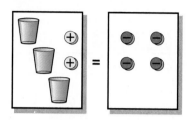

2. **Write** the inverse operation of each of the following.
   a. addition
   b. subtraction
   c. multiplication
   d. division

3. **Write** the inverse of each of five things you do to get ready for school.

4. **Draw** the graph of the temperature $-4°F$ on a number line.

Guided Practice

Name the first step in solving each equation. Then solve the equation and graph the solution.

5. $3n - 5 = 16$

6. $2t - 5 = -1$

7. $\frac{p}{3} + 6 = -2$

8. $-1 - 4y = 7$

9. $9 - 3c = 12$

10. $\frac{r}{5} + 3 = -5.5$

11. $-2r + 3.1 = 1.7$

12. $\frac{1}{4}(k - 8) = -3$

13. $\frac{t}{3} + \frac{2}{3} = 1\frac{5}{6}$

Translate each sentence into an equation. Then solve the equation and graph the solution.

14. Four times a number less five is fifteen.

15. Three more than a number divided by two is eleven.

16. The sum of a number and six, divided by eight, is negative two.

Exercises

Independent
Practice

Solve each equation and graph the solution.

17. $2x + 5 = -13$

18. $-3x - 4 = 8$

19. $\frac{w}{-2} + 5 = 11$

20. $\frac{m}{3} - 5 = -9$

21. $-\frac{1}{5}(y + 1) = 4$

22. $-12 + 8m = 36$

23. $-4m + 7.2 = -6.8$

24. $7 - 2m = -3$

25. $\frac{b}{-7} + 3 = -5$

26. $13 - 5n = -1.5$

27. $\frac{4}{3}x + 7 = -1$

28. $\frac{1}{6}(r - 3) = -5$

29. $-3m - 5 = -18$

30. $16 = 35n - 4$

31. $-\frac{7}{5}s - 3 = 11$

32. **Geometry** Solve $C = \pi d$ for d if $\pi \approx 3.14$ and $C = 100$.

33. **Geometry** Solve $P = 2\ell + 2w$ for ℓ if $w = 4$ and $P = 40$.

Translate each sentence into an equation. Then solve the equation and graph the solution.

34. Seven plus the quotient of a number and four is two.

35. Three more than the product of a number and five is negative seven.

36. Ten less than twice a number is sixteen.

Mixed Review　37. Compute mentally: $237 + 412$. *(Lesson 1-4)*

38. Solve $r = -24 \div 8$ *(Lesson 7-8)*

39. **Smart Shopping**　Jeanne finds a $52 sweater marked down $13. Express this price reduction as a percent. *(Lesson 11-8)*

40. **Diets**　Elaine Mann weighed 142 pounds when she began her diet. After several months on the diet, she weighed 125 pounds. Find the percent of decrease for Mrs. Mann's body weight. *(Lesson 12-7)*

41. Evaluate $C(6, 2)$. *(Lesson 13-8)*

Problem Solving and Applications

42. **Business**　Dave and three of his friends started a lawn mowing business during the summer. At the end of the summer, they divided the profits equally. With his share of the profits, plus a $50 gift certificate from his parents, Dave had just enough money to buy the new bike he wanted. The bike cost $210.

　a. How much did each of the boys earn in their lawn mowing business?

　b. Graph the solution.

43. **Transportation**　Ms. Jackson takes a taxi from the airport to her home. The cost of a taxi is $3, plus 60¢ for each mile traveled.

　a. If she is charged $15.60 at the end of the trip, how many miles did she travel?

　b. Graph the solution.

44. **Weather**　The graph at the right shows the low temperatures for a week in January in Muskegon, Michigan.

　a. Use the formula $C = \frac{5}{9} \times (F - 32)$ to convert the low temperature on January 21 to degrees Celsius.

　b. Convert the low temperature on January 24 to degrees Celsius.

　c. Graph both of these temperatures on a number line.

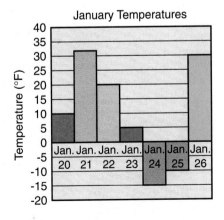

January Temperatures

45. **Critical Thinking**　Solve the equation $-3x - 4 = x + 2$. Graph the solution.

Lesson 14-2 Solving Two-Step Equations　　**539**

14-3 Equations with Two Variables

Objective

Solve equations with two variables.

Words to Learn

Ordered pair

Bob Learn, Jr., has scored more "300 games" in bowling than any other bowler. A "300 game" means all strikes, or a perfect score. Since average bowlers are not that good, they are given a "handicap" in order to compete with skilled bowlers like Mr. Learn. A handicap is an advantage given to a player in the form of extra points.

Suppose you are given a handicap of 8 points. Your final score would be $x + 8$, where x is your actual score. The table below shows how you can find the final scores given three actual scores.

Actual Score	Add the Handicap	Final Score
x	$x + 8$	y
102	$102 + 8$	110
120	$120 + 8$	128
127	$127 + 8$	135

The relationship between the actual score and the final score can be written as an equation with two variables. Let x represent your actual score and y represent your final score, as shown in the table. The equation, then, is $y = x + 8$.

Recall that solving an equation means to replace the variable so that you have a true sentence. A solution of an equation with two variables consists of two numbers, one for each variable. When you replace the variables with the numbers, the result is a true sentence.

A solution of an equation with two variables is usually written as an **ordered pair,** (x, y). Based on the table above, we know that three solutions to the equation $y = x + 8$ are (102, 110), (120, 128), and (127, 135).

Problem-Solving Hint

••••••••••••
Use a table.

Example 1

Find four solutions for $y = 2x - 3$. Write the solutions as ordered pairs.

Select any four values for x. We chose 2, 1, 0, and -1. Substitute these values for x to find y and complete the table on the next page.

x	2x − 3	y	(x, y)
2	2(2) − 3	1	(2, 1)
1	2(1) − 3	−1	(1, −1)
0	2(0) − 3	−3	(0, −3)
−1	2(−1) − 3	−5	(−1, −5)

Four solutions are $(2, 1)$, $(1, -1)$, $(0, -3)$, and $(-1, -5)$.

Example 2 *Problem Solving*

Conservation In order to conserve America's forests, lumber companies are often required to replace the trees they cut down. About half of the new trees planted are expected to survive and become full-grown trees. Find out how many pine trees are expected to survive if 154 pine trees are planted.

Let x represent the number of new trees planted. Let y represent the number of full-grown trees.

Write an equation.

$$\left(\begin{array}{c}\text{Number of new trees}\\ \text{expected to survive}\end{array}\right) = \frac{1}{2} \times \left(\begin{array}{c}\text{Number of new}\\ \text{trees planted}\end{array}\right)$$
$$y = \frac{1}{2} \times x$$

The equation is $y = \frac{1}{2}x$.

Now find the number of pine trees expected to survive.

$y = \frac{1}{2}x$

$\quad = \frac{1}{2}(154)$ *Replace x with 154.*

$\quad = 77$

About 77 pine trees are expected to survive.

Checking for Understanding

Communicating Mathematics

Read and study the lesson to answer each question.

1. **Tell** why $(1, -1)$ is a solution of $y = 3x - 4$.

2. **Write** two solutions of $y = x - 2$.

3. **Show** whether $(6, 12)$ is a solution to the equation in Example 2.

Guided Practice

Copy and complete the table for each equation. Then use the results to write four solutions for each equation. Write the solutions as ordered pairs.

4. $y = 2x + 1$

x	2x + 1	y
1	2(1) + 1	
2	2(2) + 1	
3	2(3) + 1	
4	2(4) + 1	

5. $y = 3x$

x	3x	y
−1	3(−1)	
0	3( )	
1	3( )	
2	3( )	

6. $y = -2x + 3$

x	−2x + 3	y
−1	−2( ) + 3	
0	−2( ) + 3	
1	−2( ) + 3	
2		

7. $y = 1.5x$

x	1.5x	y
1		
2		
3		
4		

Exercises

Independent Practice

Find four solutions for each equation. Write your solutions as ordered pairs.

8. $y = 2x + 3$ **9.** $y = x - 2$ **10.** $y = 3x - 1$

11. $y = 5x$ **12.** $y = 4x - 1$ **13.** $y = -2x$

14. $y = -x - 2$ **15.** $y = -2x + 2$ **16.** $y = \frac{1}{4}x$

17. $y = \frac{1}{4}x + 1$ **18.** $y = \frac{1}{2}x - 1$ **19.** $y = -2$

Translate each of the following into an equation. Then find four solutions for each equation.

20. Luis makes \$5 per hour. How much does he make in x hours?

21. A plumber charges an initial fee of \$25, plus \$35 for each hour he works. How much does he charge for x hours?

Mixed Review

22. \$3,000,000 is the winning prize in the Ohio lottery. Express this amount in scientific notation. *(Lesson 2-6)*

23. Express $\frac{5}{8}$ as a decimal. *(Lesson 4-7)*

24. Solve $2q + 6 = -20$. *(Lesson 14-2)*

Problem Solving and Applications

25. Sports Like bowling, golf is a game in which players can be given a handicap in order to compete with more skillful players. Since the goal in golf is to get the *lowest* possible score, the handicap is *subtracted* from the actual score.

   a. If a golfer is given a handicap of 18 strokes, write an equation for the final score given the actual score.

   b. If the golfer's actual scores are 104, 98, 102, and 108, what are her final scores?

   c. Write the solutions as ordered pairs. Explain what the numbers in the ordered pairs mean.

26. Consumer Awareness To rent videos from The Video Store, you must pay a one-time \$5 membership fee and then a \$2 daily fee for each tape rented.

   a. Write an equation for the cost of renting tapes on your first visit to The Video Store.

   b. How much would you have to pay if you rented three videos?

27. Critical Thinking Find four solutions for the equation $y - x = 1$.

28. Journal Entry In your own words, explain how you would find a solution for $y = -2x + 1$.

14-4 Graphing Equations with Two Variables

Objective

Graph equations by plotting points.

Words to Learn

linear equation

"Don't use up all the hot water!" Have you ever heard this when you were taking a shower? A shower uses about 5 gallons of water per minute. If you want to estimate how much water you are using in the shower, you can use the equation $y = 5x$, where x is the number of minutes you are in the shower, and y is the number of gallons of water.

This equation can be graphed on a coordinate system.

Step 1

Make a table to find at least four solutions to the equation.

x	5x	y	(x, y)
0	5(0)	0	(0, 0)
1	5(1)	5	(1, 5)
2	5(2)	10	(2, 10)
3	5(3)	15	(3, 15)

You can review the coordinate system on page 259.

Step 2

Graph the solutions on a coordinate system.

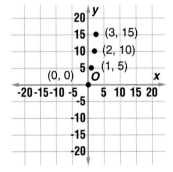

Estimation Hint

• • • • • • • • • • • • •

Graphs are useful for approximating answers to equations. For example, by looking at the graph at the right, you can estimate how many gallons of water are used in $2\frac{1}{2}$ minutes.

Step 3

Draw a line passing through the points.

The line is the graph of *all* solutions for $y = 5x$.

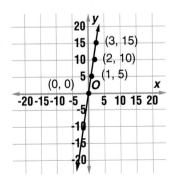

An equation like $y = 5x$ is called a **linear equation** because its graph is a straight line. Only two points are needed to graph the line. However, graph more points as a check for accuracy.

Example

Graph the equation $y = 2x - 1$.

Step 1
Make a table to find several solutions.

x	2x − 1	y	(x, y)
2	2(2) − 1	3	(2, 3)
1	2(1) − 1	1	(1, 1)
0	2(0) − 1	−1	(0, −1)
−1	2(−1) − 1	−3	(−1, −3)

Step 2
Graph the solutions on a coordinate system.

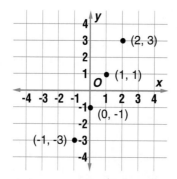

Step 3
Draw a line through the points.

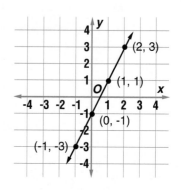

Checking for Understanding

Communicating Mathematics

Read and study the lesson to answer each question.

1. **Write**, in your own words, the three steps you take to graph an equation.

2. **Tell** why only those solutions graphed in the first quadrant for the equation $y = 5x$ on page 543 make sense when related to the use of hot water.

3. **Tell** why the equation in the example is a linear equation.

Guided Practice

Copy and complete each table. Then graph the equation.

4. $y = 3x$

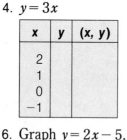

5. $y = 2x + 2$

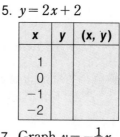

6. Graph $y = 2x - 5$.

7. Graph $y = -\frac{1}{2}x$.

Exercises

Independent Practice

Graph each equation.

8. $y = 4x$

9. $y = x - 3$

10. $y = -2x + 2$

11. $y = -x + 2$

12. $y = -3x - 1$

13. $y = 4x - 3$

14. $y = \frac{1}{2}x + 1$

15. $y = x + \frac{1}{2}$

16. $y = \frac{1}{4}x + 2$

Translate each of the following into an equation. Then graph.

17. The first number is three more than the second.

18. The first number is two times the second number less five.

19. The first number is four more than negative three times the second number.

Mixed Review 20. Complete: 36 oz = __?__ lb *(Lesson 6-6)*

21. **Geometry** Find the missing length in the triangle at the right. *(Lesson 9-4)*

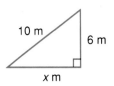

10 m

6 m

x m

22. Find four solutions of $y = -3x + 7$. Write the solutions as ordered pairs. *(Lesson 14-3)*

Problem Solving and Applications 23. **Employment** Angel gets paid $6 per hour for her job at a fast-food restaurant.
   a. Write an equation that tells how much money Angel makes in x hours.
   b. Graph the equation. Use the graph to estimate how much she will make if she works 7.5 hours.

DATA SEARCH

24. **Data Search** Refer to pages 530 and 531. Make a graph showing the total number of endangered species according to classification and location.

25. **Critical Thinking** How can you tell whether a point that appears to lie on a line on a coordinate grid is really on the line?

26. **Journal Entry** How many minutes do you spend in the shower? Use the graph on page 543 to estimate how much water you use in the shower. Then use the equation $y = 5x$ to find out how much water you use.

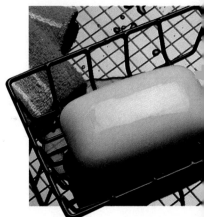

14 **Mid-Chapter Review**

1. Two whole numbers have a sum of 15 and a product of 54. What are the numbers? *(Lesson 14-1)*

Solve each equation and graph the solution. *(Lesson 14-2)*

2. $4x + 1 = -15$
3. $-\frac{1}{3}(n + 7) = 5$
4. $9 - 5x = -6$

Find four solutions for each equation. Write the solutions as ordered pairs. Then graph each equation. *(Lessons 14-3, 14-4)*

5. $y = -x + 3$
6. $y = \frac{1}{2}x + 3$
7. $y = 5x - 2$

14-5A A Function of Time

A Preview of Lesson 14-5

Objective

Use a function rule to find the output of a function.

Materials

clock with second hand
graph paper

Do you ever have to remind yourself to breathe? Probably not. Breathing is something we do naturally, without even thinking about it.

In this lab, you will learn how breathing is a *function* of time. That is, the number of times you breathe is related to time.

Try this!

- Sit quietly at your desk facing the clock. Count the number of times you breathe *out* in one minute. Record your result. Repeat four more times and record each result.

- Find the mean of the five results. This is the average number of times you breathe per minute. Copy the table below and use this average to complete it.

LOOKBACK

You can review mean on page 104.

Minutes	Minutes × Average Breaths per Minute	Total Breaths
1	1× __?__	
2	2× __?__	
3	3× __?__	
4	4× __?__	
5	5× __?__	
6	6× __?__	

DID YOU KNOW

Hiccups are usually caused by improper digestion of food. When you hiccup, the large muscle that controls your breathing, called the diaphragm, jerks quickly. It causes you to take a sudden sharp breath of air—a hiccup!

In this table, the minutes are called *input* values and the total breaths are called *output* values. The middle column contains the *function rule*. When you input a value into the function rule, you get an output value. Since the output depends on the input, we call this a function. We can say that the number of breaths is a function of time.

What do you think?

1. Graph the ordered pair (minutes, breaths).
2. How can you use the graphs to estimate the number of breaths you take in 12 minutes?

Extension

3. Repeat this lab using heartbeats per minute.

14-5 Functions

Modern banking began in the thirteenth century in Italy. The early Italian bankers carried out their business on benches in the street. These benches were called "bancos," which is where we get our word for "bank."

Banking has come a long way since then. Today, we don't even have to talk to another human being to withdraw money from the bank! Automatic teller machines (ATMs) can do that for us. We *input* the amount we need, and the machine *outputs* our money. This relationship is an example of a **function**.

In mathematics, we input a number into a function and compute to find the output. The output produced depends on the *function rule* used. When you use an ATM, the amount of money you ask for determines the amount of money you withdraw. When you use math functions, the output also depends on the input. In mathematics, we say that the output *is a function of* the input.

Example 1

Find the output, given the input and the function rule.

Input: 11, 12, 13

Function rule: The output is double the input.

We can write the function rule as:
    $2 \times$ input $=$ output.

Now make a function table to organize the information.

Input	Function Rule 2 × Input	Output
11	2 × 11	22
12	2 × 12	24
13	2 × 13	26

Place each input number into the function rule. Then compute.

The output is 22, 24, and 26.

We can write functions using algebraic notation. Let x represent the input numbers. The function rule in Example 1 can be written as $2 \times x$, or $2x$.

The output is usually represented by the notation $f(x)$. This is read *f of x*, which means *function of x*. So, the function in Example 1 can be written as $f(x) = 2x$.

Example 2

Find the output for the function $f(x) = 3x + 5$, given $x = -1$, 1, 2, and 3.

Make a function table.

Input	Function Rule	Output
x	$3x + 5$	$f(x)$
-1	$3(-1) + 5$	2
1	$3(1) + 5$	8
2	$3(2) + 5$	11
3	$3(3) + 5$	14

The output, $f(x)$, is 2, 8, 11, and 14.

Accountants keep track of how money is spent and received. They use functions to help them analyze data. Sometimes, they graph functions so they can get a better picture of what is happening with the budget. These graphs are helpful for people in management who need to make decisions on the budget.

For more information on accounting, contact: American Institute of Certified Public Accountants 1211 Ave. of the Americas New York, NY 10036.

The solutions for a function are usually written as ordered pairs, $(x, f(x))$. Four solutions for the function in Example 2 are $(-1, 2)$, $(1, 8)$, $(2, 11)$, and $(3, 14)$.

In Example 3, you will use the solutions of a function to graph the function.

Example 3

Graph $f(x) = -2x + 1$.

First make a function table. List at least three values for x.

x	$-2x + 1$	$f(x)$	$(x, f(x))$
1	$-2(1) + 1$	-1	$(1, -1)$
0	$-2(0) + 1$	1	$(0, 1)$
-1	$-2(-1) + 1$	3	$(-1, 3)$

Graph the ordered pairs. Draw a line passing through the points.

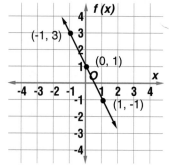

Checking for Understanding

Communicating Mathematics

Read and study the lesson to answer each question.

1. **Tell**, in your own words, what $f(x)$ means.
2. **Write** the function rule for *the output is 5 less than twice the input.*
3. **Tell** how you would label the axes for the graph of the function $f(x) = -2x + 3$.

Guided Practice

Copy and complete each function table. Then graph the function.

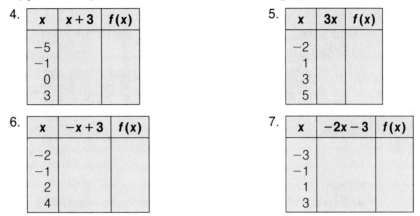

4.

x	x + 3	f(x)
−5		
−1		
0		
3		

5.

x	3x	f(x)
−2		
1		
3		
5		

6.

x	−x + 3	f(x)
−2		
−1		
2		
4		

7.

x	−2x − 3	f(x)
−3		
−1		
1		
3		

Exercises

Independent Practice

Find the output for each function, given the input and the function rule.

8. $f(x) = 3x - 1$
   $x = -3, -1, 2, 3$

9. $f(x) = 6x$
   $x = -1, 0, \frac{3}{2}, 2$

10. $f(x) = -3x + 4$
    $x = -1, 0, \frac{1}{3}, 1$

11. $f(x) = \frac{1}{2}x + 5$
    $x = -3, -2, -1, 0$

12. $f(x) = x - 1.5$
    $x = 0.5, 1, 1.5, 2$

13. $f(x) = -2x - 4$
    $x = -4.5, -3, -1.5, 0$

14. Find four solutions for the function $f(x) = -4x + 7$. Write the solutions as ordered pairs.

15. Find four solutions for the function $f(x) = \frac{1}{4}x - 2$. Write the solutions as ordered pairs.

Graph each function.

16. $f(x) = 2x - 3$

17. $f(x) = -3x$

18. $f(x) = -x + 6$

19. $f(x) = 5x + 1$

20. $f(x) = \frac{1}{2}x + 4$

21. $f(x) = 1.5x - 2$

Mixed Review

22. **Physical Fitness** Every morning, Juan walks a rectangular route that has a length of 0.5 mile and a width of 0.25 mile. How far does Juan walk each morning? *(Lesson 5-6)*

23. **Geometry** Classify the triangle at the right by its sides and by its angles. *(Lesson 8-3)*

24. Mr. McDaniel is on a business trip in Dallas, Texas. He has 3 different restaurants to choose from for breakfast, 6 different restaurants to choose from for lunch, and 2 different restaurants to choose from for dinner. How many different ways can Mr. McDaniel eat all three meals given the choices? *(Lesson 13-2)*

25. Graph the equation $y = -2x$. *(Lesson 14-4)*

26. **Computer Connection** The first step in writing a computer program is often drawing a flow chart. The flow chart below was written to determine solutions for a function.

COMPUTER

CONNECTION

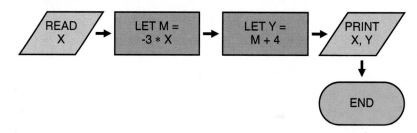

a. Study the flow chart and determine the function it represents.
b. Find the output, given an input of 2.
c. Graph the function.

27. **Geometry** The circumference (C) of a circle is a function of its diameter (d). The function can be written $C(d) = 3.14d$. *Note: This is the same as $f(x) = 3.14x$.*
a. Why would negative numbers not make sense as input?
b. Make a function table. Find the circumference (output) for four different diameters (input).
c. Draw the graph. Tell how you can use the graph to estimate d given C.

28. **Sports** In the 1992 Olympics, Bonnie Blair of the United States won the gold medal in the women's 500-meter speed skating competition. Her average speed was about 12.4 meters per second.
a. Graph the function $d(t) = 12.4t$.
b. How many seconds, t, did it take her to complete the 500-meter race?

29. **Critical Thinking** Find a function rule for the input and output numbers below.

x	−1	0	1	2	3
f(x)	3	4	5	6	7

30. **Journal Entry** Explain how a function is like an automatic teller machine.

14-6 Graphing Transformations

Objective

Graph transformations on a coordinate plane.

Words to Learn

reflection
translation
transformation

Quillwork is an art that was perfected by Native Americans. The quills of a porcupine were "embroidered" in patterns on tobacco bags, moccasins, and belts. Although this artform is rarely practiced today, the designs have been imitated and used in other forms of art.

Many of these designs consist of a variety of geometric transformations. That is, one geometric shape is used in many different positions to form a pattern or design.

◀ **LOOK BACK**

You can review reflections on page 327 and translations on page 324.

There are many ways to move a geometric shape on a coordinate plane. A figure can be flipped, turned, slid, stretched, or shrunk. When a figure is flipped, it is called a **reflection.** When it is slid, it is called a **translation.** Each of these kinds of **transformations** can be described using ordered pairs and then graphed.

Example 1

Triangle ABC has vertices $A(1, 1)$, $B(2, 4)$, and $C(5, 2)$. Graph its reflection over the x-axis.

Graph $\triangle ABC$ by graphing each ordered pair and connecting the points to form $\triangle ABC$. Label each vertex.

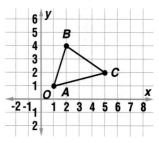

Multiply the y-coordinate of each ordered pair by -1. Write the new ordered pair.

Vertex of $\triangle ABC$	Multiply the y-coordinate by -1	New ordered pairs
$A(1, 1)$	$(1, 1(-1))$	$A'(1, -1)$
$B(2, 4)$	$(2, 4(-1))$	$B'(2, -4)$
$C(5, 2)$	$(5, 2(-1))$	$C'(5, -2)$

A′ is read "A prime,"
B′ as "B prime," and
C′ as "C prime."

Graph the new ordered pairs and label each point. Connect these points to form △A'B'C'.

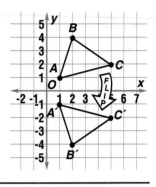

The two triangles are congruent. Triangle A'B'C' is the result of flipping △ABC over the x-axis.

The transformation in Example 1 is a reflection over the x-axis. You can also do a reflection over the y-axis by multiplying the x-coordinate by −1. *You will reflect △ABC over the y-axis in Exercise 2.*

Another transformation is a *translation*. Remember a translation slides the figure from one location to the next without changing its orientation.

Example 2

Graph △KLM with vertices K(−9, −5), L(−7, −1), and M(−1, −6). Graph its translation △K'L'M' with vertices K'(−2, −1), L'(0, 3), and M'(6, −2). Describe the movement from △KLM to △K'L'M'.

Graph each triangle. Notice that the two triangles are congruent.

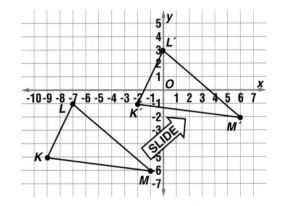

To describe the movement from △KLM to △K'L'M', look for a pattern in the ordered pairs.

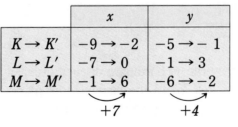

	x	y
$K \rightarrow K'$	$-9 \rightarrow -2$	$-5 \rightarrow -1$
$L \rightarrow L'$	$-7 \rightarrow 0$	$-1 \rightarrow 3$
$M \rightarrow M'$	$-1 \rightarrow 6$	$-6 \rightarrow -2$
	+7	+4

Each x-coordinate was moved 7 units to the right and each y-coordinate was moved 4 units up. The movement of each vertex can be described by (7, 4). The new ordered pairs can be written as follows: $(x, y) + (7, 4) = (x + 7, y + 4)$.

Checking for Understanding

Read and study the lesson to answer each question.

1. **Tell** what a reflection and a translation mean.

2. **Draw** the result of reflecting △*ABC* from Example 1 over the *y*-axis.

3. **Tell** what type of transformation is shown by the table at the right.

△*RST*	△*R′S′T′*
R(3, 2)	*R*′(4, 4)
S(1, 5)	*S*′(2, 7)
T(−1, 2)	*T*′(0, 4)

Guided Practice

Classify each graph as a reflection or a translation.

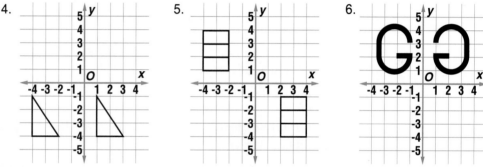

4.   5.   6.

7. a. Graph △*XYZ*, with vertices *X*(2, 1), *Y*(−4, 1), and *Z*(−1, 5).
   b. Reflect △*XYZ* over the *x*-axis.
   c. Reflect △*XYZ* over the *y*-axis.
   d. Translate △*XYZ* 4 units down and 3 units left.
   e. Translate △*XYZ* 3 units up and 2 units right.

Exercises

Independent
Practice

Graph each triangle and its transformation. Write the ordered pairs for the vertices of the new triangle.

8. △*ABC* with vertices *A*(−2, 4), *B*(2, 1), and *C*(1, 6) translated 5 units right and 3 units up

9. △*DEF* with vertices *D*(1, 3), *E*(5, 1), and *F*(5, 8) reflected over the *x*-axis

10. △*GHI* with vertices *G*(−3, −1), *H*(−5, −6), and *I*(−1, −6) translated 2 units down and 4 units right

11. △*JKL* with vertices *J*(−5, 7), *K*(−2, 5), and *L*(−7, 1) reflected over the *y*-axis

12. rectangle *PQRS* with vertices *P*(−4, 0), *Q*(−4, −3), *R*(−2, −3), and *S*(−2, 0) translated 2 units down and 5 units right

Mixed Review

13. **Geometry** Find the volume of a rectangular prism having a length of 6 meters, a width of 4 meters, and a height of 1.5 meters. *(Lesson 10-5)*

14. Find the output for the function $f(x) = 4x + 6$, given $x = -3$, 2, and 20. *(Lesson 14-5)*

15. **Art**   Describe the transformations that were used to create the quilled pattern on page 551.

16. **Geometry**
    a. Reflect the triangle at the right over the y-axis. Name the new figure formed by both triangles.
    b. Reflect the new figure over the x-axis. Name this new figure.

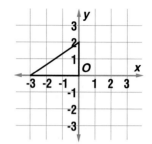

17. **Games**   In the game of chess, game pieces are slid up or down, left or right, or diagonally.
    a. What type of transformation is used in this game?
    b. Write the movement of the game piece at the right as an ordered pair.

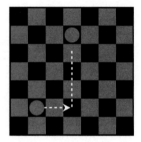

18. **Critical Thinking**
    a. Graph the image of △XYZ if both the x- and y-coordinates are multiplied by −1.
    b. Describe this transformation.

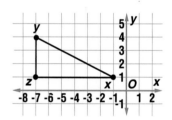

19. **Mathematics and Crafts**   Read the following paragraph.

> Quilts made in the United States during the 18th and 19th centuries are an important type of American folk art. These quilts were often made of colorful geometric forms that were either pieced together or appliqued onto a large piece of cloth. Sometimes the design on the quilt tells a story, commemorates an important historical event, or describes an important family occasion.

Use graph paper and design a quilt pattern that describes an important occasion in your family. Use reflections and translations to make your pattern.

14-6B Dilations

A Follow-Up of Lesson 14-6

Objective

Change the size of a figure on a coordinate plane.

Materials

picture or cartoon
graph paper
straightedge
colored pencil

Words to Learn

dilation

Some copy machines can reduce and enlarge images. In mathematics, the process of reducing and enlarging a figure is a transformation called a **dilation**.

Try this!

Work with a partner.

● Place a piece of graph paper over a cartoon or picture you want to enlarge. Trace the picture. *It may help to put the paper against a window pane to see the image more clearly.*

● On another piece of graph paper, use a colored pencil to draw horizontal lines every 2 squares. Then draw vertical lines every 2 squares.

● Now sketch the parts of the figure contained in each small square of your original picture onto each large square of the grid you created.

What do you think?

1. Has the figure been enlarged or reduced? By how much?
2. What type of grid would you use to *reduce* a picture?
3. **Geometry** Are your two pictures congruent or similar? Explain your answer.

Study Guide and Review

Communicating Mathematics

Choose the letter that best matches each of the following.

1. Four times x less six is 14.
2. a solution to $y = -3x + 5$
3. a term that describes $y = -2x + 6$
4. a rule that assigns an output to each input
5. a mirror image over a given line

6. In your own words, explain the steps required to graph a linear equation.

a. function
b. $(-3, 4)$
c. $4x - 6 = 14$
d. reflection
e. $(1, 2)$
f. translation
g. $6 - 4x = 14$
h. linear equation

Skills and Concepts

Objectives and Examples

Upon completing this chapter, you should be able to:

Review Exercises

Use these exercises to review and prepare for the chapter test.

- solve two-step equations *(Lesson 14-2)*

$$-6t - 5 = 19$$
$$-6t - 5 + 5 = 19 + 5$$
$$-6t = 24$$
$$\frac{-6t}{-6} = \frac{24}{-6}$$
$$t = -4$$

Solve each equation and graph the solution.

7. $3p - 4 = 8$
8. $\frac{x}{2} + 5 = 3$
9. $8 - 6w = 50$
10. $5m + 6 = -4$
11. $\frac{1}{3}(y - 4) = 5$
12. $-1.5b + 1 = 7$

- solve equations with two variables *(Lesson 14-3)*

Find four solutions for $y = 3x + 2$.

x	3x + 2	y	(x, y)
−1	3(−1) + 2	−1	(−1, −1)
0	3(0) + 2	2	(0, 2)
2	3(2) + 2	8	(2, 8)
3	3(3) + 2	11	(3, 11)

Four solutions are $(-1, -1)$, $(0, 2)$, $(2, 8)$, and $(3, 11)$.

Find four solutions for each equation. Write your solutions as ordered pairs.

13. $y = 4x - 9$
14. $y = \frac{1}{3}x$
15. $y = -2 - 3x$
16. $y = -6x + 1$
17. $y = x + 5$
18. $y = 3x + 4$
19. $y = 0$
20. $y = -x$

Objectives and Examples

- graph equations by plotting points *(Lesson 14-4)*

Graph $y = x + 3$.

x	x + 3	y	(x, y)
−2	−2 + 3	1	(−2, 1)
1	1 + 3	4	(1, 4)
3	3 + 3	6	(3, 6)

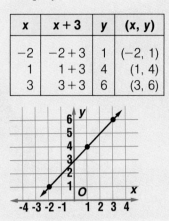

- complete function tables; graph functions *(Lesson 14-5)*

Find the output for the function $f(x) = 2x + 6$, given $x = -3, 0, 2,$ and 5.

Input	Function Rule	Output
x	2x + 6	f(x)
−3	2(−3) + 6	0
0	2(0) + 6	6
2	2(2) + 6	10
5	2(5) + 6	16

- graph transformations *(Lesson 14-6)*

Graph $\triangle QRS$ with vertices $Q(-3, 4)$, $R(-3, 1)$, and $S(4, 1)$. Then translate $\triangle QRS$ 2 units left and 3 units down.

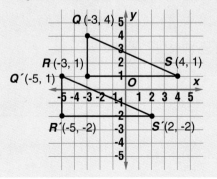

Review Exercises

Graph each equation.

21. $y = 2x$
22. $y = -\frac{1}{3}x$
23. $y = 5x + 2$
24. $y = x$

Find the output for each function, given the input and the function rule.

25. $f(x) = \frac{1}{2}x + 3$
    $x = 4, 0, -2, -4$
26. $f(x) = -4x$
    $x = -3, -1, 0, 2$
27. $f(x) = 1.5x - 4$
    $x = -4, -2, 0, 2$
28. $f(x) = -6x + 3$
    $x = -\frac{1}{2}, 0, \frac{1}{2}, 1$

Graph each triangle and its transformation. Write the ordered pairs for the vertices of the new triangle.

29. $\triangle ABC$ with $A(4, -2)$, $B(-2, -3)$, and $C(-1, 6)$ translated 3 units right and 4 units up
30. Reflect $\triangle ABC$ over the *x*-axis. Name the new figure formed by both triangles.

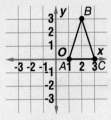

Applications and Problem Solving

31. **Stamp Collecting**   Four friends collect stamps. They are comparing how many Mexican stamps they each have. Jeff has 3 times as many as Fina. Mario has 4 fewer stamps than Danielle, but 3 more than Fina. Fina has 9 stamps. How many stamps does each friend have?

32. **Consumer Math**   A parking garage in New York City charges $3 for the first two hours and then $0.75 for each additional hour. Walter parks his car in the garage at 9:00 A.M. and when he returns must pay a $5.25 parking fee. What time did Walter return? *(Lesson 14-2)*

33. **Catering**   Christina's Catering Service charges $12.50 per person for a sit-down dinner.
    a. Write a function that represents the cost for x people.
    b. What is the cost for 40 people?

Curriculum Connection Projects

- **Travel**   Call two rental car companies and ask them for their rates for a medium-sized car. Write equations for the cost of renting a car from each company. Decide which company you would rent from if you were going to use the car for two days and travel about 150 miles.

- **Art**   Design a quillwork pattern for a belt using reflections, translations, and dilations of triangles.

Read More About It

McCauley, David. *City.*
O'Dell, Scott. *The Captive.*
Burns, Marilyn. *The I Hate Mathematics Book.*

1. **Baking** Ming baked a batch of oatmeal cookies. He kept half of them for himself and donated the other half to a charity bake sale. He packed the bake-sale cookies in 4 boxes, a dozen in each box. How many cookies did Ming bake?

Solve each equation and graph the solution.

2. $6x - 4 = 3$
3. $\frac{w}{-3} - 5 = 10$
4. $11 = \frac{1}{4}(p - 1)$

5. **School** Detentions at Morris Junior High are assigned to students who are late in the following manner: 20 minutes of detention for anyone who is up to 15 minutes late and an additional 5 minutes for every minute late beyond 15. Derrick received a 35-minute detention. How late was he?

Find four solutions for each equation. Write your solutions as ordered pairs.

6. $y = -x + 4$
7. $y = 3x - 2$
8. $y = -\frac{1}{2}x$

9. **Sales** Ms. White takes a job at a local computer store at a salary of $200 per week plus a $50 commission on every computer she sells.

   a. Write an equation for Ms. White's weekly salary if she sells x computers.

   b. What will she earn for a week in which she sells 5 computers?

Graph each equation.

10. $y = -3x$
11. $y = \frac{1}{3}x - 3$
12. $y = 2x + 5$

13. Margie charges $5 per page for report typing. Translate this information into an equation showing the cost for a report that has x pages. Graph the equation.

Find the output for each function, given the input and the function rule.

14. $f(x) = -2x - 1$; $x = -1, 2, 5, 100$
15. $f(x) = \frac{x}{10}$; $x = 20, 5, 0, -10$

Graph each function.

16. $f(x) = 2x - 5$
17. $f(x) = 3.5x + 1$

Graph each triangle and its transformation. Label all vertices.

18. $\triangle ABC$ with vertices $A(-4, 2)$, $B(3, 4)$, and $C(-1, 6)$ translated 2 units right and 4 units down

19. $\triangle XYZ$ with vertices $X(2, 5)$, $Y(2, -1)$, and $Z(5, -1)$ reflected over the x-axis

20. $\triangle QRS$ with vertices $Q(6, 4)$, $R(-1, 2)$, and $S(2, -3)$ translated 1 unit up and 5 units right

Bonus Graph the equation $y = x^2$.

Chapter 14 Academic Skills Test

Directions: Choose the best answer. Write A, B, C, or D.

1. Which numbers are factors of 1,215?

   A 2, 3, 4, and 5
   B 3, 4, and 5
   C 3, 5, and 9
   D 5, 6, and 9

2. If the number on the spinner tells the number of dollars you win, what is the expected value of a spin?

   A $1.00
   B $1.75
   C $2.00
   D $2.50

   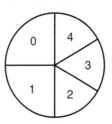

3. $-12 - 10 =$

   A -22 B -2
   C 2   D 22

4. The grading scale for a test is shown below. What score would be given for missing six problems?

-1	-2	-3	-4
98	95	93	90

   A 85   B 86
   C 87   D 88

5. Which regular polygon can be used by itself to make a tessellation?

   A pentagon   B hexagon
   C heptagon   D nonagon

6. What is the area of the trapezoid?

   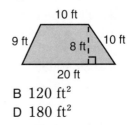

   A 80 ft^2 B 120 ft^2
   C 160 ft^2 D 180 ft^2

7. A cube has a surface area of 144 cm^2. How can you find the surface area of one face?

   A Divide 144 by 4.
   B Divide 144 by 6.
   C Divide 144 by 8.
   D None of these

8. How much water is in the pool when filled to a depth of 5 ft?

   A 800 ft^3
   B 864 ft^3
   C 1,200 ft^3
   D 1,440 ft^3

   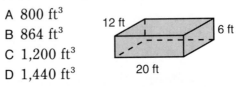

9. What is the unit price per ounce if a 14-oz can of peaches costs $1.19?

   A 1.2¢   B 8.5¢
   C 11.8¢   D 16.7¢

10. To find $12\frac{1}{2}\%$ of a number using a calculator, you can enter what decimal for $12\frac{1}{2}$?

    A 12.5   B 12.2
    C 0.125   D 0.122

11. 0.5% of $2,405 is about

    A $12   B $100
    C $120   D $1,200

12. If $592 was made in sales of sweatshirts, what was the total amount of sales (to the nearest dollar)?

A $1,850

B $1,798

C $185

D $180

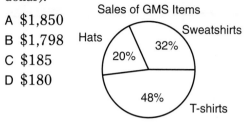

Sales of GMS Items

Hats 20%

Sweatshirts 32%

48%

T-shirts

13. 72 out of 120 students usually buy a plate lunch. If there are 500 students in the school, about how many plate lunches should be prepared?

A 80   B 210   C 300   D 400

14. There are 4 flavors of yogurt, 6 kinds of toppings, and 4 kinds of syrup. How many different combinations of yogurt, topping, and syrup (one of each) can be ordered?

A 14   B 24   C 48   D 96

15. Suppose you flip a coin twice. Which outcome has the greater probability?

A 2 heads

B 2 tails

C 1 head and 1 tail

D They are all the same.

16. In how many ways can three people be arranged in a row?

A 9   B 6   C 4   D 3

17. The cost of a taxi is $3, plus 75¢ for each mile traveled. If the total fare is $12, how many miles are traveled?

A 3.75   B 7.5   C 12   D 16

18. What is the function rule for the input and output?

A $f(x) = x - 3$

B $f(x) = x + 3$

C $f(x) = 2x - 1$

D $f(x) = 3x + 1$

x	$f(x)$
-2	-5
-1	-2
0	1
1	4
2	7

19. Which is an equation for the line graphed?

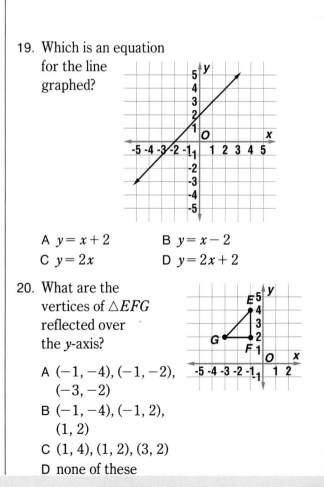

A $y = x + 2$

B $y = x - 2$

C $y = 2x$

D $y = 2x + 2$

20. What are the vertices of $\triangle EFG$ reflected over the y-axis?

A $(-1, -4), (-1, -2),$ $(-3, -2)$

B $(-1, -4), (-1, 2),$ $(1, 2)$

C $(1, 4), (1, 2), (3, 2)$

D none of these

EXTENDED PROJECTS HANDBOOK

To The Student

One of the goals of *Mathematics: Applications and Connections* is to give you the opportunity to work with the mathematics that you will likely encounter outside the classroom. This includes the mathematics demanded by many of the courses you will take in high school and by most jobs as well as the mathematics that will be required of a good citizen of the United States.

Equally important, the authors want you to approach the mathematics you will encounter in your life with curiosity, enjoyment, and confidence.

Hopefully, the **Extended Projects Handbook** reflects these goals.

Three of the most important "big" ideas you are working with throughout *Mathematics: Applications and Connections* are the following. The **Extended Projects** include "big" ideas.

1. Proportional Reasoning

You probably have a great deal of experience with proportional reasoning. One example is a straight line in which the "rise" is proportional to the "run" and their ratio is the slope of the line. Other topics include ratio, rate, percent, similarity, scale drawings, and probability.

You have made connections among the various applications of proportional reasoning and in this way have seen that the various items listed above are all part of a very big idea — proportions.

2. Multiple Representations

Mathematics provides you with many ways to present information and relationships. These include sketches, perspective drawings, tables, charts, graphs, physical models, verbalizing, and writing. You can use a computer to make graphs, data bases, spreadsheets, and simulations.

You have represented information and relationships in many different ways to completely describe various kinds of situations using mathematics.

3. Patterns and Generalizations

Mathematics has been called the science of patterns. You have experience recognizing and describing simple number and geometric patterns. You will be asked to make, test, and then use generalizations about given information in order to help you solve problems.

You may have used an algebraic expression to generalize a number pattern or the idea of similarity to make a scale drawing.

AND THE WINNER IS...

The first rock video, "Video Killed the Radio Star," performed by the Bugles was aired on MTV at midnight, August 1, 1981. Since that time, music videos have been an important part of music performers' careers as well as the music industry in general.

Each year MTV Networks recognizes outstanding achievement in the field of video music with the MTV Video Music Awards. Since their beginning in 1984, the number of awards has grown to 21. The categories include Best Male Video, Best Female Video, and Best Group Video as well as special categories for rap, heavy metal, and dance videos. The network also recognizes technical achievement, direction, and cinematography.

Who would be this year's winners if your group were hosting the show? Make up a questionnaire to find out.

Design a questionnaire that will identify the sample of people interviewed.

1. Talk about the importance of including questions on the questionnaire about the people you will be interviewing.

Categorizing your subjects

These questions will give you background information about your sample.

Examples:
- Male or female?
- Age range?
- How often do you watch MTV?
- How often do you listen to the radio?
- Do you buy cassettes and CDs of your favorite music performers regularly?
- Do you watch award shows on TV?

Awareness responses

These questions inform you of the general awareness of the people you are interviewing about past music winners.

Examples:
- Do you know who won the 1991 MTV Award for Best Female Video? (Janet Jackson, "Love Will Never Do Without You.")
- Do you know what 1950s star won a Grammy for Best Male Vocal Performance in 1990? (Roy Orbison, *Oh Pretty Woman*)

- Do you know the month the People's Choice Awards are usually televised? (March)
2. Discuss the types of questions you want to cover in your questionnaire. Your survey can include songs or videos or both. Remember, not all of the people questioned may watch videos.

This opinion poll can include the following kinds of questions.

- Who do you think should win for Best Female Video this year?
- What is your favorite song that was released this year?
- Who would you say is the best all-time group?
- What do you like most about music videos?

3. Conduct your survey with people of many age groups. Record their responses on a worksheet. When the interviewing is complete, transfer your data to a frequency table.
4. Discuss ways to interpret the information you have gathered.
   - Was there one performer or song that was an obvious favorite?
   - Did males prefer different entertainers than females?
   - Were the people who listened to the radio more informed than those who did not?
5. After conducting your interviews, evaluate your procedures.
   - Were there any questions that you wished you had included on the questionnaire? What were they?
   - Was there enough information to draw reasonable conclusions?
   - How many people did your group interview?
   - How can you make your topic more interesting?
6. Present the data you collected and your conclusions to the class in a creative way(s).

EXTENSION: Can Your Vote Be Counted?

In February of each year, nominees for the People's Choice Awards are announced. The awards are presented in March. As the name implies, the winners are chosen by the public, not peers in the industry. Other award shows such as the Soap Opera Digest Awards, created by the magazine *Soap Opera Digest,* surveys its subscribers who choose the winners after the editors select the nominees. Have you ever wondered what it would be like to participate in the voting process and see if your favorite performers are winners?

Research an award show where the American public chooses the winners. You can begin by writing to networks, contacting *TV Guide,* or a production company such as Dick Clark Productions. Discuss other ways you could obtain the information.

Write a story for your school newspaper detailing the steps you took to find out how award shows such as People's Choice Awards determine the winners.

Project

2

The Question Is . . .

In the late 1980s and early 1990s, trivia was very popular in this country. Trivia is knowledge of unimportant facts. Trivia Bowls were held on college campuses and game shows. Board games focused on the tidbits of knowledge that are stored in our minds. Have you ever wondered who wrote these questions and how they knew the answers?

In this project, your group will act as creative consultants and write trivia questions based on averages. Work in groups of 4.

Each person in your group should be assigned a task based on the following roles. You may wish to change roles after one week.

The Organizer

This person will ensure that the group's information is organized neatly in a file folder. All worksheets, charts, and other information for the group should be contained in one place.

The Investigator

This person will divide the work among group members and try to assist in helping to locate difficult-to-find information.

The Recorder

This person writes a report of the group's progress, decides how often this report is needed, and gives it to the Organizer.

The Question and Answer Person

This person will record all of the facts collected and put it in a question and answer format that can be distributed to the teacher.

All group members will gather facts and data.

One possible way to write each question is to begin with:

Each day in the United States about. . .

Make a list of trivia or fact questions that follow this format. You can group questions by theme if you wish, or write on a variety of topics.

Examples:

1. . . . how many gallons of water do people use in the home?
2. . . . how many times does a 7th grader blink?
3. . . . how many couples get married?
4. . . how many times does a 7th grader's heart beat?
5. . . . how many people immigrate to the United States from other nations?
6. . . . how many magazines are sold?
7. . . . how many disposable diapers make it to landfills?
8. . . . . how many gallons of soft drinks are consumed?

9. . . . how many children go to school?
10. . . . how many ounces of wood do 40,000 termites eat?
11. . . . how many hours of television does a 7th grader watch?
12. . . . how many compact discs are purchased?
13. . . . how many people go to the movies?
14. . . . how many people attend a high school sporting event?
15. . . . how much money do teenagers spend on clothes?

After you have developed your list of questions, be sure to keep a record of your sources and how you arrived at your answers. You can use almanacs and other library resources. Some of your questions may require personal interviews with other students or people in the community. To find answers to some questions, you may need to get results from several people and average them.

When your group is finished, discuss your findings with the class. How did you use what you know about averages to find your answers? Were you surprised by any of your answers?

Project 3 Where In The World?

The National Geographic Society defines geography as "a field of knowledge that deals with the Earth and all the life on it." Geography encompasses many areas of physical, cultural, political, economic, historical, and environmental facts.

In this project, you will make a map of the world and label it with facts about different countries, oceans, and animals found around the world.

Work in a group. Decide how large your map should be. What information do you want to include on it? Will it be something you want to display on a bulletin board when you are finished? How can you use proportions to make your map as accurate as possible?

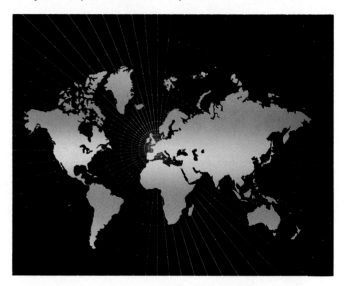

Look at an atlas before you begin drawing. Be as creative as possible. For example, you may want to draw penguins on Antarctica or maple leafs on Canada. You can include a legend to show mountains or bodies of water. You can use color to distinguish countries from each other. Discuss these and other issues with your group to develop a plan on how you will start the project.

Interesting But True

Assign group members different countries to research. Remember, you want to make your map as unique and interesting as possible. So, you will want to find out factual information about different countries that other classmates might not know.

Examples:

More than 40% of the world's oil imports come from the Middle East, where nearly two-thirds of the world's known oil resources are found.

Zambia is a country located in Africa. There are more than 73 ethnic groups speaking more than 80 languages in Zambia.

Share the facts you found with your group members. Choose the most interesting facts and decide how you will label these on your map. If you are not skilled at drawing, you can cut pictures out of magazines and position them on your map.

Once your map is complete, you are now ready to challenge other groups with questions about the world. Have your classmates assume the role of a detective who is trying to locate a missing person.

Your group should make up questions to assist the detectives in finding your missing person.

Examples:

Someone told you that the person you are trying to find is in Seattle, Denver, or Houston. Then you receive an anonymous tip that he is in the city with the highest elevation. Where should you look?

Your missing person decides to head toward the most populated country in the world. Where should you look?

Your missing person was spotted in a country that has the largest movie industry in the world. Where should you look?

Take turns solving each question. You could use push pins or stickers to mark the places once each mystery is solved. Some clues may take longer to solve than others so you may want to assign the questions ahead of time and give the detectives time to make an educated guess.

EXTENSION: Where Did It Come From?

You may want to extend this project by examining things you use at home or school to find out in which country they were made. Make a chart of your findings. You can include food, clothing, appliances, games, or electronics on your list. Then locate these countries on your map.

Project 4

The Big Debate

How do people learn? Do you have different learning strategies for different subjects? Can you only learn by memorization? Do you understand more fully if you see it done first?

There are many different opinions on how people learn. One difference of opinion is whether hands-on materials such as tiles and counters should be used to teach students at the middle school or high school level. These materials are called manipulatives because you can manipulate them.

In this project, your group will conduct a debate on this issue.

A debate is between two individuals, teams, or groups who present arguments to support opposing sides of a question, according to a set form or procedure.

Work in groups of five. Two group members will take the position that students at the middle and high school level should use manipulatives to learn mathematics. They will be Team A. The other team, Team B, will argue against this point. One group member will serve as the moderator.

Before you begin, you may want to agree on certain rules and procedures. For example, you may want to limit the amount of time each team speaks. Or, you may want the side speaking against the issue to speak first. Make a list of your agreements and follow them during your debate. The moderator will make sure that the rules are followed.

Getting Started

Each team should meet and discuss a strategy. Outline all of the arguments for teaching with manipulatives as well as the arguments against regardless of which team you are on. It is wise to think about what the opposition will say so that you will have a response that addresses their point.

You are now ready to begin researching your topic. In a debate, it is important to support your position with as much convincing evidence as you can gather to sway the audience. You may want to conduct interviews, distribute a questionnaire, or conduct experiments of your own.

For example, suppose you are on Team A. Develop a problem that you believe is best solved using manipulatives. You may want to use a problem involving probability. Ask classmates to participate. Present the same problem to one set of students and provide them with the manipulative(s) they need to solve the problem. Have the other set of students solve the problem without the manipulative(s).

To check their understanding, ask students to explain how they solved the problem. Record your results. Do they support your position? If they do, you can use this information in your debate. Make a chart or some other visual display to present as part of your argument.

Team B can conduct a similar experiment and use the results if the findings support their position.

Remember, the more data you can gather, the more convincing your argument will be. You can also make a list of situations where using manipulatives enhances learning.

If you are on Team B, you may want to show how using manipulatives can be hard to manage with big graphs or in a limited time

span. You can conduct an experiment where the solution to a problem could be found more quickly using mental math, estimation, or a calculator. Record your findings in a chart and present that as part of your argument.

Regardless of which team you are on, you should try to be as objective as possible and not allow your personal feelings or preferences get in the way of your position.

The Debate

After you have gathered your data, you should write out your arguments on index cards and keep them with you during your presentation. You may want to write a short paragraph summarizing your position before you begin presenting your data.

The moderator should begin the debate by asking the first team to present their position by saying: "Resolved: Manipulatives Should Be Used to Teach Mathematics." Then Team A begins their arguments. After the preset time has elapsed, Team B gets the same amount of time to respond to what Team A said.

After the debate has been completed, your classmates should vote to see who presented the better argument.

Extra Practice

Lesson 1-2   Estimate. Use front-end estimation.

1. $216 + 492$
2. $1,235 + 5,645$
3. $6,478 - 2,345$
4. $1,510 + 523$
5. $298 + 109$
6. $8,710 - 610$
7. $15,269 + 14,795$
8. $958,752 - 125,568$
9. $145 + 256 + 872$
10. $23,456 + 758,567$
11. $6,915 - 3,854$
12. $7,785 + 563 + 1,973$
13. $1,085,365 - 523,894$
14. $1,616 + 5,897 + 2,439$
15. $\$665 + \$900 + \$213$
16. $45,648 - 44,896$

Lesson 1-3   Estimate. Use compatible numbers.

1. $601 \div 6$
2. $364 \div 6$
3. $699 \div 9$
4. $658 \div 8$
5. $821 \div 90$
6. $786 \div 80$
7. $398 \div 13$
8. $243 \div 6$
9. $269 \div 20$
10. $1,423 \div 7$
11. $1,499 \div 50$
12. $735 \div 8$
13. $1,602 \div 2$
14. $672 \div 65$
15. $198 \div 9$
16. $410 \div 7$
17. $8,620 \div 5$
18. $1,410 \div 7$
19. $41 \div 6$
20. $149 \div 5$

Lesson 1-4   Add or subtract mentally.

1. $44 + 56$
2. $12 + 36$
3. $19 + 61$
4. $56 - 23$
5. $45 + 25$
6. $27 + 39$
7. $18 + 62$
8. $92 - 45$
9. $108 + 204$
10. $916 - 465$
11. $622 + 104$
12. $376 - 213$
13. $1,410 + 562$
14. $462 + 37$
15. $768 - 246$
16. $99 + 11$
17. $1,620 + 2,710$
18. $2,569 + 4,298$
19. $5,235 - 4,945$
20. $\$456 + \$1,324$

Lesson 1-7

Evaluate each expression.

1. $14 - 5 + 7$
2. $12 + 10 - 5 - 6$
3. $50 - 6 + 12 + 4$
4. $12 - 2 \times 3$
5. $16 + 4 \times 5$
6. $5 + 3 \times 4 - 7$
7. $2 \times 3 + 9 \times 2$
8. $6 \times 8 + 4 \div 2$
9. $7 \times 6 - 14$
10. $8 + 12 \times 4 \div 8$
11. $13 - 6 \times 2 + 1$
12. $80 \div 10 \times 8$
13. $1 + 2 + 3 + 4$
14. $1 \times 2 \times 3 \times 4$
15. $6 + 6 \times 6$
16. $14 - 2 \times 7 + 0$
17. $156 - 6 \times 0$
18. $30 - 14 \times 2 + 8$

Lesson 1-8

Evaluate each expression if $a = 3$, $b = 4$, and $c = 12$.

1. $a + b$
2. $c - a$
3. $a + b + c$
4. $b - a$
5. $c - a \times b$
6. $a + 2 \times b$
7. $b + c \div 2$
8. ab
9. $a + 3b$
10. $a + c \div 6$
11. $25 + c \div b$
12. abc
13. $2(a + b) \div 7$
14. $2c \div b$
15. $144 - abc$
16. $2ab$
17. $c \div a + 10$
18. $9b \div 3$
19. $2b - a$
20. ac

Lesson 1-9

Write each product using exponents.

1. $2 \cdot 2 \cdot 2 \cdot 2 \cdot 2$
2. $6 \cdot 6 \cdot 6 \cdot 7 \cdot 7$
3. $9 \cdot 9 \cdot 9 \cdot 9 \cdot 9 \cdot 9 \cdot 10$
4. $k \cdot k \cdot k \cdot \ell \cdot \ell \cdot \ell$
5. $14 \cdot 14 \cdot 6$
6. $3 \cdot 3 \cdot 3 \cdot 3 \cdot y \cdot y$

Write each power as a product.

7. 13^4
8. 9^6
9. $2^3 \cdot 3^2$
10. x^5
11. 169^3
12. $13{,}410^2$

Evaluate each expression.

13. 5^6
14. 17^3
15. 2^{12}
16. $3^5 \cdot 2^3$
17. $6^4 \cdot 3$
18. $2^2 \cdot 3^2 \cdot 4^2$
19. 176^2
20. $6 \cdot 4^3$
21. five squared
22. 2 to the fifth power
23. 4 cubed

Lesson 1-10 Solve each equation.

1. $b + 7 = 12$
2. $a + 3 = 15$
3. $s + 10 = 23$
4. $9 + n = 13$

5. $20 = 24 - n$
6. $4x = 36$
7. $2y = 10$
8. $15 = 5h$

9. $j \div 3 = 2$
10. $14 = w - 4$
11. $24 \div k = 6$
12. $b - 3 = 12$

13. $c \div 10 = 8$
14. $y \div 2 = 8$
15. $6 = t \div 5$
16. $42 = 6n$

17. $14 + m = 24$
18. $g - 3 = 10$
19. $7 + a = 10$
20. $3y = 39$

21. $\frac{f}{2} = 12$
22. $16 = 4v$
23. $81 = 80 + a$
24. $9 = \frac{72}{x}$

Lesson 2-1 Replace each ● with $<$, $>$, or $=$.

1. 0.36 ● 0.63
2. 1.74 ● 1.7
3. 4.03 ● 4.003

4. 0.06 ● 0.066
5. 10.5 ● 10.05
6. 3.0 ● 3

7. 5.632 ● 5.623
8. 0.423 ● 0.5
9. 2.020 ● 2.202

10. 0.93 ● 0.9
11. 0.205 ● 0.025
12. 0.46 ● 0.49

13. 13.100 ● 13.1
14. 6.25 ● 6.20
15. 9.99 ● 9.099

16. 0.030 ● 0.03
17. 0.062 ● 0.62
18. 1.14 ● 1.09

19. 10.1 ● 100.0
20. 0.02 ● 0.002
21. 2.101 ● 2.11

Lesson 2-2 Round each number to the underlined place-value position.

1. $5.\underline{6}4$
2. $0.2\underline{6}25$
3. $0.45\underline{6}95$
4. $6.\underline{2}49$

5. $15.\underline{2}98$
6. $0.002\underline{6}325$
7. $758.9\underline{9}9$
8. $\underline{4}.25$

9. $32.65\underline{8}2$
10. $\underline{0}.025$
11. $1.004\underline{4}9$
12. $9.\underline{2}5$

13. $67.4\underline{9}2$
14. $25.\underline{1}9$
15. $26.\underline{9}6$
16. $4.0\underline{0}65$

17. $26.96\underline{6}6$
18. $1.\underline{2}499999$
19. $2.0\underline{1}2$
20. $1\underline{6}.569$

Lesson 2-3

Estimate. Use an appropriate strategy.

1.	$\begin{array}{r} 0.245 \\ + 0.256 \end{array}$	**2.**	$\begin{array}{r} 2.45698 \\ - 1.26589 \end{array}$	**3.**	$\begin{array}{r} 0.5962 \\ + 1.2598 \end{array}$	**4.**	$\begin{array}{r} 17.985 \\ - 9.001 \end{array}$

5.	$\begin{array}{r} 12.6589 \\ - 6.3874 \end{array}$	**6.**	$\begin{array}{r} 0.005698 \\ + 0.015963 \end{array}$	**7.**	$\begin{array}{r} 1.26589 \\ + 0.76589 \end{array}$	**8.**	$\begin{array}{r} 15.986325 \\ - 12.965236 \end{array}$

9.	$\begin{array}{r} 8.5 \\ \times 9.1 \end{array}$	**10.**	$\begin{array}{r} 12.9568 \\ \times 6.1563 \end{array}$	**11.**	$\begin{array}{r} 9.652 \\ \times 6.2 \end{array}$	**12.**	$\begin{array}{r} 25.49862 \\ \times 4.2136 \end{array}$

13. $1.12 + 0.9865 + 1.023 + 0.89 + 0.99 + 1.03569$

14. $3\overline{)11.75}$

15. $82.1 + 79.3 + 81.5 + 79 + 80 + 81.256$

16. $4.1\overline{)16.123}$

Lesson 2-4

Multiply.

1. 9.6×10.5

2. 3.2×0.1

3. 10.5×9.6

4. 5.42×0.21

5. 7.42×0.2

6. 0.001×0.02

7. 0.6×542

8. 6.7×5.8

9. 3.24×6.7

10. 9.8×4.62

11. 7.32×9.7

12. 0.008×0.007

13. 0.0001×56

14. 4.5×0.2

15. 9.6×2.3

16. 5.63×8.1

17. 10.35×9.1

18. 28.2×3.9

19. 102.13×1.221

20. 2.02×1.25

21. 8.37×89.6

Lesson 2-5

Multiply mentally.

1. 1.2×10

2. 0.23×100

3. $1.235 \times 1,000$

4. 1.2×10^3

5. 0.002×100

6. 3.56×10^2

7. 0.000012×10^{10}

8. 95.23×10^1

9. 76.425×10^4

10. $1.0056 \times 10,000$

11. 4.7×1

12. 9.6×10^0

Lesson 2-6   Write each number in scientific notation.

1. 720
2. 7,560
3. 892
4. 1,400
5. 91,256
6. 51,000
7. 145,600
8. 90,100
9. 123,580,000,000

Write each number in standard form.

10. 4.5×10^3
11. 2×10^4
12. 1.725896×10^6
13. 9.61×10^2
14. 1×10^7
15. 8.256×10^8
16. 5.26×10^4
17. 3.25×10^2
18. 6.79×10^5

Lesson 2-7   Divide.

1. $9\overline{)0.036}$
2. $13\overline{)39.39}$
3. $45\overline{)0.585}$
4. $8\overline{)0.024}$
5. $6\overline{)0.312}$
6. $7\overline{)0.161}$
7. $7\overline{)7.21}$
8. $3\overline{)9.18}$
9. $0.72 \div 12$
10. $0.36 \div 9$
11. $0.56 \div 14$
12. $32.2 \div 8$
13. $0.3869 \div 5.3$
14. $0.39 \div 7.8$
15. $0.0426 \div 7.1$
16. $0.1185 \div 7.9$
17. $0.84 \div 12$
18. $4.544 \div 64$
19. $0.384 \div 9.6$
20. $0.2262 \div 8.7$

Lesson 2-8   Divide. Round to the nearest tenth.

1. $26.5 \div 4$
2. $46.25 \div 8$
3. $19.38 \div 9$
4. $8.5 \div 2$
5. $90.88 \div 14$
6. $23.1 \div 4$
7. $19.5 \div 27$
8. $26.5 \div 19$
9. $46.23 \div 25$
10. $46.25 \div 25$
11. $4.26 \div 9$
12. $18.74 \div 19$

Divide. Round to the nearest hundredth.

13. $17.9 \div 21$
14. $57.9 \div 14$
15. $21.555 \div 6$
16. $6.435 \div 7$
17. $15.23 \div 8$
18. $1.2356 \div 3$
19. $156.8 \div 25$
20. $19.563 \div 6$
21. $0.125 \div 1$

Lesson 2-9   Complete.

1. $400 \text{ mm} = \underline{\hspace{0.8cm}} \text{ cm}$
2. $4 \text{ km} = \underline{\hspace{0.8cm}} \text{ m}$
3. $660 \text{ cm} = \underline{\hspace{0.8cm}} \text{ m}$

4. $0.3 \text{ km} = \underline{\hspace{0.8cm}} \text{ m}$
5. $30 \text{ mm} = \underline{\hspace{0.8cm}} \text{ cm}$
6. $84.5 \text{ m} = \underline{\hspace{0.8cm}} \text{ km}$

7. $\underline{\hspace{0.8cm}} \text{ m} = 54 \text{ cm}$
8. $18 \text{ km} = \underline{\hspace{0.8cm}} \text{ cm}$
9. $\underline{\hspace{0.8cm}} \text{ mm} = 45 \text{ cm}$

10. $4 \text{ kg} = \underline{\hspace{0.8cm}} \text{ g}$
11. $632 \text{ mg} = \underline{\hspace{0.8cm}} \text{ g}$
12. $4,497 \text{ g} = \underline{\hspace{0.8cm}} \text{ kg}$

13. $\underline{\hspace{0.8cm}} \text{ mg} = 21 \text{ g}$
14. $61.2 \text{ mg} = \underline{\hspace{0.8cm}} \text{ g}$
15. $61 \text{ g} = \underline{\hspace{0.8cm}} \text{ mg}$

16. $\underline{\hspace{0.8cm}} \text{ mg} = 0.51 \text{ kg}$
17. $0.63 \text{ kg} = \underline{\hspace{0.8cm}} \text{ g}$
18. $\underline{\hspace{0.8cm}} \text{ kg} = 563 \text{ g}$

Lesson 3-3   Find the range for each set of data. Choose an appropriate scale and intervals. Draw a number line to show the scale and intervals.

1. 25, 26, 27, 25, 28, 27, 26, 28, 25

2. 110, 210, 156, 174, 135, 198, 127, 160

3. 85, 76, 91, 81, 97, 74, 77, 82, 93

4. 0.5, 0.6, 0.1, 0.4, 0.8, 0.6, 0.3, 0.55

5. 3, 4, 5, 1, 8, 5, 2, 3, 6, 4, 7, 8, 2, 1

Lesson 3-4   Make a line plot for each set of data. Circle any outliers on the line plot.

1. 25, 26, 27, 25, 28, 27, 21, 26, 28, 25

2. 110, 210, 156, 174, 125, 198, 165, 185

3. 600, 650, 700, 600, 625, 675, 450, 650

4. 0.5, 0.6, 0.1, 0.4, 0.8, 0.6, 0.7, 0.5

5. 3, 4, 5, 2, 6, 1, 2, 4, 3, 6, 9, 1, 2, 3

Lesson 3-5

Find the mode(s), median, and mean for each set of data. Round answers to the nearest tenth.

1. 1, 5, 9, 1, 2, 5, 8, 2

2. 2, 5, 8, 9, 7, 6, 3, 5

3. 1, 2, 1, 2, 2, 1, 2

4. 12, 13, 15, 12, 12, 11

5. 256, 265, 247, 256

6. 957, 562, 462, 847, 721

7. 46, 54, 66, 54, 46, 66

8. 81, 82, 83, 84, 85, 86, 87

Lesson 3-6

Make a stem-and-leaf plot for each set of data.

1. 23, 15, 39, 68, 57, 42, 51, 52, 41, 18, 29

2. 5, 14, 39, 28, 14, 6, 7, 18, 13, 28, 9, 14

3. 189, 182, 196, 184, 197, 183, 196, 194, 184

4. 71, 82, 84, 95, 76, 92, 83, 74, 81, 75, 96

Lesson 3-7

1. Darlene's quiz scores in science have been steadily going up since her parents hired a tutor for her. Based on the graph at the right, predict what Darlene's score will be on the next quiz.

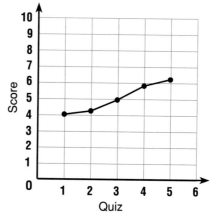

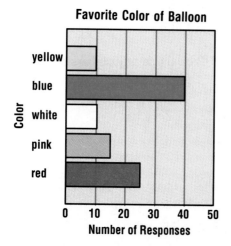

Favorite Color of Balloon

2. A balloon maker asked 100 kids what their favorite color of balloon is. The graph at the left shows their responses. What color of balloon should he make the most of?

Lesson 4-1   Determine whether the first number is divisible by the second number.

1. 279; 3
2. 1,240; 6
3. 3,250; 5
4. 835; 4
5. 5,550; 10
6. 315; 9
7. 777; 6
8. 4,214; 3
9. 3,012; 2
10. 244; 4
11. 984; 6
12. 1,000; 5

Determine whether each number is divisible by 2, 3, 4, 5, 6, 9, or 10.

13. 453
14. 2,225
15. 504
16. 4,300
17. 672
18. 8,240
19. 111
20. 6,232
21. 999
22. 5,200
23. 3,217
24. 804

Lesson 4-2   Determine whether each number is composite or prime.

1. 32
2. 417
3. 5,212
4. 2,111
5. 71
6. 1,005
7. 239
8. 3,215

Write the prime factorization of each number.

9. 81
10. 525
11. 245
12. 1,120
13. 750
14. 2,400
15. 914
16. 975

Use your calculator to find the prime factors of each number. Then write the prime factorization of each number.

17. 423
18. 972
19. 144
20. 72

Lesson 4-3   Identify each sequence as arithmetic, geometric, or neither. Then find the next three terms in each sequence.

1. 5, 9, 13, 17, …
2. 3, 6, 12, 24, …
3. 10, 15, 25, 40, …
4. 4.5, 5.4, 6.3, 7.2, …
5. 90, 100, 91, 99, 92, …
6. 0.5, 0.4, 0.3, …
7. 64, 16, 4, 1, …
8. 10, 50, 250, …
9. 8, 24, 72, 216, …
10. 16, 8, 4, 2, …
11. 49, 7, 1, $\frac{1}{7}$, …
12. 500, 400, 300, …
13. 40, 42, 46, 52, 60, …
14. 75, 15, 3, $\frac{3}{5}$, …
15. 27, 9, 3, 1, …
16. 1, 2, 4, 7, 11, …
17. 10, 100, 1,000, …
18. 1, 10, 19, 28, 37, …
19. 225, 250, 275, …
20. 2.5, 3.0, 3.5, 4.0, …

Lesson 4-5   Find the GCF of each set of numbers.

1. 12, 16   **2.** 63, 81   **3.** 225, 500   **4.** 37, 100

5. 240, 32   **6.** 640, 412   **7.** 36, 81   **8.** 350, 140

9. 72, 170   **10.** 255, 51   **11.** 48, 72

12. 86, 200   **13.** 24, 56, 120   **14.** 48, 60, 84

15. 32, 80, 96   **16.** 49, 14, 70   **17.** 6, 8, 12

18. 33, 55, 77   **19.** 27, 15, 300   **20.** 45, 150, 225

Lesson 4-6   Express each fraction in simplest form.

1. $\dfrac{14}{28}$ **2.** $\dfrac{15}{25}$ **3.** $\dfrac{100}{300}$ **4.** $\dfrac{14}{35}$

5. $\dfrac{9}{51}$ **6.** $\dfrac{54}{56}$ **7.** $\dfrac{75}{90}$ **8.** $\dfrac{24}{40}$

9. $\dfrac{180}{270}$ **10.** $\dfrac{312}{390}$ **11.** $\dfrac{240}{448}$ **12.** $\dfrac{71}{82}$

13. $\dfrac{333}{900}$ **14.** $\dfrac{85}{255}$ **15.** $\dfrac{84}{128}$ **16.** $\dfrac{640}{960}$

Lesson 4-7   Express each fraction as a decimal. Use bar notation if necessary.

1. $\dfrac{16}{20}$ **2.** $\dfrac{25}{100}$ **3.** $\dfrac{7}{8}$ **4.** $\dfrac{48}{60}$

5. $\dfrac{11}{40}$ **6.** $\dfrac{13}{50}$ **7.** $\dfrac{55}{300}$ **8.** $\dfrac{18}{12}$

Express each decimal as a fraction in simplest form.

9. 0.38   **10.** 2.05   **11.** 0.075   **12.** 0.18

13. 0.675   **14.** 15.33   **15.** 0.64   **16.** 6.04

Lesson 4-8   The spinner shown at the right is equally likely to stop on each of the regions numbered 1 to 8. Find the probability that the spinner will stop on each of the following.

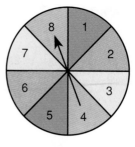

1. an even number
2. a prime number
3. a factor of 12
4. a composite number
5. a number less than 5
6. a factor of 36

A package of balloons contains 5 green, 3 yellow, 4 red, and 8 pink balloons. If you reach in the package and choose one balloon at random, what is the probability that you will select each of the following? Express each ratio as both a fraction and a decimal.

7. a red balloon
8. a green balloon
9. a pink balloon
10. a yellow balloon
11. a red or yellow balloon

Lesson 4-9   Find the LCM of each set of numbers.

1. 4, 9
2. 6, 16
3. 3, 8, 14
4. 24, 36

5. 48, 84
6. 12, 18, 28
7. 8, 9
8. 49, 56

9. 42, 66
10. 15, 39
11. 32, 80, 96
12. 56, 64

13. 24, 42
14. 250, 80
15. 26, 169
16. 5, 18, 45

17. 11, 22, 33
18. 56, 14, 70
19. 16, 24
20. 13, 14

Lesson 4-10   Find the LCD for each pair of fractions.

1. $\frac{3}{8}, \frac{2}{3}$
2. $\frac{5}{9}, \frac{7}{12}$
3. $\frac{4}{9}, \frac{8}{15}$
4. $\frac{11}{24}, \frac{17}{42}$
5. $\frac{12}{36}, \frac{15}{42}$
6. $\frac{25}{27}, \frac{43}{81}$
7. $\frac{32}{64}, \frac{15}{48}$
8. $\frac{2}{6}, \frac{14}{15}$

Replace each ⬤ with $<$, $>$, or $=$ to make a true statement.

9. $\frac{7}{9}$ ⬤ $\frac{3}{5}$
10. $\frac{14}{25}$ ⬤ $\frac{3}{4}$
11. $\frac{8}{24}$ ⬤ $\frac{20}{60}$
12. $\frac{5}{12}$ ⬤ $\frac{4}{9}$
13. $\frac{18}{24}$ ⬤ $\frac{10}{18}$
14. $\frac{4}{6}$ ⬤ $\frac{5}{9}$
15. $\frac{11}{49}$ ⬤ $\frac{12}{42}$
16. $\frac{5}{14}$ ⬤ $\frac{2}{6}$

Lesson 5-1

Change each improper fraction to a mixed number in simplest form or a whole number.

1. $\frac{5}{4}$ 2. $\frac{10}{7}$ 3. $\frac{6}{3}$ 4. $\frac{9}{4}$ 5. $\frac{3}{2}$ 6. $\frac{9}{3}$

7. $\frac{16}{10}$ 8. $\frac{7}{3}$ 9. $\frac{3}{1}$ 10. $\frac{8}{6}$ 11. $\frac{21}{20}$ 12. $\frac{21}{7}$

13. $\frac{12}{4}$ 14. $\frac{5}{2}$ 15. $\frac{7}{4}$ 16. $\frac{20}{6}$ 17. $\frac{10}{3}$ 18. $\frac{26}{5}$

Lesson 5-2

Round each fraction to 0, $\frac{1}{2}$, or 1.

1. $\frac{3}{8}$ 2. $\frac{1}{9}$ 3. $\frac{6}{7}$ 4. $\frac{7}{12}$ 5. $\frac{1}{6}$ 6. $\frac{10}{12}$

Round to the nearest whole number.

7. $5\frac{7}{8}$ 8. $3\frac{7}{12}$ 9. $7\frac{1}{10}$ 10. $2\frac{5}{12}$ 11. $2\frac{4}{9}$ 12. $8\frac{3}{4}$

Estimate.

13. $\frac{3}{7} + \frac{6}{8}$ 14. $\frac{3}{10} + \frac{4}{7}$ 15. $\frac{3}{9} + \frac{7}{8}$ 16. $\frac{1}{8} + \frac{8}{9}$

17. $1\frac{1}{2} + 2\frac{1}{4}$ 18. $3\frac{1}{8} + 7\frac{6}{7}$ 19. $4\frac{2}{3} + 6\frac{7}{8}$ 20. $3\frac{2}{3} \times 2\frac{1}{3}$

21. $\frac{4}{5} \times 3$ 22. $9\frac{7}{8} - 6\frac{2}{3}$ 23. $\frac{3}{7} - \frac{1}{15}$ 24. $\frac{26}{17} \times \frac{37}{38}$

Lesson 5-3

Add or subtract. Write each sum or difference in simplest form.

1. $\frac{5}{11} + \frac{9}{11}$ 2. $\frac{5}{8} - \frac{1}{8}$ 3. $\frac{7}{10} + \frac{7}{10}$ 4. $\frac{9}{12} - \frac{5}{12}$

5. $\frac{2}{9} + \frac{1}{3}$ 6. $\frac{1}{2} + \frac{3}{4}$ 7. $\frac{1}{4} - \frac{3}{12}$ 8. $\frac{3}{7} + \frac{6}{14}$

9. $\frac{1}{4} + \frac{3}{5}$ 10. $\frac{4}{9} + \frac{1}{2}$ 11. $\frac{5}{7} + \frac{4}{6}$ 12. $\frac{3}{4} - \frac{1}{6}$

13. $\frac{3}{5} + \frac{3}{4}$ 14. $\frac{2}{3} - \frac{1}{8}$ 15. $\frac{9}{10} + \frac{1}{3}$ 16. $\frac{8}{15} + \frac{2}{9}$

17. $\frac{6}{7} + \frac{6}{9}$ 18. $\frac{3}{7} + \frac{3}{4}$ 19. $\frac{5}{7} + \frac{5}{9}$ 20. $\frac{7}{8} + \frac{5}{6}$

Lesson 5-4

Add or subtract. Write each sum or difference in simplest form.

1. $2\frac{1}{3} + 1\frac{1}{3}$
2. $5\frac{2}{7} - 2\frac{3}{7}$
3. $6\frac{3}{8} + 7\frac{1}{8}$
4. $2\frac{3}{4} - 1\frac{1}{4}$

5. $5\frac{1}{2} - 3\frac{1}{4}$
6. $2\frac{2}{3} + 4\frac{1}{9}$
7. $7\frac{4}{5} + 9\frac{3}{10}$
8. $3\frac{3}{4} + 5\frac{5}{8}$

9. $10\frac{2}{3} + 5\frac{6}{7}$
10. $17\frac{2}{9} - 12\frac{1}{3}$
11. $6\frac{5}{12} + 12\frac{5}{12}$
12. $7\frac{1}{4} + 15\frac{5}{6}$

13. $6\frac{1}{8} + 4\frac{2}{3}$
14. $7 - 6\frac{4}{9}$
15. $8\frac{1}{12} + 12\frac{6}{11}$
16. $7\frac{2}{3} + 8\frac{1}{4}$

17. $12\frac{3}{11} + 14\frac{3}{13}$
18. $21\frac{1}{3} + 15\frac{3}{8}$
19. $19\frac{1}{7} + 6\frac{1}{4}$
20. $9\frac{2}{5} - 8\frac{1}{3}$

Lesson 5-5

Multiply. Write each product in simplest form.

1. $\frac{2}{3} \times \frac{3}{5}$
2. $\frac{1}{6} \times \frac{2}{5}$
3. $\frac{4}{9} \times \frac{3}{7}$
4. $\frac{5}{12} \times \frac{6}{11}$

5. $\frac{3}{8} \times \frac{8}{9}$
6. $\frac{3}{5} \times \frac{1}{12}$
7. $\frac{2}{5} \times \frac{5}{8}$
8. $\frac{7}{15} \times \frac{3}{21}$

9. $\frac{5}{6} \times \frac{15}{16}$
10. $\frac{6}{14} \times \frac{12}{18}$
11. $\frac{2}{3} \times \frac{3}{13}$
12. $\frac{4}{9} \times \frac{1}{6}$

13. $3 \times \frac{1}{9}$
14. $5 \times \frac{6}{7}$
15. $\frac{3}{5} \times 15$
16. $3\frac{1}{2} \times 4\frac{1}{3}$

17. $3\frac{5}{8} \times 4\frac{1}{2}$
18. $\frac{4}{5} \times 2\frac{3}{4}$
19. $6\frac{1}{8} \times 5\frac{1}{7}$
20. $2\frac{2}{3} \times 2\frac{1}{4}$

Lesson 5-6

Find the perimeter of each figure shown below.

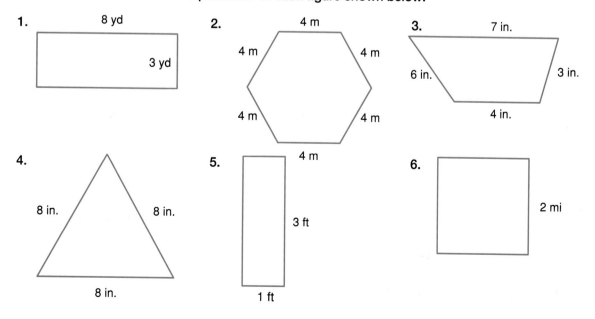

1. 8 yd, 3 yd

2. 4 m, 4 m, 4 m, 4 m, 4 m, 4 m

3. 7 in., 6 in., 3 in., 4 in.

4. 8 in., 8 in., 8 in.

5. 4 m, 3 ft, 1 ft

6. 2 mi

Extra Practice **583**

Lesson 5-7

Find the circumference of each circle.

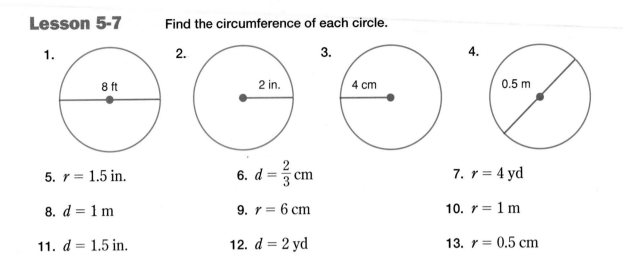

1. 8 ft

2. 2 in.

3. 4 cm

4. 0.5 m

5. $r = 1.5$ in.

6. $d = \frac{2}{3}$ cm

7. $r = 4$ yd

8. $d = 1$ m

9. $r = 6$ cm

10. $r = 1$ m

11. $d = 1.5$ in.

12. $d = 2$ yd

13. $r = 0.5$ cm

Lesson 5-9

Name the property shown by each statement.

1. $\frac{4}{5} \times \frac{2}{3} = \frac{2}{3} \times \frac{4}{5}$

2. $\frac{3}{10} \times 3\frac{1}{3} = 1$

3. $\frac{24}{27} \times 1 = \frac{24}{27}$

4. $\left[\frac{1}{2} + \frac{3}{4}\right] + \frac{5}{6} = \frac{1}{2} + \left[\frac{3}{4} + \frac{5}{6}\right]$

5. $\frac{2}{3} \times \left[\frac{1}{2} + \frac{5}{6}\right] = \frac{2}{3} \times \frac{1}{2} + \frac{2}{3} \times \frac{5}{6}$

6. $\frac{2}{3} \times \frac{3}{2} = 1$

Compute mentally.

7. $5 \times 5\frac{1}{5}$

8. $3 \times 1\frac{2}{5}$

9. $3\frac{2}{7} \times 7$

10. $\frac{1}{2} \times 2\frac{1}{2}$

11. $3\frac{3}{4} \times \frac{1}{2}$

12. $2\frac{1}{3} \times 6$

13. $\frac{1}{5} \times 10$

14. $\frac{1}{3} \times 3\frac{2}{5}$

Lesson 5-10

Divide. Write each quotient in simplest form.

1. $\frac{2}{3} \div \frac{3}{2}$

2. $\frac{3}{5} \div \frac{2}{5}$

3. $\frac{7}{10} \div \frac{3}{8}$

4. $\frac{5}{9} \div \frac{2}{5}$

5. $4 \div \frac{2}{3}$

6. $8 \div \frac{4}{5}$

7. $9 \div \frac{5}{9}$

8. $\frac{2}{7} \div 2$

9. $\frac{1}{14} \div 7$

10. $\frac{2}{13} \div \frac{5}{26}$

11. $\frac{4}{7} \div \frac{6}{7}$

12. $\frac{7}{8} \div \frac{1}{3}$

13. $15 \div \frac{3}{5}$

14. $\frac{9}{14} \div \frac{3}{4}$

15. $\frac{8}{9} \div \frac{5}{6}$

16. $\frac{4}{9} \div 36$

17. $\frac{3}{5} \div \frac{2}{3}$

18. $\frac{8}{9} \div \frac{4}{5}$

19. $\frac{3}{4} \div \frac{15}{16}$

20. $6 \div \frac{1}{5}$

Lesson 6-1

Solve each equation by using the inverse operation. Round to the nearest tenth.

1. $q - 7 = 7$
2. $g - 3 = 10$
3. $b + 7 = 12$
4. $a + 3 = 15$
5. $4x = 36$
6. $39 = 3y$
7. $4z = 16$
8. $54 = 9w$
9. $0.011 + h = 5.0$
10. $63 + f = 71$
11. $7 = 91 - g$
12. $9 = 19 - j$
13. $\frac{x}{6} = 6$
14. $\frac{x}{7} = 8$
15. $8 = \frac{c}{10}$
16. $4 = \frac{x}{2}$
17. $z + 0.34 = 3.1$
18. $23 = n - 0.09$
19. $2g = 0.6$
20. $r - 3 = 4$
21. $\frac{t}{3} = 1.2$

Lesson 6-2

Solve each equation. Check your solution.

1. $r - 3 = 14$
2. $t + 3 = 21$
3. $s + 10 = 23$
4. $7 + a = 10$
5. $14 + m = 24$
6. $9 + n = 13$
7. $s - 0.4 = 6$
8. $x - 1.3 = 12$
9. $y + 3.4 = 18$
10. $0.013 + h = 4.0$
11. $6 + f = 71$
12. $7.2 + g = 9.1$
13. $z - 12.1 = 14$
14. $w - 0.1 = 0.32$
15. $v - 18 = 13.7$
16. $s + 1.3 = 18$
17. $t + 3.43 = 7.4$
18. $x + 7.4 = 23.5$
19. $p + 3.1 = 18$
20. $q - 2.17 = 21$
21. $w - 3.7 = 4.63$

Lesson 6-3

Solve each equation. Check your solution.

1. $2m = 18$
2. $42 = 6n$
3. $72 = 8k$
4. $20r = 20$
5. $420 = 5s$
6. $325 = 25t$
7. $14 = 2p$
8. $18q = 36$
9. $40 = 10a$
10. $100 = 20b$
11. $416 = 4c$
12. $45 = 9d$
13. $\frac{m}{7} = 5$
14. $\frac{n}{3} = 6$
15. $4 = \frac{p}{4}$
16. $4 = \frac{x}{2}$
17. $\frac{s}{9} = 8$
18. $6 = \frac{t}{5}$
19. $\frac{w}{7} = 8$
20. $\frac{c}{8} = 2$

Lesson 6-4 Translate each phrase into an algebraic expression.

1. six less than p

2. twenty more than c

3. the quotient of a and b

4. Ann's age plus 6

5. x increased by twelve

6. $1,000 divided by z

7. 3 divided into y

8. the product of 7 and m

9. the difference of f and 9

10. twenty-six less q

11. 19 decreased by z

12. two less than x

Lesson 6-6 Complete.

1. 4,000 lb = _____ tons

2. 5 tons = _____ lb

3. 2 lb = _____ oz

4. 12,000 lb = _____ tons

5. $\frac{1}{4}$ lb = _____ oz

6. 6 lb 2 oz = _____ oz

7. 3 gal = _____ pt

8. 24 fl oz = _____ c

9. 8 pt = _____ c

10. 10 pt = _____ qt

11. $2\frac{1}{4}$ c = _____ fl oz

12. 12 pt = _____ c

13. 4 gal = _____ qt

14. 4 qt = _____ fl oz

15. 4 pt = _____ c

16. 9 lb = _____ oz

17. 15 qt = _____ gal

18. 6 lb = _____ oz

19. 2 gal = _____ fl oz

20. 3 tons = _____ lb

21. 18 qt = _____ pt

Lesson 6-7 Find the area of each figure shown or described below.

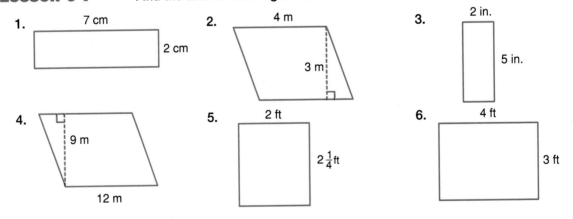

7. rectangle: $\ell = 19$ m, $w = 6$ m

8. parallelogram: $b = 0.2$ m, $h = 0.3$ m

9. rectangle: $\ell = 0.2$ m, $w = 0.3$ m

Lesson 7-1   Write an integer for each situation.

1. a gain of 14 points
2. a $25 withdrawal
3. six degrees below zero
4. a loss of 3 pounds
5. a loss of 20 yards
6. a profit of $16

Write the integer represented by the point for each letter. Then find its opposite and its absolute value.

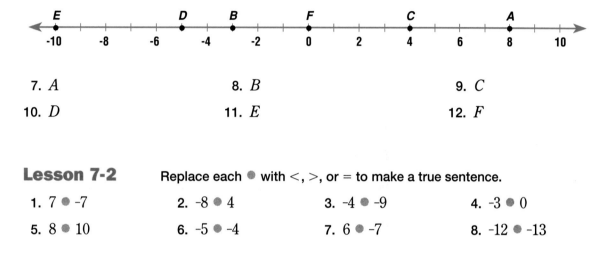

7. A
8. B
9. C
10. D
11. E
12. F

Lesson 7-2   Replace each ● with $<$, $>$, or $=$ to make a true sentence.

1. $7 ● -7$
2. $-8 ● 4$
3. $-4 ● -9$
4. $-3 ● 0$
5. $8 ● 10$
6. $-5 ● -4$
7. $6 ● -7$
8. $-12 ● -13$

Order the integers from least to greatest.

9. $-2, -8, 4, 10, -6, -12$
10. $19, -19, -21, 32, -14, 18$
11. $18, 23, 95, -95, -18, -23, 2$
12. $46, -48, -47, -52, -18, 12$

Lesson 7-3   Name the x-coordinate and y-coordinate for each point labeled at the right. Then tell in which quadrant the point lies.

1. A
2. B
3. C
4. D
5. E
6. F
7. G
8. H
9. I

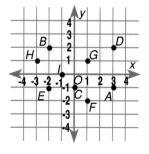

On graph paper, draw a coordinate plane. Then graph and label each point.

10. $N (-4, 3)$
11. $K (2, 5)$
12. $W (-6, -2)$
13. $X (5, 0)$
14. $Y (4, -4)$
15. $M (0, -3)$
16. $Z (-2, 0.5)$
17. $S (-1, -3)$

Lesson 7-4   Solve each equation.

1. $a = -4 + 8$

2. $14 + 16 = b$

3. $-7 + (-7) = h$

4. $g = -9 + (-6)$

5. $-18 + 11 = d$

6. $k = -36 + 40$

7. $42 + (-18) = f$

8. $-42 + 29 = r$

9. $m = 18 + (-32)$

10. $-33 + (-12) = w$

11. $h = -13 + (-11)$

12. $47 + 12 = y$

13. $-96 + (-18) = g$

14. $x = 95 + (-5)$

15. $y = -69 + (-32)$

16. $-100 + 98 = a$

17. $-120 + 2 = b$

18. $-120 + (-2) = c$

19. $5 + (-7) = y$

20. $w = 25 + (-25)$

21. $x = -56 + (-4)$

Lesson 7-5   Solve each equation.

1. $3 - 7 = y$

2. $-5 - 4 = w$

3. $a = -6 - 2$

4. $r = 8 - 13$

5. $6 - (-4) = b$

6. $12 - 9 = x$

7. $-2 - 23 = c$

8. $z = 63 - 78$

9. $a = 0 - (-14)$

10. $-20 - 0 = d$

11. $-5 - (-9) = h$

12. $a = 58 - (-10)$

13. $55 - 33 = k$

14. $m = 72 - (-19)$

15. $n = -41 - 15$

16. $84 - (-61) = a$

17. $-51 - 47 = x$

18. $c = -81 - 21$

19. $z = -4 - (-4)$

20. $-99 - 1 = p$

21. $26 - (-14) = y$

Lesson 7-7   Solve each equation.

1. $5(-2) = d$

2. $a = 6(-4)$

3. $4(21) = y$

4. $-11(-5) = c$

5. $x = -6(5)$

6. $a = -50(0)$

7. $-5(-5) = z$

8. $-4(8) = q$

9. $b = 3(-13)$

10. $x = -12(5)$

11. $3(-16) = y$

12. $a = 2(2)$

13. $b = 2(-2)$

14. $c = -2(2)$

15. $d = -2(-2)$

16. $-3(2)(-4) = j$

17. $6(3)(-2) = k$

18. $x = 5(-12)$

19. $a = (-4)(-4)$

20. $y = -3(12)$

21. $2(2)(-2) = b$

Lesson 7-8 Solve each equation.

1. $a = 4 \div (-2)$

2. $16 \div (-8) = x$

3. $-14 \div (-2) = c$

4. $d = 32 \div 8$

5. $g = 18 \div (-3)$

6. $h = -18 \div 3$

7. $8 \div (-8) = y$

8. $t = 0 \div (-1)$

9. $-25 \div 5 = k$

10. $c = -14 \div (-7)$

11. $-32 \div 8 = m$

12. $n = -56 \div (-8)$

13. $-81 \div 9 = y$

14. $81 \div (-9) = w$

15. $x = 81 \div 9$

16. $q = -81 \div (-9)$

17. $18 \div (-2) = a$

18. $-55 \div 11 = c$

19. $25 \div (-5) = r$

20. $x = -21 \div 3$

21. $-42 \div (-7) = y$

Lesson 7-9 Solve each equation. Check your solution.

1. $-4 + b = 12$

2. $z - 10 = -8$

3. $-7 = x + 12$

4. $m + (-2) = 6$

5. $r - (-8) = 14$

6. $a + 6 = -9$

7. $3m = -15$

8. $0 = 6r$

9. $r \div 7 = -8$

10. $\frac{c}{-4} = 10$

11. $\frac{y}{12} = -6$

12. $-2a = -8$

Write an equation for each problem below. Then solve.

13. The sum of 5 and a number g is 12. Find g.

14. The quotient when 16 is divided by a number x is -4. Find x.

15. When -6 is multiplied by a number f, the product is -36. Find f.

Lesson 7-10 Write each number in standard form.

1. 3×10^{-5}

2. 8×10^{-2}

3. 6×10^{-6}

4. 4×10^{-3}

5. 7×10^{-1}

6. 5×10^{-4}

Write each decimal in scientific notation.

7. 0.002

8. 0.00008

9. 0.00000005

10. 0.06

11. 0.00000000009

12. 0.7

Lesson 8-1   Classify each angle as acute, obtuse, right, or straight.

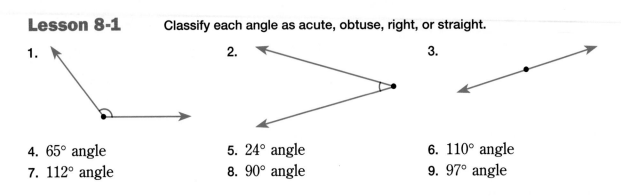

1.   **2.**   **3.**

4. 65° angle   **5.** 24° angle   **6.** 110° angle
7. 112° angle   **8.** 90° angle   **9.** 97° angle

Lesson 8-2   Determine which figures are polygons. If a figure is not a polygon, explain why.

1. 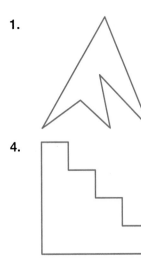   **2.**   **3.**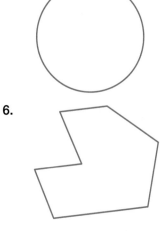

4.   **5.** 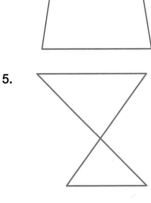   **6.**

Lesson 8-3   Classify each triangle by its sides and by its angles.

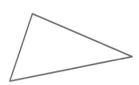

1.   **2.**   **3.**

Name every quadrilateral that describes each figure. Then underline the name that best describes the figure.

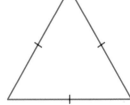

4.   **5.**   **6.**

Lesson 8-4   Tell whether each polygon is a regular polygon. If not, tell why.

1.

2.

3.

4.

5.

6.

Lesson 9-2   Find the square of each number.

1. 6   2. 12   3. 7   4. 15

5. 20   6. 14   7. 24   8. 1

9. 11   10. 40   11. 25   12. 9

Find each square root.

13. $\sqrt{49}$ 14. $\sqrt{64}$ 15. $\sqrt{169}$ 16. $\sqrt{324}$

17. $\sqrt{900}$ 18. $\sqrt{225}$ 19. $\sqrt{2,500}$ 20. $\sqrt{81}$

21. $\sqrt{289}$ 22. $\sqrt{576}$ 23. $\sqrt{8,100}$ 24. $\sqrt{676}$

Lesson 9-3   Estimate.

1. $\sqrt{15}$ 2. $\sqrt{35}$ 3. $\sqrt{112}$ 4. $\sqrt{75}$

5. $\sqrt{27}$ 6. $\sqrt{249}$ 7. $\sqrt{88}$ 8. $\sqrt{1,500}$

9. $\sqrt{612}$ 10. $\sqrt{340}$ 11. $\sqrt{495}$ 12. $\sqrt{264}$

13. $\sqrt{350}$ 14. $\sqrt{834}$ 15. $\sqrt{3,700}$ 16. $\sqrt{298}$

17. $\sqrt{101}$ 18. $\sqrt{800}$ 19. $\sqrt{58}$ 20. $\sqrt{750}$

21. $\sqrt{1,200}$ 22. $\sqrt{1,000}$ 23. $\sqrt{5,900}$ 24. $\sqrt{999}$

25. $\sqrt{374}$ 26. $\sqrt{512}$ 27. $\sqrt{3,750}$ 28. $\sqrt{255}$

29. $\sqrt{83}$ 30. $\sqrt{845}$ 31. $\sqrt{200}$ 32. $\sqrt{10,001}$

Lesson 9-4

Use the Pythagorean Theorem to find the length of each hypotenuse given the lengths of the legs. Round answers to the nearest tenth.

1. 4 ft, 6 ft   **2.** 12 cm, 25 cm   **3.** 15 yd, 24 yd   **4.** 8 mm, 11 mm

Find the missing lengths. Round decimal answers to the nearest tenth.

5. *a*: 14 cm; *c*: 18 cm   **6.** *b*: 15 ft; *c*: 24 ft   **7.** *a*: 5 yd; *b*: 8 yd

Given the following lengths, determine whether each triangle is a right triangle. Write *yes* or *no*.

8. 6 mm, 8 mm, 10 mm   **9.** 12 ft, 15 ft, 20 ft   **10.** 300 m, 400 m, 500 m

Lesson 9-6

Estimate the area of each figure.

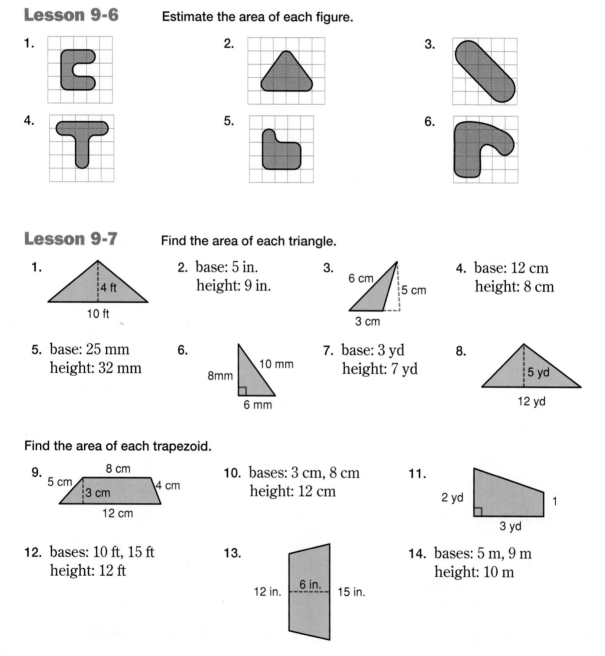

1.   **2.**   **3.**

4.   **5.**   **6.**

Lesson 9-7

Find the area of each triangle.

1. 4 ft 10 ft

2. base: 5 in.
  height: 9 in.

3. 6 cm 5 cm 3 cm

4. base: 12 cm
  height: 8 cm

5. base: 25 mm
  height: 32 mm

6. 8mm 10 mm 6 mm

7. base: 3 yd
  height: 7 yd

8. 5 yd 12 yd

Find the area of each trapezoid.

9. 8 cm 5 cm 3 cm 4 cm 12 cm

10. bases: 3 cm, 8 cm
  height: 12 cm

11. 2 yd 3 yd 1

12. bases: 10 ft, 15 ft
  height: 12 ft

13. 12 in. 6 in. 15 in.

14. bases: 5 m, 9 m
  height: 10 m

Lesson 9-8

Find the area of each circle shown or described below. Round answers to the nearest tenth.

1. radius, 8 in.

2. 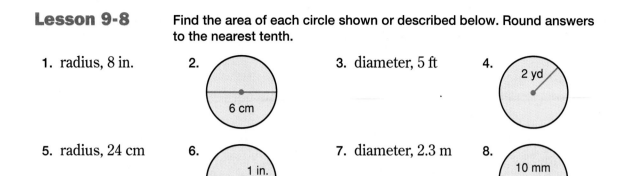 6 cm

3. diameter, 5 ft

4. 2 yd

5. radius, 24 cm

6. 1 in.

7. diameter, 2.3 m

8. 10 mm

Find the length of the radius of each circle given the following areas. Round answers to the nearest tenth.

9. 15 cm^2

10. 24 ft^2

11. 125 in^2

12. 36 yd^2

13. 100 m^2

14. 200 mm^2

15. 72 ft^2

16. 142 in^2

Lesson 10-3

Find the surface area of each rectangular prism. Round answers to the nearest tenth.

1. length, 8 ft
   width, 6.5 ft
   height, 7 ft

2. length, $4\frac{1}{2}$ cm
   width, 10 cm
   height, $8\frac{3}{4}$ cm

3. length, 9.4 yd
   width, 2 yd
   height, 5.2 yd

4. length, 20 mm
   width, 15 mm
   height, 25 mm

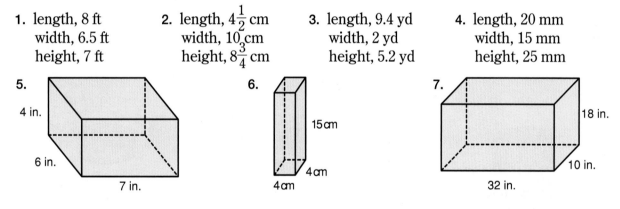

5. 4 in. 6 in. 7 in.

6. 15 cm 4 cm 4 cm

7. 18 in. 10 in. 32 in.

Lesson 10-4

Find the surface area of each cylinder. Use 3.14 for π. Round answers to the nearest tenth.

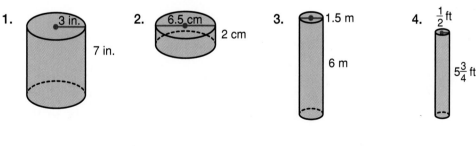

1. 3 in. 7 in.

2. 6.5 cm 2 cm

3. 1.5 m 6 m

4. $\frac{1}{2}$ ft $5\frac{3}{4}$ ft

Find the surface area of each cylinder. Use $\frac{22}{7}$ for π.

5. height, 6 cm
   radius, 3.5 cm

6. height, $5\frac{1}{2}$ in.
   diameter, 3 in.

7. height, 16.5 mm
   diameter, 18 mm

8. height, 22 yd
   radius, 10.5 yd

Lesson 10-5

Find the volume of each rectangular prism.

1. length, 1.5 in.
   width, 3 in.
   height, 6 in.

2. length, 4.5 cm
   width, 6.75 cm
   height, 2 cm

3. length, 3 ft
   width, 10 ft
   height, 2 ft

4. length, 16 mm
   width, 0.7 mm
   height, 12 mm

5. length, 18 cm
   width, 23 cm
   height, 15 cm

6. length, $3\frac{1}{2}$ ft
   width, 10 ft
   height, 6 ft

7. length, 25 mm
   width, 32 mm
   height, 10 mm

8. length, 12 in.
   width, $5\frac{1}{2}$ in.
   height, $3\frac{3}{8}$ in.

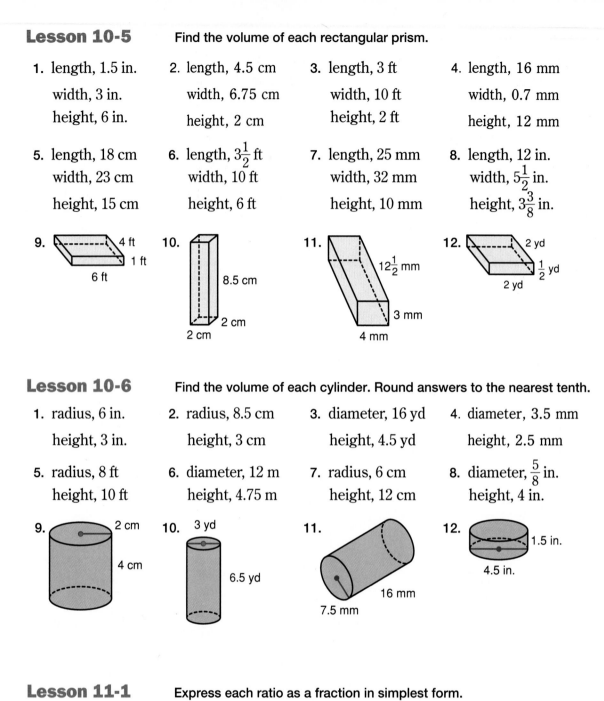

9. 6 ft, 4 ft, 1 ft

10. 8.5 cm, 2 cm, 2 cm

11. $12\frac{1}{2}$ mm, 3 mm, 4 mm

12. 2 yd, 2 yd, $\frac{1}{2}$ yd

Lesson 10-6

Find the volume of each cylinder. Round answers to the nearest tenth.

1. radius, 6 in.
   height, 3 in.

2. radius, 8.5 cm
   height, 3 cm

3. diameter, 16 yd
   height, 4.5 yd

4. diameter, 3.5 mm
   height, 2.5 mm

5. radius, 8 ft
   height, 10 ft

6. diameter, 12 m
   height, 4.75 m

7. radius, 6 cm
   height, 12 cm

8. diameter, $\frac{5}{8}$ in.
   height, 4 in.

9. 2 cm, 4 cm

10. 3 yd, 6.5 yd

11. 7.5 mm, 16 mm

12. 4.5 in., 1.5 in.

Lesson 11-1

Express each ratio as a fraction in simplest form.

1. 45 to 15
2. 64:128
3. 12 weeks out of 15
4. 14 to 49
5. 125:25
6. 18 to 81
7. 33 minutes:60 minutes
8. 16:40
9. 120 to 180
10. 32:64
11. 10 ft to 8 yd
12. 90 to 100

Tell whether the ratios in each pair are equivalent. Show your answer by simplifying.

13. 14 to 77 and 8 to 44
14. $\frac{48}{16}$ and $\frac{1}{3}$
15. 65:13 and 500:100
16. 72 to 90 and 20 to 16
17. 250:100 and 5:2
18. $\frac{32}{2}$ and $\frac{3}{48}$
19. 8 hours to 5 days and 24 hours to 15 days

Lesson 11-2 Express each rate as a unit rate.

1. $240 for 4 days
2. 250 people in 5 buses
3. 500 miles in 10 hours
4. 18 cups for 24 pounds
5. 32 people in 8 cars
6. 3 dozen for $4.50
7. 245 tickets in 5 days
8. 12 classes in 4 semesters
9. 60 people in 4 rows
10. 48 ounces in 3 pounds
11. 20 people in 4 groups
12. 1.5 pounds for $3.00
13. 45 miles in 60 minutes
14. $5.50 for 10 disks
15. 360 miles for 12 gallons
16. $8.50 for 5 yards
17. 24 cups for $1.20
18. 160 words in 4 minutes
19. $60 for 5 books
20. $24 for 6 hours

Lesson 11-3 Solve each proportion.

1. $\frac{4}{9} = \frac{x}{3}$
2. $\frac{12}{m} = \frac{15}{10}$
3. $\frac{36}{90} = \frac{16}{t}$
4. $\frac{g}{32} = \frac{8}{64}$
5. $\frac{5}{14} = \frac{10}{a}$
6. $\frac{k}{18} = \frac{5}{3}$
7. $\frac{120}{150} = \frac{p}{20}$
8. $\frac{15}{w} = \frac{60}{4}$
9. $\frac{81}{90} = \frac{y}{20}$
10. $\frac{14}{s} = \frac{8}{4}$
11. $\frac{h}{3} = \frac{36}{9}$
12. $\frac{44}{8} = \frac{150}{t}$
13. $\frac{42}{8} = \frac{36}{d}$
14. $\frac{125}{v} = \frac{35}{5}$
15. $\frac{u}{72} = \frac{2}{4}$
16. $\frac{45}{80} = \frac{j}{3}$

Lesson 11-4 Tell whether each pair of polygons is similar. Justify your answer.

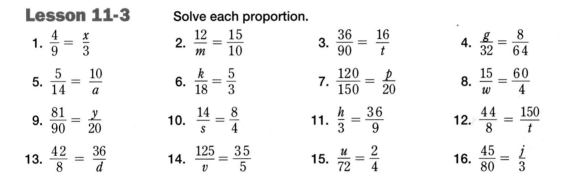

Find the value of *x* in each pair of similar polygons.

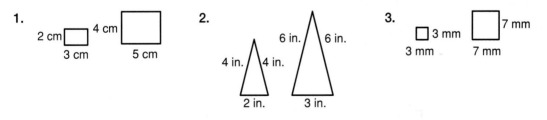

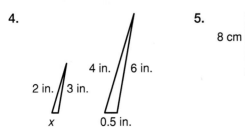

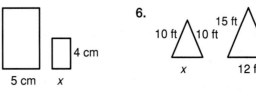

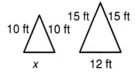

Lesson 11-5

On a map, the scale is 1 inch:50 miles. For each map distance, find the actual distance.

1. 5 inches   2. 12 inches   3. $3\frac{1}{2}$ inches   4. $2\frac{3}{8}$ inches

5. $\frac{4}{5}$ inch   6. $6\frac{3}{4}$ inches   7. $2\frac{5}{6}$ inches   8. 8 inches

On a scale drawing, the scale is $\frac{1}{2}$ inch:2 feet. Find the dimensions of each room in the scale drawing.

9. 14 feet by 18 feet   10. 32 feet by 6 feet

11. 3 feet by 5 feet   12. 20 feet by 30 feet

13. 8 feet by 15 feet   14. 25 feet by 80 feet

Lesson 11-7

Express each ratio as a percent.

1. $\frac{32}{100}$ 2. 48 out of 100   3. 25 hundredths   4. $85\frac{1}{2}$:100

5. $\frac{16}{100}$ 6. 23 out of 100   7. 58.5 hundredths   8. 67:100

9. $\frac{28}{100}$ 10. 54 out of 100   11. 3 hundredths   12. 89.25:100

Write a percent to represent the shaded area. If necessary, round answers to the nearest percent.

13.   14.   15.

Lesson 11-8

Express each fraction as a percent.

1. $\frac{14}{25}$ 2. $\frac{28}{50}$ 3. $\frac{14}{20}$ 4. $\frac{9}{12}$

5. $\frac{4}{6}$ 6. $\frac{3}{8}$ 7. $\frac{7}{10}$ 8. $\frac{17}{17}$

9. $\frac{9}{16}$ 10. $\frac{80}{125}$ 11. $\frac{8}{9}$ 12. $\frac{3}{16}$

Express each percent as a fraction in simplest form.

13. 32%   14. 18.5%   15. 89%   16. 72%

17. $52\frac{1}{4}$%   18. $33\frac{1}{3}$%   19. 11%   20. 1%

21. 28%   22. 55%   23. $26\frac{1}{4}$%   24. $3\frac{1}{3}$%

Lesson 11-9

Express each decimal as a percent.

1. 0.41 2. 0.375 3. 0.916 4. 0.09
5. 1 6. 0.425 7. 0.895 8. 0.0455

Express each percent as a decimal.

9. 67% 10. 43.5% 11. 2.5% 12. $55\frac{1}{5}$%
13. 28.3% 14. 100% 15. 9.05% 16. 39%

Replace each ● with <, >, or = .

17. 24% ● 0.024 18. 0.1 ● 10% 19. $66\frac{2}{3}$% ● 0.66 20. 0.4525 ● 4.525%
21. 38% ● 0.38 22. 1 ● 1% 23. 0.695 ● 695% 24. 2.08% ● 0.028

Lesson 11-10

Express each percent as a decimal.

1. 125% 2. 0.045% 3. 895% 4. 0.000075%
5. 200% 6. 0.001% 7. 0.01345% 8. 555%

Express each number as a percent.

9. $4\frac{1}{4}$ 10. $7\frac{9}{10}$ 11. 3.245 12. 0.003
13. 25 14. 16.74 15. $2\frac{3}{5}$ 16. 900

Replace each ● with <, >, or = .

17. 3.25 ● 325% 18. 2,000% ● 2 19. 45 ● 4.5% 20. 245% ● 2.45
21. $24 \times \frac{1}{4}$ ● 24 × 25% 22. $16 \times 1\frac{1}{3}$ ● $133\frac{1}{3}$% × 16

Lesson 12-1

Use a proportion to solve each problem. Round answers to the nearest tenth.

1. What number is 25% of 280?
2. What number is 32% of 54?
3. 90% of 72 is what number?
4. Find 45% of 125.5.
5. What number is 80% of 500?
6. 12% of 120 is what number?
7. Find 68% of 50.
8. What number is 23% of 500?
9. 20% of $58\frac{1}{2}$ is what number?
10. Find 75% of 1.
11. What number is $33\frac{1}{3}$% of 66?
12. 50% of 350 is what number?
13. Find 80% of 8.
14. What number is $37\frac{1}{2}$% of 32?
15. 95% of 40 is what number?
16. Find 30% of 26.

Lesson 12-3

Write the fraction, decimal, mixed number, or whole number equivalent of each percent that could be used to estimate.

1. 28%
2. 99%
3. 450%
4. 0.09%
5. $\frac{3}{4}$%
6. 65.5%
7. $15\frac{3}{5}$%
8. 39.45%
9. $8\frac{1}{2}$%
10. 48.2%
11. 0.009%
12. 287%

Estimate.

13. 50% of 37
14. 18% of 90
15. 60.5% of 60
16. 300% of 245
17. 0.7% of 200
18. 1% of 48
19. 7% of 24
20. 400% of 13
21. $5\frac{1}{2}$% of 100
22. 40.01% of 16
23. 70% of 300
24. 35% of 35

Lesson 12-4

Write a proportion for each problem. Then solve. Round answers to the nearest tenth.

1. What number is 24% of 60?

2. 38 is what percent of 50?

3. 54 is 25% of what number?

4. What percent of 300 is 50?

5. What number is 65% of 200?

6. $24\frac{1}{2}$% of what number is 15?

7. 99 is what percent of 150?

8. What percent of 240 is 32?

9. 20% of what number is 6?

10. What number is 15.5% of 45?

11. 54 is 40% of what number?

12. What percent of 150 is 30?

13. 68 is $33\frac{1}{3}$% of what number?

14. What number is 85% of 1,000?

15. What percent of 450 is 50?

16. 42 is what percent of 126?

Lesson 12-5

Write an equation for each problem. Then solve. Round answers to the nearest tenth.

1. 12% of what number is 50?
2. Find 45% of 50.
3. 38 is what percent of 62?
4. $28\frac{1}{2}$% of 64 is what number?
5. 5% of what number is 12?
6. 80 is what percent of 90?
7. $66\frac{2}{3}$% of what number is 40?
8. Find 46.5% of 75.
9. 90 is what percent of 95?
10. Find 22% of 22.
11. 16% of what number is 2?
12. 75 is what percent of 300?
13. 75% of 80 is what number?
14. Find 60% of 45.
15. What number is 55.5% of 70?
16. 80.5% of what number is 80.5?

Lesson 12-6

Use the information in the following charts to make a circle graph.

1.

Car Sales by Body Style	
Style	Percent
Sedan	45
Station Wagon	22
Pickup Truck	9
Sports Car	13
Compact Car	11

2.

Favorite Flavor of Ice Cream	
Flavor	Percent
Vanilla	28
Chocolate	35
Strawberry	19
Mint Chip	12
Coffee	6

Lesson 12-7

Find the percent of change. Round to the nearest whole percent.

1. old: $75
   new: $50

2. old: 450
   new: 675

3. old: 3.25
   new: 2.95

4. old: $5.75
   new: $6.25

5. old: 180
   new: 160

6. old: 32.5
   new: 44

7. old: 1.5
   new: 1.0

8. old: 450
   new: 400

9. old: $1,500
   new: $1,200

10. old: 750
    new: 600

11. old: $65
    new: $75

12. old: 380
    new: 320

13. old: 0.75
    new: 1.0

14. old: $3.95
    new: $4.25

15. old: 350
    new: 420

16. old: 500
    new: 100

Lesson 12-8

Find the sales tax or discount to the nearest cent.

1. $45 sweater; 6% tax

2. $18.99 CD; 15% off

3. $39 shoes; $5\frac{1}{2}$% tax

4. $199 ring; 10% off

5. $29 shirt; 7% tax

6. $55 plant; 20% off

Find the total cost or sale price to the nearest cent.

7. $19 purse; 25% off

8. $150 clock; 5% tax

9. $2 notebook; 15% off

10. $145 coat; $6\frac{1}{4}$% tax

11. $89 radio; 30% off

12. $300 table; $\frac{1}{3}$ off

Find the rate of discount to the nearest percent.

13. regular price, $45
    sale price, $40

14. regular price, $250
    sale price, $200

15. regular price, $89
    sale price, $70

Lesson 12-9

Find the interest to the nearest cent for each principal, interest rate, and time.

1. $2,000, 8%, 5 years
2. $500, 10%, 8 months
3. $750, 5%, 1 year
4. $175.50, $6\frac{1}{2}$%, 18 months
5. $236.20, 9%, 16 months
6. $89, $7\frac{1}{2}$%, 6 months
7. $800, 5.75%, 3 years
8. $5,500, 7.2%, 4 years
9. $245, 6%, 13 months

Find the interest to the nearest cent on credit cards for each credit card balance, interest rate, and time.

10. $750, 18%, 2 years
11. $1,500, 19%, 16 months
12. $300, 9%, 1 year
13. $4,750, $19\frac{1}{2}$%, 30 months
14. $2,345, 17%, 9 months
15. $689, 12%, 2 years
16. $390, 18.75%, 15 months
17. $1,250, 22%, 8 months
18. $3,240, 18%, 14 months

Lesson 13-1

Make a tree diagram to show all the outcomes in the sample space. Then give the total number of outcomes.

1. rolling 2 number cubes
2. choosing an ice cream cone from waffle, plain, or sugar and a flavor of ice cream from chocolate, vanilla, or strawberry
3. making a sandwich from white, wheat, or rye bread, cheddar or swiss cheese and ham, turkey, or roast beef
4. flipping a penny twice
5. choosing one math class from algebra and geometry and one foreign language class from French, Spanish, or Latin

Lesson 13-2

Find the total number of outcomes in each situation.

1. choosing a local phone number if the exchange is 234 and each of the four remaining digits is different
2. choosing a way to drive from Millville to Westwood if there are 4 roads that lead from Millville to Miamisburg, 2 roads that connect Miamisburg to Hathaway, and 4 highways that connect Hathaway to Westwood
3. tossing a quarter, rolling a number cube, and tossing a dime
4. spinning the spinners shown below

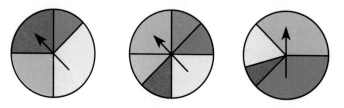

Lesson 13-5

Of 42,000 registered voters, the voting preferences of a random sample of 2,000 are listed in the table at the right.

Candidate	Number of Votes
Brown	540
Kim	380
Andrews	620
Undecided	460

1. How many voters out of the 42,000 might you expect to vote for Kim?
2. How many voters out of the 42,000 might you expect to be undecided at election time?
3. If the undecided voters choose Brown, how many votes might Brown expect to receive?

Lesson 13-6   Find each probability.

1. Two evenly-balanced nickels are flipped. Find the probability that one head and one tail result.

2. A wallet contains four $5 bills, two $10 bills, and eight $1 bills. Two bills are selected without the first selection being replaced. Find $P(\$5, \text{then } \$5)$.

3. Two chips are selected from a box containing 6 blue chips, 4 red chips, and 3 green chips. The first chip selected is not replaced before the second is drawn. Find $P(\text{red, then green})$.

4. A blue die and a red die are rolled. Find the probability that an odd number is rolled on the blue die and a multiple of 3 is rolled on the red die.

Lesson 13-7   Find the value of each expression.

1. $3!$
2. $0!$
3. $6!$
4. $P(5, 3)$
5. $P(6, 6)$
6. $P(10, 2)$
7. $P(5, 0)$
8. $P(3, 2)$

9. How many different five-digit zip codes can be formed if no digit can be repeated?

10. Eight runners are competing in a 100-meter sprint. In how many ways can the gold, silver, and bronze medals be awarded?

11. In a lottery for which 30 tickets were sold (all to different people), in how many ways can the grand prize, second prize, and third prizes be awarded?

Lesson 13-8

Find the value of each expression.

1. $\dfrac{8!}{3!}$

2. $\dfrac{4!}{0!}$

3. $\dfrac{7!}{6!}$

4. $\dfrac{10!}{5!}$

5. $C(5,3)$

6. $C(9,4)$

7. $C(3,3)$

8. $C(10,2)$

9. List all of the possible combinations of Andrew, Jonathon, Megan, Rebecca, and Jeffrey taken 2 at a time. If all of the combinations are equally likely, find the probability that the combination chosen will consist of 2 males.

10. List all the combinations of the digits 1, 3, 5, 7, 9 taken 3 at a time. If all of the combinations are equally alike, find the probability that the combination chosen will contain a 1.

Lesson 14-2

Solve each equation and graph the solution.

1. $3x + 6 = 6$

2. $\dfrac{p}{4} + 5 = 7$

3. $-10 + 2d = 8$

4. $\dfrac{3}{5}k + 2 = 8$

5. $12 - 5w = 3$

6. $5t - 4 = 6$

7. $2q - 6 = 4$

8. $\dfrac{g}{6} + 3 = 9$

9. $15 = 6y + 2$

10. $3s - 4 = 9$

11. $18 - 7f = 4$

12. $13 + 3p = 7$

13. $\dfrac{1}{2}(x - 3) = 2$

14. $4.2 + 7z = 2.8$

15. $-9m - 9 = 9$

16. $32 + 0.2c = 1$

17. $14 - 5t = 14$

18. $\dfrac{r}{6} - 4 = 8$

19. $-\dfrac{1}{4}(r - 2) = 4$

20. $4d - 3 = 9$

21. $16 - 2w = 9$

22. $4k + 13 = 20$

23. $7 = 5 - 2w$

24. $8x + 15 = 14$

25. $92 - 16b = 12$

26. $14e + 14 = 28$

27. $1.1j + 2 = 7.5$

28. $16 - \dfrac{1}{2}k = 10$

29. $4r + 3 = 25$

30. $16 - 5t = 3$

31. $3.5 + 1.5w = 7$

32. $-\dfrac{1}{3} + 5s = -\dfrac{4}{3}$

Lesson 14-3

Find four solutions for each equation. Write your solutions as ordered pairs.

1. $y = 3x + 2$

2. $y = -5x - 3$

3. $y = -\dfrac{1}{2}x - 1$

4. $y = 7x + 1$

5. $y = -3.5x - 0.5$

6. $y = 2x + 8$

7. $y = -2x - 16$

8. $y = -8x$

9. $y = \dfrac{1}{3}x - 1$

10. $y = -2$

11. $y = x$

12. $y = -5x - 5$

13. $y = 13$

14. $y = 3x - \dfrac{3}{2}$

15. $y = 4x$

16. $y = 5x + 15$

17. $y = \dfrac{1}{2}x + 5$

18. $y = -9x + 9$

19. $y = -x$

20. $y = 6x + 2$

21. $y = 6x$

22. $y = -2x + 18$

23. $y = 5x + 1$

24. $y = -8$

25. $y = 4x + 4$

26. $y = -6x + 5$

27. $y = \dfrac{1}{4}x - 3$

28. $y = 7x + \dfrac{1}{2}$

29. $y = 12x + 4$

30. $y = 4 - x$

31. $y = 2x$

32. $y = 5x + 100$

Lesson 14-4

Graph each equation.

1. $y = 3x$

2. $y = 2x + 3$

3. $y = -x$

4. $y = 4x + 2$

5. $y = \frac{1}{2}x + 2$

6. $y = -x + 3$

7. $y = \frac{1}{4}x + 6$

8. $y = -3x + 6$

9. $y = 5x + 2$

10. $y = \frac{1}{3}x$

11. $y = -6$

12. $y = 3x - \frac{1}{3}$

13. $y = 2x + 7$

14. $y = -5x + 1$

15. $y = 13 + x$

16. $y = 5 - \frac{1}{2}x$

17. $y = x - 6$

18. $y = 5x + \frac{3}{2}$

19. $y = 16 - 4x$

20. $y = 4x + 5$

21. $y = -5x + 7$

22. $y = 13 - 7x$

23. $y = -x - 4$

24. $y = 2$

25. $y = 5 - 2x$

26. $y = 0$

27. $y = 11x + 5$

28. $y = -7 - 3x$

29. $y = \frac{1}{4}x$

30. $y = 14 + \frac{1}{2}x$

31. $y = -7x$

32. $y = 5$

Lesson 14-5

Find the output for each function, given the input and the function rule.

1. $f(x) = 2x + 4$
   $x = -2, 0, 1, 5$

2. $f(x) = 16 - 3x$
   $x = -4, -1, 3, 5$

3. $f(x) = \frac{1}{2}x$
   $x = -4, 0, 3, 8$

4. $f(x) = 5x$
   $x = -6, -4, 1, 7$

5. $f(x) = -2$
   $x = -10, -5, 0, 6$

6. $f(x) = 1.5 + 3x$
   $x = 0.5, 1.2, 3, 5$

7. $f(x) = -\frac{1}{3}x - 1$
   $x = -3, -1, 2, 6$

8. $f(x) = 2.5x$
   $x = -4, -1, 3, 6$

9. $f(x) = 3x - 6$
   $x = -15, -7, 8, 20$

10. $f(x) = -x$
    $x = -4, 3, 7, 15$

11. $f(x) = \frac{5}{4}x + 2$
    $x = -6, -4, 3, 7$

12. $f(x) = 0.25x - 0.5$
    $x = -1.5, -0.25, 1, 4$

Lesson 14-6

Graph each triangle and its transformation. Write the ordered pairs for the vertices of the new triangle.

1. $\triangle ABC$ with vertices $A(-4, 3)$, $B(2, -1)$, and $C(0, 5)$ translated 3 units left and 4 units down

2. $\triangle DEF$ with vertices $D(5, 2)$, $E(-1, -1)$, and $F(3, 4)$ reflected over the x-axis

3. $\triangle GHI$ with vertices $G(0, 7)$, $H(5, 0)$, and $I(-2, -4)$ translated 2 units right and 3 units up

4. $\triangle JKL$ with vertices $J(-4, -4)$, $K(4, -4)$, and $L(0, 0)$ reflected over the y-axis

5. Rectangle $PQRS$ with vertices $P(3, 5)$, $Q(-4, 5)$, $R(-4, -1)$, and $S(3, -1)$ translated 1 unit down and 4 units left

Glossary

absolute value (254)   The number of units a number is from zero on the number line.

acute angle (297)   Any angle that measures between 0° and 90°.

addition property of equality (225)   If you add the same number to each side of an equation, the two sides remain equal. If $a = b$, then $a + c = b + c$.

additive inverse (265)   Two integers that are opposites of each other are called additive inverses. The sum of any number and its additive inverse is zero, $a + (-a) = 0$.

algebra (28)   A mathematical language that uses letters along with numbers. The letters stand for numbers that are unknown. $10n - 3 = 17$ is an example of an algebra problem.

algebraic expression (28)   A combination of variables, numbers, and at least one operation.

area (243)   The number of square units needed to cover a surface.

arithmetic sequence (136)   A sequence of numbers in which you can find the next term by adding the same number to the previous term.

associative property of addition (204)   For any numbers a, b, and c, $(a + b) + c = a + (b + c)$.

associative property of multiplication (204)   For any numbers a, b, and c, $(a \times b) \times c = a \times (b \times c)$.

average (104)   The sum of two or more quantities divided by the number of quantities.

base (32)   The number used as a factor. In 10^3, 10 is the base.

base (244)   Any side of a parallelogram.

base (383)   The faces on the top and the bottom of a three-dimensional figure.

base (454)   In a percent proportion, the number to which the percentage is compared.

bisect (298)   To divide something into two congruent parts.

cell (36)   Each section of a spreadsheet. A cell can contain data, labels, or formulas.

center (197)   The middle point of a circle or sphere. The center is the same distance from all points on the circle or sphere.

circle (197)   The set of all points in a plane that is the same distance from a given point called the center.

circumference (197)   The distance around a circle.

cluster (101)   Data that are grouped closely together.

clustering (54)   A method used to estimate decimal sums and differences by rounding a group of closely related numbers to the same whole number.

combination (522)   An arrangement of objects in which order is unimportant.

common denominator (164)   A common multiple of the denominators of two or more fractions.

commutative property of addition (204)   For any numbers a and b, $a + b = b + a$.

commutative property of multiplication (204)   For any numbers a and b, $a \times b = b \times a$.

compatible numbers (11)   Two numbers that are easy to divide mentally. They are often members of fact families.

compensation (14)   A strategy for computing exact sums or differences mentally.

composite number (132)   Any whole number greater than one that has more than two factors.

congruent angles (298)   Two angles that have the same measure.

congruent sides (307)   Sides that have the same length.

coordinate system (259) Two perpendicular number lines that intersect at their zero points form a coordinate system.

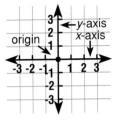

cross products (417)   If the cross products in a ratio are equal then the ratio forms a proportion. In the proportion $\frac{2}{3} = \frac{8}{12}$, the cross products are 2×12 and 3×8.

cup (239)   A customary unit of capacity equal to 8 fluid ounces.

cylinder (388)   A three-dimensional figure with two parallel congruent circular bases.

D **data base** (96)   A collection of data that is organized and stored on a computer for rapid search and retrieval.

decagon (304)   A polygon having ten sides.

degree (297)   The most common unit of measurement for angles.

dependent event (511)   Two or more events in which the outcome of one event does affect the outcome of the other event or events.

diameter (197)   The distance across a circle through its center.

diameter

discount (479)   The amount deducted from the original price.

distributive property (205)   The sum of two addends multiplied by a number is the sum of the product of each addend and the number. $a \times (b + c) = a \times b + a \times c$.

divisible (129)   A number is divisible by another if the quotient is a whole number and the remainder is zero.

division property of equality (228)   If each side of an equation is divided by the same nonzero number, then the two sides remain equal. If $a = b$, then $\frac{a}{c} = \frac{b}{c}$, $c \neq 0$.

dodecagon (304)   A polygon having twelve sides.

E **equation** (38)   A mathematical sentence that contains an equals sign, $=$.

equiangular (313)   A polygon with equal angles.

equilateral triangle (308) A triangle with three congruent sides.

equivalent equations (225)   Two or more equations with the same solution. $x + 3 = 5$ and $x = 2$ are equivalent equations.

evaluate (28)   To find the value of an expression by replacing variables with numerals.

event (157)   A specific outcome or type of outcome.

expected value (201)   The average value that one would expect to get over many attempts.

experimental probability (500)   An estimated probability based on the relative frequency of positive outcomes occurring during an experiment.

exponent (32)   The number of times the base is used as a factor. In 10^3, the exponent is 3.

exterior angle (314) If you extend a side of a polygon, an exterior angle is formed.

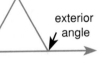
exterior angle

F **face** (383) Any surface that forms a side or a base of a prism.

factor (32, 129)   When two or more numbers are multiplied, each number is a factor of the product.

factorial (519)   The expression $n!$ is the product of all counting numbers beginning with n and counting backwards to 1.

field (96)   The elements within each record of a computer data base file.

file (96)   A collection of data within a computer data base about a particular subject.

frequency table (93)   A table for organizing a set of data that shows the number of times each item or number appears.

front-end-estimation (8)   A method used to estimate decimal sums and differences by adding or subtracting the front-end digits, then adjusting by estimating the sum or difference of the remaining digits, then adding the two values.

function (547)   A relationship in which the output value depends upon the input according to a specified rule. For example, with a function $f(x) = 2x$, if the input is 5, the output is 10.

Fundamental Counting Principle (497)   If there are m ways of selecting an item from set A and n ways of selecting an item from set B, then there are $m \times n$ ways of selecting an item from set A and an item from set B.

G **gallon** (239)   A customary unit of capacity equal to 4 quarts.

geometric sequence (136)   A sequence of numbers in which you can find the next term by multiplying the previous term by the same number.

gram (78)   The basic unit of mass in the metric system.

greatest common factor (GCF) (145)   The greatest of the common factors of two or more numbers. The greatest common factor of 18 and 24 is 6.

H **height** (244)   The vertical distance from the base of a parallelogram to its other side.

heptagon (304)   A polygon having seven sides.

hexagon (304)   A polygon having six sides.

hypotenuse (344)   In a right triangle, the side opposite the right angle is called the hypotenuse.

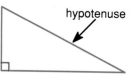

identity property of addition (204)   For any number a, $a + 0 = a$.

identity property of multiplication (204)   For any number a, $a \times 1 = a$.

improper fraction (174)   A fraction that has a numerator that is greater than or equal to the denominator.

independent event (510)   Two or more events in which the outcome of one event does not affect the outcome of the other event or events.

integers (254)   The whole numbers and their opposites. . . . , $-3, -2, -1, 0, 1, 2, 3, \ldots$

interest (482)   The amount charged or paid for the use of money.

interval (98)   The difference between successive values on a scale.

inverse property of multiplication (204)   The product of a number and its multiplicative inverse is 1. For all fractions $\frac{a}{b}$, where $a, b \neq 0$, $\frac{a}{b} \times \frac{b}{a} = 1$.

inverse operation (220)   Pairs of operations that undo each other. Addition and subtraction are inverse operations. Multiplication and division are inverse operations.

irregular figures (351)   Figures that do not necessarily have straight sides and square corners.

isosceles triangle (308)   A triangle that has at least two congruent sides.

L **leaf** (109)   The second greatest place value of data in a stem-and-leaf plot.

least common denominator (LCD) (164)   The least common multiple of the denominators of two or more fractions.

least common multiple (LCM) (161)   The least of the common multiples of two or more numbers, other than zero. The least common multiple of 2 and 3 is 6.

leg (344)   A leg of a right triangle is either of the two sides that form the right angle.

line plot (101)   A vertical graph showing a picture of information on a number line.

linear equation (543)   An equation for which the graph is a straight line.

line symmetry (327)   Figures that match exactly when folded in half have line symmetry.

line of symmetry (327)
A fold line on a figure that shows symmetry. Some figures can be folded in more than one way to show symmetry.

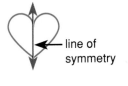
line of symmetry

liter (78)   The basic unit of capacity in the metric system. A liter is a little more than a quart.

mean (104)   The arithmetic average; the sum of the numbers in a set of data divided by the number of pieces of data.

median (104)   The middle number when a set of data are arranged in numerical order. When there are two middle numbers, the median is their mean.

meter (78)   The basic unit of length in the metric system.

metric system (78)   A system of weights and measures based on tens. The meter is the basic unit of length, the kilogram is the basic unit of weight, and the liter is the basic unit of capacity.

mixed number (174)   A number that shows the sum of a whole number and a fraction. $6\frac{2}{3}$ and $8\frac{3}{4}$ are mixed numbers.

mode (104)   The number or item that appears most often in a set of data.

multiplication property of equality (228)   If each side of an equation is multiplied by the same number, then the two sides remain equal. If $a = b$, then $ac = bc$.

multiplicative inverse (204)   A number times its multiplicative inverse is equal to 1. The multiplicative inverse of $\frac{2}{3}$ is $\frac{3}{2}$.

negative integer (254)   Whole numbers to the left of zero on the number line or numbers less than zero.

nonagon (304)   A polygon having nine sides.

obtuse (297)   Any angle that measures between 90° and 180°.

octagon (304)   A polygon having eight sides.

opposite (254)   Two integers are opposites if they are represented on the number line by points that are the same distance from zero, but on opposite sides of zero. The sum of opposites is zero.

ordered pair (259)   A pair of numbers where order is important. An ordered pair, which is graphed on a coordinate plane, is written in this form: (*x*-coordinate, *y*-coordinate).

order of operation (24)   The rules to follow when more than one operation is used. 1. Do all operations within grouping symbols first. 2. Do multiplication and division from left to right. 3. Do addition and subtraction from left to right.

origin (259)   The point of intersection of the *x*-axis and *y*-axis in a coordinate system.

ounce (238)   A customary unit of weight. 16 ounces equals 1 pound.

outliers (101)   Data that are far apart from the rest of the data.

parallelogram (244)   A quadrilateral that has both pairs of opposite sides parallel.

pentagon (304)   A polygon having five sides.

percent (433, 454)   A ratio that compares a number to 100.

percentage (454)   In a percent proportion, a number (P) that is compared to another number called the base (B).

$$\frac{\text{Percentage}}{\text{Base}} = \text{Rate or } \frac{P}{B} = \frac{r}{100}.$$

percent of decrease (477)

$$\frac{\text{Amount of Decrease}}{\text{Original Value}} \times 100$$

percent of increase (477)

$$\frac{\text{Amount of Increase}}{\text{Original Value}} \times 100$$

perfect square (338)   Squares of whole numbers.

perimeter (194)   The distance around a geometric figure.

permutation (517)   An arrangement or listing of objects in which order is important.

pint (239)   A customary unit of capacity equal to 2 cups.

polygon (303)   A simple closed figure in a plane formed by three or more line segments.

population (506)   The entire group of items or individuals from which the samples under consideration are taken.

population density (414)   The population per square mile.

positive integers (254)   Whole numbers greater than zero or to the right of zero on the number line.

pound (238)   A customary unit of weight equal to 16 ounces.

power (32)   A number expressed using an exponent. The power 7^3 is read *seven to the third power*, or *seven cubed*.

prime factorization (132)   A composite number that is expressed as the product of prime numbers. The prime factorization of 12 is $2 \times 2 \times 3$.

prime number (132)   A number that has exactly two factors, 1 and the number itself.

principal (482)   The amount of an investment or a debt.

prism (383)   A three-dimensional figure that has two parallel and congruent bases in the shape of polygons.

probability (157)   The ratio of the number of ways an event can occur to the number of possible outcomes; how likely it is that an event will occur.

proper fraction (174)   A fraction that has a numerator that is less than the denominator.

proportion (417)   A proportion is an equation that shows that two ratios are equivalent, $\frac{a}{b} = \frac{c}{d}, b \neq 0, d \neq 0$.

Pythagorean Theorem (344)   In a right triangle, the square of the length of the hypotenuse is equal to the sum of the squares of the lengths of the legs. $a^2 + b^2 = c^2$

Q **quadrant** (259)   One of the four regions into which two perpendicular number lines separate a plane.

	y - axis
Quadrant III	Quadrant I
O	x - axis
Quadrant III	Quadrant IV

quadrilateral (303)   A polygon having four sides.

quart (239)   A customary unit of capacity equal to 2 pints.

R **radical sign** (339)   The symbol used to represent a nonnegative square root is $\sqrt{}$.

radius (197)   The distance from the center of a circle to any point on the circle.

radius

random (157)   Outcomes occur at random if each outcome is equally likely to occur.

random (506)   A sample is called random if the members of the sample are selected purely on the basis of chance.

range (98)   The difference between the greatest number and the least number in a set of data.

rate (414)   A ratio of two measurements with different units.

rate (454)   In a percent proportion, the ratio of a number to 100.

rate (482)   The percent charged or paid for the use of money.

ratio (411)   A comparison of two numbers by division. The ratio comparing 2 to 3 can be stated as 2 out of 3, 2 to 3, 2:3, or $\frac{2}{3}$.

reciprocal (204)   Another name for a multiplicative inverse.

record (96)   The subject or sub unit within a computer data base file.

rectangle (243)   A parallelogram with all angles congruent.

rectangular prism (383) A prism with rectangles as bases.

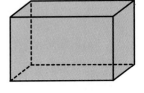

reflection (327)   A mirror image of a figure across a line of symmetry.

regular polygon (313)   A polygon that is both equiangular and equilateral.

repeating decimal (154)   A decimal whose digits repeat in groups of one or more. Examples are 0.181818. . . and 0.83333. . .

rhombus (308)   A parallelogram with all sides congruent.

right angle (297)   Any angle that measures exactly 90°.

S **sales tax** (479)   A tax based on the amount received for articles sold.

sample (506)   A small part or piece of anything that shows what the whole is like.

sample space (492)   The set of all possible outcomes.

scale (98)   The set of all possible values of a given measurement, including the least and greatest numbers in the set, separated by the intervals used.

scale drawing (426)   A representation of something that is too large or too small to be drawn to actual size.

scalene triangle (307)   A triangle with no congruent sides.

scientific notation (67)   A way of expressing numbers as the product of a number that is at least 1, but less than 10, and a power of ten. In scientific notation 5,500 is 5.5×10^3.

sequence (136)   A list of numbers in a specific order.

similar polygons (422)   Two polygons are similar if their corresponding angles are congruent and their corresponding sides are in proportion. They have the same shape but may not be the same size.

simplest form (150)   The form of a fraction when the GCF of the numerator and denominator is 1. The fraction $\frac{1}{4}$ is in simplest form because the GCF of 1 and 4 is 1.

solution (38)   Any number that makes an equation true. The solution for $n + 10 = 15$ is 5.

solve (38)   To replace a variable with a number that makes an equation true.

spreadsheet (36)   A computer spreadsheet organizes numerical data into rows and columns. It is used for organizing and analyzing data and formulas.

square (338)   The product of a number and itself. $7^2 = 7 \times 7 = 49$.

square root (339)   One of the two equal factors of a number. If $a^2 = b$, then a is the square root of b. The square root of 144 is 12 because $12^2 = 144$.

stem (109)   The greatest place value of data in a stem-and-leaf plot.

stem-and-leaf plot (109)   A system used to condense a set of data where the greatest place value of the data forms the stem and the next greatest place value forms the leaves.

straight angle (297)   Any angle that measures exactly 180°.

subtraction property of equality (225)   If you subtract the same number from each side of an equation, then the two sides remain equal. If $a = b$, then $a - c = b - c$.

surface area (383)   The sum of the areas of all the faces of a three-dimensional figure.

T

terminating decimal (154)   A quotient in which the division ends with a remainder of zero. 0.25 and 0.125 are terminating decimals.

tessellation (321)   A repetitive pattern of polygons that fit together with no holes or gaps.

theoretical probability (500)   The long-term probability of an outcome based on mathematical principles.

tiling (321)   Covering a surface with regular figures.

time (482)   When used to calculate interest, time is given in years.

ton (238)   A customary unit of weight equal to 2,000 pounds.

translation (324)   A method used to make changes in the polygons of tessellations by sliding a pattern to create the same change on opposite sides.

trapezoid (307, 355)   A quadrilateral with exactly one pair of parallel sides.

tree diagram (492)   A diagram used to show the total number of possible outcomes in a probability experiment.

triangle (355)   A polygon that has three sides.

U

undecagon (304)   A polygon having eleven sides.

unit price (414)   The cost of an item and its unit rate.

unit rate (414)   A rate in which the denominator is 1 unit.

V

variable (28)   A symbol, usually a letter, used to represent a number in mathematical expressions or sentences. In $3 + a = 6$, a is a variable.

vertex (297)   A vertex of an angle is the common endpoint of the rays forming the angle.

vertex

volume (394)   The number of cubic units needed to fill a space.

X

***x*-axis** (259)   The horizontal line of the two perpendicular number lines in a coordinate plane.

***x*-coordinate** (259)   The first number of an ordered pair.

Y

***y*-axis** (259)   The vertical line of the two perpendicular number lines in a coordinate plane.

***y*-coordinate** (259)   The second number of an ordered pair.

Selected Answers

1 Tools for Problem Solving

Pages 6-7 Lesson 1-1
4. 1,609 miles   5. yes   7. 58 tables
9. $191.40   11. 498 miles
13. Sample answer: about 530 million radios

Pages 9-10 Lesson 1-2
4. 1,560   5. 310   6. 1,900   7. 2,200   8. $105
9. 12,500   10. 12,300   11. 600   12. Yes, it is
reasonable.   13. 10,500   15. 9,200
17. 14,900   19. $18   21. 8,800   23. 24,300
25. false   26. 20 days   27. 30-ounce bottle; $6
as opposed to $5.   28. no   29. $190   31. no

Pages 12-13 Lesson 1-3
4. 200   5. 100   6. 90   7. 80   8. 20   9. 20
10. 10   11. 30   12. 200   13. 80   14. 40
15. 50   16. about 40 miles per gallon   17. 100
19. 300   21. 10   23. 20   25. 8   27. 600
29. 60   31. 300   33. 200   35. 16,350 seats
36. 200 tons   37. 9,600
38. about 220 campers   39. about $2,860
41. about 400 pounds

Pages 15-16 Lesson 1-4
4. $70 + 23$ 5. $6.00 + $3.25   6. $150 + 64$
7. $189 - 55$ 8. $149 - 46$ 9. $6.99 - $4.27
10. 91   11. 193   12. $8.25   13. 27   14. 335
15. $3.63   17. 111   19. 22   21. 823
23. 952   25. 154   27. 279   29. 4,584
31. $1.03   33. 1,163   35. 458   36. 48 miles
37. about 110 representatives   38. Yes, since
the estimate is $71.   39. about 6   40. about 7
hours   41. 225 squares

Pages 18-19 Lesson 1-5
4. exact   5. estimate   6. estimate   7. exact
8. exact   9. estimate   11. 36 miles per gallon
13. Sample answer for 1, 2, 3, 4, 5: 245×13
15. about 16 cubic miles

Page 19 Mid-Chapter Review
1. Sample answer: 120 pounds ÷ 6 = 20 quarts
3. 3,500   5. 90   7. 216   9. 108 meters

Page 21 Lesson 1-6
3. Not enough facts; the cost of each video.
4. $40.95   5. Not enough facts; whether or not
it is a leap year.   7. $1.26   9. Not enough facts;
which socks he bought.

Pages 25-26 Lesson 1-7
4. multiplication   5. addition   6. subtraction
7. addition   8. multiplication   9. addition
10. 23   11. 4   12. 28   13. 24   14. 13
15. 12   17. subtraction   19. multiplication
21. addition   23. 2   25. 8   27. 6   29. 5
31. 7   33. 25   35. 9   37. $(16 + 5) \times 4 \div 2 =$
42   39. $(36 \div 3 - 9) \div 3 = 1$ 41. 3,840 cans
42. about $17,100   43. about 4   44. 415
45. 18 pounds   47. $48   49.a. 19   b. 20
c. 32   d. 3   e. 12   f. 225

Pages 30-31 Lesson 1-8
4. 8   5. 1   6. 2   7. 18   8. 10   9. 5   10. 12
11. 9   12. 5   13. 6   14. 5   15. 20   17. 21
19. 7   21. 4   23. 16   25. 2   27. 12
29. 101 discs   31. 10   33.a. $740m$
b. 3,700 mph

Pages 33-35 Lesson 1-9
4. $2 \cdot 2 \cdot 2 \cdot 2$ 5. $7 \cdot 7 \cdot 7 \cdot 7 \cdot 7$ 6. $12 \cdot 12 \cdot 12$
7. $9 \cdot 9 \cdot 9 \cdot 9 \cdot 9 \cdot 9 \cdot 9$ 8. 6^3 9. 15^4 10. a^6
11. 625   12. 64   13. 256   14. 512
15. $100,000,000   17. $9 \cdot 9 \cdot 9$ 19. $n \cdot n \cdot n \cdot n$
$\cdot n \cdot n \cdot n \cdot n$ 21. 12^2 23. 49   25. 243
27. 81   29. 81   31. 36   33. 125   35. true.
37. true   39. 625   41. 121   43. 128
44. about 30 miles per gallon   45. 34 years old
46. 39   47. 56 pens and pencils   48. 36
49.a. $7y$ b. 84 years old   51. 10^{18}
55. $999,800,000,000

Pages 40-41 Lesson 1-10
4. false   5. true   6. true   7. false   8. 13
9. 41   10. 81   11. 20   12. 7   13. 18
14. 143   15. 44   16. 80   17. 37   18. 50
mph   19. 4   21. 7   23. 18   25. 96   27. 56
29. 86   31. 48   33. 173   35. 112   37. 146

39. 72 **41.** 22 **43.** 17 **45.** 154 **46.** 238
47. $246 **48.** false **49.a.** $75 + ($1)n
b. $120 **50.** 4^4 **51.** 216,000 seconds
53. 84 cm

Pages 42-44　Study Guide and Review
7. 420 miles **9.** 24,400 **11.** 650
13. 310,000 **15.** 200 **17.** 70 **19.** 123
21. 9,334 **23.** 7,326 **25.** 25 **27.** 10 **29.** 49
31. 10 **33.** 40 **35.** 13 **37.** 729 **39.** 256
41. 47 **43.** 100 **45.** Conchita, Emma,
Barbara, Doris, Anna

2　Applications with Decimals

Pages 49-50　Lesson 2-1
4. >

0.51　0.56
0.5　0.55　0.6

6. <

0.97　1.06
0.95　1.0　1.05　1.1

7. > **8.** = **9.** <

11. <

0.23　0.29
0.20　0.25　0.30

15. >

2.49　2.50
2.48　2.50　2.52

17. > **19.** = **21.** < **23.** < **25.** <
27. > **29.** 5.009, 5.07, 5.13 **31.** 0.087, 0.901,
1.001, 2 **33.** $250 **34.** 485 **35.** Maria is 18
years old. **36.** 86,698 **39.a.** 19.924, 19.875,
19.837, 19.837 **39.b.** Daniela Silivas
39.c. Phoebe Mills and Gabriela Potorac
39.d. They are tied.

Pages 52-53　Lesson 2-2
4. 0.3 **5.** 0.25 **6.** 7.038 **7.** 17.5 **8.** 8
9. 12.13 **10.** 23 **11.** 0.24 **12.** 16.5
13. 0.22 **14.** 709.1 **15.** 0.085 **17.** 0.4
19. 1.0 **21.** 0.2 **23.** 15.5 **25.** 4.530
27. 0.8 **29.** 60 **31.** 1.70
33.

3.67
3.0　3.5　4.0

34. $730 **35.** 93 items **36.** 64 **37.** 18
38. 6.32, 8.75, 8.78, 9, 9.15, 10.29 **39.a.** 9 cm
39.b. 9.2 cm **41.** 7.22 in^3

Pages 55-57　Lesson 2-3
4. 15 **5.** 30 **6.** 28 **7.** 7 **8.** 240 **9.** 60
10. 159 **11.** 5 **12.** 12.40 **13.** 3 **14.** 25
15. 6 **16.** 34% **17.** $19 **19.** 10 **21.** 64
23. 9 **25.** 600 **27.** 2,400 **29.** 200 **31.** 600
33. 7,200 **35.** 244 cards **36.** $500 + 5n$
37. 15 **38.** < **39.** 5.7 miles **41.** about 5
times faster **43.** about 12,000 miles

Page 59　Review
1. 6.4 **3.** 0.24 **5.** $11.77 **7.** 1.747
9. 180.75 **11.** 8.64 **13.** 11.18 **15.** 37.632
17. 51.525 **19.** 13.59 **21.** 10.72 **23.** 12.43
25. 28.601 **27.** 168 **29.** 3.64
31. 268.3 million people

Pages 62-63　Lesson 2-4
4. 5.28 **5.** 0.77 **6.** 19.728 **7.** 0.28 **8.** 26.52
9. 0.1845 **10.** 5.405 **11.** 0.009 **12.** 0.53248
13. 1.14 **14.** 1.26 **15.** 15.12 **17.** 0.27
19. 0.0736 **21.** 1.53 **23.** 8.814 **25.** 9.075
27. 21.93 **29.** 0.0001 **31.** 0.1053 **33.** 0.72
35. 0.00589 **37.** $0.06 **38.** 32
39.

3.7　3.77
3.7　3.75　3.8

40. 26 miles **41.** 2.21 **43.** $3.38
45. 4382.88 days

Pages 65-66　Lesson 2-5
3. 234 **4.** 0.8 **5.** 2.8 **6.** 140,000 **7.** 125.3
8. 605 **9.** 315.9 **10.** 20,310 **11.** 78
12. 1,320 **13.** 5 **15.** 2,780 **17.** 92.5
19. 780 **21.** 6,894 **23.** 9,300 **25.** $73.00
26. 2,000 **27.** 0.03, 0.3, 0.33, 3, 3.03, 3.33
28. 63 **29.** 60 mph **31.** $12.50
33.a. 1×10^6 **33.b.** 6.07×10^2
33.c. 3.9256×10^7

Page 66　Mid-Chapter Review
1. > **3.** > **5.** 6.5 **7.** 3.1 **9.** $2 **11.** 7
13. 0.63 **15.** 21.812 **17.** $290

Pages 68-69   Lesson 2-6

3. 8.9×10^2 **4.** 4.3×10^3 **5.** 6.235×10^3
6. 5.2×10^4 **7.** 8.2×10^5 **8.** 1.264×10^8
9. 9,870   **10.** 600   **11.** 17,500   **12.** 23,000
13. 495,000,000   **14.** 570,000   **15.** 7.5×10^3
17. 4.07×10^4 **19.** 4×10^5 **21.** 7.9×10^6
23. 1.6×10^8 **25.** 14,200   **27.** 547,000
29. 27,100,000   **31.** 602,400,000
33. 1.1×10^6 **35.** 315 cards   **36.** 13
37. 112   **38.** 18.17 min   **39.** 18,700
41.a. Mauna Kea   **41.b.** 4,500 feet

Pages 72-74   Lesson 2-7

5. $3.6 \div 4$ **6.** $10.5 \div 7$ **7.** $44 \div 11$
8. $2,940 \div 84$ **9.** $18.9 \div 9$ **10.** $5,040 \div 56$
11. 5   **12.** 3.5   **13.** 1.5   **14.** 8.2   **15.** 63.75
16. 0.35   **17.** $8.2 \div 4$ **19.** $26 \div 13$
21. $14.88 \div 31$ **23.** 8   **25.** 70   **27.** 140
29. 0.91   **31.** 0.046   **33.** 0.5   **35.** 0.088
37. 12   **39.** 3.4   **41.** 0.65   **42.** no   **43.** 10
44. $147.27   **45.** 120   **46.** 6.35×10^5
47. 2.6 inches   **49.** 1,000 bacteria

Pages 76-77   Lesson 2-8

3. 41.9   **4.** 21.7   **5.** 400.0   **6.** 400,000
7. 10,000   **8.** 300,000   **9.** $1.38   **10.** $11.24
11. $2.13   **12.** 645.65   **13.** 0.26   **15.** 4.36
17. 50,000   **19.** $913.40   **21.** $185.58
23. 4,000,000   **24.** about 30,000 inches
25. 4.25 years old   **26.** 30,750
27. $0.05 per ounce   **29.** 541 times larger

Page 80   Lesson 2-9

3. 55   **4.** 43,800   **5.** 0.814   **6.** 16,500
7. 5,000   **8.** 32,000   **9.** 89,000   **10.** 67,100
11. 600   **12.** 1,010 grams   **13.** 56 cm
15. 580   **17.** 9   **19.** 6.7   **21.** 0.080
23. 73,800   **25.** 8,100   **27.** 0.047   **29.** 70 mL
31. 14,000 mg   **32.** 256   **33.** 3.19   **34.** 6.48
35. $42.67 per second

Pages 82-83   Lesson 2-10

3. No, $3,000 \times 7 = 21,000$ **4.** $20   **5.** no
7. 7 miles   **9.** 43.9 seconds   **11.** 3.5 miles
13. 923 points   **15.** 3 quarts   **17.** 34¢
19. about 16 times

Pages 84-86   Study Guide and Review

7. 3.04, 3.15, 3.7, 3.9, 4.2   **9.** 0.015, 0.105, 0.149,
0.15, 0.501   **11.** 0.257, 2.04, 2.046, 25.7, 26.04
13. 13.27   **15.** 0.1   **17.** 257.20   **19.** 0.002
21. 350   **23.** 36   **25.** 30   **27.** 0.26
29. 22.725   **31.** 13,700   **33.** 63.7   **35.** 6×10^3
37. 13,700   **39.** 10   **41.** 0.004   **43.** 1,000
45. 2.7   **47.** 0.027   **49.** 6,850   **51.** 160
53. 0.043   **55.** Thomas is lower.   **57.** no

3 Statistics and Data Analysis

Pages 91-92   Lesson 3-1

3. 400 thousand women   **4.** 1988   **5.** Sample
answer: Yes, until the recession ends.
7. $279.55   **9.** 11 cans   **11.** more   **13.** $3.20

Pages 94-95   Lesson 3-2

3. 1, 5, 10, 1, 6, 3   **4.b.** 25   **5.a.** 30 seconds
5.b. 60 seconds   **7.** $93.72   **9.** 16 people
11. Sample answer: 245×13 **13.** 1.5 inches

Pages 99-100   Lesson 3-3

4. 2 to 10, 2   **5.** 1 to 21, 4   **6.** 50 to 70, 5
7. 100 to 160, 20   **8.** 8   **9.** 19   **10.** 50
11. 444   **12.** 5.5   **13.** 10   **15.** 100   **17.** 1,000
19.a. Sample answer: 30 to 150, 20   **20.** $26.00
21. 9.66 minutes   **22.** 2,330 m   **23.** Range
must remain constant; scale and interval can vary
somewhat depending on an individual's choice.
25.a. Walt Disney

Pages 102-103   Lesson 3-4

14. 570 miles   **15.** Sample answer: scale, 0 to
40; interval, 5; range, 33

Pages 106-107   Lesson 3-5

3. 12, 13, 15, 17, 17, 18, 20; 17, 17, 16   **4.** 2, 3, 4,
5, 6, 7, 7, 8; 7, 5.5, 5.25   **5.** 90, 90, 91, 92, 94, 94,
95, 98; 90 and 94, 93, 93   **7.** 65, 65, 65.5
9. 1,755 and 1,805; 1,780; 1,780   **11.** 3.25×10^4
12.

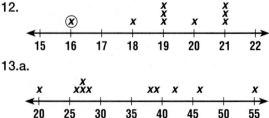

13.a.

13.b. 27, 33, 34.8 **15.a.** 89, 84, 84.7
15.b. the mode, 89

Page 107   Mid-Chapter Review

1. greater **3.** 9
5.

```
                          x
              x   x   x   x
      x       x   x   x   x   x
  x       x   x   x   x   x   x   x
◄──┼───┼───┼───┼───┼───┼───┼───┼───┼───┼──►
  11  12  13  14  15  16  17  18  19  20
```

7. Mean, it includes all scores.

Pages 110-111   Lesson 3-6

3. 1, 2, 3, 4, 5 **4.** 0, 1, 2, 3, 4
5. 8, 9, 10, 11, 12, 13

6.a.
```
1 | 6      6.b. 4   6.c. 30's
2 | 3 5 9
3 | 2 5 5 6
4 | 1 5
5 |
6 | 7      1 | 6 means 16 years.
```

7.
```
2 | 0 4 7
3 | 4 5 6 6 8
4 | 3 5 7
5 | 3 4 8 8
7 | 8      2 | 7 means 27.
```

9. 36°, 6° **11.** 110 **12.** $12, $12, $10

13.a.
```
0 | 2 2 3 4 4 6 9
1 | 1 5
2 | 2      1 | 5 means 15 points.
```

13.b. 3 **13.c.** 2 and 4, 5, 7.8; median

Pages 114-115   Lesson 3-7

3. Ana **4.** about 30 **5.** black **7.** 21.6, 23, 25.9, 27.4, 30.2

8.
```
6 | 0
7 | 2 5 6
8 | 1 3 6 8
9 | 14     6 | 0 means 60
```

9.a. about 60 million **9.b.** Sample answer: popularity of baseball compared to other sports.

Pages 117-119   Lesson 3-8

3. A, it makes the scores look higher. **4.** It starts at 80 instead of 0. **5.** No, this is not a representative group. **7.a.** $5, $10, $11.05
7.b. mean **7.c.** median **9.** rock

Pages 120-122   Study Guide and Review

15. none, 24, 24.4 **17.** none; 70,500; 74,166.7
21. The 29 brought the mean down, the other six scores were well above 81.6. **23.** greater
25. $3.90, $3.95, $4.02

4   Patterns and Number Sense

Pages 130-131   Lesson 4-1

5. no **6.** yes **7.** yes **8.** no **9.** 2, 3, 4, 5, 6, 9, 10 **10.** 2, 4 **11.** 5 **12.** 2, 3, 4, 5, 6, 9, 10
13. 2, 3, 4, 6, 9 **14.** none **15.** yes **17.** no
19. yes **21.** 2, 3, 5, 6, 10 **23.** none **25.** 5
27. 10, 20 **29.** 30, 60 **31.** 30, 60 **32.** 10
33. 0.01
34.

```
◄─┼─●─┼─────┼─────●─┼──●●──┼─────┼──●──┼──●──┼─►
  15   20   25   30   35   40   45   50
```

35. $42,100 is considerably higher than all the other salaries. **37.** 12 desks

Pages 133-135   Lesson 4-2

6. composite **7.** prime **8.** composite
9. composite **10.** $2^2 \times 3^2 \times 7$ **11.** $2 \times 3 \times 11$
12. $2^4 \times 5 \times 11$ **13.** $2 \times 3^3 \times 5$ **14.** $2^4 \times 3^2$
15. $2^3 \times 11$ **16.** 2×73 **17.** 13×17
18. 2×5^3 **19.** $2^2 \times 3 \times 5^2$ **21.** prime
23. composite **25.** $2^2 \times 3^2 \times 5 \times 7$
27. $2^2 \times 5^2 \times 13$ **29.** $2 \times 3^2 \times 5$ **31.** $5^2 \times 7$
33. $2^6 \times 5$ **35.** 10 **37.** 78 **38.** 45 **39.** 0.9

40.
```
1 | 9
2 | 5 7
3 | 1 1 2 3 4 6
4 | 1      4 | 1 means 41
```

41. 135 is divisible by 3, 5, 9 **43.a.** $4 = 2 \cdot 2$, $9 = 3 \cdot 3$, $16 = 2 \cdot 2 \cdot 2 \cdot 2$, $25 = 5 \cdot 5$, $36 = 2 \cdot 2 \cdot 3 \cdot 3$, $49 = 7 \cdot 7$, $64 = 2 \cdot 2 \cdot 2 \cdot 2 \cdot 2 \cdot 2$
43.b. Squares of prime numbers have only two factors in their prime factorization.
43.c. The hypothesis holds. **47.** 3, 5; 5, 7; 11, 13; 17, 19; 29, 31; 41, 43; 59, 61; 71, 73

Pages 138-139   Lesson 4-3

5. arithmetic: 49, 56, 63 **6.** geometric: 162, 486, 1,458 **7.** neither: 21, 28, 36
8. arithmetic: 75, 90, 105

9. neither: 15, 13, 18 **10.** geometric: $\dfrac{1}{27}, \dfrac{1}{81}, \dfrac{1}{243}$

15. geometric: 0.125, 0.0625, 0.03125
17. arithmetic: 5.4, 6.5, 7.6
19. geometric: $\frac{1}{64}$, $\frac{1}{256}$, $\frac{1}{1,024}$
21. arithmetic: 55, 66, 77   **23.** neither: 125, 216, 343   **25.** 1, 9, 25, 49: neither   **27.** 21
28. 0.0023 L   **29.** 3, 3, 3   **30.** $2 \times 3^2 \times 5 \times 7$
31. 5, 10, 20, 40, 80, 160, 320;

no, 320 min = $5\frac{1}{3}$ hours

Pages 143-144   Lesson 4-4
3. 1, 1, 2, 3, 5, 8, 13, 21, 34, 55, 89, 144, 233, 377, 610, 987, 1,597, 2,584, 4,181, 6,795
4. number under total = number under number of As + previous number of As
5.a. 1, 0.5, 0.667, 0.6, 0.625, 0.615, 0.619, 0.618, 0.618   **b.** The quotient, rounded to the nearest thousandth, eventually equals 0.618; also, the quotients are alternately greater than 0.618 and less than 0.618.   **7.** 832,040   **9.** 27, 29
11. yes; 1, 144; 1, 8   **13.** 6.35 cm

Page 147   Lesson 4-5
4. 4   **5.** 6   **6.** 11   **7.** 28   **8.** 3, 5; 15   **9.** 3, 3; 9   **10.** 15   **11.** 10   **12.** 6   **13.** 13   **15.** 12
17. 1   **19.** 10   **21.** 1   **23.** 5   **25.** 18   **27.** no
28. about 6
29.
64 66 68 70 72 74 76 78 80 82 84 86 88 90 92 94
30. multiply by 2; 208, 416, 832   **31.** 15
33. No, the factor of any number is not greater than the number.   **31.** 15   **33.** No, it must divide into all numbers.

Pages 151-153   Lesson 4-6
4. $\frac{3}{7}$ **5.** $\frac{3}{7}$ **6.** $\frac{5}{9}$ **7.** $\frac{3}{2}$ **8.** $\frac{9}{10}$ **9.** $\frac{11}{40}$ **10.** $\frac{4}{5}$
11. $\frac{3}{4}$ **12.** $\frac{25}{35}, \frac{15}{21}$ **13.** $\frac{1}{2}$ **15.** $\frac{5}{9}$ **17.** $\frac{9}{11}$
19. $\frac{7}{9}$ **21.** $\frac{7}{16}$ **23.** $\frac{2}{3}$ **25.** $\frac{4}{6}, \frac{6}{9}, \frac{8}{12}$ **27.** $\frac{6}{8}$,
$\frac{9}{12}, \frac{12}{16}$ **29.** 148   **30.** 2.7   **31.** red   **32.** 18
33. $\frac{7}{15}$ **35.** $\frac{1}{3}, \frac{1}{12}, \frac{1}{24}, \frac{7}{24}, \frac{1}{8}, \frac{1}{12}, \frac{1}{24}$ **37.** $\frac{10}{11}$

Page 153   Mid-Chapter Review
1. no   **3.** yes   **5.** $2^2 \times 7$ **7.** $2^3 \times 3^2$
9. arithmetic: 40, 47, 54

11. geometric: $\frac{1}{256}$, $\frac{1}{2,048}$, $\frac{1}{16,384}$ **12.** $\frac{143}{11} =$
13; $\frac{231}{11} = 21$ **13.** 4   **15.** 1   **17.** $\frac{2}{7}$ **19.** $\frac{19}{28}$

Pages 155-156   Lesson 4-7
3. 0.8   **4.** 0.34   **5.** 0.24   **6.** 0.028   **7.** 1.12
8. 0.56   **9.** 0.175   **10.** 1.375   **11.** $0.\overline{7}$
12. $0.\overline{36}$ **13.** $\frac{17}{20}$ **14.** $\frac{5}{8}$ **15.** $\frac{1}{2}$ **16.** $\frac{83}{100}$
17. $\frac{3}{40}$ **19.** 0.55   **21.** 0.12   **23.** 0.875
25. 0.375   **27.** $0.\overline{72}$ **29.** $0.\overline{6}$ **31.** 0.875
33. $\frac{9}{100}$ **35.** $\frac{3}{8}$ **37.** $2\frac{1}{2}$ **39.** $\frac{12}{25}$ **41.** $12\frac{41}{200}$
43. 81   **44.** 0.22, 0.23, 1.6, 2.29, 2.3, 23
45. 63, 63   **46.** 5×7^2 **47.** $\frac{13}{27}$

Pages 159-160   Lesson 4-8
4. $\frac{1}{2}$ **5.** $\frac{1}{2}$ **6.** $\frac{1}{4}$ **7.** $\frac{5}{12}$ **8.** $\frac{2}{9}, 0.\overline{2}$ **9.** $\frac{5}{18}$,
$0.2\overline{7}$ **10.** $\frac{1}{3}, 0.\overline{3}$ **11.** $\frac{1}{6}, 0.1\overline{6}$ **13.** $\frac{1}{4}$ **15.** $\frac{23}{24}$
17. $\frac{1}{24}$ **19.** $\frac{1}{4}, 0.25$ **21.** $\frac{9}{20}, 0.45$ **23.** 2
24. $2.15   **25.** 0.314   **26.** $0.\overline{4}$ **27.** $\frac{2}{9}$
29. an event that will definitely happen

Pages 162-163   Lesson 4-9
3. 60   **4.** 30   **5.** 30   **6.** 300   **7.** 36   **8.** 102
9. 44   **10.** 1,225   **11.** 15   **13.** 180   **15.** 24
17. 900   **19.** 240   **21.** 36   **23.** 2,460
24. 8.19

25.
0	2 5 9
1	0 2 3 6 7
2	3 5 5
3	1 3	1 means 31

26. 0.01   **27.** 3 years   **29.** when the smaller number is a factor of the larger number

Pages 166-167   Lesson 4-10
3. 36   **4.** 55   **5.** 26   **6.** 112   **7.** <   **8.** <
9. >   **10.** <   **11.** 30   **13.** 8   **15.** 30   **17.** 68
19. 30   **21.** 21   **23.** >   **25.** =   **27.** <
29. <   **31.** >   **33.** >   **35.** 55   **36.** 1.03×10^4
37.
1.0   1.5   2.0   2.5   3.0   3.5   4.0   4.5
38. $3 \times 5 \times 17$ **39.** 270   **41.** $\frac{4}{6}, \frac{7}{10}, \frac{7}{9}, \frac{9}{11}, \frac{8}{9}, \frac{7}{9}$
45. brass section

Pages 168-170   Study Guide and Review

9. 5   11. 2, 4, 5, 10   13. none   15. 2, 3, 6, 9

17. 3   19. $2^4 \times 3^2$ 21. 7×11

23. $2^2 \times 3 \times 5^2$ 25. $2 \times 5^2 \times 29$

27. geometric: 1,024; 4,096; 16,384

29. geometric: 100,000; 1,000,000; 10,000,000

31. 5   33. 3   35. $\frac{4}{5}$ 37. $\frac{2}{3}$ 39. $\frac{7}{11}$ 41. 0.4

43. 0.375   45. $0.\overline{5}$ 47. $\frac{1}{2}, 0.5$ 49. 30

51. 80   53. 3,969   55. $<$ 57. $<$ 59. 2

5 Applications with Fractions

Pages 176-177   Lesson 5-1

4. improper   5. mixed number   6. proper

7. improper   8. improper

9. ● ●   10. ◕ ◔

11. $1\frac{2}{3}$ 12. $4\frac{1}{2}$ 13. 3   14. $2\frac{1}{3}$ 15. 1

16. $4\frac{1}{2}$ 17. $\frac{15}{8}$ 18. $\frac{11}{4}$ 19. $\frac{3}{1}$ 20. $\frac{41}{9}$ 21. $\frac{41}{3}$

23. mixed number   25. improper   27. $1\frac{2}{7}$

29. $2\frac{2}{5}$ 31. $2\frac{1}{2}$ 33. 3   35. $4\frac{1}{3}$ 37. $\frac{19}{5}$

39. $\frac{5}{1}$ 41. $\frac{31}{8}$ 43. $\frac{35}{8}$ 45. $\frac{27}{10}$ 47. $\frac{39}{7}$

49. $9,300   50. 17.25 minutes   51. 0.09

52. ◀─◆─┼─┼─┼─◆─┼●┼─●●┼─●┼●●─▶
45 50 55 60 65 70 75 80 85 90 95 100

53. 78.125, 195.3125, 488.28125   54. $\frac{1}{16}, \frac{1}{2}, \frac{2}{3},$
$\frac{5}{6}, \frac{7}{8}$ 55. $3\frac{1}{2}$ pies   57. $2\frac{1}{2}$ pages

Pages 179-181   Lesson 5-2

4. 1   5. 0   6. $\frac{1}{2}$ 7. 1   8. $\frac{1}{2}$ 9. 6   10. 9

11. 3   12. 4   13. 7   14. $1\frac{1}{2}$ 15. $\frac{1}{2}$ 16. 2

17. 11   18. $\frac{1}{8}$ 19. 1   21. 1   23. 1   25. 0

27. $\frac{1}{2}$ 29. 1   31. 4   33. 4   35. 6   37. 9

39. 7   41. $\frac{1}{2}$ 43. 8   45. 9   47. 11   49. 8

51. $\frac{1}{2}$ 53. 3   55. no, 124   56. 9   57. 10

58. 0.036 liters

59.
```
0 | 0 1 3 3 4 7 8 9 9
1 | 0 3 4 5
2 | 4
3 | 1        3 | 1 means 31
```

60. $2^2 \times 3^2$ 61. $\frac{29}{8}$ 65. 7,000 pounds

67. about 5 cups

Pages 184-185   Lesson 5-3

4. $\frac{3}{5}$ 5. $\frac{1}{6}$ 6. $\frac{7}{24}$ 7. $1\frac{1}{10}$ 8. $\frac{11}{15}$ 9. $1\frac{1}{45}$

11. $\frac{1}{4}$ 13. $\frac{2}{3}$ 15. $1\frac{8}{35}$ 17. $1\frac{7}{18}$ 19. $1\frac{1}{6}$

21. $\frac{13}{24}$ 23. $\frac{35}{72}$ 25. 31, 31.5, no mode

26. ◀─┼─**x x**─┼**x x**┼─┼──**x**─┼──**x** **x**─▶
0   200   400   600   800   1000   1200

27. 80   28. $2 \times 12 = 24$ 29. $\frac{5}{18}$

Pages 187-188   Lesson 5-4

3. 7   4. 3   5. 8   6. 3   7. $8\frac{1}{2}$ 8. $4\frac{1}{2}$ 9. $1\frac{3}{4}$

10. $17\frac{5}{24}$ 11. $1\frac{4}{5}$ 12. $29\frac{23}{40}$ 13. 3   15. 1

17. 11   19. 13   21. $8\frac{1}{3}$ 23. $12\frac{1}{4}$ 25. $2\frac{1}{6}$

27. $4\frac{4}{15}$ 29. $7\frac{1}{20}$ 31. $11\frac{17}{40}$ 33. $3\frac{7}{9}$

35. 9 feet   36. 30   37. 32.5, 35, 37.5

38. 12 hours   39. $1\frac{1}{2}$ 41. $1\frac{3}{4}$ inches

Pages 192-193   Lesson 5-5

3. $\frac{3}{10}$ 4. $\frac{5}{9}$ 5. $\frac{1}{4}$ 6. $1\frac{1}{2}$ 7. $6\frac{2}{3}$ 8. $\frac{1}{6}$

9. $\frac{1}{10}$ 10. $1\frac{3}{5}$ 11. $8\frac{2}{3}$ 13. $\frac{1}{10}$ 15. $\frac{3}{10}$

17. $\frac{1}{10}$ 19. $\frac{2}{7}$ 21. $\frac{5}{16}$ 23. $\frac{1}{2}$ 25. $7\frac{1}{3}$

27. 6   29. 6   30. It is high because of the

outlier, $49,500.   31. 2, 3, 4, 5, 6, 10   32. $16\frac{5}{8}$

33. 16 inches

Page 193   Mid-Chapter Review

1. $\frac{5}{2}$ 3. $\frac{10}{3}$ 5. 1   7. 20   9. $15\frac{5}{24}$ 11. $10\frac{1}{9}$

13. $\frac{5}{21}$

Pages 195-196   Lesson 5-6

3. 56 ft   4. $17\frac{7}{8}$ in.   5. $21\frac{3}{4}$ in.   6. 32 feet

7. $3\frac{1}{4}$ in. 8. $2\frac{3}{4}$ in.   9. 50 miles   11. 54 in.

13. 10.4 miles   15. $42\frac{1}{2}$ feet   17. $3\frac{1}{2}$ inches

18. 7,689 pennies   19. $\frac{3}{4}$ 20. $\frac{5}{12}$

21.b. 103 bushels

Pages 199-200   Lesson 5-7

5. 25.12 ft   6. 21.98 m   7. 28.26 yd
8. 37.68 cm   9. 43.96 in.   10. 20.096 m
11. 65.94 in.   12. 20.41 in.   13. 3 meters
14. 15 meters   15. 11 ft   17. 29.83 km
19. 55 mi   21. 9.42 km   23. 38.936 cm
25. $27\frac{1}{2}$ ft   27. 26 feet   29. about 200 cases
30. 1.9, 2.3, 2.6, 3.4, 3.6, 3.8   31. 0.875
32. $50\frac{4}{5}$ feet   33. 200.96 in.   35.a. 6.28, 12.56,
25.12, 50.24, 100.48, 200.96   b. When the
diameter is doubled, the circumference doubles.

Pages 202-203   Lesson 5-8

3. $\frac{1}{2}$ 4. $\frac{1}{4}$ 5. $\frac{1}{4}$ 6. $0.75   7. win   9. lose

11. $3.50   13. Coke, Pepsi, Slice

14. $30\frac{9}{14}$ inches   15. $1\frac{1}{2}$ girls

Pages 205-206   Lesson 5-9

3. identity of $+$ 4. associative of $\times$ 5. no

6. yes   7. yes   8. yes   9. $2\frac{1}{3}$ 10. $2\frac{1}{5}$

11. $2\frac{1}{8}$ 13. $\frac{8}{7}$ 15. $\frac{1}{3}$ 17. multiplicative

inverse   19. distributive of $+$ over $\times$ 21. 11

23. $3\frac{1}{6}$ 24. 65 words/minute   25. 3.5

26. $\frac{5}{36}$ 27. 500   29. $1\frac{1}{6}$ in^2

Pages 208-209   Lesson 5-10

3. $\frac{5}{3}$ 4. $\frac{1}{2}$ 5. 3   6. $\frac{2}{9}$ 7. $1\frac{1}{2}$ 8. $3\frac{1}{2}$ 9. $\frac{2}{3}$

10. $\frac{5}{14}$ 11. $\frac{3}{8}$ 12. $5\frac{1}{3}$ 13. $\frac{2}{5}$ 14. $1\frac{1}{2}$

15. $\frac{6}{5}$ 17. $\frac{5}{4}$ 19. $\frac{5}{18}$ 21. $2\frac{2}{5}$ 23. $\frac{2}{3}$

25. $\frac{7}{16}$ 27. $\frac{2}{3}$ 29. $\frac{4}{15}$ 31. $3\frac{3}{8}$ 33. $\frac{2}{3}$

35. $\frac{2}{3}$ 37. $\frac{17}{210}$ 38. 75   39. 10.3   40. $\frac{7}{12}$

41. 241.78 feet   42. $2\frac{6}{7}$ 43. 16 lots

Page 213   Lesson 5-11
3. 80   4. $2.70   5. $10.65   7. $38

Pages 214-216   Study Guide and Review

9. $\frac{25}{7}$ 11. $15\frac{3}{4}$ 13. $2\frac{5}{8}$ 15. 18   17. $\frac{1}{2}$

19. 90   21. 3   23. $\frac{1}{4}$ 25. $\frac{19}{40}$ 27. $8\frac{1}{4}$

29. $13\frac{9}{14}$ 31. $18\frac{1}{5}$ 33. $34\frac{4}{7}$ 35. $75\frac{13}{21}$ ft

37. $23\frac{9}{35}$ yd   39. $20\frac{26}{35}$ ft   41. $12\frac{3}{4}$ 43. $4\frac{1}{18}$

45. $3\frac{3}{5}$ 47. $5\frac{11}{24}$ cups

6 An Introduction to Algebra

Pages 221-222   Lesson 6-1

4. $5 = 8 - 3$ 5. $9 = 4 + 5$ 6. $m = 19 - 7$

7. $n = 21 + 12$ 8. $5 = 30 \div 6$ 9. $54 = 6 \times 9$

10. $a = 13 \times 2$ 11. $c \div 3 = 14$ 12. $e = 48 \div$
4   13. 8   15. 7   17. 31   19. 114   21. 300

23. 11.6   25. 13.4   27. $1\frac{1}{4}$ 29. 22   31. 9.2

33. 13.4   35. 0.8   37. about $800   38. $8.59

39.
```
0 | 0 2 2 4 5 7 7 9
1 | 0 2 3 3 8 9
2 | 0 3        1 | 2 means 12
```

40. $\frac{3}{4}$ 41. $8\frac{4}{5}$ 43. 98   45. 4.4 points

Pages 226-227   Lesson 6-2

5. 25   6. 36   7. 39   8. 9   9. 11.7   10. 21

11. $4\frac{1}{4}$ 12. 9   13. $7\frac{3}{4}$ 15. 16   17. 25

19. 38   21. 73   23. $47\frac{5}{12}$ 25. 127   27. 1.3

29. 10.7   31. $76\frac{1}{15}$ 33. 20.2   34. 9.8×10^3

35.

36. 24   37. 48   39.a. $4\frac{3}{4}$ hours   b. $3\frac{1}{4}$ hours

Pages 230-231   Lesson 6-3

4. 5   5. 36   6. 294   7. 7   8. 24.025   9. 6
10. $1\frac{3}{5}$ 11. 864   12. 96   13. 7   15. 13
17. 18   19. 51   21. 44   23. 768   25. 6
27. 32.4   29. 40   31. 243   32. mean = 14,
median = 14   33. $2 \cdot 2 \cdot 2 \cdot 3$ 34. $3\frac{5}{12}$
35. 100   37. $3,200   39.a. $9,600 = 4m$
b. 2,400 bits per second

Page 231   Mid-Chapter Review

1. 48   **3.** 157   **5.** 68   **7.** 0.45   **9.** 306

Pages 234-235   Lesson 6-4

5. $t + 7$ **6.** $r + 2$ **7.** $p - 8$ **8.** $g - 4$

9. $12 + s$ **10.** $18 - y$ **11.** $\frac{c}{4}$ **12.** $3a$

13. $\frac{b}{2}$ **14.** $7a$ **15.** t minus 10, 10 less than t

16. the quotient of 4 and d, 4 divided by d

17. 10 times n, the product of 10 and n

18. 14 plus h, 14 more than h **19.** $y + 5$

21. $2p$ **23.** $\frac{a}{3}$ **25.** $\frac{a}{6}$ **27.** $19 - r$

29. $j + 1{,}110$ **31.** $8 - d$ **33.** $\frac{6}{k}$ **35.** $s - 8$

37. $x - 15$ **41.** $\frac{1}{4}$ **42.** Bill   **43.** 12   **44.** $\frac{22}{75}$

45. \$136.50   **47.a.** brushing teeth   **b.** $\frac{t}{2}$

49. 17.3, 26, 40.3, 41.8, 47.3, 56.2, 60.7

Page 237   Lesson 6-5

3. 3   **4.** 110   **5.** not enough information

7. $\frac{5}{9}$

Pages 239-240   Lesson 6-6

3. 6   **4.** 4,000   **5.** 80   **6.** 8   **7.** 3   **8.** 3
9. 7.5   **10.** 6   **11.** 8   **13.** yes   **15.** 10,000

17. 10   **19.** 3   **21.** 4   **23.** 5   **25.** 9   **27.** $\frac{1}{4}$

29. 9 pints   **31.** $\frac{1}{4}$ lb   **32.** 4.5   **33.** \$61.02

34.

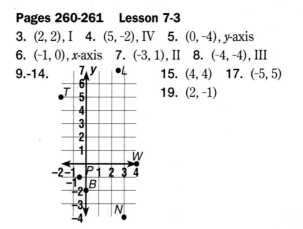

35. $3\frac{7}{12}$ **36.** $12 + d$ **39.** $68°, 76°$

Pages 244-245   Lesson 6-7

4. $24\,\text{in}^2$ **5.** $14\,\text{m}^2$ **6.** $32.5\,\text{ft}^2$ **7.** $6\,\text{in}^2$
8. $216\,\text{cm}^2$ **9.** $40\,\text{cm}^2$ **11.** $28\,\text{mm}^2$

13. $19\frac{1}{8}\,\text{m}^2$ **15.** $414\,\text{ft}^2$ **17.** $142.38\,\text{ft}^2$

19. $18\,\text{cm}^2$ **21.** about 210   **22.** 3,000   **23.** 32
servings   **25.a.** yes   **b.** $126\,\text{ft}^2$

Pages 246-248   Study Guide and Review

9. $1\frac{1}{2}$ **11.** 0   **13.** $\frac{3}{8}$ **15.** 14.26   **17.** $6\frac{1}{2}$

19. 63   **21.** 1.5   **23.** 8   **25.** $\frac{5}{6}$ **27.** 4

29. 36   **31.** $x + 5$ **33.** $13 - r$ **35.** $9 + q$

37. $b - 23$ **39.** $\frac{c}{100}$ **41.** 80   **43.** $3\frac{3}{8}$

45. 11.5 or $11\frac{1}{2}$ **47.** $45\,\text{yd}^2$ **49.** $36.72\,\text{in}^2$

51. $6\frac{1}{24}\,\text{in}^2$ **53.** no   **55.** $250\,\text{in}^2$

7 Integers

Pages 255-256   Lesson 7-1

5. $+4$ **6.** -5 **7.** $+12$ **8.** $+6$ **9.** -6
10. $+10$ **11.** $-5, 5, 5$ **12.** $-1, 1, 1$ **13.** $-8, 8, 8$
14. $8, -8, 8$ **15.** $0, 0, 0$ **16.** $3, -3, 3$ **17.** 1
18. $63°\text{F}, -21°\text{F}$ **19.** $+6$ **21.** $+2$ **23.** -25
25. $-6, 6, 6$ **27.** $-9, 9, 9$ **29.** $-1, 1, 1$ **31.** -1
33. 178
34.

35. $11\frac{1}{9}$ **36.** $30{,}000\,\text{ft}^2$ **37.** $20{,}320; -282$

Page 258   Lesson 7-2

3. $>$ **4.** $<$ **5.** $<$ **6.** $>$ **7.** $<$ **8.** $>$
9. $-98, -76, -1, 14, 31, 56$ **11.** $>$ **13.** $>$
15. $<$ **17.** $-91, -76, -9, -6, 2, 18, 32$
19. $-14, -7, -1, 0, 5, 13$ **20.** \$2, 245
21. Sample answer: 24, 48   **22.** 21, 21
23. $21°\text{F}$ **25.** -3

Pages 260-261   Lesson 7-3

3. $(2, 2)$, I   **4.** $(5, -2)$, IV   **5.** $(0, -4)$, y-axis
6. $(-1, 0)$, x-axis   **7.** $(-3, 1)$, II   **8.** $(-4, -4)$, III
9.-14.   **15.** $(4, 4)$ **17.** $(-5, 5)$
19. $(2, -1)$

21., 23., 25., 27., 29.

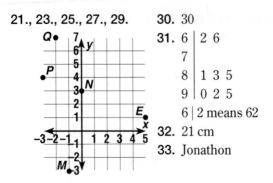

30. 30

31.

```
6 | 2 6
7 |
8 | 1 3 5
9 | 0 2 5
```
6 | 2 means 62

32. 21 cm

33. Jonathon

Pages 265-266   Lesson 7-4

5. 0   **6.** −   **7.** +   **8.** +   **9.** +   **10.** −
11. −   **12.** +   **13.** -5   **14.** -10   **15.** 11
16. 7   **17.** 0   **18.** 9   **19.** gained 4 yards
21. -4   **23.** 6   **25.** 25   **27.** 100   **29.** -13
31. -6   **33.** 0   **35.** -14   **37.** $6/pound
39. $\frac{1}{16}, \frac{1}{32}, \frac{1}{64}$ **40.** $t = 8$
41.

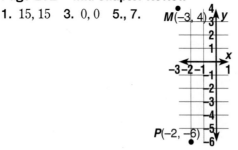
$B(-4, 4)$

43. T + 45 seconds

Pages 269-271   Lesson 7-5

5. 19   **6.** -12   **7.** 27   **8.** -7   **9.** 45   **10.** -8
11. -3   **13.** -27   **15.** -68   **17.** 20   **19.** 0
21. -83   **23.** 0   **25.** 11   **27.** 1   **29.** -18
31. -10   **33.** -4   **35.** -12   **37.** -2   **39.** -6
40. 3,500   **41.** 42   **42.** $1.40   **43.** $w − 65$
44. -5   **45.** 59°, 840°, 5°   **47.a.** normal rainfall
b. -3, -2, -4, 2   **47.c.** August; it has the lowest
rainfall compared to normal   **47.d.** There was a
drought this summer.

Page 271   Mid-Chapter Review

1. 15, 15   **3.** 0, 0   **5., 7.**

$M(-3, 4)$
$P(-2, -6)$

9. -12   **11.** -7   **13.** 8

Pages 275-276   Lesson 7-6

3. 1 and 3 alternating; 1, 3, 1   **4.** squares of
ordered whole numbers; add 3, 5, 7, 9...
5. 54, 48   **7.** 255 people

9.

11. 9,567 + 1,085 = 10,652

Pages 279-280   Lesson 7-7

5. -42   **6.** 20   **7.** -18   **8.** -81   **9.** -45   **10.** 9
11. -18   **12.** -18   **13.** 9   **14.** -108   **15.** -324
16. 54   **17.** 90   **19.** 121   **21.** 36   **23.** -96
25. -68   **27.** -70   **29.** 49   **31.** 189   **33.** 49
35. 150   **37.** -525   **39.** 45   **40.** 1.5
41. 0.375   **42.** $\frac{11}{36}$ **43.** 84   **44.** 87 feet

45. No; at -18 feet the pressure dropped 3(2.7),
so it is 22.8 pounds per sq. in.

Page 282   Lesson 7-8

3. -3   **4.** -10   **5.** 5   **6.** -20   **7.** 17   **8.** -7
9. 31   **10.** 24   **11.** 8   **13.** 4   **15.** -5   **17.** -3
19. -56   **21.** -8   **23.** -8   **25.** -1   **27.** 24
29. 64   **30.** 2.5, 2, 2   **31.** -60   **33.** -2°F

Pages 285-286   Lesson 7-9

5. -4   **6.** -28   **7.** -7   **8.** -12   **9.** 13   **10.** -54
11. -10,100   **12.** 32   **13.** 420   **14.** -8
15. -56   **17.** 43   **19.** -401   **21.** -190   **23.** 46
25. -168   **27.** -5   **29.** 6   **31.** 24.5 cm
32. $\frac{35}{24}$ **33.** 2.5 lb   **34.** 40 ft/min
35. 110 feet; −50 − (−160) = 110   **37.** 17 cm
39. 24 years

Pages 288-289   Lesson 7-10

3. 0.0003   **4.** 0.6   **5.** 0.00007   **6.** 2×10^{-5}
7. 5×10^{-3} **8.** 9×10^{-7} **9.** 0.000000001
11. 0.7   **13.** 0.00003   **15.** 0.0000008
17. 1×10^{-3} **19.** 9×10^{-1} **21.** 3×10^{-7}
23. 0.00006   **24.** 3 P.M.   **25.** $\frac{1}{4} < \frac{2}{7}$ **26.** $\frac{5}{9} < \frac{3}{4}$
27. 32   **29.** 0.002 cm   **31.** 0.04 oz.

Pages 290-292   Study Guide and Review

9. 42   **11.** 75   **13.** -5   **15.** -1, 1, 1   **17.** <
19. >   **21.** >   **23.** I   **25.** III   **27.** y-axis

23., 25., 27.

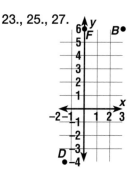

29. 4 **31.** 7 **33.** -25 **35.** -2 **37.** 8 **39.** -6
41. -8 **43.** 9 **45.** -25 **47.** -5 **49.** 6
51. -3 **53.** -5 **55.** -480 **57.** 5×10^{-9}
59. 0.002 **61.** They gained 5 yards.

8 Investigations in Geometry

Pages 299-300   Lesson 8-1
4. obtuse **5.** acute **6.** right **7.** obtuse
8. right **9.** acute **10.** straight **11.** 42°
13. acute **15.** acute **17.** obtuse **19.** acute
21. acute **23.** 20 **25.** 72° **26.** 0.75
27. 1.9 miles **28.** 6×10^{-5} **29.a.** A **b.** C

Pages 304-305   Lesson 8-2
3. No, not closed. **4.** yes **5.** yes **7.** No,
sides do not meet at vertex. **9.** yes **11.** yes
13. 70 **14.** acute **15.a.** 5 **15.b.** 9
17.a. octagon, stop **b.** square, curve
c. triangle, yield **d.** rectangle, speed limit 50

Pages 309-310   Lesson 8-3
4. scalene, right **5.** scalene, obtuse
6. isosceles, acute **7.** parallelogram, rectangle
8. trapezoid **9.** parallelogram, rectangle,
rhombus, square **11.** equilateral, acute
13. scalene, obtuse **15.** scalene, acute
17. parallelogram, rectangle
19. parallelogram, rhombus
21. parallelogram, rectangle, square, rhombus

23. 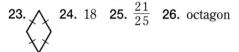 **24.** 18 **25.** $\frac{21}{25}$ **26.** octagon

Pages 315-316   Lesson 8-4
5. No, not equiangular **6.** yes **7.** No, not
equilateral or equiangular **8.** 90°, square
9. yes **11.** yes **13.** 36°, 144° **15.** rhombus
17. 24, 18, 16, 16 **18.** 23 **21.a.** 40

Page 316   Mid-Chapter Review
1. obtuse **3.** parallelogram

Page 320   Lesson 8-5
3. Q, octagon; R, square; S, hexagon
5. Susan Sales **7.** about 10 yards

Page 323   Lesson 8-6
3. yes **4.** no **5.** no **7.** no **11.** 1 square,
2 octagons **13.** No, not equilateral or
equiangular. **15.** Use one of each.

Page 326   Lesson 8-7
3.

 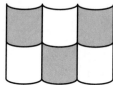

5.

 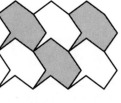

7.

 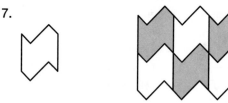

9. No, the patterns on the top and bottom will
not tessellate. **10.** $1\frac{7}{8}$ **11.** no
13. No; no opposite side. **15.** 1.3 miles

Pages 328–329   Lesson 8-8
3.

4.

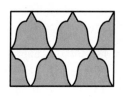

5. **7.**

9.

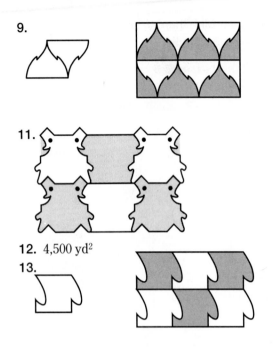

11.

12. 4,500 yd²

13.

Pages 330-332   Study Guide and Review

9. acute   **11.** straight   **13.** obtuse   **15.** yes
17. yes   **19.** scalene, right
21. parallelogram, <u>rhombus</u>   **23.** No, not
equilateral or equiangular.   **25.** No, not
equilateral or equiangular.   **27.** yes

31.

35. isosceles, right

9 **Area**

Pages 336-337   Lesson 9-1

2. 7 tables   **3.** 45, 62   **4.** 16 years old
5. 2 packages of each   **7.** 24 ways   **9.** b

Pages 339-340   Lesson 9-2

4. 4   **5.** 9   **6.** 49   **7.** 100   **8.** 121   **9.** 144
10. 2   **11.** 6   **12.** 9   **13.** 10   **15.** 25
17. 196   **19.** 400   **21.** 900   **23.** 8   **25.** 12
27. 20   **29.** 40   **31.** 21   **33.** 225 m²   **35.** yes
37. yes   **39.** yes   **40.** about $1,500
41. 14, 12   **42.** 245   **43.** 7.6   **45.** 64 squares

Page 342   Lesson 9-3

3. 3   **4.** 5   **5.** 5   **6.** 6   **7.** 8   **8.** 8   **9.** 10
10. 10   **11.** 3   **13.** 7   **15.** 10   **17.** 12
19. 19   **21.** 27   **23.** $\sqrt{14}$ **24.** 5   **25.** $\frac{19}{20}$
26. 14   **29.** 15 feet

Pages 346-347   Lesson 9-4

4. legs 3, 4; hypotenuse 5   **5.** legs 5, 12;
hypotenuse 13   **6.** legs 7, 24; hypotenuse 25
7. $6^2 + 8^2 = c^2$; 10 ft   **8.** $5^2 + 12^2 = c^2$; 13 cm
9. $(9\frac{1}{2})^2 + (4\frac{2}{3})^2 = c^2$; 10.6 m   **10.** $8.2^2 + 15.6^2$
$= c^2$; 17.6 yd   **11.** $10^2 + b^2 = 25^2$; 23 m
12. $12^2 + a^2 = 15^2$; 9 yd   **13.** $5^2 + b^2 = 13^2$;
12 cm   **14.** $15^2 + b^2 = 30^2$; 26 ft   **15.** no
16. yes   **17.** no   **19.** 7.6 in.   **21.** 16.1 ft
23. 20.2 yd   **25.** 16.6 ft   **27.** 12.4 in.
29. 23.7 cm   **31.** no   **33.** no   **35.** no
37. no   **39.** December 31   **40.** $2^4 \times 3^2 \times 5$
41. 75   **42.** 18   **43.** about 3 miles

Pages 349-350   Lesson 9-5

3. 23.6 m   **4.** 16.5 cm   **5.** 11.5 ft   **6.** 9.4 miles
7. about 102.5 ft   **9.** $3/pound   **10.** no

Page 350   Mid-Chapter Review

1. 36   **3.** 484   **5.** ≈4   **7.** ≈12   **9.** ≈18.6 in.

Pages 352-353   Lesson 9-6

3. 28 square units   **4.** 18 square units   **5.** 22
square units   **6.** 52 square units   **7.** 18 square
units   **8.** 56 square units   **9.** 42 square units
11. 63 square units   **13.** 52 square units
17. scalene, right   **18.** ≈29.7 feet
19.a. Alabama, Colorado, Montana, Oregon
b. Maryland, Maine   **c.** Montana, Colorado,
Oregon, Alabama, Maine, Maryland
21. 48 square units

Pages 357-358   Lesson 9-7

4. 12 m²   **5.** 900 ft²   **6.** 27 in²   **7.** 30 ft²
8. 238 in²   **9.** 248.37 m²   **10.** 57 cm²
11. 80 yd²   **13.** 56 km²   **15.** 0.77 in²
17. $2\frac{5}{36}$ yd²   **19.** 150 yd²   **21.** $127\frac{1}{6}$ ft²
23. 25; scale: 5 to 35; intervals of 5   **24.** $7\frac{17}{24}$
25. about 32 square units

Pages 361-362   Lesson 9-8

4. 12.6 ft^2 5. 153.9 m^2 6. 78.5 in^2
7. 1.5 cm^2 8. 2.0 cm 9. 2.9 ft 10. 4.9 m
11. 6.0 in. 13. 153.9 cm^2 15. 615.4 yd^2
17. 754.4 ft^2 19. 113.0 cm^2 21. $1,962.5 \text{ ft}^2$
23. $1,194.0 \text{ m}^2$ 25. 5.0 m 27. 0.6 km
29. 1.8 yd 31. 5.3 m 33. $\$2.36$
34. $-8, -4, -3, 0, 1, 4, 6$ 35. square, rhombus
36. 60 in^2 37. about 706.5 m^2

Pages 366-367   Lesson 9-9

3. $\frac{1}{16}$ 4. $\frac{3}{20}$ 5. $\frac{1}{6}$ 7. $\frac{3}{14}$ 9. $\frac{1}{3}$ 11. $\frac{5}{26}$
13. obtuse   14. about 2.83 cm^2, about 3.46
cm^2, about 2.54 cm^2, about 4.52 cm^2 15. $\frac{1}{45}$
17. Gulf of Mexico, Sea of Okhotsk, Sea of Japan

Pages 368-370   Study Guide and Review

9. 1 11. 13 13. 100 15. 6 17. 6 19. 20
21. $\approx 11.2 \text{ yd}$ 23. $\approx 17.0 \text{ m}$ 25. $\approx 5.8 \text{ mi}$
27. 35 square units   29. 60 yd^2 31. 30 ft^2
33. 254.34 mm^2 35. 94.99 yd^2 37. $\frac{1}{3}$
39. $\frac{1}{20}$

10 Surface Area and Volume

Pages 379-380   Lesson 10-1

7.   15.

17. Yes, everyone is using the same figures. There is only one top, front, and side view.

18. 49 sick days   19. 3.68 cm^2 20. $\frac{17}{50}$

Page 382   Lesson 10-2

3. $20 \text{ in.} \times 8 \text{ in.} \times 8 \text{ in.}$ 4. $5, 3$ 5. 55 boxes
7. 4 prisms   9. 6 bricks

Pages 384-386   Lesson 10-3

4. $4,200 \text{ mm}^2$ 5. 94 ft^2 6. 192 in^2 7. 88 yd^2
8. 167.4 m^2 9. 310 cm^2 11. 181.3 cm^2
13. $308\frac{1}{4} \text{ ft}^2$ 15. 784 cm^2 17. $3,804 \text{ m}^2$
19. 48 in^2 21. about $\$3.00$ 22. 1 25.a. 252 in^2

Pages 389-391   Lesson 10-4

4.a. 12.56 cm^2 b. 100.48 cm^2 c. 125.6 cm^2
5. 125.9768 m^2 6. 282.6 ft^2 7. 942 cm^2
9. 602.9 mm^2 11. 324.9 ft^2 13. 904.3 m^2
15. 415.8 cm^2 17. $2,313\frac{1}{7} \text{ m}^2$ 19. $1,168\frac{3}{4} \text{ yd}^2$
21. 678.24 in^2 23. quadrilateral, parallelogram, rectangle, square   24. 6 ft^2 27. 1 can

Page 391   Mid-Chapter Review

3. 88 cm^2 5. 71.8432 ft^2

Pages 395-397   Lesson 10-5

3. 64 cm^2 4. 135 in^3 5. 120 mm^3
6. 53.24 cm^3 7. 150 in^3 8. 24 cm^3
9. $236\frac{1}{4} \text{ in}^3$ 11. 226.8 ft^3 13. 28 in^3
15. 350 mm^3 17. $1,051.732 \text{ cm}^3$
19.a. 343 in^3 b. $V = s^3$ 20. $2, 3, 4, 5, 6, 9, 10$
21. about 13 22. 196.25 m^2 23. $V = x^3$
27. 480 cubic meters

Pages 399-400   Lesson 10-6

5. 100.5 in^3 6. 129.8 in^3 7. 169.6 m^3
8. 183.7 cm^3 9. 235.5 in^3 10. 37.7 mm^3
11. 76 cm^3 13. 401.9 cm^3 15. 89.5 yd^3
17. 282.6 in^3 19. 161.9 ft^3 21. 39 inches
22. -60 23. 152.88 cm^3 25. about 2.5 cans
27. 132 soft drinks

Page 403   Lesson 10-7

3. 351.68 cm^2 4. $\$27$ 5. -9 7. 2 yd^3
9. No, the box has a surface area of $5,200 \text{ cm}^2$.

Pages 404-406   Study Guide and Review

7.

11. 23.9 yd^2 13. 195.1 in^2 15. $1,406.7 \text{ cm}^2$
17. $168\frac{3}{4} \text{ ft}^2$ 19. 9.4 ft^3 21. 47.2 in^3
23.a. 231 in^3 b. about 18 ounces
25. 196.25 in^3

11 Ratio, Proportion, and Percent

4. $\frac{3}{4}$ 5. $\frac{1}{2}$ 6. $\frac{6}{1}$ 7. $\frac{3}{4}$ 8. $\frac{4}{3}$

9. no, $\frac{65}{100} = \frac{13}{20}$ 10. yes   11. no, $\frac{12}{9} = \frac{4}{3}$

12. yes   13. $\frac{9}{5}$ 15. $\frac{7}{2}$ 17. $\frac{1}{3}$ 19. $\frac{1}{9}$ 21. $\frac{4}{3}$

23. $\frac{16}{7}$ 25. yes, 32:24 = 4:3 and 96:72 = 4:3

27. yes, 4:72 = 1:18 and 12:216 = 1:18

29. no, 6:39 = 2:13   31. 100   32. multiply by 3; 324, 972, 2,916   33. -5, -1, 3, 5

34. 117.75 in³   35.a. $\frac{6}{8} = \frac{9}{12} = \frac{30}{40}$

35.b. $\frac{3}{4} = \frac{3}{4} = \frac{3}{4}$ 35.c. $\frac{54}{96}$ or $\frac{9}{16}$; $\frac{9}{16} = \left(\frac{3}{4}\right)^2$

Pages 415-416   Lesson 11-2

4. $0.79 per pound   5. 40 mph   6. 2 cups per pound   7. $1.19 per disk   8. $70 per day
9. 4 people per car   10. 4 pounds per week
11. 30 tickets per day   13. 80 miles per day
15. $1.45 per disk   17. $\frac{1}{2}$ cup per pound
19. 3 people per car   21. 0.003 g   22. $8\frac{3}{4}$
23. hexagon   24. 5:9   25. 47 people per square mile   27. 305 people per square mile
29. 13,919 people per square mile

Pages 419-420   Lesson 11-3

4. 6   5. 6   6. 15   7. 2   8. 22.5   9. $1\frac{2}{3}$
10. 4.8   11. 16   12. $0.25   13. 12   15. 6
17. 3   19. 75   21. 3   23. 200   25. no, 7 × 15 = 105 and 8 × 13 = 104   26. 2.0×10^6
28. 25   29. $0.31 per ounce   31. 5.6 pounds

Pages 423-425   Lesson 11-4

4. yes, $\frac{2}{6} = \frac{1}{3}$ and $\frac{3}{9} = \frac{1}{3}$ 5. no, $\frac{10}{5} = \frac{2}{1}$ and $\frac{12}{7} = \frac{12}{7}$ 6. 8 m   7. 3 in.   8. 125 feet
9. no, $\frac{2}{2} = \frac{1}{1}$ and $\frac{3}{6} = \frac{1}{2}$ 11. no, $\frac{3}{6} = \frac{1}{2}$ and $\frac{4}{7} = \frac{4}{7}$ 12. yes, $\frac{5}{2.5} = \frac{2}{1}$ and $\frac{8}{4} = \frac{2}{1}$ 13. 2.5 ft
15. 6 m   17. 75 inches   18. about 2 seconds
19. skiing   20. $10\frac{2}{3}$ 21. 74 cm

4. 266 miles   5. 114 miles   7. 25 inches
8. $2\frac{1}{2}$ inches   9. $1\frac{13}{32}$ inches   10. 71 inches
11. 480 mi   13. 105 mi   15. 390 miles
17. 30 mi   19. 1.6 cm   21. 9 in. by 6 in.
23. 5 in. by $5\frac{1}{2}$ in.   24. $\frac{4}{5}$
25. 3 points per day   26. no

Page 429   Mid-Chapter Review

1. $\frac{1}{5}$ 3. $\frac{24}{7}$ 5. $1.19 per pound
7. 6 inches per day   9. 12   11. 60
13. $5\frac{1}{4}$ cm   15. $2\frac{1}{2}$ inches

Pages 431-432   Lesson 11-6

3. after 7 stops   4. 10 handshakes
5. 4 T-shirts and 3 gym shorts
7. 1,024 recipes   9. 10 chairs
11. 11 tapes and 33 compact discs

Pages 434-435   Lesson 11-7

5. 45%   6. 37%   7. 13%   8. 18.5%   9. 12.5%
10. 98.5%   11. 42%   12. 30%   13. 50%
14. 28%   15. 60%   16. 79%   17. 22%
19. 98%   21. 11%   23. 62%   25. 40%
27. 25%   29. 36%   30. 125.1   31. $17\frac{1}{2}$

Pages 437-439   Lesson 11-8

3. 90%   4. 45%   5. $33\frac{1}{3}$%   6. $87\frac{1}{2}$%   7. $31\frac{1}{4}$%
8. $\frac{3}{10}$ 9. $\frac{1}{100}$ 10. $\frac{9}{20}$ 11. $\frac{23}{100}$ 12. $\frac{9}{40}$
13.a. 60%   b. 40%   15. 86%   17. 40%
19. 30%   21. 25%   23. 12.5%   25. 43.75%
27. 32%   29. $\frac{1}{4}$ 31. $\frac{18}{25}$ 33. $\frac{7}{10}$ 35. $\frac{4}{5}$
37. $\frac{1}{8}$ 39. $\frac{2}{3}$ 41. $\frac{7}{40}$ 43. $\frac{1}{16}$ 45. $\frac{27}{50}$
46. $\frac{15}{24} > \frac{17}{32}$ 47. 5×10^{-5}
48. isosceles triangle, obtuse triangle
49. 8 feet   50. 120 cm³   51. 34%   53. $\frac{11}{20}$

Pages 442-443   Lesson 11-9

3. 46%   4. 5%   5. 60%   6. 56.5%   7. 0.39
8. 0.04   9. 0.7   10. 0.2325   11. 1.7   13. 75%
15. 32.5%   17. 3%   19. 1%   21. 99.9%

23. 100% 25. 0.89 27. 0.02 29. 0.9
31. 0.134 33. 0.625 35. 1 37. > 39. <
41. = 43. > 45. 0.63 47. 125
48. $3\frac{1}{2}$ pounds 49. 21 50. 12 51. 352 in²
52. $\frac{7}{20}$ 53.a. 34.4% b. about 34 hits

Pages 445-447 Lesson 11-10
5. 4 6. 1.8 7. 1.3 8. 1.45 9. 0.0075
10. 0.0024 11. 0.002 12. 0.00125 13. 180%
14. 110% 15. 0.5% 16. 0.35% 17. 925%
18. 750% 19. 0.92% 20. 0.116%
21. Yes, people moved to California between
1980-1990 so the population increased.
22. No, he cannot have more goals than shots
taken. 23. 169% 25. 0.00068 27. 2
29. 0.00012 31. 0.00032 33. 500%
35. 525% 37. 190% 39. 28,500% 41. 0.1%
43. 0.9% 45. 310% 47. 400% 49. =
51. > 53. =
55. Yes, an antique car can be worth more now
than was originally paid for it.
57. No, John can't give away more than all the
coins in his collection.
59. Yes, a pine tree's height can increase.
61. $2^2 \cdot 5 \cdot 7$ 62. 34 in² 64. 42 cm²
65. 207.24 in² 66. 0.45 67. 309%

Pages 448-450 Study Guide and Review
7. $\frac{5}{2}$ 9. $\frac{1}{6}$ 11. $\frac{30}{11}$ 13. $\frac{7}{1}$
15. 4 cups per person 17. $4.75 per pound
19. $9.50 per hour 21. 75 23. 1,750
25. yes, $\frac{5}{10} = \frac{8}{16}$ 27. no, $\frac{4}{6} \neq \frac{7}{14}$
29. 144 km 31. 360 km 33. 432 km
35. 3,600 km 37. 63% 39. 60% 41. $\frac{27}{200}$
43. 62.5% 45. 47% 47. 0.75 49. 99.5%
51. 1.25 53. 0.2% 55. 475%
57. 2,444 people per square mile

12 Applications with Percent

Pages 455-456 Lesson 12-1
5. 93, 100; 200 6. $\frac{P}{88}$; 56.3 7. $\frac{P}{220} = \frac{40}{100}$;
88 8. $\frac{P}{16.5} = \frac{12}{100}$; 2.0 9. 140 11. 64

13. 72.4 15. 28.8 17. 8 19.a. $9.10
b. $149.10 20. 0.78 21. 16 22. 56.52 in²
23. 0.065% 25.a. about 47 million
b. about 1,468,750 pounds 27. 676
immigrants 29. $224.75

Pages 457-458 Lesson 12-2
2. 18 people 3. $3,600 5. 4 parts 7. 4

Pages 460-461 Lesson 12-3
3. 36% 4. 52% 5. 30% 6. 100 7. 1.2
8. $68 9. 0.4 11. 2 13. 0.1 15. 0.01
17. 0.01 19. $\frac{1}{8}$ 21. 24 23. 6 25. 10
27. 1.1 29. 90 31. 0.1 33. about $3
36. 41 37. acute 38. 560 miles 39. 720
41. about 25 million Americans

Pages 463-464 Lesson 12-4
4. b 5. c 6. a 7. 27 8. 7.5% 9. 106.7
10. 50% 11. 20.4 12. 125 13. 140 students
15. 20%, 18 17. 105%, 36 19. 30, 15
21. 28.6% 23. 3.6 26. 192 27. 31.6
29. 10% 31. 60 33. about $10,800
34. 2, 5, 10 35. $\frac{3}{8}$ 36. 4 37. about 48
39. about 12.2% 41. about 413 books

Pages 466-468 Lesson 12-5
4. 40 5. 44% 6. 5.1 7. 25% 8. 90 9. 9.0
10. 30.9 11. 47.5% 12. 42.5 13. 65.5%
14. 12.5 16. 96.2 17. 125% 19. 2.3
21. 13.6 23. 64 25. 124 27. 500 29. 84.7
31. 6.9 33. 66.7% 35. Copying any percent
under 100 is reducing. 36. 6 37. 52
38. It is not a closed figure. 39. 326.56 cm³
40. 76.67% 43. 16 free throws

Page 468 Mid-Chapter Review
1. $10 3. 6 5. about 23,018 children 7. 3
9. 7.8 11. 80 13. 125 15. 33.3%

Pages 472-473   Lesson 12-6

3. $\frac{7}{31}$ 4. $\frac{5}{31}$ 5. $\frac{8}{31}$ 6. $\frac{4}{31}$ 7. $\frac{5}{31}$ 8. $\frac{2}{31}$

9. 83° 10. 58° 11. 94° 12. 47° 13. 58°

14. 22° 19. $2.21 20. no 21. 7

Pages 477-478   Lesson 12-7

4. 50% 5. 20% 6. 50% 7. 37% 8. 10%

9. 18% 11. 20% 13. 41% 15. 19%

17. 100% 19. 59% 21. 50% 22. $w - 6$

23. 405.6 cm²

Pages 480-481   Lesson 12-8

3. $1.40 4. $0.96 5. $5.78 6. $7.50

7. $90.85 8. $186.51 9. $10.60 10. $1.87

11. 17% 12. 20% 13. $2.22 15. $17.33

17. $12.79 19. $4.27 21. 14% 23. 30%

25. $55.12 26. 63 27. 15.7 inches 28. 128

29. 12.8% 31.a. $18.59 b. $2.79

33.a. $507 33.b. $8,307

Pages 482-483   Lesson 12-9

3. $24 4. $61.22 5. $3.23 6. $1,165.50

7. $60.13 8. $171 9. $472.50 11. $43.75

13. $637.20 15. $205 17. $38 19. $28.80

20. 4, 3.5, 3 21. 113 in² 22. 30%

25. $1,120

Pages 484-486   Study Guide and Review

9. 357 11. 352.8 13. 450 15. 30 17. 650

19. 12 21. 5.3% 23. 52.5% 25. 217

27. 750 29. 55% 31. 15% 33. $43.75

35. $16.67 37. $3.75 39. $324

41. 190 days

13 Discrete Math and Probability

Pages 493-494   Lesson 13-1

4. 4 outcomes 5. 12 outcomes

6. 16 outcomes 7. 27 outcomes

9. 9 outcomes 11. 12 outcomes

13. 45, 42, no mode 14. $630

15. 27 numbers

Pages 498-499   Lesson 13-2

4. 30 outcomes 5. 12 outcomes

6. 260 outcomes 7. 16 outcomes

9. 18 outcomes 10. 12 11. 12

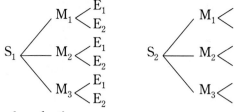

13. *ghn* selections

Pages 501-502   Lesson 13-3

3. $\frac{2}{3}$ 4. $\frac{5}{12}$ 5.a. $\frac{8}{15}$

b. The experimental probability is slightly

greater. 7. $\frac{1}{36}$ 9. 0 10. about 15

11. about 33% 13.a. 0.38 b. no

Page 505   Lesson 13-4

3. 1 4. at least 8 packages 5. at least 6 times

7. $\frac{3}{4}$

Page 505   Mid-Chapter Review

1. 6 outcomes
orange — wheat, rice, corn
apple — wheat, rice, corn

3. $\frac{1}{10}$

Pages 507-509   Lesson 13-5

5. about 22,260 voters 6. about 3,360 voters

7. about 19,740 votes 9. 16% 11. 32%

13. about 11,000 voters 15. Sample answer:
No, because those called may not be registered

voters. 16. 400 17. obtuse, scalene 18. $\frac{1}{2}$

19.a. about 600,000 b. about 900,000

c. about 1,200,000 23. 336 days

Pages 512-513   Lesson 13-6

4. dependent 5. independent 6. $\frac{1}{12}$ 7. $\frac{7}{69}$

8. $\frac{1}{4}$ 9. independent 11. $\frac{1}{15}$ 13. $\frac{1}{6}$

15. $\frac{45}{343}$ 16. $\frac{33}{40}$

17.

18. 39 games 19. 0.3

Pages 519-520 Lesson 13-7

4. 6. 5. 1 6. 4 7. 20 8. 6 ways
9. 120 permutations 10. 60 flags 11. 1
13. 360 15. 120 words 17. 5,040 ways
19. 27,216 zip codes 20. 108 21. -18
22. 0.7744 23. 6 ways 25. $\frac{1}{6}$

Pages 524-525 Lesson 13-8

5. 60 6. 56 7. 15 8. 4 9. RST, RSU, RSV, RTU, RTV, RUV, STU, STV, SUV, TUV 11. $\frac{1}{14}$

13. 1,680 15. 28 17. $\frac{1}{252}$

19. permutation; 24 ways 21. Sample answer is: 1.9 million farms, 500 acres 22. 1.8

23. 120 signals 25. $\frac{1}{20}$ 27.a. damage is total, ground becomes wavy 27.b. once every 5 to 10 years

Pages 526-528 Study Guide and Review

7. 6 outcomes 9. 216 outcomes 11. $\frac{3}{10}$

13. $\frac{7}{10}$ 15. about 2,000 voters 17. $\frac{5}{51}$ 19. 1

21. 720 23. 1 25. 70 27. 6,840 ways

14 Functions and Graphs

Pages 533-534 Lesson 14-1

2. 3 ounces 3. Jim, 18; Phyllis, 6; Ed, 10; Matsu, 2 5. 30 gallons 7. $1.35
11. 2,048 cartons 13. 15,132,158.6 times

Pages 538-539 Lesson 14-2

5. 7 7. -24

9. -1 11. 0.7

13. $3\frac{1}{2}$ 15. $3 + \frac{n}{2} = 11$; 16

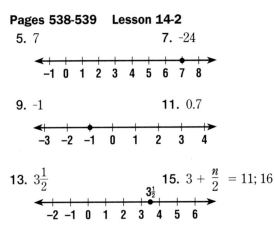

17. -9 19. -12

21. -21 23. 3.5

25. 56 27. -6

29. $4\frac{1}{3}$ 31. -10

33. 16 35. $5x + 3 = -7$; $x = -2$

37. 649 38. -3 39. 25% 40. 12% decrease
41. 15 43.a. 21 miles
b.

Pages 541-542 Lesson 14-3

4.

x	2x + 1	y	
1	2(1) + 1	3	(1, 3)
2	2(2) + 1	5	(2, 5)
3	2(3) + 1	7	(3, 7)
4	2(4) + 1	9	(4, 9)

5.

x	3x	y	
-1	3(-1)	-3	(-1, -3)
0	3(0)	0	(0, 0)
1	3(1)	3	(1, 3)
2	3(2)	6	(2, 6)

6.

x	-2x + 3	y	
-1	-2(-1) + 3	5	(-1, 5)
0	-2(0) + 3	3	(0, 3)
1	-2(1) + 3	1	(1, 1)
2	-2(2) + 3	-1	(2, -1)

7.

x	1.5x	y	
1	1.5 (1)	1.5	(1, 1.5)
2	1.5 (2)	3	(2, 3)
3	1.5 (3)	4.5	(3, 4.5)
4	1.5 (4)	6	(4, 6)

9. {(-1, -3), (0, -2), (1, -1), (2, 0)}

11. {(-1, -5), (0, 0), (1, 5), (2, 10)}

13. {(-1, 2), (0, 0), (1, -2), (2, -4)}

15. {(-1, 4), (0, 2), (1, 0), (2, -2)}

17. {(-1, $\frac{3}{4}$), (0, 1), (1, $1\frac{1}{4}$), (2, $1\frac{1}{2}$)}

19. {(-1, -2), (0, -2), (1, -2), (2, -2)}

21. $25 + 35x = y$; {(1, 60), (2, 95), (3, 130), (4, 165)} **22.** 3.0×10^6 **23.** 0.625 **24.** -13

25.a. $f = g - 18$ **b.** 86, 80, 84, 90

c. (104, 86), (98, 80), (102, 84), (108, 90); (original score, handicap score)

Pages 544-545　Lesson 14-4

4.

x	y	(x, y)
2	6	(2, 6)
1	3	(1, 3)
0	0	(0, 0)
-1	-3	(-1, -3)

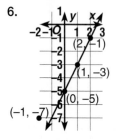

6.

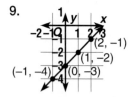

9.

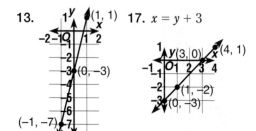

13.

17. $x = y + 3$

20. $2\frac{1}{4}$ **21.** 8 m **22.** (-1, 10), (0, 7), (1, 4), (2, 1) **23.a.** $y = 6x$

b.

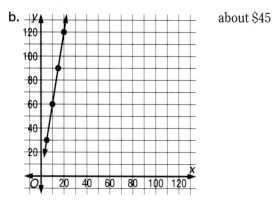

about $45

Page 545　Mid-Chapter Review

1. 6 and 9 **3.** -22

7. (2, 8), (1, 3), (0, -2), (-1, -7)

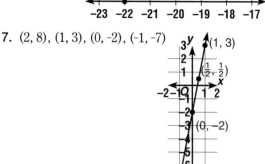

Pages 549-550　Lesson 14-5

4.

x	x + 3	f(x)
-5	-5 + 3	-2
-1	-1 + 3	2
0	0 + 3	3
3	3 + 3	6

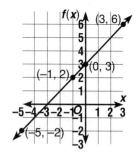

6.

x	-x + 3	f(x)
-2	-(-2) + 3	5
-1	-(-1) + 3	4
2	-2 + 3	1
4	-4 + 3	-1

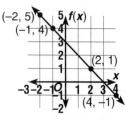

9. -6, 0, 9, 12 **11.** $3\frac{1}{2}, 4, 4\frac{1}{2}, 5$

13. 5, 2, -1, -4

15. (-4, -3), (0, -2), (4, -1), (8, 0)

17.

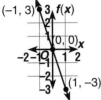

21.

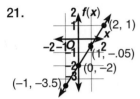

22. $1\frac{1}{2}$ miles   **23.** isosceles, right

24. 36 ways

25.

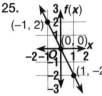

27.a. no negative diameters

b.

d	3.14d	C(d)
1	3.14 (1)	3.14
2	3.14 (2)	6.28
3	3.14 (3)	9.42
4	3.14 (4)	12.56

c.

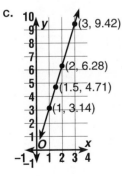

Pages 553-554   Lesson 14-6

4. translation   **5.** translation   **6.** reflection

7.a.

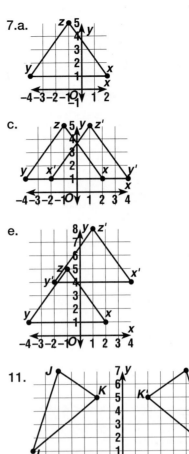

c.

e.

11.

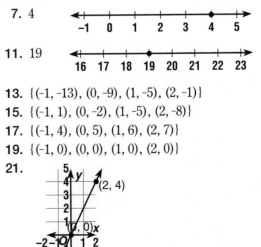

13. 36 m³   **14.** -6, 14, 86

15. on the lid: reflections; on the side: translations and reflections

17.a. translation   **b.** (2, 4)

Pages 556-558   Chapter 14 Study Guide and Review

7. 4

11. 19

13. {(-1, -13), (0, -9), (1, -5), (2, -1)}

15. {(-1, 1), (0, -2), (1, -5), (2, -8)}

17. {(-1, 4), (0, 5), (1, 6), (2, 7)}

19. {(-1, 0), (0, 0), (1, 0), (2, 0)}

21.

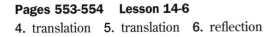

23.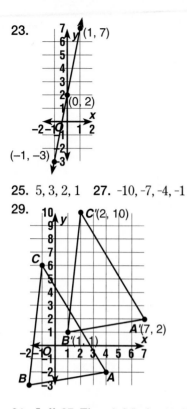

25. 5, 3, 2, 1   **27.** -10, -7, -4, -1

29.

31. Jeff, 27; Fina, 9; Mario, 12; Danielle, 16

33.a. $f(x) = 12.50x$ b. $500

Index

Photo Credits

Cover: (tl) Tom Mareschal/The Image Bank, (tr) S. Achernar/The Image Bank, (ml) Doug Armand/Tony Stone Worldwide, (mr) Michael Smith Studios, (b) Comstock Inc./Jack Elness

iii, Robert Mullenix; **viii,** (t) Konard Wothe/The Image Bank, (b) Aaron Haupt; **ix,** Michael A. Keller/The Stock Market; **x,** (t) Duomo/Rick Rickman, (m) Marti Pie/The Image Bank, (b) Pictures Unlimited, (bkgd) Mark Gibson; **xi,** (t) Caroline Kroeger/Animals Animals, (b) Hank Morgan/Photo Researchers, Inc.; **xii,** (t) Pictures Unlimited, (b) Janet Adams; **xiii,** (t) KS Studio, (b) Mike Luque/Photo Researchers, Inc.; **xiv,** (t) file photo, (m) Brownie Harris/The Stock Market, (b) Skip Comer, (bkgd) Glencoe photo; **xv,** (t) KS Studio, (b) Bud Fowle; **xvi,** (t) Skip Comer, (b) Ken Frick; **xvii,** (t) Fotex/Shooting Star, (bl) Intek Imagineering/Masterfile, (br) Tony Stone Worldwide, (bkgd) Glencoe photo; **2,** (t) Richard Price/Westlight, (b) Reprinted with special permission of King Features Syndicate, Inc.; **3,** (t) Allsport/Vandystadt/Jean-Marc Barey, (b) Robert Mullenix; **6,** Tom & Pat Leeson/Photo Researchers, Inc; **7,** (t) Ken Frick, (b) ©FPG International, Inc.; **10,** Walter Iooss Tr./The Image Bank; **11,** Archive Photos; **13,** Randy Trine; **15,** Ken Frick; **16,** Duomo/David Madison; **19,** Doug Martin; **20,** Skip Comer; **22-23,** J.Messerschmidt/Westlight; **24,** Ken Frick; **26,** courtesy Chrysler Corporation; **28,30,** Ken Frick; **31,** NASA/Science Source/Photo Researchers, Inc; **32,** Skip Comer; (tl) Skip Comer, (tr) Doug Martin, (b) Ken Frick; **36-37,** Skip Comer; **38,** Steve Lissau; **39,** Doug Martin; **40,** Gabe Palmer/The Stock Market; **41,** Doug Martin; **44,** Ken Frick; **46,** (l) courtesy of the Girl Scouts of the United States of America, (r) Historical Pictures Service; **47,** (t) Sobell/Klonsky/The Image Bank, (m) Bob Daemmrich/Stock Boston, (b) The Bettmann Archive; **48,** Doug Martin; **51,** Cincinnati Convention & Visitor's Bureau; **53,** Skip Comer; **54,** Gabe Palmer/The Stock Market; **55,** Ken Frick; **56,** Shooting Star; **57,** Skip Comer; **59,** Duomo/Paul Sutton; **63,** (t) Konard Wothe/The Image Bank, (b) Tim Courlas; **65,** Paul W. Nesbit; **66,** Comstock, Inc.; **69,** Rick Golt/Photo Researchers, Inc; **71,** Scala/Art Resource, New York; **72,** (l) Ken Brate/Photo Researchers, Inc.,(r) Duomo/David Madison; **74,** Historical Pictures Service; **75,** Skip Comer; **76,** NASA; **81,83,** Doug Martin; **86,** Skip Comer; **88,** (l) Harald Sund/The Image Bank, (r) Used by permission. Merriam-Webster, Inc.; **89,** (t) George Holton/Science Source/Photo Researchers Inc., (ml) ©David Bartruff/FPG International, Inc., (mr) R.Ian Lloyd/Westlight, (bl) Archive Photos/Lambert, (br) Culver Pictures, Inc.; **90,** Ken Frick; **92,** Doug Martin; **93,** Jon Provost/Movie Still Archives, (b) ©William Read Woodfield/FPG International, Inc.; **103,** Skip Comer; **104,** KS Studios; **107-108,** Doug Martin; **111,** Andy Caulfield/The Image Bank; **112,114,** Skip Comer; **116,** Latent Image; **118-119,** Doug Martin; **122,** MAK-I; **126,** (t) courtesy of the USDA Forest Service, (b) W.Cody/Westlight; **127,** (t) Tom Ulrich/Tony Stone Worldwide, (b) Calvin and Hobbes ©1989, Universal Press Syndicate. Reprinted with permission. All rights reserved., (br) Doug Wilson/Westlight; **129,** MAK-I; **131,** Doug Martin; **132,** Welzenbach/The Stock Market; **134,** MAK-I; **136,** Comstock, Inc; **139,** Matt Meadows; **142-143,** MAK-I; **144,** Joe Towers/The Stock Market; **145,** ©FPG International, Inc.; **147,** (t) Skip Comer, (b) Tim Courlas; **148,** (t,b) Robert Mullenix, (m) Skip Comer; **150,** Doug Martin; **152,** MAK-I; **154,** Dave Hogan/LGI; **157,** Michael Kevin Daly/The Stock Market; **159,** MAK-I; **160,** H.Armstrong Roberts; **161,** Alex Webb/Magnum; **163,165,** MAK-I; **167,** Doug Martin; **170,** MAK-I; **172,** (t) Grant V. Faith/The Image Bank, (bl) Larry Lefever/Grant Heilman Photography Inc., (br) Dr. E.R.Degginger; **173,** (t) Comstock, Inc., (ml) The Bettmann Archive, (mr) C.Ursillo/H.Armstrong Roberts,Inc., (b) courtesy of Texas Instruments; **174,** KS Studios; **177,** Porterfield/Chickering/Photo Researchers, Inc.; **178,** Doug Martin; **181,** (t) B. Bartholomew/Black Star, (b) The Granger Collection; **183,** Ed Hille/The Stock Market; **184,** Turner Entertainment; **186,** Skip Comer; **191,** Comstock, Inc.; **193,** Skip Comer; **197,** Hank Morgan/Photo Researchers, Inc.; **199,** Tom Braise/The Stock Market; **201,** Skip Comer; **203,** (l) MAK-I, (r) Ken Frick; **204,** R.B. Sanchez/The Stock Market; **205,** Margot Conts/Animals Animals; **207,** Doug Martin; **208,** Latent Image; **209,** Skip Comer; **210,** (l) Latent Image, (r) Doug Martin; **211,** Doug Martin; **212,** (t) Skip Comer, (b) Doug Martin; **213,** (l) Pictures Unlimited, (r) Doug Martin; **218,** (t) John A. Sawyer/Profiles West, (b) PEANUTS reprinted by permission of United Features Syndicate, Inc.; **219,** (bl) Robert Mullenix, (br) Rob Mustard; **220,** Comstock, Inc.; **221,** ©Alan Nyiri/FPG International,Inc.; **222,** Pictures Unlimited; **223,** Doug Martin; **225,** Mary Evans Picture Library/Photo Researchers, Inc; **226,** Latent Image; **228,** Vic Bider/The Stock Market; **230,** Doug Martin; **233,** Flip & Debra Schulke/Black Star; **236-237,** Doug Martin; **238,** Caroline Kroeger/Animals Animals; **239,** Pictures Unlimited; **240,** Don C. Nieman; **242,** Skip Comer; **243,** Ken Frick; **245,** Roy Scheider/The Stock Market; **248,** Doug Martin; **252,** A.J.Verkaik/The Stock Market; **253,** (t) GARFIELD reprinted by permission of UFS, Inc., (m) Comstock, Inc./Stuart Cohen, (b) Comstock, Inc.; **254,** Clark Mishler/The Stock Market; **256,** Johnny Johnson; **258,** Skip Comer; **261,** Werner Stoy/The Image Bank; **263,266,** NASA; **269,** (l) NASA, (r) California Institute of